Noise and vibration from high-speed trains

Edited by

V. V. Krylov

Department of Civil and Structural Engineering
Nottingham Trent University

Thomas Telford

Published by Thomas Telford Publishing, Thomas Telford Ltd, 1 Heron Quay, London E14 4JD
URL: http://www.t-telford.co.uk

Distributors for Thomas Telford books are
USA: ASCE Press, 1801 Alexander Bell Drive, Reston, VA 20191-4400
Japan: Maruzen Co. Ltd, Book Department, 3–10 Nihonbashi 2-chome, Chuo-ku, Tokyo 103
Australia: DA Books and Journals, 648 Whitehorse Road, Mitcham 3132, Victoria

First published 2001

A catalogue record for this book is available from the British Library
ISBN: 0 7277 2963 2

© V. V. Krylov and Thomas Telford Limited 2001

All rights, including translation, reserved. Except as permitted by the Copyright, Designs and Patents Act 1988, no part of this publication may be reproduced, stored in a retrieval system or transmitted in any form or by any means, electronic, mechanical, photocopying or otherwise, without the prior written permission of the Publishing Director, Thomas Telford Publishing, Thomas Telford Ltd, 1 Heron Quay, London E14 4JD

This book is published on the understanding that the authors are solely responsible for the statements made and opinions expressed in it and that its publication does not necessarily imply that such statements and/or opinions are or reflect the views or opinions of the publishers. While every effort has been made to ensure that the statements made and the opinions expressed in this publication provide a safe and accurate guide, no liability or responsibility can be accepted in this respect by the authors or publishers

Typeset by Helius, Brighton
Printed and bound in Great Britain by MPG Books, Bodmin

Contents

Preface		xi
Part 1.	**Generation and propagation of railway noise**	**1**
1.	**Theory of generation of wheel/rail rolling noise**	**3**
	D. J. Thompson	
1.1.	Introduction	3
1.2.	Wheel dynamics	6
	1.2.1. Modes of vibration of a railway wheel	6
	1.2.2. Frequency response functions	9
	1.2.3. Effects of rotation	10
1.3.	Track dynamics	11
	1.3.1. Models for track vibration	11
	1.3.2 Frequency response functions	11
	1.3.3. Propagation along the track	12
	1.3.4. Sleeper response	13
	1.3.5. Effects of preload	13
1.4.	Roughness and interaction	14
	1.4.1. Equations of wheel/rail interaction	14
	1.4.2. Contact receptances	15
	1.4.3. Wheel and rail roughness	16
	1.4.4. Roughness modification at the contact zone	17
	1.4.5. Effective damping of a rolling wheel	18
1.5.	Radiation of sound	18
	1.5.1. Radiation from the wheel	18
	1.5.2. Radiation from the rail	20
	1.5.3. Radiation from the sleepers	21
	1.5.4. Aerodynamic sources	21
	1.5.5. Contribution of various sources	22
1.6.	Validation	23
	1.6.1. Experimental set-up	23
	1.6.2. Results	23
	1.6.3. Sine wheel tests	25
1.7.	Summary	25
1.8.	References	25

2.	**Wheel and rail excitation from roughness**		**27**
	P. J. Remington		
	2.1.	Introduction	27
	2.2.	Roughness modelling	29
		2.2.1. Average roughness model	31
		2.2.2. Distributed point-reacting spring model	33
		2.2.3. Full elastic-interaction model	40
	2.3.	Roughness measurement	45
		2.3.1. Accelerometer-based devices	45
		2.3.2. Displacement-based devices	46
	2.4.	Wheel and rail roughness characteristics	48
	2.5.	Controlling wheel/rail noise at the source	53
		2.5.1. Roughness amplitude reduction	54
		2.5.2. Contact stiffness reduction and contact area increase	56
	2.6.	Summary and conclusions	61
	2.7.	References	62
3.	**High-speed train noise barrier tests at reduced scale**		**65**
	J. D. van der Toorn		
	3.1.	Modelling outdoor sound propagation	65
	3.2.	Scale modelling	65
		3.2.1. Similarity	65
		3.2.2. Measurable quantities	66
		3.2.3. Sound sources	66
		3.2.4. Receiver	70
		3.2.5. Atmospheric absorption	70
		3.2.6. Ground plane	71
		3.2.7. Barriers	73
	3.3.	Scale modelling of railway noise	73
		3.3.1. An acoustical 1:32 scale model of a high-speed train	73
		3.3.2. An acoustical 1:32 scale model of a railway track	75
	3.4.	Design of sound-absorbing barriers at a scale of 1:32	76
		3.4.1. Reference absorption curve	76
		3.4.2. Absorption extracted from excess attenuation	79
	3.5.	Barrier tests	79
	3.6.	Concluding remarks	81
	3.7.	Acknowledgements	82
	3.8.	References	82
4.	**Generic prediction models for environmental railway noise**		**85**
	J. J. A. van Leeuwen		
	4.1.	Introduction	85
	4.2.	Noise indicators	85
		4.2.1. Annoyance	85
		4.2.2. The noise level and the A-frequency-weighted noise level	86

	4.2.3.	Root mean square average	86
	4.2.4.	The maximum sound level $L_{A,\,max}$	87
	4.2.5.	The long-time average sound level and the equivalent sound level	87
	4.2.6.	Statistical indicators	87
	4.2.7.	The basic indicators: $L_{A,\,day}$, $L_{A,\,evening}$, $L_{A,\,night}$ and $L_{A,\,24\,h}$	87
	4.2.8.	The composite indicator L_{den}	88
4.3.	Background to environmental-noise predictions		88
	4.3.1.	Why noise predictions?	88
	4.3.2.	Noise predictions for where?	88
	4.3.3.	What do we want to calculate?	89
	4.3.4.	When to use prediction models	91
	4.3.5.	How do you provide your input?	92
	4.3.6.	Sequence of noise predictions	93
4.4.	What is a noise prediction model?		94
4.5.	Noise prediction methodology		95
4.6.	Source description model		96
	4.6.1.	Sound radiation characteristics	98
4.7.	Propagation models		98
	4.7.1.	Geometrical spreading	100
	4.7.2.	Atmospheric absorption	101
	4.7.3.	Absorption by the ground	101
	4.7.4.	Attenuation due to a barrier or another obstacle	102
	4.7.5.	Additional types of attenuation	104
	4.7.6.	Reflections	105
	4.7.7.	Meteorological correction	105
4.8.	Calculation of the noise level		106
	4.8.1.	Calculating the noise level with monopole or dipole noise sources	107
4.9.	The determination of the sound propagation paths		109
4.10.	Accuracy of a generic prediction model		112
4.11.	Conclusions		113
4.12.	References		114

Part 2. Measurements and control of railway noise — 117

5. Measurements of railway noise — 119
M. T. Kalivoda

5.1.	Introduction		119
5.2.	Exterior noise		120
	5.2.1.	Diagnostics	122
	5.2.2.	Type testing	126
	5.2.3.	Monitoring	144
	5.2.4.	Non-acoustic factors influencing exterior rail noise	149
5.3.	Interior noise		158
	5.3.1.	Diagnostics	158
	5.3.2.	Type testing	160
5.4.	References		160

6. Means of controlling rolling noise at source — 163
C. J. C. Jones and D. J. Thompson

    6.1. Introduction — 163
    6.2. Wheel noise — 164
        6.2.1. Damping treatments — 164
        6.2.2. Wheel shape optimization — 166
        6.2.3. Resilient wheels — 168
        6.2.4. Reduced wheel radiation — 169
    6.3. Track noise — 170
        6.3.1. Rail pad stiffness — 170
        6.3.2. Damping treatments — 173
        6.3.3. Rail shape optimization — 174
        6.3.4. Track mobility — 176
        6.3.5. Ballastless track forms — 177
    6.4. Roughness — 177
        6.4.1. Effects of braking system — 177
        6.4.2. Rail corrugation — 179
        6.4.3. Changes to the contact zone — 180
    6.5. Shielding — 180
    6.6. Measures in combination — 180
    6.7. Summary — 182
    6.8. References — 182

Part 3. Bursting noise associated with non-linear pressure waves in tunnels — 185

7. Micropressure waves radiating from a Shinkansen tunnel portal — 187
T. Maeda

    7.1. Introduction — 187
    7.2. Generation of a compression wave by a train — 189
    7.3. The propagation of the compression wave through the tunnel — 192
    7.4. Radiation of the micropressure wave out of the tunnel portal — 198
    7.5. Measures to decrease the micropressure waves — 203
        7.5.1. Measures applied to Shinkansen tunnels — 204
        7.5.2. Measures applied to Shinkansen trains — 206
    7.6. References — 210

8. Emergence of an acoustic shock wave in a tunnel and a concept of shock-free propagation — 213
N. Sugimoto

    8.1. Introduction — 213
    8.2. Overview of the problem — 216
    8.3. Analysis of the near field — 219
        8.3.1. Linear acoustic theory — 219
        8.3.2. Evaluation of the pressure field — 220

	8.4.	Analysis of the far field	223
		8.4.1. Formulation	223
		8.4.2. Non-linear wave equation for the far field	226
		8.4.3. Evolution of the pressure wave into a shock	228
	8.5.	Shock-free propagation	229
		8.5.1. Linear dispersion characteristics	229
		8.5.2. Effects of the array of Helmholtz resonators	234
		8.5.3. Suppression of shock formation	236
	8.6.	Experimental verification	241
		8.6.1. Experimental set-up	241
		8.6.2. Experimental results	243
	8.7.	Conclusion	244
	8.8.	References	245

Part 4.   Generation of ground vibrations by surface trains   **249**

9.   Generation of ground vibration boom by high-speed trains   **251**
V. V. Krylov

	9.1.	Introduction	251
	9.2.	Quasi-static pressure mechanism of generating ground vibrations	252
		9.2.1. Dynamic properties of the track	253
		9.2.2. Forces applied from sleepers to the ground	255
	9.3.	Green's function for the problem	256
		9.3.1. Homogeneous elastic half-space	257
		9.3.2. Effect of layered ground structure	258
	9.4.	Calculation of generated ground vibrations	262
		9.4.1. Vibrations from a single axle load	262
		9.4.2. Vibrations from a complete train	262
	9.5.	Trans-Rayleigh trains	263
		9.5.1. General discussion	263
		9.5.2. Ground vibrations from TGV and Eurostar trains	265
		9.5.3. High-speed trains travelling underground	270
		9.5.4. Waveguide effects of embankments on generated	
		ground vibration fields	277
	9.6.	Conclusions	281
	9.7.	Acknowledgements	282
	9.8.	References	282

10.   Free-field vibrations during the passage of a high-speed train: experimental results and numerical predictions   **285**
G. Degrande

	10.1.	Introduction	285
	10.2.	The *in situ* measurements	287
		10.2.1. The train	287
		10.2.2. The track	288

	10.2.3.	The soil	288
	10.2.4.	The experimental set-up	290
10.3.	Experimental results		291
	10.3.1.	The passage of a Thalys HST at a speed $v = 314$ km/h	291
	10.3.2.	The influence of the train speed	293
10.4.	Krylov's analytical prediction model		298
	10.4.1.	The force transmitted by a sleeper due to a single axle load	300
	10.4.2.	The forces transmitted by all sleepers due to a train passage	301
	10.4.3.	Response of the soil	302
10.5.	Analytical predictions		303
	10.5.1.	Track response	303
	10.5.2.	Green's functions	305
	10.5.3.	Free-field response	306
10.6.	Conclusion		312
10.7.	Acknowledgements		313
10.8.	References		313

11. High-speed trains on soft ground: track–embankment–soil response and vibration generation 315
C. Madshus and A. M. Kaynia

11.1.	Introduction		315
11.2.	Case study		315
	11.2.1.	Test site and test programme	317
	11.2.2.	Observations	317
11.3.	Measurements		323
11.4.	Dynamic properties of soil and embankment materials		326
11.5.	Numerical simulation		333
	11.5.1.	Simulations and comparisons	336
11.6.	Countermeasures		337
11.7.	Physical model		339
11.8.	Environmental vibration		342
11.9.	Conclusions		343
11.10.	Acknowledgements		344
11.11.	References		344

12. Ground vibrations alongside tracks induced by high-speed trains: prediction and mitigation 347
H. Takemiya

12.1.	Introduction		347
12.2.	Basic theory		349
	12.2.1.	Solution method for a moving load	349
	12.2.2.	Track–ground dynamic interaction	351
	12.2.3.	Modelling of a loading by train	354
	12.2.4.	Ground vibration due to a quasi-static moving load	355
	12.2.5.	Elastodynamic analysis	357

12.3.	Features of the response for a moving load		363
	12.3.1.	Dispersion characteristics of layers	363
	12.3.2.	Transient responses	364
	12.3.3.	Ground surface motions	374
	12.3.4.	Response of track–ground system	375
12.4.	Field measurements, theoretical prediction and mitigation		377
	12.4.1.	Measurement data	377
	12.4.2.	Wave propagation at the site	380
	12.4.3.	Prediction of ground motions	383
	12.4.4.	Vibration mitigation measures – WIBs	385
12.5.	Conclusion		387
12.6.	Appendix: layer stiffness matrix		389
	12.6.1.	The layer stiffness matrix with respect to stresses acting on the z plane $\{\sigma_{12}\ \sigma_{22}\ \sigma_{32}\}$	389
	12.6.2.	The stiffness matrix for a half-space with respect to stresses acting on the z plane	391
12.7.	References		391

Part 5. Ground vibrations generated by underground trains   395

13. Prediction and measurements of ground vibrations generated from tunnels built in water-saturated soil   397
S. A. Kostarev, S. A. Makhortykh and S. A. Rybak

13.1.	Introduction	397
13.2.	Waves radiated by a cylindrical oscillating shell	398
13.3.	Transmission of vibrations to the ground surface	405
13.4.	Two-level elastic system for vibration reduction	407
13.5.	Method of estimation of the elastic parameters and damping of layered ground	411
13.6.	Discussion	418
13.7.	Acknowledgements	421
13.8.	References	421

14. Measures for reducing ground vibration generated by trains in tunnels   423
H. E. M. Hunt

14.1.	Introduction	423
14.2.	Tunnels with floating slabs	424
14.3.	Vibration from railway tunnels	425
14.4.	Conclusions	430
14.5.	References	430

Index   431

Preface

During the last decade, high-speed railways have become one of the most advanced and fast-developing branches of transportation. The reasons for this are the relatively low air pollution per passenger, compared with road vehicles, and the very high speeds achievable by the most advanced modern trains – French TGV, Eurostar, Thalys, the German ICE, British high-speed trains, the Italian Pendolino, the Swedish X2000, the Japanese Shinkansen, etc. For example, for French TGV trains a maximum speed of more than 515 km/h was achieved in May 1990, and speeds close to 300 km/h are now typical for commercially used TGV and Eurostar trains. Prospective plans for the year 2010 assume that the New European Trunk Line will have connected Paris, London, Brussels, Amsterdam, Cologne and Frankfurt by a high-speed railway service that will provide fast and more convenient passenger communications within Europe. Similar plans are being developed in the USA and Japan. All these make high-speed railways increasingly competitive with air and road transport at short and medium distances.

Unfortunately, when train speeds increase, the intensity of railway-generated noise and vibration generally becomes higher. And this represents a major environmental problem for nearby residents, schools and hospitals. Railway operators and local authorities need to be familiar with those new aspects of railway noise and vibration which are associated with high-speed trains. Almost all known mechanisms of generation of railway noise and vibration are speed dependent. These include both wheel/rail rolling noise and aerodynamic noise, the latter being important for train speeds higher than 300 km/h. This applies even more so for generated ground vibrations. For example, when train speeds exceed certain critical velocities of elastic waves propagating in the ground or in the track/ground system, new mechanisms of generation of ground vibrations may appear, in addition to those already known for conventional trains. In particular, a very large increase in generated ground vibrations may occur if train speeds exceed the velocity of Rayleigh surface waves in the ground. If this happens, a *ground vibration boom* takes place, similar to the sonic boom normally associated with supersonic aircraft. The first observation of a ground vibration boom took place on the recently opened high-speed railway line in Sweden. This line was built on very soft soil, with Rayleigh wave velocities as low as 45 m/s. This is why an increase in train speed from 140 to

180 km/h was sufficient for the phenomenon to be observed, thus indicating that 'supersonic' or (more precisely) 'trans-Rayleigh' trains have become today's reality.

There are many other new physical effects and mechanisms of generation of noise and vibration which are specific to high-speed trains, for example the effects of train-induced non-linear pressure wave propagation in long tunnels, resulting in bursting noise radiated from the exit tunnel portals. In addition to these new effects, the 'traditional' mechanisms of generation of railway noise and vibration and their propagation from the source to a receiver demonstrate interesting new features and sometimes behave in a different way as train speeds increase. An example of this may be seen in the design of noise barriers for high-speed railway lines. Such barriers should take into account the spatial redistribution of noise generation mechanisms as train speeds increase.

Although some of the problems of noise and vibration from high-speed trains are being addressed in an increasing number of journal papers and conference proceedings, there is still no general reference book which could help a reader starting to study this problem to find answers to numerous theoretical and practical questions. The existing reviews concerning railway-generated noise and vibration deal largely with conventional trains and do not reflect specific high-speed problems. The present book, which consists of 14 chapters grouped into five parts, aims to fill this gap. It represents the views of leading international experts on the current status of the problems of generation and propagation of noise and vibration from high-speed trains and suggests possible ways of reducing their environmental impact. The book describes mainly the results of recent academic research and is pitched largely at an advanced level. In the light of this, it is assumed that the ideal reader will have a university background in engineering, physics or applied mathematics. At the same time, several chapters of the book have been written by railway noise and vibration practitioners. These chapters contain a lot of experimental data with interesting illustrations and can be understood by a less well-prepared audience.

The intended readership of the book is rather wide. It includes scientists and engineers working on the prediction and remediation of railway noise and vibration, environmental consultants investigating particular situations associated with the environmental impact of railways, local authorities, designers of new railway lines, etc. The book will also be useful to university students, railway enthusiasts and for members of the general public concerned with topical environmental issues.

Victor V. Krylov

Part I
Generation and propagation of railway noise

1. Theory of generation of wheel/rail rolling noise

D. J. Thompson
Institute of Sound and Vibration Research, University of Southampton, Highfield, Southampton SO17 1BJ, UK

1.1. Introduction

As transport needs grow, the environmental impact of different modes of transport is coming under critical review. Railways are usually seen as an environmentally friendly option for large volumes of both passenger and freight traffic. Expansion of the high-speed railway network can help to curb the growth in road and air journeys; if freight is transferred back onto the rails this can remove heavy lorries from the roads; in densely populated areas new light rail and tramway schemes are being proposed to help solve traffic congestion. Railways are therefore entering a new era of higher speeds and higher capacities for intercity and urban systems, and for freight as well as passenger traffic.

Unfortunately, noise and vibration are often perceived as an environmental weakness of railways. The prospect of new railway construction has led to resistance from residents, partly based on noise and vibration issues, which promoters are then required to take into account. For example, there are many new high-speed railway projects in Europe affected in this way. Moreover, in recent years, in response to growing public concerns, a number of countries have introduced regulations limiting railway noise and Europe-wide legislation seems certain to follow. Without reducing the noise from individual trains, a reduction in the permissible level can imply a significant restriction on the train service or speed. Noise barriers, commonly seen as a routine solution to excessive noise from roads, are already widely used in some countries along railway lines. However, these have the disadvantage that they can be visually intrusive for the lineside residents as well as the passengers, and they are also expensive. Moreover, the acoustic effect of such barriers is limited and may be insufficient in some European countries to achieve compliance with new national noise legislation.

There is therefore a growing awareness in the railway community that methods of reducing noise at source are needed. Since the rolling stock and track have an expected life in excess of 30 years, solutions are required which can be applied to existing vehicles and infrastructure as well as to new systems. For effective solutions to be developed, there is a requirement for increased fundamental understanding of how the noise is generated. Detailed computational models are required for the generation of rolling noise, which can be used to design low-noise wheels and tracks. This chapter describes the mechanisms of noise generation and how it can be modelled. The options for noise reduction at source are dealt with in Chapter 6.

The main source of noise from railway operations on open line is rolling noise. This is a broadband noise emitted by the wheel and rail during rolling on straight, unjointed track. It contains significant energy in the frequency range 200 to 5000 Hz with a peak in the A-weighted sound emitted around 1–2 kHz, depending on the speed. Typical sound pressure levels at 25 m from the track, taken from [1.1–1.4], are indicated in Fig. 1.1. As the speed, V, increases, the sound pressure level due to rolling noise increases at a rate of about $30 \log_{10} V$. At speeds over about 300 km/h, aerodynamic sources become dominant. These have a much greater speed dependence, which causes the curves to steepen. It is also apparent from Fig. 1.1 that the vehicles with cast-iron tread brakes (BR Mk II and TGV-PSE) produce considerably more noise at a given speed than vehicles with purely disc or drum braking.

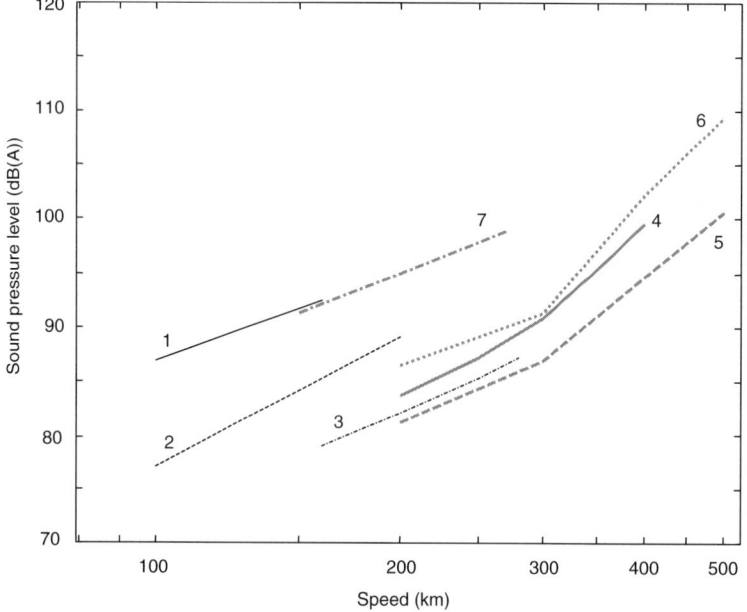

Fig. 1.1. Sound pressure levels at 25 m from the track as a function of train speed for various types of train: 1, BR Mk II coaches (tread-braked); 2, BR Mk III coaches (disc-braked); 3, Talgo (drum braked); 4, ICE (disc-braked); 5, TGV-A trailer (disc-braked); 6, TGV-A power cars; 7, TGV-PSE (disc and tread-braked)

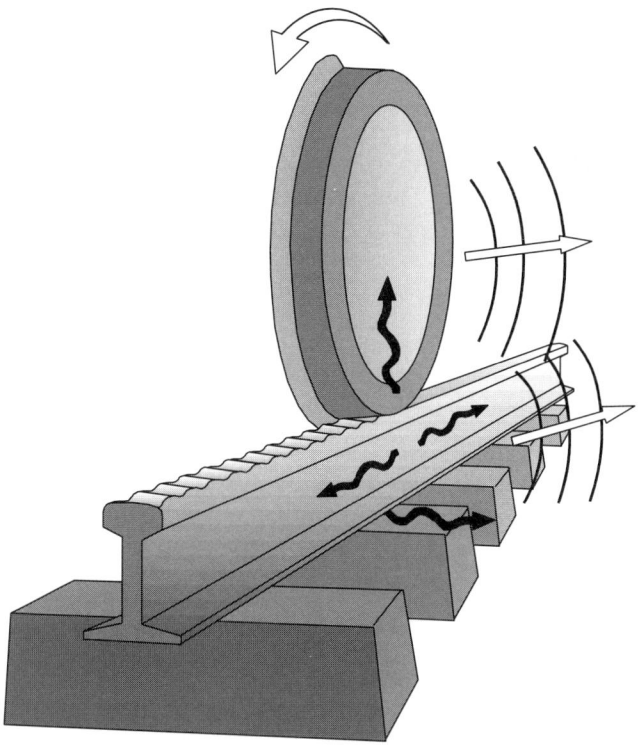

Fig. 1.2. Schematic diagram showing how rolling noise is generated by the wheel/rail interaction

Figure 1.2 shows the process of rolling-noise generation schematically. The wheel and rail surfaces are not perfectly smooth, but contain small-amplitude unevennesses (roughness). This causes the wheel and rail to vibrate relative to each other, and these vibrations radiate noise to the wayside. Early theoretical models of rolling noise, produced by Remington [1.5–1.8], were based on this premise of roughness excitation. These models were developed further and extended by Thompson [1.9–1.14]. Subsequent research funded by the European Rail Research Institute (ERRI) resulted in implementation of the prediction model in a computer program, TWINS (Track–Wheel Interaction Noise Software) [1.15]. Extensive full-scale validation experiments [1.16] have been carried out and will be discussed in Section 1.6.

The process of rolling-noise generation is shown in Fig. 1.3 in the form of a flow chart. This represents the various stages of calculation included in the TWINS theoretical model. At the heart of the calculation is a model for the forces generated at the wheel/rail contact owing to the interaction of the wheel and the track during rolling. The wheel and the track are represented in terms of frequency response functions calculated using separate models – the displacement divided by force, as a function of frequency, is known as the receptance. These models must contain sufficient complexity to reproduce accurately the dynamic behaviour of these structures over the whole frequency range of interest. The dynamic forces calculated by the interaction model then form the excitation to the structural models in order to determine the vibration responses over the surface areas of the wheel and track. Models for the radiation of sound from the wheel, the rail and the sleeper complete the calculation of the total sound radiated due to the rolling action of a single wheel. The sound from a complete train is assumed to be formed of a series of incoherent sources related to each wheel. The various elements of Fig. 1.3 will be discussed in the following sections, after which results of the complete model will be presented.

1.2. Wheel dynamics

1.2.1. Modes of vibration of a railway wheel

A railway wheel is a lightly damped, resonant structure. When struck it rings like a bell, a structure which it strongly resembles. As with all structures, the frequencies at which it rings are its resonance frequencies, and the associated vibration pattern is called the mode shape.

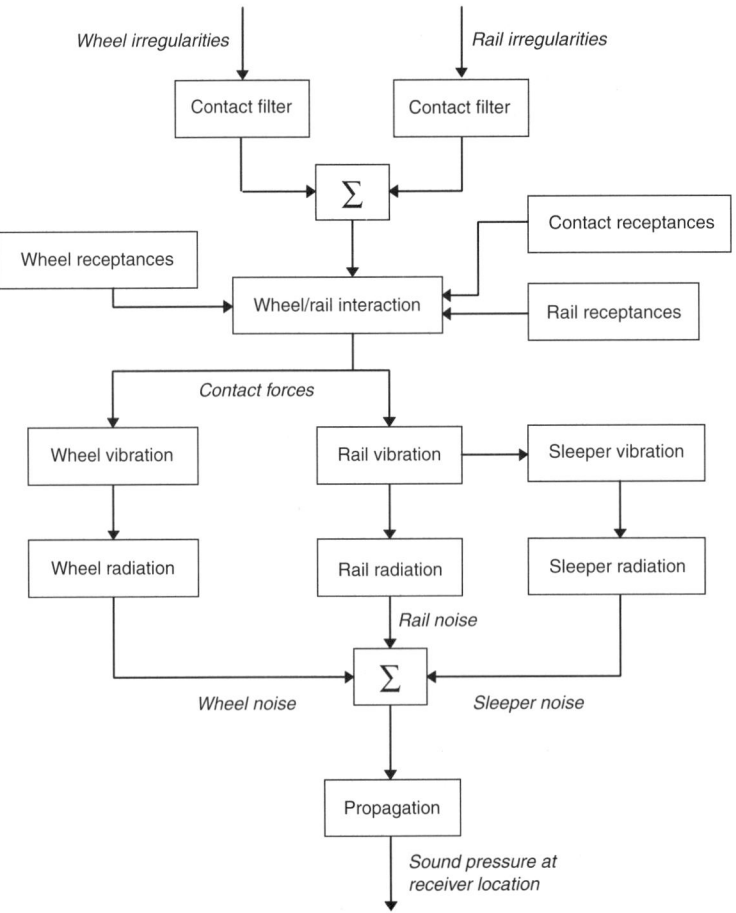

Fig. 1.3. Flow chart of the TWINS calculation model for rolling noise

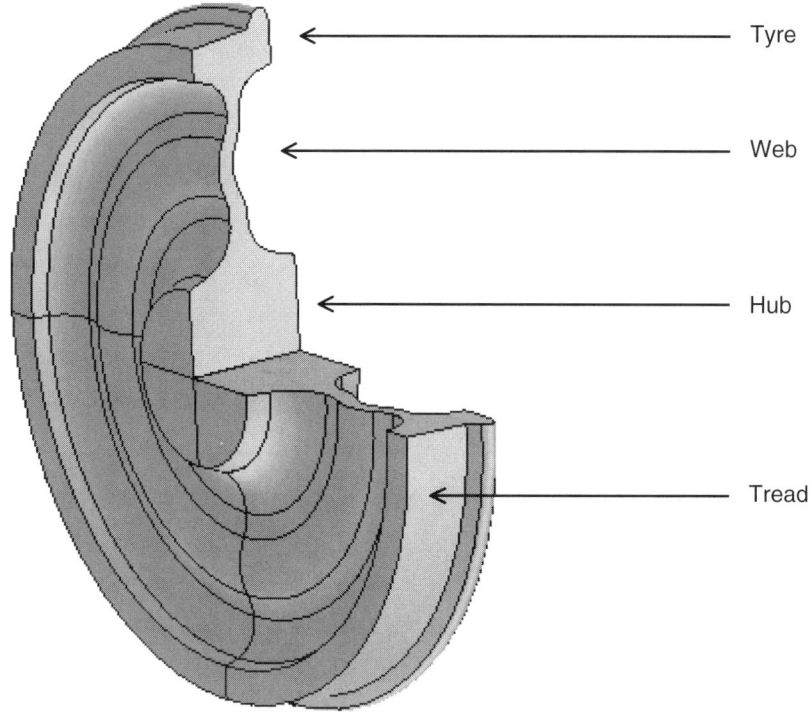

Fig. 1.4. A typical railway wheel

Wheels are usually axisymmetric structures. Their normal modes of vibration can therefore be described in terms of the number of diametral lines at which the vibration pattern has a zero (node lines). The modes of a flat disc, to which a wheel can be approximated, can be categorized in the following way: its out-of-plane modes can be described by two integers, n, the number of nodal diameters, and m, the number of nodal circles. A perfectly flat disc also has a series of purely in-plane modes: radial modes with n nodal diameters and circumferential modes with n nodal diameters. Within the frequency range of interest for railway wheels, no in-plane modes with nodal circles occur.

A railway wheel differs from a flat disc, as can be seen in Fig. 1.4, which shows a typical railway wheel. Important features are the thick tyre region at the perimeter, and the thick hub at the centre connecting the wheel to the axle. Another important difference between a wheel and a disc is that a railway wheel is not symmetric about a plane perpendicular to its axis. The tyre region is asymmetric owing to the flange, and the web is usually also asymmetric, at least on wheels designed for tread braking, the curved web being designed to allow for thermal expansion of the tyre region. An important consequence of this asymmetry is that radial and out-of-plane (axial) modes are coupled.

The finite-element method [1.17] can be used quite effectively to calculate the natural frequencies and mode shapes of a railway wheel. Figure 1.5 shows an

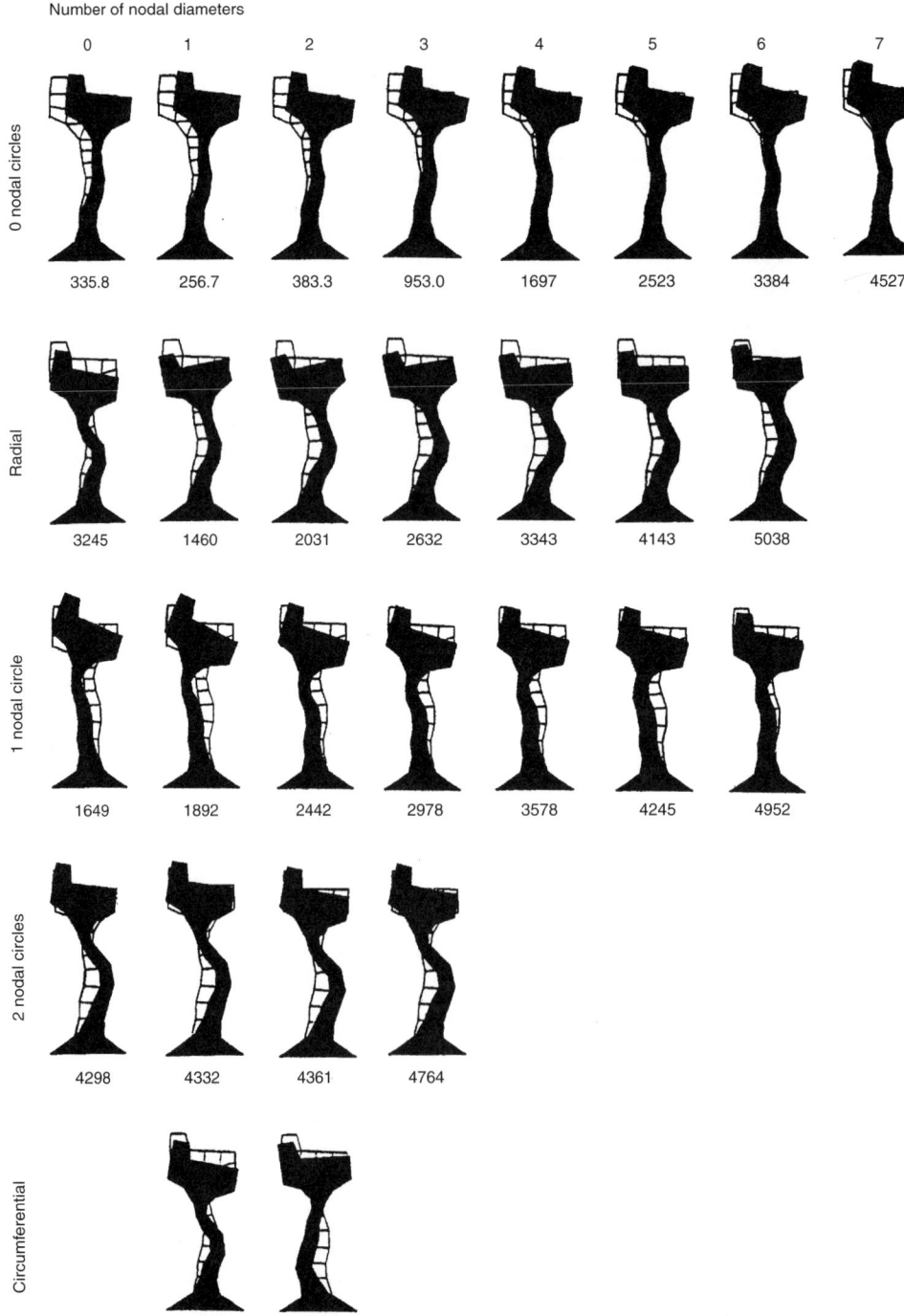

Fig. 1.5. Modes of vibration of a TGV wheel [1.18]

example of results for a TGV wheel [1.18]. The cross-section through the wheel is shown, along with an exaggerated form of the deformed shape in each mode of vibration. Each column contains modes of a particular number of nodal diameters, n. The first row contains axial modes with no nodal circle. These have their largest out-of-plane vibration at the running surface of the wheel. These modes form the set that are usually excited in curve squeal but are not excited significantly in rolling noise. The second and third rows contain the set of one-nodal-circle axial modes and radial modes. Owing to the asymmetry of the wheel cross-section and their proximity in frequency, these two sets of modes are strongly coupled; that is, both contain axial *and* radial motion. It is these two sets of modes that are most strongly excited by roughness during rolling on straight track, owing to their radial component at the wheel/rail contact point.

The modes shown in Fig. 1.5 are of the wheel alone, constrained rigidly at the inner edge of its hub. The first column of modes, $n = 0$, are in practice coupled to extensional motion in the axle, and the second set, $n = 1$, are coupled to bending motion in the axle. As a result of this coupling with the axle, which is constrained by the roller bearings within the axle boxes, these sets of modes experience greater damping than the modes with $n \geq 2$. The latter do not involve deformation of the axle and therefore have very low modal damping; their damping ratios are typically 10^{-4}.

1.2.2. Frequency response functions

In order to couple the wheel to the track in a theoretical model, the frequency response functions of the wheel at the interface point are required. These may be expressed in terms of receptance, the vibration displacement due to a unit force as a function of frequency, mobility, the velocity divided by force, or acceleration, the acceleration divided by force.

A standard technique in vibration analysis is to construct such frequency response functions from a modal summation. For each mode, the natural frequency f_{mn} is written as a circular frequency $\omega_{mn} = 2\pi f_{mn}$. Then the response at circular frequency ω, in the form of a receptance α_{jk}^{W}, is

$$\alpha_{jk}^{W} = \sum_{n,m} \frac{\psi_{mnj} \psi_{mnk}}{m_{mn}(\omega_{mn}^2 - \omega^2 + 2i\zeta_{mn}\omega\omega_{mn})} \quad (1.1)$$

where ψ_{mnj} is the mode shape amplitude of mode (m, n) at position j (the response position), ψ_{mnk} is the mode shape amplitude of mode (m, n) at position k (the force position), m_{mn} is the modal mass of mode (m, n) (a normalization factor for the mode shape amplitude), ζ_{mn} is the modal damping ratio of mode (m, n) and i is the square root of -1.

Figure 1.6 shows the radial point mobility of a TGV wheelset calculated using the normal modes from a finite-element model such as those shown in Fig. 1.5. This is based on equation (1.1), multiplied by $i\omega$ to convert from receptance to mobility. At low frequencies the mobility falls inversely proportional to frequency, corresponding to mass-like behaviour. Around 500 Hz an antiresonance trough appears, and above

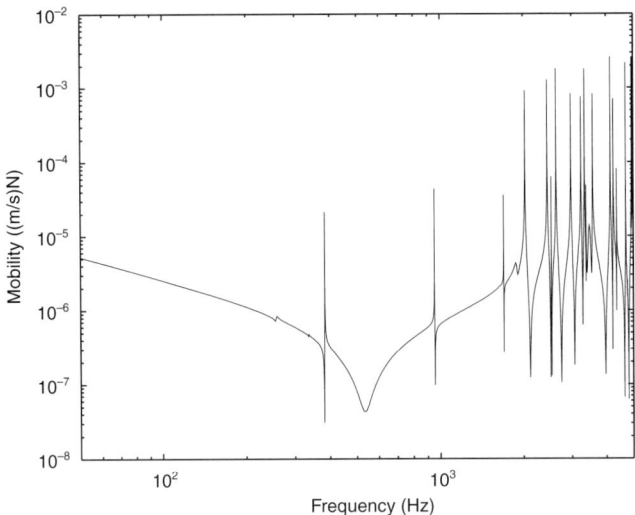

Fig. 1.6. Predicted magnitude of point mobility of a TGV wheelset in the radial direction [1.18]

this frequency the curve rises in stiffness-like behaviour until a series of sharp resonance peaks are reached at around 2 kHz. These peaks are the axial one-nodal-circle and radial sets of modes identified in Fig. 1.5.

1.2.3. Effects of rotation

The modes of the wheel with $n > 0$ nodal diameters have an angular dependence of $\cos n\theta$ or $\sin n\theta$, where θ is the angle around the wheel. These can also be represented as the sum of two rotating waves with angular dependence $e^{\pm in\theta + i\omega t}$, each with half the amplitude. The interference between these two waves leads to a standing-wave pattern, which is the mode shape.

When the wheel rotates, however, each of these waves travels at a different rotational speed relative to the wheel/rail contact point; the wave travelling with the wheel travels faster, and the other wave slower, than do waves in a stationary wheel. Consequently, the frequency at which resonance occurs in a particular wave is altered and each resonance peak is split into two peaks, each corresponding to a wave rotating in one direction [1.14]. These two resonant waves are thus excited by different frequency components at the wheel/rail interface.

If the behaviour is considered in a frame of reference rotating with the wheel, both peaks still occur at the original resonance frequency. As will be seen, the excitation is a random process caused by the roughness of the wheel and rail. Consequently, the two waves are excited by different frequency components of the interaction force. Therefore no fixed interference can occur between them in the wheel, either in the frame of reference rotating with the wheel or in the frame fixed relative to the contact point.

In summary, although the vibration behaviour of the wheel is described using a modal basis, there are no node lines on a rotating wheel that is excited by roughness.

1.3. Track dynamics

1.3.1. Models for track vibration

In comparison with the wheel, which is a finite, highly resonant structure, the track is effectively an infinite structure, characterized by much greater damping. Most modern tracks comprise continuously welded flat-bottomed rails attached to concrete sleepers via a resilient pad and held in place by spring clips. Traditionally, track is laid on a layer of consolidated ballast, approximately 30 cm deep below the sleeper. The space between and around the sleepers is filled to form a 'ballast shoulder'. Slab track construction, in which the rails are laid on a concrete base formed directly on the consolidated foundation, i.e. having no ballast layer, is also becoming popular for high-speed railway applications in some countries.

A detailed review of track vibration models is given in [1.19]. Three alternative track models are included in TWINS [1.15]. References [1.20] and [1.21] give a fuller description of these models and a comparison with the results of field measurements. The simplest of these models represents the rail as a Timoshenko beam supported on a continuous spring–mass–spring foundation, representing the rail pad, the sleeper and the ballast respectively. The use of a Timoshenko beam, including shear deformation and rotational inertia, is necessary for frequencies above about 500 Hz. This model is simple to use but lacks detail compared with the other models (see below). The sleeper can also be represented by a beam to include its bending resonances, although in practice the ballast and ground introduce a high degree of damping to the sleeper.

The second model uses the same beam for the rail but it is now supported on a periodic foundation. The effects of periodicity can be important for vibration in the vertical direction, particularly for stiff pads. The third model, from [1.12], includes cross-sectional deformation of the rail in combination with a continuous track foundation. This model gives more reliable results for the lateral direction, where torsion and web bending are important as well as lateral bending.

1.3.2 Frequency response functions

Figure 1.7 shows the predicted and measured vertical mobility of a track with concrete twin-block sleepers, from [1.21]. Two broad peaks can be discerned, at around 100 Hz and 450 Hz. The first of these corresponds to the whole track mass bouncing on the stiffness of the ballast and the second to the rail bouncing on the stiffness of the rail pad, in antiphase with the sleeper. The frequencies of these peaks are determined by the stiffness of the ballast and pad. Between them, at around 250 Hz, is an antiresonance, which is the result of the sleeper mass and pad stiffness forming a resonant system. At the resonance frequency of this system, energy is absorbed from the rail. This is similar to the action of a 'tuned absorber'.

The periodic support due to the sleepers leads to another 'resonance' phenomenon, seen here at about 1 kHz. When the bending wavelength in the rail equals twice the span length between two sleepers, a standing wave can occur. This is known as the 'pinned–pinned' mode and gives rise to a maximum in the frequency response function between sleepers and a minimum above the sleeper. This phenomenon is strongest in tracks with stiff pads and regular sleeper spacing. A moderate amount of irregularity in the sleeper spacing, typical of tracks in reality, leads to it having a much reduced effect [1.22].

1.3.3. Propagation along the track

The noise radiated by the rail is determined not only by its vibration at the contact point but also by the rate of decay of vibration along its length. For a low decay rate, a large length of rail vibrates and the noise radiation is high. For ballasted track, at low frequencies, below typically 300–500 Hz, the sleepers are strongly coupled to the rail. Since they have a larger area and higher radiation efficiency than the rail, it is important to consider the vibration of the sleepers along with that of the wheel and rail. For track in which the rail is fastened directly to a concrete slab, tunnel or bridge deck, the use of special fasteners with a low stiffness ('resilient baseplates') can lead to lower decay rates in the rail and uncoupling between the rail and its foundation at a lower frequency.

Figure 1.8 shows the rate of decay of vibration along the rail in dB/m for the track corresponding to Fig. 1.7. The sound power radiated by the rail in a given frequency band is approximately inversely proportional to the decay rate in this band, i.e. a doubling of decay rate would lead to a reduction in radiated sound by 3 dB. At low frequencies the decay rate is high (above 10 dB/m) as the vibration in the rail is localized and waves do not propagate in the rail. At high frequencies, where wave propagation occurs, the decay rates are much lower. A broad peak in the decay rate

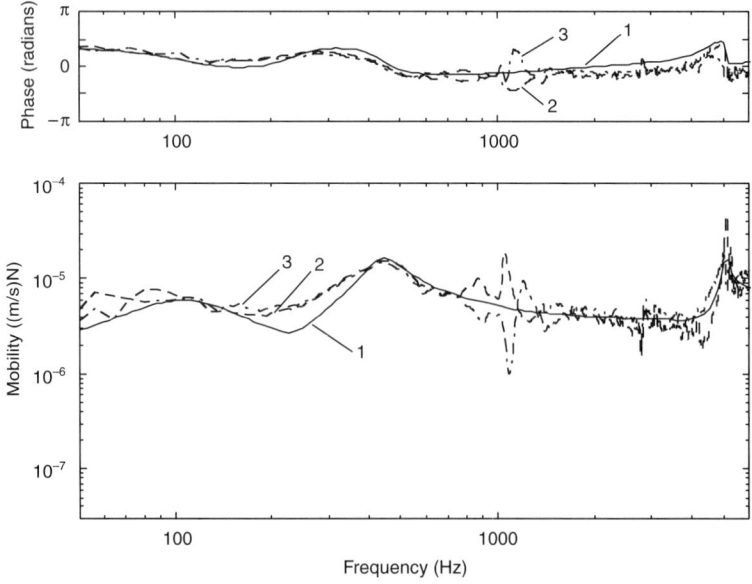

Fig. 1.7. Predicted and measured vertical point mobility of a track with twin-block sleepers and moderately soft pads [1.21]: 1, predicted; 2, measured between sleepers; 3, measured above sleepers

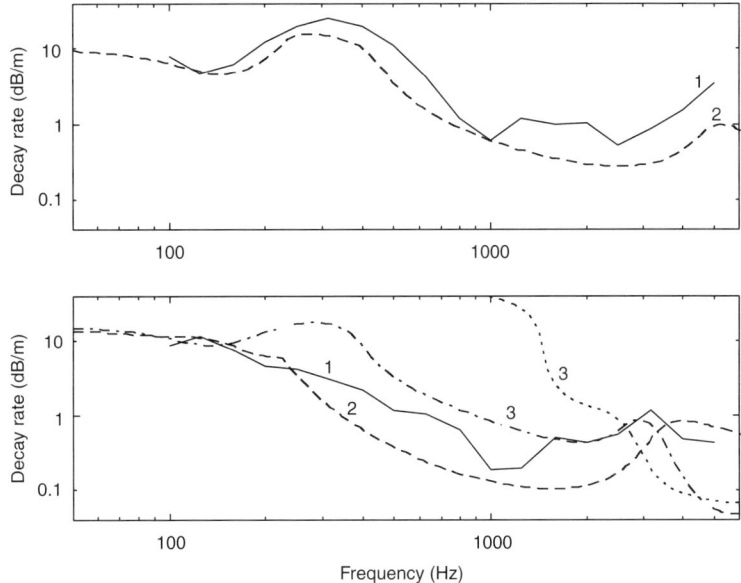

Fig. 1.8. Decay rates in the rail for track with twin-block sleepers and moderately soft pads [1.21]: 1, measured; 2, predicted for dominant wave; 3, predicted for other lateral waves

curve for vertical vibration occurs around 300 Hz, close to the antiresonance frequency at which the sleeper acts as a mass–spring dynamic absorber.

1.3.4. Sleeper response

As noted already, the sleeper is uncoupled from the rail at high frequencies. Figure 1.9 shows the predicted ratio of vibration on the sleeper to that on the rail. Around the frequency of the antiresonance in the rail mobility this ratio has a peak, and above this it falls. For monoblock sleepers, bending resonances of the sleeper lead to a modification of these results, although the high damping of the ballast limits the effect. The stiffness of the ballast layer experienced by the sleeper increases considerably with frequency and this should be taken into account in the model.

1.3.5. Effects of preload

Wu and Thompson [1.23] have developed a methodology for studying the effect of the local preload close to the wheel. This preload leads to a local increase in the ballast and pad stiffnesses, which reach their unloaded values by about the fifth or sixth sleeper from the excitation. The static load distribution is modelled with a non-linear model and the appropriate stiffnesses are then used in a dynamic model. The point receptances of the track are found to be dominated by the local loaded stiffnesses, whereas the wave propagation and decay rates are dominated by the unloaded region. As, in practice, multiple wheel loads are present on the track, the decay of vibration is enhanced at each wheel position. In cases where the rail pad is strongly progressive, this can lead to an effect of up to 4 dB in radiated noise, i.e. the preloaded track radiates less noise than an unloaded track would do [1.24].

1.4. Roughness and interaction
1.4.1. Equations of wheel/rail interaction

The wheel and rail are coupled together at their point of contact, as shown in Fig. 1.10. This coupling can be written in terms of the equilibrium of forces,

$$F^W = F^R \tag{1.2}$$

where F^W is the force acting upwards on the wheel and F^R is the force acting downwards on the rail, and the continuity of displacement,

$$u^W = u^R - u^C + r \tag{1.3}$$

where u^W is the (upward) displacement of the wheel, u^R is that of the rail, u^C is the local compression occurring in the contact zone and r is the relative displacement introduced between the wheel and rail owing to profile variations (positive for a peak in the profile). The contact spring is shown in Fig. 1.10 on the wheel side of the roughness, but in practice it makes no difference to the derivation whether it is located on one side or divided into two parts either side of the roughness. Writing the receptance of the wheel as $\alpha_{jk}^W = u_j^W/F_k^W$, where j is the direction of the response and k is the direction of the force, and α_{jk}^R for the receptance of the rail and α_{jk}^C for the

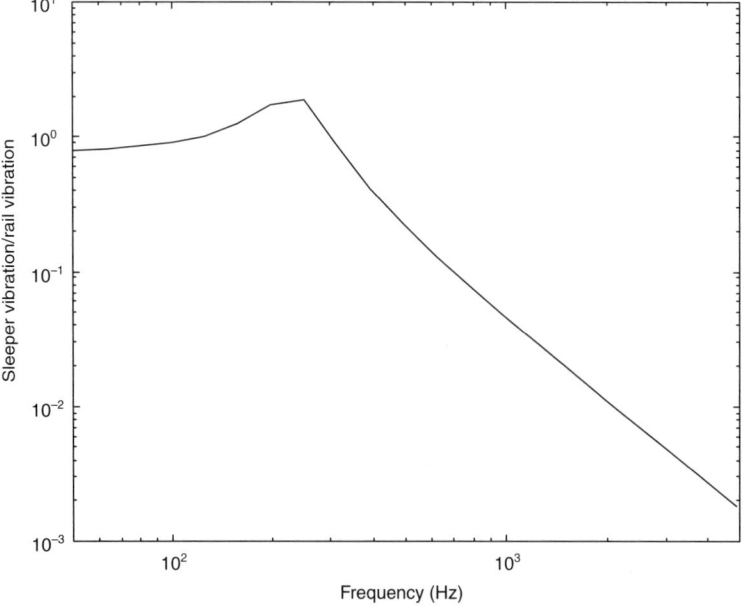

Fig. 1.9. Ratio of sleeper vibration to rail vibration for track with twin-block sleepers and moderately soft pads [1.21]

THEORY OF GENERATION OF WHEEL/RAIL ROLLING NOISE

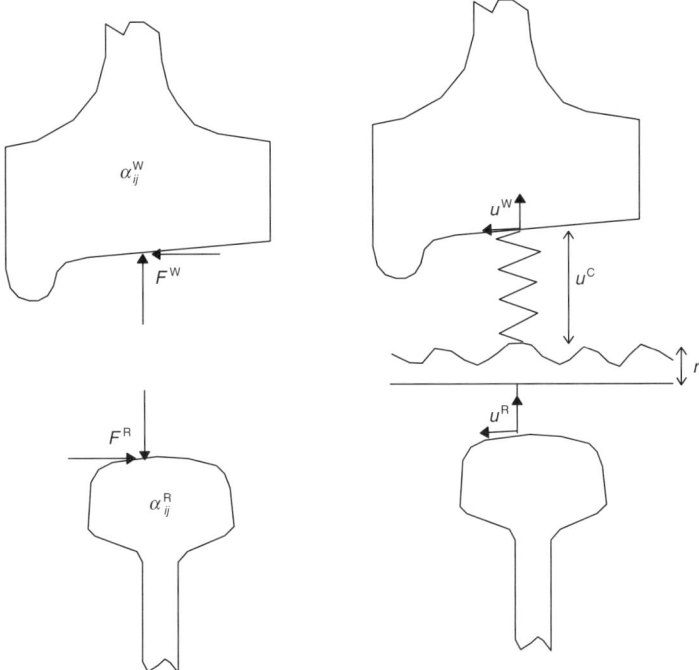

Fig. 1.10. Wheel/rail contact showing (a) equilibrium of forces and (b) continuity of displacements and excitation by roughness amplitude r

receptance of local deformations of the contact zone, the above equations can be combined in matrix form to give

$$\{u^W\} = \{\alpha^W\}[\alpha^W + \alpha^R + \alpha^C]^{-1}\{r\} \quad (1.4)$$

and

$$\{u^R\} = -\{\alpha^R\}[\alpha^W + \alpha^R + \alpha^C]^{-1}\{r\} \quad (1.5)$$

where the vector $\{r\}$ contains zeros in all directions except the vertical, which contains the roughness amplitude. The minus sign in equation (1.5) derives from the directions adopted as positive in Fig. 1.10.

1.4.2. Contact receptances

In the above expressions, a receptance has been given for the contact zone. This refers to the relative displacement across the contact zone divided by the load. For the vertical direction, Hertzian deformation occurs, which has a non-linear load–deflection curve:

$$F = Au^{3/2} \quad (1.6)$$

where the constant A depends on the radii of curvature and material properties of the wheel and rail. For small amplitudes relative to the quasi-static deflection, this can be linearized to give an approximate stiffness for the 'contact spring':

$$k_H = \frac{2}{\xi}\left(\frac{3E^{*2}r_e F_0}{2}\right)^{1/3} \quad (1.7)$$

where $E^* = E/(1 - \nu^2)$ is the plane strain elastic modulus (E is Young's modulus and ν is Poisson's ratio), F_0 is the static load, ξ is a dimensionless quantity dependent on the radii of curvature of the wheel and rail (see e.g. [1.9]), and r_e is the equivalent radius of curvature at the contact, given by

$$\frac{1}{r_e} = \frac{1}{2}\left(\frac{1}{r_W} + \frac{1}{r_{RT}} + \frac{1}{r_{WT}}\right) \quad (1.8)$$

Here r_W is the radius of the wheel, r_{WT} is its transverse radius of curvature and r_{RT} is the transverse radius of curvature of the rail. From equation (1.7), the contact receptance in the vertical direction $\alpha^C = 1/k_H$ can be derived.

For other directions contact receptances also exist. Those in the lateral direction are derived from the creep force–creepage relationships. More details can be found in [1.9, 1.13].

1.4.3. Wheel and rail roughness

As indicated in Fig. 1.2, the dominant cause of excitation of rolling noise is now widely agreed to be surface unevennesses and irregularities, usually referred to as 'roughness'. This is demonstrated qualitatively by the differences between the noise produced by wheels equipped with cast-iron block brakes, which have been found to

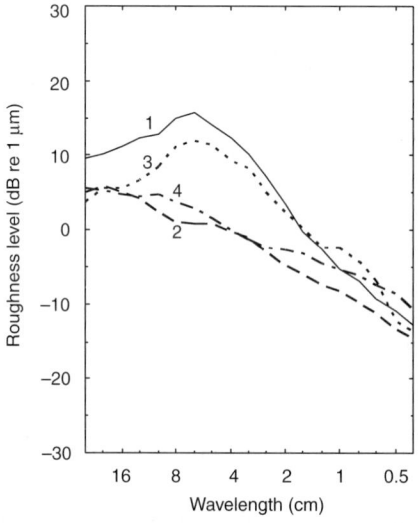

Fig. 1.11. Typical wheel roughness spectra [1.25]: 1, disk brakes plus cast-iron blocks; 2, disk brakes plus sinter blocks; 3, cast-iron blocks only; 4, disk brakes only

THEORY OF GENERATION OF WHEEL/RAIL ROLLING NOISE

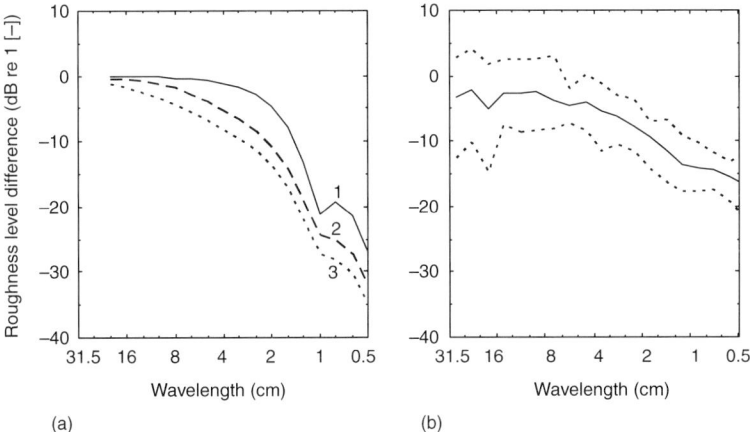

Fig. 1.12. Contact filter effect for a contact patch length of 11 mm. (a) From theory for different values of correlation parameter α [1.7]: 1, high correlation α = 1; 2, medium correlation α = 5; 3, low correlation α = 10. (b) Derived from time-domain analysis of roughness data: mean, maximum and minimum per one-third octave band of six sets of data (from [1.26])

exhibit corrugation, and those with disc brakes, which are smooth (see Fig. 1.1), and also between smooth and corrugated track.

The wavelengths responsible are typically 5 mm to 200 mm, with amplitudes from below 1 μm up to 50 μm or more. These induce a relative vertical motion between the wheel and rail, as seen in equation (1.3), the frequency f of which is related to the wavelength λ by

$$f = v/\lambda \tag{1.9}$$

where v is the speed in metres per second for wavelengths in metres and frequencies in hertz. Typical wheel roughness spectra are shown in Fig. 1.11, taken from [1.25].

1.4.4. Roughness modification at the contact zone

The wheel/rail contact does not occur at a point but over an area, known as the contact patch. This is typically 10 to 15 mm long and about the same width. When roughness wavelengths are short compared with the contact path length, their effect on the wheel/rail system is attenuated because of averaging across the contact patch. This effect is known as the contact filter. This is significant from about 2–3 kHz for speeds of 300 km/h, and at lower frequencies for lower speeds. In determining this effect a weighted average of the roughness needs to be calculated, taking into account the precise profiles of the contacting surfaces.

In early analytical models for this effect [1.6, 1.7], the extent of the correlation of the roughness across the width of the contact had to be assumed, as very detailed roughness data were not available. Figure 1.12(a) shows results from this model for a contact patch length of 11 mm. More recently, Remington has developed a distributed point-reacting spring (DPRS) model (see Chapter 2). This model is

Table 1.1. Damping of wheel modes during rolling [1.27]

n	Mode	Natural frequency (Hz)	Predicted loss factor	Measured loss factor
2	Radial	1722	0.007	0.018
3	Radial	2426	0.004	0.006
0	Radial	2761	0.005	0.006
4	Radial	3186	0.003	–
5	Radial	4089	0.002	–
2	One-nodal-circle	2387	0.003	0.008
3	One-nodal-circle	3030	0.002	–
4	One-nodal-circle	3707	0.001	0.0025

intended to be used with roughness measurements performed on multiple parallel lines a few millimetres apart. Such measurements have been made as part of a number of experimental exercises. Figure 1.12(b), from [1.26], shows, using a series of such measurements in combination with the DPRS model, that the filtering effect is not as severe at high frequencies as the analytical model of [1.6, 1.7] indicated. This topic is covered in more detail in Chapter 2.

1.4.5. Effective damping of a rolling wheel

It has been seen in Section 1.2 that a free wheel has very low damping. When the wheel is in contact with the rail, the track introduces damping to the wheel since its mobility has a phase that is close to 0°. This is characteristic of a damper, although it occurs owing to structural wave propagation in the track which radiates energy away from the wheel/rail contact point.

For a stationary wheel, modes with an antinode at the contact point will be damped by this mechanism but those with a node at the contact point will not. However, owing to wheel rotation the resonant behaviour of a wheel consists of rotating waves, as described in Section 1.2.3, and both the forward- and backward-rotating waves are damped by the rail.

The damping of the modes of the wheel while it is rolling can be derived from an analysis of the bandwidth of the peaks, from either experimental or predicted data. Table 1.1 lists such results, from [1.27]. This shows that the damping experienced by a rolling wheel on the track is considerably greater than for a free wheel. Consequently, if, in order to produce a noise-reducing wheel design, its structural damping is increased a little, the effect on rolling noise may be negligible even though the effect on a free wheel in the laboratory may be significant. This is discussed further in Chapter 6.

1.5. Radiation of sound

1.5.1. Radiation from the wheel

The radiated sound power from a vibrating structure can be written as

$$W_{\text{rad}} = \rho c S \sigma \langle v^2 \rangle \tag{1.10}$$

where S is the surface area of the structure, ρc is the characteristic specific impedance of air (density multiplied by speed of sound) and $\langle v^2 \rangle$ is the spatially averaged mean square normal velocity of the vibrating surface. The term σ is the radiation ratio, or radiation efficiency, and depends on the shape of the structure and the velocity distribution, in particular its wavelength. At high frequencies σ tends to 1, whereas at low frequencies σ is much less than 1.

For a wheel, the radiation ratio depends at low frequencies on the number of nodal diameters, n; the same result applies for both standing and propagating waves. Figure 1.13 shows radiation ratios predicted using an axisymmetric boundary element model for axial vibration. For $n = 0$ modes the low-frequency behaviour follows a trend in which σ is proportional to f^4, which is characteristic of a dipole radiator. For increasing values of n, the power of the frequency dependence increases according to $2n + 4$. Similar results apply to radial motion, except that the low-frequency behaviour has a frequency dependence of f^{2n+2}.

The frequency at which σ tends to 1 is dependent on the wheel radius and also increases slightly with increasing n. Comparing the curves of radiation ratio with the natural frequencies of the wheel shown in Fig. 1.5, it is found that only a small number of modes have their natural frequencies in the region where σ is much less than 1, the most important radial and one-nodal-circle modes occurring in the high-frequency region where σ tends to 1.

In order to use the sound power to estimate the sound pressure at a particular location, the directivity is also required. The directivity of the wheel radiation is a complex function of the wheel shape, mode shape and frequency. It has been found in laboratory experiments [1.18] that the directivity associated with axial (one-nodal-circle) modes can be approximated by that of a dipole, and the directivity for radial motion by a monopole (omnidirectional). These do not predict a minimum on the

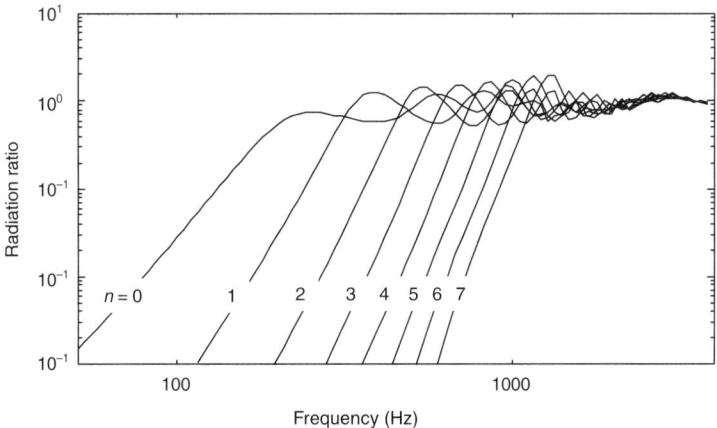

Fig. 1.13. *Radiation ratio of a wheel of diameter 0.92 m predicted using boundary elements for various numbers of nodal diameters*

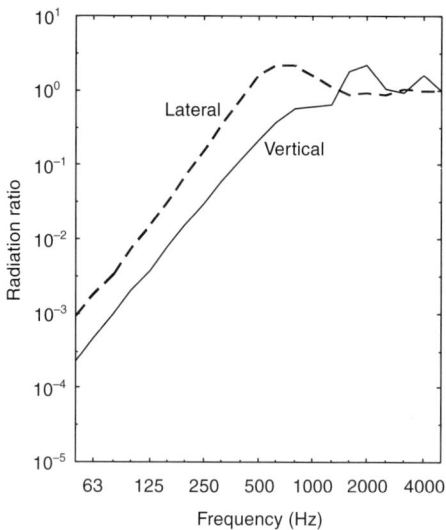

Fig. 1.14. Radiation ratio for a rail of type UIC60 vibrating vertically or laterally

axis of the wheel, as is the case in reality, nor the presence of other lobes in the radiation pattern, but if the purpose is to predict the average sound during the passage of a wheel past a trackside observer, they are sufficient.

1.5.2. Radiation from the rail

The sound radiated by the rail can be predicted using similar techniques to those for the wheel. Figure 1.14 shows the radiation ratio of the rail produced by a vertical or a lateral vibration of the rail along its whole length, with no deformation of the cross-section. At high frequencies this tends to a constant power, as given by equation (1.10) with σ tending to 1. At low frequencies, as for the wheel, σ is much less than 1, in this case depending on the cube of the frequency, typical of a line dipole. The sound radiated by the vertical vibration is lower than that from the lateral vibration at low frequencies owing to partial cancellation of the sound radiated by the top and bottom of the head and foot of the rail.

Another aspect that must be considered for the rail is the effect of the wavenumber and rate of decay of the vibration in the rail. The model just described is two-dimensional. This is valid when the wavelength of waves in the rail is large compared with that of sound in air and the spatial decay is small so that the vibration extends over a large distance. This is the case for frequencies over about 250 Hz for a very wide range of tracks [1.28]. At lower frequencies neither of these assumptions is valid; it has been seen in Fig. 1.8 that the decay rate is high at low frequencies and the wavelength also becomes similar to that in air.

As for the wheel, the sound power prediction needs to be combined with an estimate of the directivity in order to give the sound pressure at a particular position. Here, as for the wheel, simple monopole and dipole directivities are found to be acceptable.

1.5.3. Radiation from the sleepers

Owing to their large area compared with that of the rail, the radiation efficiency of sleepers is close to 1 over most of the frequency range of interest. Since they are well coupled to the rail at low frequencies (see Fig. 1.9), the sleepers dominate the sound radiation at low frequencies. A simple model based on a baffled piston can be used to predict the radiated power, and this is used with a uniform directivity to estimate pressures.

1.5.4. Aerodynamic sources

It has already been stated that aerodynamic noise becomes comparable to wheel/rail rolling noise only at train speeds of around 300 km/h (see Fig. 1.1). Aerodynamic sources of noise are therefore of importance only for trains travelling at the current highest speeds, and these are normally achieved only in locations outside towns and cities, where typical distances to dwellings and other receivers are greater. However, a number of considerations give the study of aerodynamic sources importance. Train speeds will undoubtedly be increased in the future as trains are required to compete with other modes of transport over longer and longer distances. Additionally, the use of barriers for reduction of noise propagation from the sources at wheel and rail height leaves the aerodynamic sources at higher level, above the barrier, untreated. This is well known to be the case for pantograph noise in Japan, where the use of lineside barriers is widespread. Reduction of the mechanical sources of noise results in a greater comparative significance of the aerodynamic noise sources at lower speeds; these will eventually form a limit below which noise reduction is not possible without tackling aerodynamic noise.

Turbulent air flow, which can be caused by many different parts of the vehicle, may become a significant source of aerodynamic noise at high train speed [1.29]. The locations of a number of sources have been identified and their strengths have been quantified in studies using specialized microphone array measurement techniques [1.30]. Important sources have been found to fall broadly into two categories. The first category consists of air flow over structural elements: the pantograph and the electrical isolators protruding at the base of the pantograph, the recess on the roof of the vehicle into which the pantograph drops, the recess formed at the intercoach connections, the bogies, and ventilators on the roofs of the carriages. The second category is where noise is created owing to the turbulent boundary layer, for example passing over the surfaces of the carriages. In addition, the flow over the succession of cavities presented by the louvred openings that allow for heat dissipation from the locomotive are a source of various types of aerodynamic noise depending on the length and depth of the cavity.

The sound power from various simple aeroacoustic sources is dependent on the train speed v according to a power law:

$$W \sim v^\alpha \tag{1.11}$$

For an aeroacoustic monopole source, such as the pulsating flow from an exhaust pipe, α equals 4. For a dipole source, such as the tones generated by vortex shedding from a cylinder, α has a value of 6. For a quadrupole source such as wind shear, α equals 8.

Empirically based models for each source of aerodynamic noise from trains can be derived if the source strengths are determined. The source classification in terms of monopole, dipole and quadrupole components and the component source strengths for a real train rely on empirical results for the source terms, measured using advanced experimental techniques such as microphone arrays either at the lineside or in wind tunnels. Such measurement studies may be complemented by the use of computational-fluid-dynamics (CFD) models that are also currently undergoing comparatively rapid development, particularly in terms of their applicability to problems of realistic complexity.

While a working model for the aerodynamic sources from trains, analogous to the TWINS model for rolling noise, is some way off, it is the objective of some current research [1.31].

1.5.5. Contribution of various sources

Figure 1.15 shows a typical example of the sound powers radiated by the sleepers, rail and wheel, as predicted by the TWINS model. The case modelled represents a TGV wheel at 300 km/h on a track composed of UIC60 rail, concrete twin-block sleepers and relatively soft pads (see Figs 1.7–1.9). As noted above, the sleeper dominates the sound radiation at low frequencies. The wheel radiation is only significant at high frequencies, here at 2 kHz and above. This frequency range is determined by the natural frequencies of the radial and one-nodal-circle modes; see

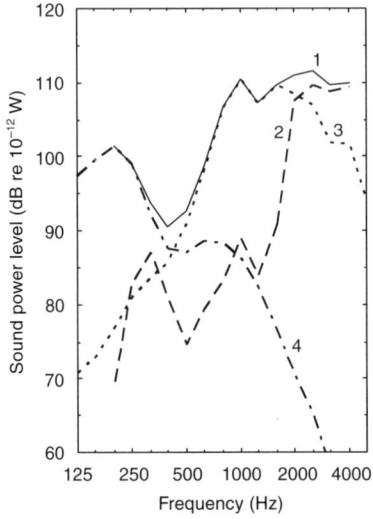

Fig. 1.15. Predicted components of sound power from a single wheel and the associated track vibration, TGV wheel on track with moderately soft pads: 1, total; 2, wheel; 3, rail; 4, sleeper

Table 1.2. Combinations of wheel, track and speed (km/h) used in the TWINS validation tests

	SNCF Corail wheel	SBB EW4 wheel	DB optimized wheel
UIC54 rails, hard pads, bibloc sleepers	50, 80, 125, 160	50, 80, 125, 160	–
UIC54 rails, no pads, wooden sleepers	50, 80, 125, 160	50, 80, 125, 160	–
UIC60 rails, medium pads, monoblock sleepers	80, 125, 160	80, 125, 160	80, 125, 160

Fig. 1.5. In the midfrequencies the rail is the dominant source; its contribution at high frequencies also being important, although less than that of the wheel.

In the case illustrated, the region between 1 and 4 kHz has most effect on the overall A-weighted level. The rail and wheel contribute equally in this case. However, in other situations this balance of wheel to rail will be different. Factors that influence it include the wheel and rail designs, the shape of the roughness spectrum (see Fig. 1.11) and the train speed. As speed increases, the peak in the sound spectrum will tend to move towards higher frequencies, which will increase the relative importance of the wheel. For the track considered here, the relatively soft pads included in the model lead to a higher rail component owing to the low decay rates found in the rail.

1.6. Validation

1.6.1. Experimental set-up

A series of experiments were performed in 1992–1993 to validate the TWINS models [1.15, 1.16]. These included three different types of wheel and track, and tests were carried out at up to four speeds, as indicated in Table 1.2. Extensive static measurements were performed to ensure that the wheel and track were correctly modelled, for example the resonance frequencies and damping of the wheel, and the frequency response functions and decay rates of the track. Very detailed roughness measurements were performed, with many parallel measurement lines across the running surface for input to the DPRS model referred to in Section 1.4.4 above.

1.6.2. Results

Figure 1.16 shows the overall results for these tests. This shows the predicted overall A-weighted sound level at 2 m from the track plotted against the measured result. Despite a range of values of 30 dB, the results only vary from a straight line by a standard deviation of 2 dB. This straight line is slightly offset, indicating an average overprediction of about 2 dB.

Figure 1.17 summarizes the spectral results of the 25 cases in terms of difference spectra. If the predicted noise were to correspond precisely to the measured noise, a value of 0 dB would be obtained. The mean over all 25 cases is close to 0, although the standard deviation in each band is up to 5 dB, yielding a rather large confidence interval. It has been found that the uncertainties in the input data, i.e. the roughness,

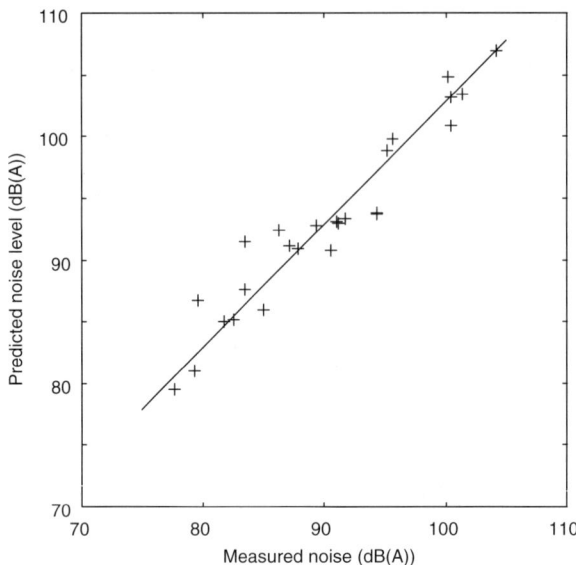

Fig. 1.16. Predicted versus measured noise for a range of three track types, three wheel types and up to four speeds [1.16]

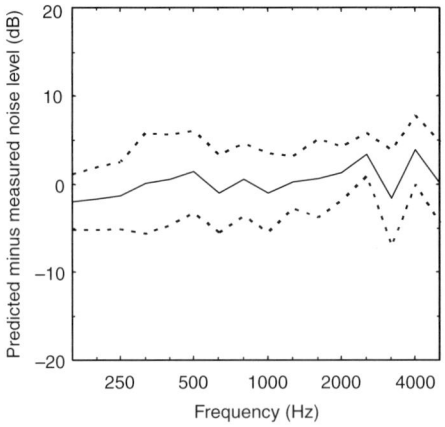

Fig. 1.17. Predicted minus measured noise spectra for the 25 combinations shown in Fig. 1.16. The mean and a range of one standard deviation are shown for each one-third octave band [1.16]

are large enough to account for this wide confidence interval. Furthermore, predictions of noise based on measured vibration had a standard deviation of only 2 dB in each band, which confirms that the greater uncertainties are contained in the first part of the model relating vibration to roughness.

Overall, despite the remaining uncertainties, the level of agreement from the validation tests gives confidence that the model can be used to design quieter wheels and track, as will be discussed further in Chapter 6. Further tests, not shown here, confirmed the conclusions for a freight wheel on two different types of track.

1.6.3. Sine wheel tests

In another series of tests, performed in 1990, some of the problems with defining the excitation were overcome by using wheels machined with a special sinusoidal profile [1.27]. By gradually increasing the train speed, the single-frequency excitation was allowed to sweep through the frequency range. Six wheels with profiles of different wavelengths were used to cover a wide range of excitation frequencies. These tests allowed some specific aspects of the theoretical models to be verified. It was confirmed that peaks in the wheel vibration occur at frequencies up to 20 Hz above the resonance frequencies of the free wheel owing to coupling with the rail and that the damping experienced by a rolling wheel is significantly greater than that of a free wheel, as predicted in [1.14] (see also Table 1.1).

1.7. Summary

Models for the generation of rolling noise show that the wheel, the rail and the sleeper can all be significant radiators of noise in different parts of the frequency range. These models have been demonstrated to be sufficiently refined to allow their use in detailed engineering design of low-noise wheels and track. This is the subject of Chapter 6.

At high speed, aerodynamic sources become increasingly important. Theoretical models for these are still at an earlier stage of development than those for the mechanical sources of noise.

1.8. References

1.1. HEMSWORTH, B. Recent developments in wheel/rail noise research. *Journal of Sound and Vibration*, 1979, **66**, 297–310.
1.2. MAUCLAIRE, B. Noise generated by high speed trains. New information acquired by SNCF in the field of acoustics owing to the high speed test programme. *Proceedings of Internoise*. Gothenburg, 1990, pp. 371–374.
1.3. KING, W. F., III. The components of wayside noise generated by high-speed tracked vehicles. *Proceedings of Internoise*. Gothenburg, 1990, pp. 375–378.
1.4. WETTSCHUREK, R. and HAUCK, G. Geräusche und Erschütterungen aus dem Schienenverkehr. In: M. Heckl and H. A. Müller (eds) *Taschenbuch der Technischen Akustik*, Springer, Berlin, 1994, 2nd edn, ch. 16.
1.5. REMINGTON, P. J. Wheel/rail noise, part I: characterization of the wheel/rail dynamic system. *Journal of Sound and Vibration*, 1976, **46**, 359–379.
1.6. REMINGTON, P. J. Wheel/rail noise, part IV: rolling noise. *Journal of Sound and Vibration*, 1976, **46**, 419–436.
1.7. REMINGTON, P. J. Wheel/rail rolling noise, I: theoretical analysis. *Journal of the Acoustical Society of America*, 1987, **81**, 1805–1823.
1.8. REMINGTON, P. J. Wheel/rail rolling noise, II: validation of the theory. *Journal of the Acoustical Society of America*, 1987, **81**, 1824–1832.
1.9. THOMPSON, D. J. Wheel/rail noise – theoretical modelling of the generation of vibrations. PhD thesis, University of Southampton, 1990.
1.10. THOMPSON, D. J. Wheel–rail noise generation, part I: introduction and interaction model. *Journal of Sound and Vibration*, 1993, **161**, 387–400.
1.11. THOMPSON, D. J. Wheel–rail noise generation, part II: wheel vibration. *Journal of Sound and Vibration*, 1993, **161**, 401–419.

1.12. THOMPSON, D. J. Wheel–rail noise generation, part III: rail vibration. *Journal of Sound and Vibration*, 1993, **161**, 421–446.
1.13. THOMPSON, D. J. Wheel–rail noise generation, part IV: contact zone and results. *Journal of Sound and Vibration*, 1993, **161**, 447–466.
1.14. THOMPSON, D. J. Wheel–rail noise generation, part V: inclusion of wheel rotation. *Journal of Sound and Vibration*, 1993, **161**, 467–482.
1.15. THOMPSON, D. J., HEMSWORTH, B. and VINCENT, N. Experimental validation of the TWINS prediction program for rolling noise, part 1: description of the model and method. *Journal of Sound and Vibration*, 1996, **193**, 123–135.
1.16. THOMPSON, D. J., FODIMAN, P. and MAHÉ, H. Experimental validation of the TWINS prediction program for rolling noise, part 2: results. *Journal of Sound and Vibration*, 1996, **193**, 137–147.
1.17. PETYT, M. *Introduction to finite element vibration analysis*, Cambridge University Press, Cambridge, 1990.
1.18. THOMPSON, D. J. and DITTRICH, M. G. *Wheel Response and Radiation – Laboratory Measurements of Five Types of Wheel and Comparisons with Theory*. ORE, Utrecht, 1991, Technical Document DT248 (C163).
1.19. KNOTHE, K. and GRASSIE, S. L. Modeling of railway track and vehicle/track interaction at high frequencies. *Vehicle System Dynamics*, 1993, **22**, 209–262.
1.20. THOMPSON, D. J. and VINCENT, N. Track dynamic behaviour at high frequencies. Part 1: theoretical models and laboratory measurements. *Vehicle System Dynamics*, 1995, Supplement 24, 86–99.
1.21. VINCENT, N. and THOMPSON, D. J. Track dynamic behaviour at high frequencies. Part 2: experimental results and comparisons with theory. *Vehicle System Dynamics*, 1995, Supplement 24, 100–114.
1.22. WU, T. X. and THOMPSON, D. J. The influence of random sleeper spacing and ballast stiffness on the vibration behaviour of railway track. *Acustica–Acta Acustica*, 2000, **86**(2), 313–321.
1.23. WU, T. X. and THOMPSON, D. J. The effects of local preload on the foundation stiffness and vertical vibration of railway track. *Journal of Sound and Vibration*, 1999, **219**, 881–904.
1.24. THOMPSON, D. J., JONES, C. J. C., WU, T. X. and DE FRANCE, G. The influence of the non-linear stiffness behaviour of rail pads on the track component of rolling noise. *Proceedings of the Institution of Mechanical Engineers, Journal of Rail and Rapid Transit*, 1999, **213F**, 233–241.
1.25. DITTRICH, M. G., BIEGSTRAATEN, F. J. W., DINGS, P. C. G. J. and THOMPSON, D. J. Wheel roughness and rolling noise – the influence of braking systems and mileage. *Rail Engineering International* 1994, **23**(3), 17–20.
1.26. THOMPSON, D. J. On the relationship between wheel and rail surface roughness and rolling noise. *Journal of Sound and Vibration*, 1996, **193**, 149–160.
1.27. THOMPSON, D. J., VINCENT, N. and GAUTIER, P. E. Validation of a model for railway rolling noise using field measurements with sinusoidally profiled wheels. *Journal of Sound and Vibration*, 1999, **223**, 587–609.
1.28. THOMPSON, D. J., JONES, C. J. C. and TURNER, N. Comparison of 2D and 3D rail radiation models. ISUR Contract Report No. 99/28, Southampton, 1999.
1.29. KING, W. F., III. A precis of developments in the aeroacoustics of fast trains. *Journal of Sound and Vibration*, 1996, **193**, 349–358.
1.30. BARSIKOW, B. Experiences with various configurations of microphone arrays used to locate sound sources on railway trains operated by the DB-AG. *Journal of Sound and Vibration*, 1996, **193**, 283–293.
1.31. TALOTTE, C. Aerodynamic noise, a critical survey. *Journal of Sound and Vibration*, 2000, **231**(3), 549–562.

2. Wheel and rail excitation from roughness

P. J. Remington
BBN Technologies, 70 Fawcett Street, Cambridge, MA 02138, USA

2.1. Introduction

Environmental noise generated by high-speed rail lines is of significant concern to transportation planners throughout the world. In the vicinity of these lines the noise produced by the interaction between the wheel and rail often dominates the overall noise signature. The forces that are a consequence of that interaction are generated in a small contact region, throughout which the wheel and rail are in intimate contact. The size of this contact region is typically about 1 cm^2, and its shape and the distribution of normal stresses that occur there depend on the static wheel load and the geometry of the wheel and rail. As irregularities on the running surfaces of the wheel and rail, which we term *roughness*, enter the region of contact, they modify the normal stresses in the region and cause fluctuations in the force between wheel and rail as shown in Fig. 2.1. It has come to be accepted in recent years that these fluctuating forces generated by roughness are the source of wheel/rail rolling noise [2.1–2.10].

Analytical models relating the fluctuating interaction forces in the contact region have become increasingly more sophisticated over the years. Early models [2.6–2.9] relied on the average roughness in the contact region times a nominal contact stiffness to estimate those forces. In recent years numerical models [2.1–2.3] based on the local elastic interaction between the wheel and rail in the contact region have been developed that account for irregularities in the wheel and rail profiles as well as the two-dimensional characteristics of the roughness distribution in estimating the interaction forces.

Early efforts at measuring wheel and rail roughness [2.7–2.9] relied on probes that typically used accelerometers and measured the roughness along a single line. Later, as the importance of the contact region and of the statistics of the roughness distribution in two dimensions began to be appreciated, more sophisticated techniques were employed that involved non-contact displacement-measuring devices [2.11] and displacement-measuring devices that measured the roughness along multiple parallel paths [2.12, 2.13].

As has already been mentioned, the characteristics of the roughness distribution in two dimensions affects the generation of forces in the contact region. Figure 2.2 shows the frequencies and associated inverse wavelengths for various train speeds. In this chapter we shall often use the inverse wavelength, or $k/2\pi$, where k is the wavenumber, to characterize the spatial frequency of roughness. In general, we are interested in wheel/rail noise in the 300 to perhaps 5000 Hz range. From the figure it is apparent that as the train speed increases, the range of inverse wavelengths important for the generation of rolling noise shifts to lower values (longer wavelengths). However, for audible rolling noise from trains operating at normal speeds inverse wavelengths from below 0.01 mm^{-1} (100 mm wavelength) to nearly 1 mm^{-1} (1 mm wavelength) are of interest.

Roughness, for the longest wavelengths in this range, has amplitudes on the order of millimetres, while for the shortest wavelengths the amplitudes are on the order of micrometres. The effectiveness of roughness in exciting wheel/rail noise depends on, in addition to wavelength, the amplitude of the roughness and the degree of correlation of the roughness in the direction transverse to rolling. In the case of wheels the magnitude of the roughness and the degree of correlation depend strongly on the type of braking system, e.g. disc or tread brake [2.15].

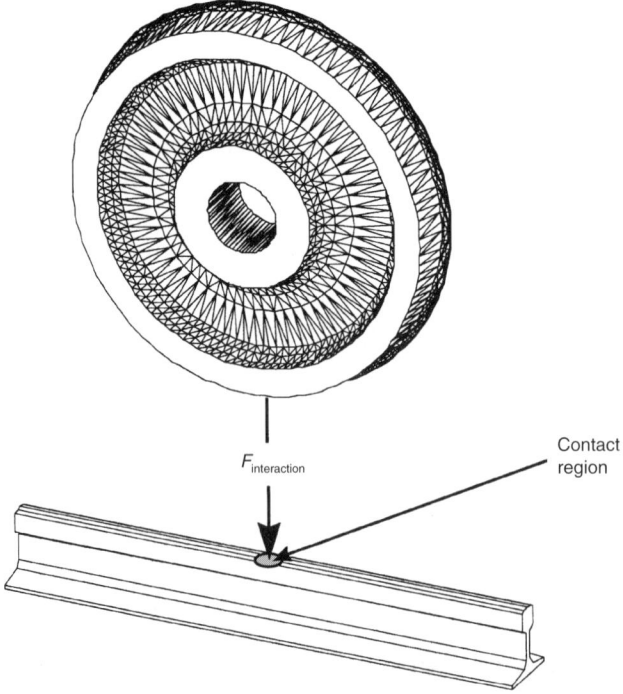

Fig. 2.1. The interaction between wheel and rail

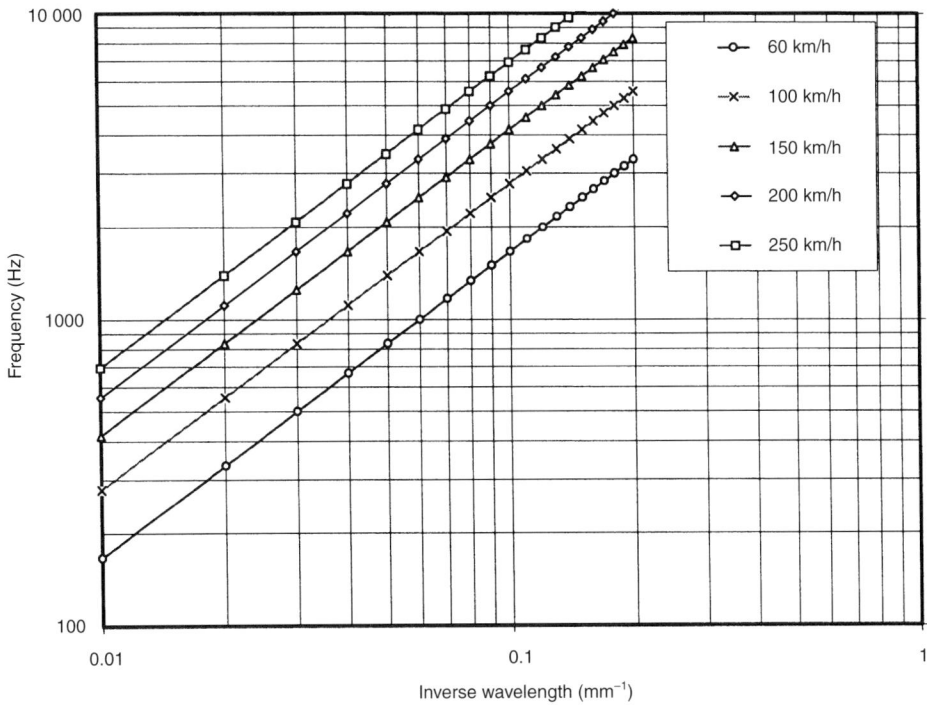

Fig. 2.2. Relationship between frequency and inverse roughness wavelength

A variety of efforts, over the years, have been undertaken to control wheel/rail noise using treatments that focus on interdicting the paths for vibration transmission and sound radiation, e.g. barriers, resilient wheels, rail damping and resilient rail fasteners [2.14, 2.16]. Efforts focused on controlling the generation of the interaction forces between wheel and rail in the contact zone, or, as we term it here, *controlling rolling noise at the source*, have been far fewer. The paucity of source control treatments is a consequence of a number of very special constraints imposed on any treatment in the contact zone. First of all, the very high stresses in the contact zone impose severe durability requirements on any treatment. In addition, the contact zone is where the traction and braking forces, needed to accelerate and stop the train, are generated. Consequently, contact zone treatments must not reduce wheel/rail adhesion. Furthermore, creep forces also generated in the contact zone play an important role in rail vehicle stability (hunting) at high speed. Therefore source treatments must avoid introducing rail vehicle instabilities or reducing the critical speed for the onset of hunting.

2.2. Roughness modelling

To be useful for the estimation of wheel/rail noise, the roughness measurements must be transformed into estimates of the interaction force between wheel and rail.

Figure 2.3 shows two equivalent circuits illustrating to first order how the wheel and rail interact in the vertical direction [2.17]. Two circuits are shown because the roughness can be characterized by either a *blocked force*, F_B, illustrated as a voltage source in the figure, or a *roughness velocity*, v_{rough}, shown as a current source. The two are equivalent and are related by the simple formula

$$F_B = Z_c v_{rough}$$

where Z_c is the contact impedance. The other variables in the figure, Z_W and Z_R, are the wheel and rail impedances, respectively.

The interaction force arises from the fact that the roughness irregularities on the running surfaces of the wheel and rail enter the contact region and alter the stress distribution. This altering of the stress distribution results in fluctuating forces between the wheel and rail, causing both local (Hertzian) and dynamic structural deformation of both wheel and rail. The local deformation is accounted for in the block diagrams by the contact impedance, which is typically modelled as a spring with a stiffness equal to the contact stiffness under the nominal wheel load. The contact stiffness is the ratio of force to deflection when the wheel is forced onto the rail and only local deformation is allowed. The force–deflection relationship is non-linear [2.18] and consequently the contact stiffness depends on the wheel load. In both equivalent circuits the contact impedance can be thought of as the source impedance. For typical wheel/rail systems, the contact impedance is much greater than the wheel or rail impedance at low frequency. Consequently, at low frequency the roughness can be thought of as a velocity source. On the other hand, at higher frequencies the wheel, rail and contact impedances are of comparable magnitude and the source characterization is more complicated.

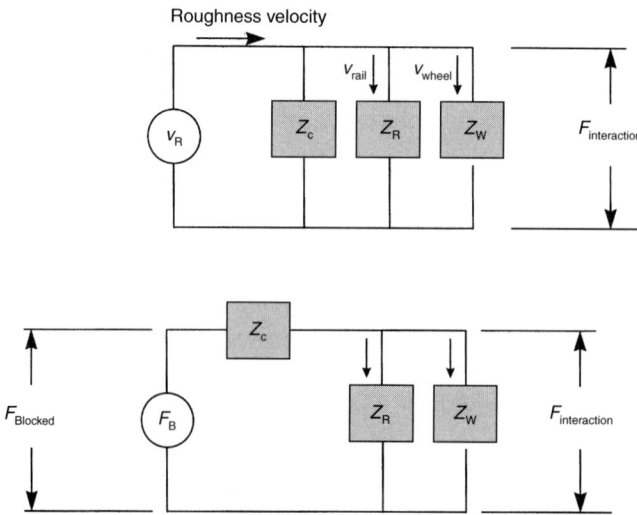

Fig. 2.3. Equivalent circuits for the interaction between wheel and rail

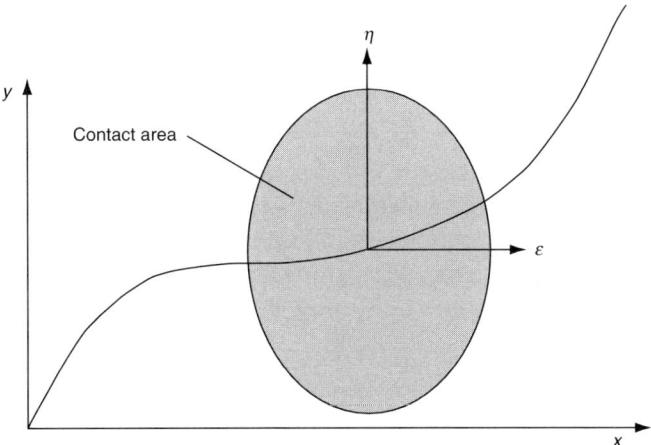

Fig. 2.4. Contact zone geometry

2.2.1. Average roughness model

In the early days of rail noise research the average roughness in the contact zone between wheel and rail multiplied by the nominal contact stiffness was used to estimate the blocked force. Figure 2.4 illustrates the contact zone geometry. One can approximate the force as a function of the position of the contact zone using the following equation:

$$f_B(x, y) = K_c \frac{1}{A_c} \int_{\text{contact area}} h_c(x-\xi, y-\eta) u_R(\xi, \eta) \, d\xi \, d\eta \tag{2.1a}$$

where f_B is the blocked force, A_c is the contact zone area, $u_R(x, y)$ is the roughness amplitude and h_c is a function that weights the roughness as a function of position in the contact zone. If the weighting function is uniform, this approach for estimating the force is referred to as the *average-roughness* model. While not correct in all circumstances, the average roughness model is simple and does provide some useful insights. If we Fourier transform equation (1a) into the wavenumber domain we obtain

$$F(k_x, k_y) = \frac{K_c}{A_c} H_c(k_x, k_y) U(k_x, k_y) \tag{2.1b}$$

where $F(k_x, k_y)$ and $U(k_x, k_y)$ are the spatial Fourier transforms of the blocked force and the roughness, respectively, and

$$H_c(k_x, k_y) = \int_{-\infty}^{+\infty} \int_{-\infty}^{+\infty} h_c(x, y) e^{j(k_x x + k_y y)} \, dx \, dy \tag{2.2}$$

The function H_c plays a role analogous to a frequency response in the time domain, and consequently the wavenumber spectrum (analogous to the frequency spectrum) of the blocked force, $S_F(k_x, k_y)$, can be related to the wavenumber spectrum of the roughness $S_R(k_x, k_y)$ by

$$S_F(k_x, k_y) = \left(\frac{K_c}{A_c}\right)^2 |H_c(k_x, k_y)|^2 S_R(k_x, k_y) \tag{2.3}$$

This expression is analogous to a similar one relating the frequency spectra of the input and output of a linear system by the frequency response of the system. If we restrict the contact area to moving at a uniform velocity along a line parallel to the x axis, it is easy to show [2.19] that the wavenumber spectrum of the blocked force is given by

$$S_F(k_x) = \left(\frac{K_c}{A_c}\right)^2 \int_{-\infty}^{+\infty} |H_c(k_x, k_y)|^2 S_R(k_x, k_y) dk_y \tag{2.4}$$

From this expression, the frequency spectrum of the blocked force can be readily obtain by noting that $\omega = k_x V$, where V is the velocity of the wheel.

Assuming now that the weighting function $h_c(x, y)$ in equations (2.1) is uniform, we can compute the wavenumber response function H_c from equation (2.2) to be

$$H_c(k_x, k_y) = \iint_{\text{contact area}} e^{jk_x x} e^{jk_y y} \, dx \, dy$$

which for a circular area of radius a becomes

$$H_c(k_x, k_y) = \frac{2J_1(ka)}{(ka)} \tag{2.5}$$

where $k = k_x^2 + k_y^2$). Substituting equation (2.5) into equation (2.4), we obtain a general expression for the blocked force based on the *average roughness*:

$$S_F(k_x) = \left(\frac{K_c}{A_c}\right)^2 \int_{-\infty}^{+\infty} \left|\frac{2J_1(ka)}{ka}\right|^2 S_R(k_x, k_y) dk_y \tag{2.6}$$

To proceed any further we need to know the dependence of the roughness wavenumber spectrum on k_y. For illustrative purposes, assume that the k_y dependence is given by

$$S_R(k_x, k_y) = S_0(k_x, k_y) \frac{1}{2\alpha k_x}; \quad -\alpha k_x \leq k_y \leq \alpha k_x$$
$$S_R(k_x, k_y) = 0; \quad |k_y| \geq \alpha k_x \tag{2.7}$$

If we Fourier transform this y dependence to obtain the spatial correlation in the y direction we obtain a result of the form $\sin(\alpha k_x \delta)/\alpha k_x \delta$. This result implies that as α increases, the correlation length in the y direction decreases. Substituting equation (2.7) into equation (2.6) and making the substitution $k_y = k_x \tan \theta$, we obtain [2.19]

$$S_F(k_x) = \frac{4}{\alpha} \left(\frac{K_c}{A_c k_x a}\right)^2 S_0(k_x) \int_0^{\tan^{-1} \alpha} J_1^2(k_x a \sec \theta) d\theta \tag{2.8}$$

WHEEL AND RAIL EXCITATION FROM ROUGHNESS

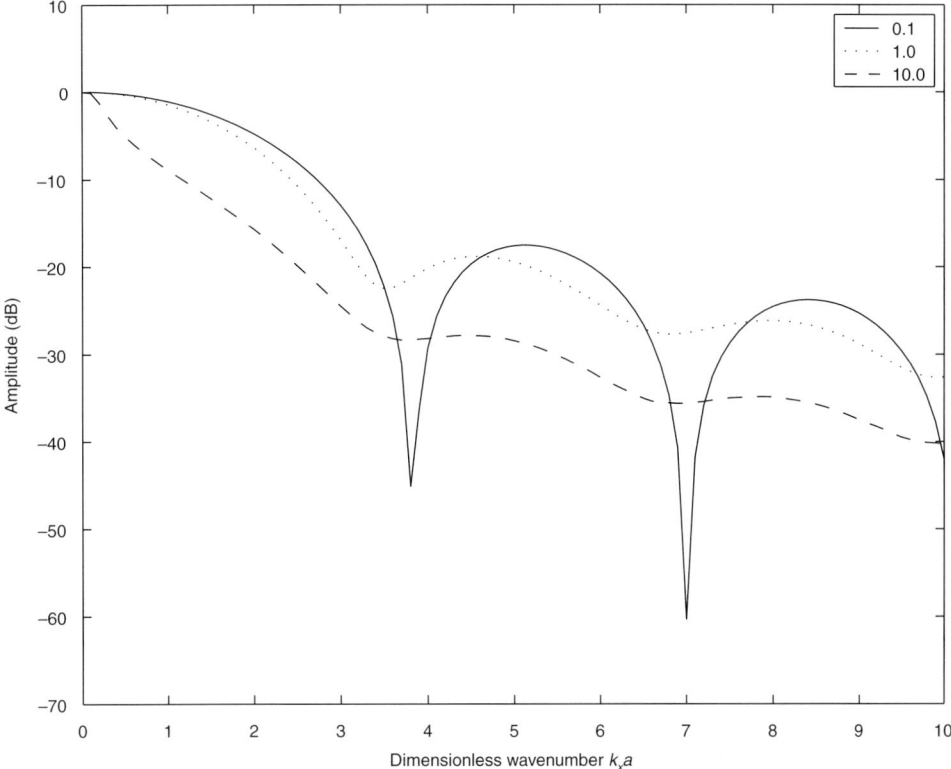

Fig. 2.5. Contact filter gain versus wavenumber

In Fig. 2.5 we evaluate this expression for various values of α. The figure clearly shows that the attenuation of the contact filter is strongly dependent on the correlation of the roughness in the direction transverse to rolling. Poorly correlated roughness (α large) is more strongly attenuated by the contact area.

2.2.2. Distributed point-reacting spring model

A more sophisticated, but still approximate, approach to the estimation of the forces in the contact region due to roughness, originally suggested by Heckl [2.20], can be obtained by covering the surfaces of the wheel and rail with distributions of point-reacting springs. By point-reacting we mean that the deflection of one spring does not induce any deflection in adjacent springs. Figure 2.4 shows the geometry of contact for a wheel and rail. The x direction is in the direction of rolling and the y direction is transverse to rolling. The deflection of a spring at distances ξ and η from the centre of the contact region is given by

$$u = u_0 - \frac{\xi^2}{2R_W} - \frac{\eta^2}{2R_R} + \frac{\eta^2}{2R_{WT}} \tag{2.9}$$

where u_0 is the displacement of the wheel relative to the rail, R_W is the wheel radius,

R_R is the transverse rail radius and R_{WT} is the transverse wheel radius, assumed in the equation to be concave relative to the rail.

Hertz [2.18] has shown that for smooth bodies in contact, the normal stress distribution in the contact region is given by

$$\sigma = \sigma_0 \left(1 - \frac{\xi^2}{a^2} - \frac{\eta^2}{b^2}\right)^{1/2} \tag{2.10}$$

where a and b are the semi-major axes of the elliptical area of contact. Consequently, to agree with the results of Hertz, our individual springs must be non-linear such that the stress distribution is given by

$$\sigma = Ku_0^{1/2}\left[1 - \frac{\xi^2}{2R_W u_0} - \frac{\eta^2}{2u_0}\left(\frac{1}{R_R} - \frac{1}{R_{WT}}\right)\right]^{1/2} \tag{2.11}$$

By making the springs generate forces proportional to the square root of the deflection, we are assured of getting the proper stress distribution in the contact region, provided

$$a = (2R_W u_0)^{1/2}$$

$$b = \left[2\left(\frac{R_R R_{WT}}{R_{WT} - R_R}\right)u_0\right]^{1/2} \tag{2.12}$$

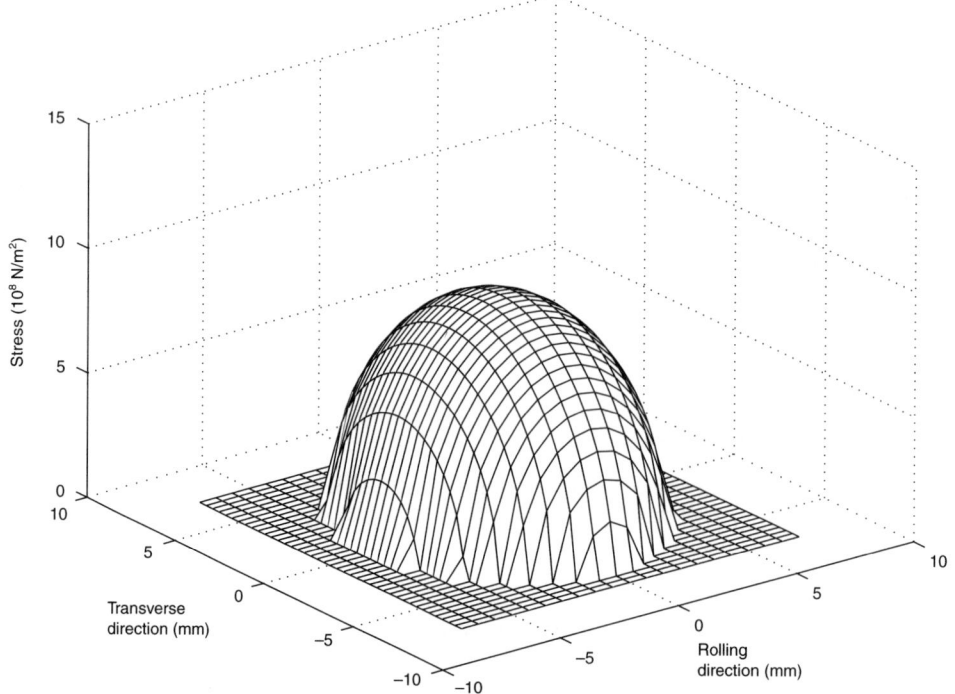

Fig. 2.6. *Stress distribution in the contact region*

Table 2.1. Parameter values for wheel/rail system

Parameter	Symbol	Value
Wheel load	Q	63 738 N
Wheel radius	R_W	508 mm
Rail radius	R_r	304 mm
Wheel transverse radius	R_{WT}	Infinite
Modulus of wheel and rail	E	2×10^{11} N/m²
Poisson's ratio	ν	0.3
Static deflection	u_0	76 µm

In general, equation (2.12) will overestimate the dimensions of the contact region. We can, however, make equation (2.12) correct if we define a new set of *equivalent* wheel and rail radii $\tilde{R}_W$ and $\tilde{R}_R$ such that equation (2.11) becomes

$$\sigma = K u_0^{1/2} \left(1 - \frac{x^2}{2\tilde{R}_W u_0} - \frac{y^2}{2\tilde{R}_R u_0} \right)^{1/2}$$

To define $\tilde{R}_W$ and $\tilde{R}_R$ we require the exact formulation of Hertz, defining the dimensions of the contact area and the static deflection under load. The required equations are given by [2.21]

$$a = a^* (3Qk/r)^{1/3}$$
$$b = b^* (3Qk/r)^{1/3} \qquad (2.13)$$
$$u_0 = \frac{u^*}{2} (3Qk/r)^{2/3} r$$

where Q is the load, $k = (1 - \nu^2)/E$ (ν is Poisson's ratio and E is the modulus of the wheel and rail material), $r = 1/R_W + 1/R_R - 1/R_{WT}$, and a^*, b^* and u^* are constants defined by Hertz and tabulated by Harris [2.21]. Combining equations (2.12) and (2.13), we find that

$$\tilde{R}_W = \frac{(a^*)^2}{u^*} \frac{1}{1/R_W + 1/R_R - 1/R_{WT}}$$
$$\tilde{R}_R = \frac{(b^*)^2}{u^*} \frac{1}{1/R_W + 1/R_R - 1/R_{WT}} \qquad (2.14)$$

We now estimate the interaction force between wheel and rail, when the contact region is centred at (x, y) by integrating the normal stress over the area of contact,

$$f_B(x, y) = \iint_{\substack{\text{contact} \\ \text{region}}} K u_0^{1/2} \left(1 - \frac{\xi^2}{2\tilde{R}_W u_0} - \frac{\eta^2}{2\tilde{R}_R u_0} \right)^{1/2} d\xi\, d\eta \qquad (2.15)$$

and use equation (2.14) to calculate the *equivalent radii*. Figure 2.6 shows the stress distribution in the contact area for the wheel and rail parameters given in Table 2.1.

The stress is simply the integrand in equation (2.15) in which K has been chosen such that the force, predicted by equation (2.15), is the same as that predicted by Hertz's theory to produce the static deflection u_0. The constant K is independent of load, but will change if the wheel and rail geometries are changed. The stress distribution in the figure has the ellipsoidal shape predicted by Hertz's theory, with zero stress outside of the contact region. The major axis of the contact area is 12.8 mm in the rolling direction and 10.6 mm in the transverse direction, just as predicted by Hertz's theory. Finally, Fig. 2.7 shows that the relationship between force and deflection predicted by equation (2.15) is the same as that predicted by Hertz's theory.

Equation (2.15) is for smooth bodies in contact. It can be extended to include roughness by simply adding the roughness at x and y, $u_R(x,y)$, to the static deflection, u_0, as follows:

$$f_B(x,y) = \iint_{\substack{\text{contact} \\ \text{region}}} K u_0^{1/2} \left(1 + \frac{u_R(x+\xi, y+\eta)}{u_0} - \frac{\xi^2}{2\tilde{R}_W u_0} - \frac{\eta^2}{2\tilde{R}_R u_0} \right)^{1/2} d\varepsilon \, d\eta \qquad (2.16)$$

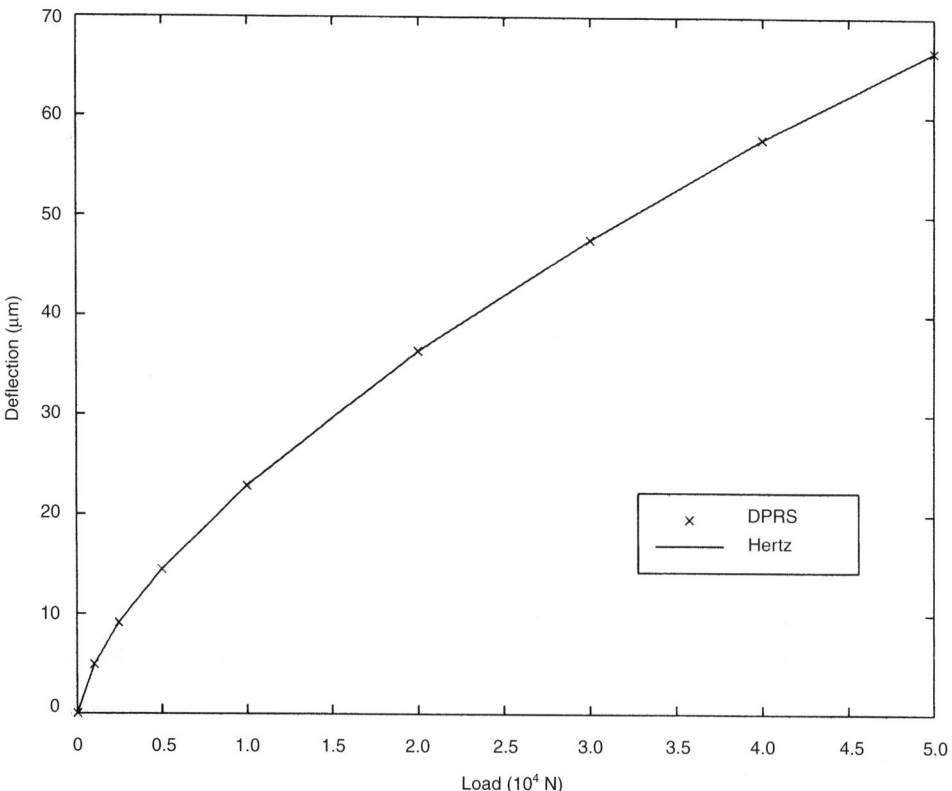

Fig. 2.7. Force–deflection relationship predicted by the DPRS model compared with the predictions of Hertz's theory

WHEEL AND RAIL EXCITATION FROM ROUGHNESS

Equations (2.14) and (2.16) together define the *distributed point-reacting spring* (DPRS) model of roughness excitation.

It is interesting to examine the effect of the roughness wavelength in the rolling direction and of the degree of correlation of the roughness in the direction transverse to rolling on the interaction force, as was done for the average roughness model in Fig. 2.5. To do so we need to determine numerically the contact filter $H_c(k_x, k_y)$. If we define the roughness to be sinusoidally distributed, such that

$$u_R(x, y) = u_{R0} \cos[k_x(x+\xi)]\cos[k_y(y+\eta)] \tag{2.17}$$

where x and y define the location of the centre of the contact region, ξ and η are the coordinates within the region, as illustrated previously in Fig. 2.4, and k_x and k_y are the wavenumbers in the rolling direction and the transverse direction, respectively. By substituting this equation into equation (2.16), one can compute the blocked force. By repeating the calculation at a number of points along the path of travel, the blocked force as a function of position can be computed. In general, we will fix y and allow x to vary. If the amplitude of the sinusoidal roughness in equation (2.17) u_{R0} is small compared with the static deflection u_0, f_B the blocked force will also be approximately sinusoidal. By carrying out the indicated calculations in equation (2.16) at discrete intervals along the path of travel, we can calculate the amplitude of the blocked force for any chosen k_x and k_y. By repeating the calculation for a number of different wavenumbers k_x and k_y, the blocked force as a function of wavenumber can be computed. The contact filter for each wavenumber combination can then be obtained from equation (2.1b) as

$$H_c(k_x, k_y) = \frac{A_c F(k_x, k_y)}{K_c u_{R0}} \tag{2.18}$$

If we carry out the above calculation for $k_y = 0$, using the same wheel/rail parameters as before, the result is closely equivalent to the average-roughness model case in Fig. 2.5 for $\alpha = 0.1$, where the roughness is well correlated in the direction transverse to rolling. Figure 2.8 compares the result of the DPRS model for the case $k_y = 0$ with the result of the average-roughness model shown in Fig. 2.5 for $\alpha = 0.1$. The DPRS calculations were carried out using a roughness amplitude of 10% of the static deflection. As the figure shows, the two results are comparable, indicating that the sophistication of the DPRS model is not needed for low-amplitude, well-correlated roughness.

For the case of roughness poorly correlated in the direction transverse to rolling, the picture is somewhat different. In order to compare the blocked-force predictions of the two approaches for roughness poorly correlated in the direction transverse to rolling, we have plotted the result of evaluating equation (2.8) with $\alpha = 10$ in Fig. 2.9. We have also plotted the result from evaluation of equation (2.16) using the roughness spectrum of equation (2.7). The curve for the average-roughness model shows less attenuation due to contact area filtering than does the curve for the DPRS model. The reason for the discrepancy between the average-roughness model and the DPRS model for uncorrelated roughness can best be illustrated by examining Fig. 2.10. There, the incremental normal stress distribution in the contact zone due to roughness is shown. The incremental stress distribution is the stress distribution with roughness on the wheel or rail minus the stress distribution that would be obtained on the same wheel and rail if they were perfectly smooth.

The distribution has been computed using roughness the wavelengths of which are much greater than the dimensions of the contact region. The figure shows a significant rise in the stress difference near the edges of the contact zone. For the average-roughness model, the equivalent figure would show a uniform distribution of stress difference. Consequently, with the distributed point-reacting spring model, the roughness near the edges of the contact zone would be expected to contribute most to the blocked force. If the roughness is not well correlated in the direction transverse to rolling, the DPRS model adds up contributions from roughness components which are separated further and hence are more poorly correlated than in the average-roughness model. Consequently, the distributed

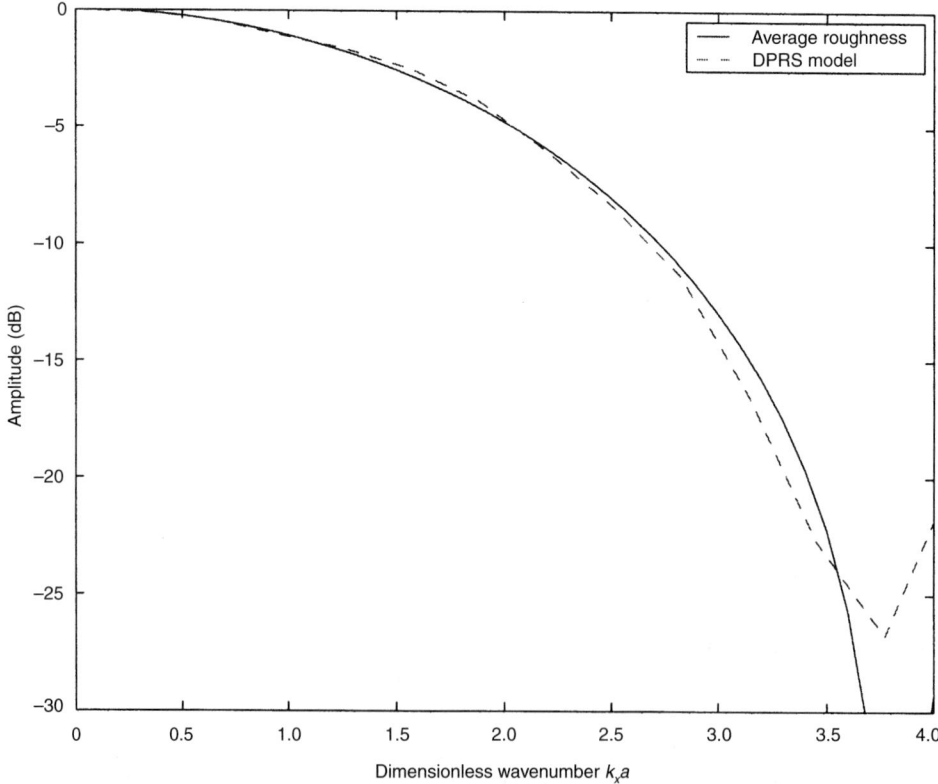

Fig. 2.8. Comparison of the contact filter for the DPRS model and average-roughness model for low-amplitude well-correlated roughness

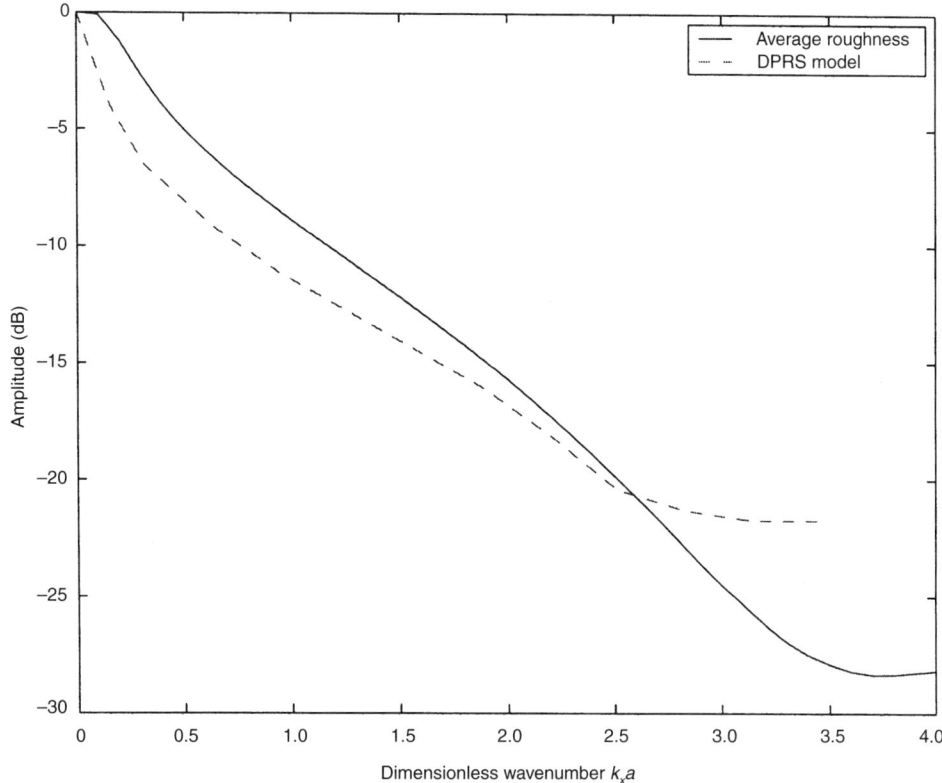

Fig. 2.9. Comparison of the contact filter for the DPRS model and average-roughness model for low-amplitude poorly correlated roughness

point-reacting spring model will, in general, compute a lower blocked force than will the average-roughness model for roughness poorly correlated in the direction transverse to rolling. The distributed point-reacting spring model acts as if the effective area of the contact region were increased. The increased incremental stress at the edges of the contact zone is a consequence of the non-linear force–deflection characteristic of the distributed springs in the DPRS model. The dependence of the distributed spring forces on the square root of the deflections means that at low deflection the force increases more rapidly with deflection that it does at high deflection. Consequently, at the edges of the contact region, where the deflections due to the static load are small, roughness will create greater changes in stress than in the interior of the contact region, where the static deflections are large. Thus, the concentration of incremental stress near the boundaries of the zone of contact is a direct result of the force–deflection characteristic of the springs in the DPRS model. This is not, however, an artefact of the model, as we shall see in the next section, where a third analytical model that accounts for the full elastic interaction in the contact region confirms the DPRS predictions.

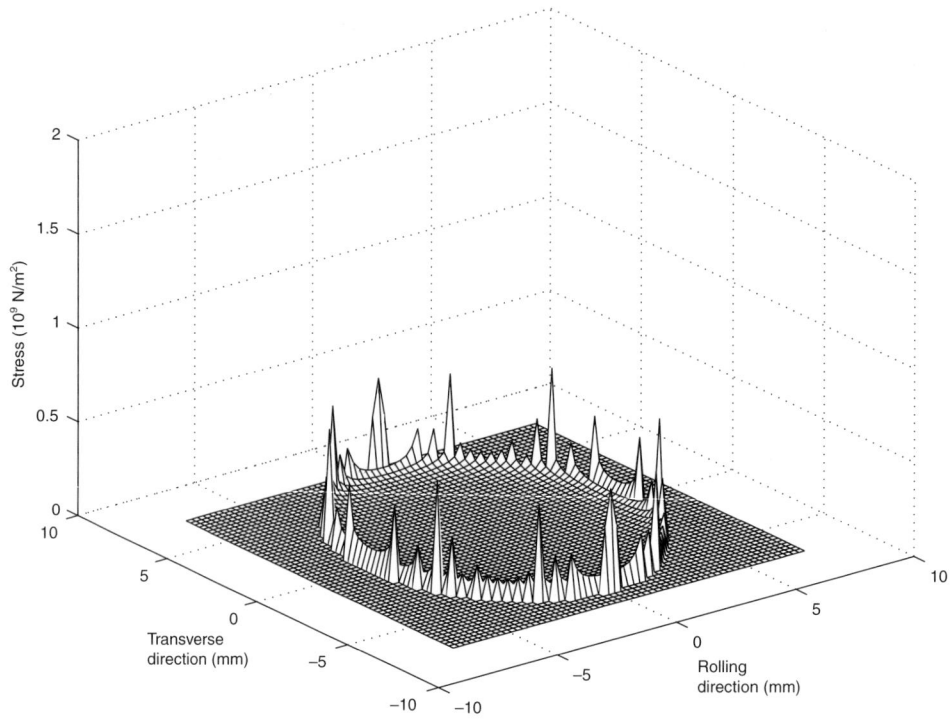

Fig. 2.10. Incremental stress distribution in the contact region due to roughness

2.2.3. Full elastic-interaction model

The final model that has been developed for estimating the forces in the contact region accounts for the full elastic interaction between wheel and rail. The *full elastic-interaction* model, first utilized by Paul and Hashemi [2.22, 2.23] for examining contact stresses in wheels and rails with worn profiles, begins by assuming that both the wheel and the rail can be modelled locally as elastic half-spaces. In general this is a reasonable assumption, since the effects of forces applied to the surface of an elastic half-space decay rapidly with distance from the point of application of the force. If a point force f is applied normal to the surface of an elastic half-space, the normal displacement u of the surface at a distance r from the forces is given by [2.24]

$$u = \frac{f(1-\nu^2)}{\pi E r} \qquad (2.19)$$

where E is the Young's modulus of the material, ν is Poisson's ratio and r is the distance between the force and the point of deflection. Directly under the point force, the displacement is infinite. However, if the force is distributed over an area, the displacement is given by

$$u = \frac{mf(1-\nu^2)}{EA^{1/2}} \qquad (2.20)$$

where m is a constant defined by the aspect ratio of the area over which the force is distributed ($m = 0.95$ for a square [2.24]) and A is the area over which the force is distributed.

For a known force distribution equation (2.20) can be used to calculate the deflection both within and outside the force distribution. For interaction problems, however, the force distribution is unknown and the area over which it is distributed is also unknown. Hertz solved the problem by setting up an integral equation for the deflection at the interface between the two interacting bodies in terms of the unknown pressure distribution in the contact area:

$$u(x,y) = \int_{\substack{\text{contact}\\\text{area}}} \frac{f(\xi,\eta)(1-\nu^2)}{\pi E(\xi^2+\eta^2)^{1/2}} d\xi\, d\eta \tag{2.21}$$

For smooth bodies of simple geometry, he was able to obtain an exact solution. Here the problem is somewhat more difficult, as the bodies may not be simple in geometry (constant radii of curvature that are large compared with the deflections and contact regions) and they are not smooth.

Therefore, we apply here a numerical approach to allow us to deal with the more complex geometries and with roughness. We break the surface where the wheel and rail come into contact into a number of small areas. The total area must be greater than the anticipated contact area. We then use a discretized version of equation (2.21) given by

$$u_i = \sum_j H_{ij} f_j \tag{2.22}$$

where

$$H_{ij} = \frac{1-\nu^2}{\pi E[(\xi_i - \xi_j)^2 + (\eta_i - \eta_j)^2]^{1/2}}; \quad i \neq j$$

$$= \frac{m(1-\nu^2)}{EA^{1/2}}; \quad i = j$$

Here u_i is the deflection at location (ξ_i, η_i), f_j is the force at location (ξ_j, η_j) and ξ is the distance from the nominal centre of the contact region in the rolling direction and η is the corresponding distance in the direction transverse to rolling. The result is a matrix equation

$$\boldsymbol{u} = [\mathbf{H}]\boldsymbol{f} \tag{2.23}$$

in which $\boldsymbol{u}$ is the vector of displacements at all locations within the contact region, $\boldsymbol{f}$ is the vector of forces at those locations and $[\mathbf{H}]$ is the matrix formed from H_{ij} defined above. The solution for the forces in the contact region is obtained by using equation (2.23) in an iterative fashion. We begin by solving equation (2.23) to obtain the forces, assuming the displacements at each point to be given by the vector defined by

$$u_i = u_0 + u_R(x + \xi_i, y + \eta_i) - \frac{\xi_i^2}{2R_W} - \frac{\eta_i^2}{2R_R} + \frac{\eta_i^2}{2R_{WT}} \tag{2.24}$$

where x and y are the coordinates of the nominal centre of the contact area. The force vector is then obtained from

$$f_0 = [\mathbf{H}_0]^{-1} u_0 \tag{2.25}$$

where u_0 is given by equation (2.24) for all of the incremental areas and $\mathbf{H}_0$ is given by equation (2.22). The resulting force vector will show that a number of the incremental areas in the initial guess of the contact region will be in tension. Since tension cannot occur in the contact region, the rows and columns corresponding to these areas are removed from the matrix $\mathbf{H}_0$ and a new matrix equation is formed,

$$f_1 = [\mathbf{H}_1]^{-1} u_1 \tag{2.26}$$

in which the subscript 1 indicates that only those elements of f and u and rows and columns of $\mathbf{H}_0$ for which the corresponding incremental areas are in compression are retained. The process continues until the force vector contains only compression forces.

When the calculations are carried out as indicated for smooth bodies, the stress distribution is identical to that for the DPRS model shown in Fig. 2.6. When roughness with wavelengths large compared with the dimensions of the contact zone is added, the differential stress distribution (the difference between the stress distribution with and without roughness) is the same as that predicted by the DPRS model as shown in Fig. 2.10. Even for high-amplitude roughness, as large as 50% of the static deflection under the nominal wheel load, the two models provide nearly identical predictions of the stress distribution [2.17, 2.25]. Only when the roughness wavelengths become short compared with the dimensions of the contact zone do discrepancies between the two models become apparent, and then the differences are primarily in the details of the stress distribution and not in the overall interaction force, which is of primary interest here.

To compare the results of the two models using measured roughness requires some modifications to deal with the large dynamic range of typical roughness data. In particular, roughness with large-amplitude long-wavelength components, as is typical of most wheel and rail roughness, cannot be handled unless the wheel is allowed to move relative to the rail. In practice the wheel suspension and rail foundation allow the wheel and rail to move relative to one another and thereby reduce any forces generated by these long-wavelength roughness components. The DPRS and full elastic-interaction models as described so far keep the wheel and rail fixed relative to one another and allow only for local or Hertzian deformation in the contact region. Consequently, when large-amplitude long-wavelength roughness is encountered, the resulting large deformations in the contact zone result in anomalous predictions of the blocked force for the shorter-wavelength roughness of most interest for wheel/rail noise.

To alleviate this problem, an approach has been employed [2.17, 2.25] in which the rail is fixed and the wheel is allowed to move as a rigid body, as illustrated in Fig. 2.11. As the figure shows, a damper has been attached to the wheel to prevent ringing at the contact resonance. By properly choosing the wheel mass, a wavelength

WHEEL AND RAIL EXCITATION FROM ROUGHNESS

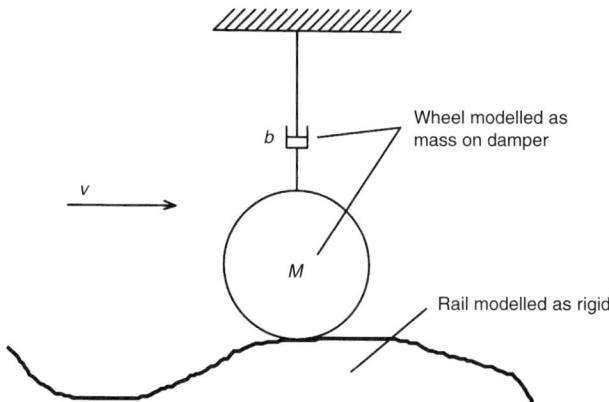

Fig. 2.11. Wheel model used to deal with high-amplitude long-wavelength roughness components

can be selected below which the wheel follows the roughness and generates little force and above which the wheel is essentially fixed relative to the rail and the resulting force is the blocked force. This wavelength is referred to as the cut-off wavelength λ_c. By modifying the calculation scheme in this manner we introduce what is essentially a high-pass filter, but a high-pass filter that mimics the actual physics of the wheel/rail interaction.

The equation of motion for the wheel is given by

$$M\frac{d^2u_0}{dt^2} + b\frac{du_0}{dt} = f_B - f_s \tag{2.27}$$

where M is the wheel mass, b is the damping coefficient of the damper in the figure, f_B is the force between the wheel and rail and f_s is the nominal static wheel load. Taking the nominal contact stiffness to be K_c, we can rewrite the above equation as

$$\frac{d^2u_0}{dx^2} + \frac{2\pi\eta}{\lambda_c}\frac{du_0}{dx} = (2\pi)^2\frac{(f_B - f_s)}{\lambda_c^2 K_c} \tag{2.28}$$

where x is the distance in the rolling direction and λ_c is the cut-off wavelength described above. Equation (2.28) is typically solved using finite-difference techniques, with the force f_B calculated from the DPRS or full elastic-interaction model by integrating the normal stresses over the contact area.

A comparison of the blocked force between the wheel and rail for roughness measured on a French National Railways (SNCF) rail [2.25] is shown in Fig. 2.12. The parameters for the wheel/rail system and the roughness data are given in Table 2.2. The force, as a function of inverse wavelength in the axial direction, predicted by the two models is seen to agree quite closely even when the wavelengths are less than the dimensions of the contact region.

Because computation times for the DPRS model can be orders of magnitude less than for the full elastic-interaction model, the DPRS model is the model of choice for most calculations. The full elastic-interaction model may be preferred when the

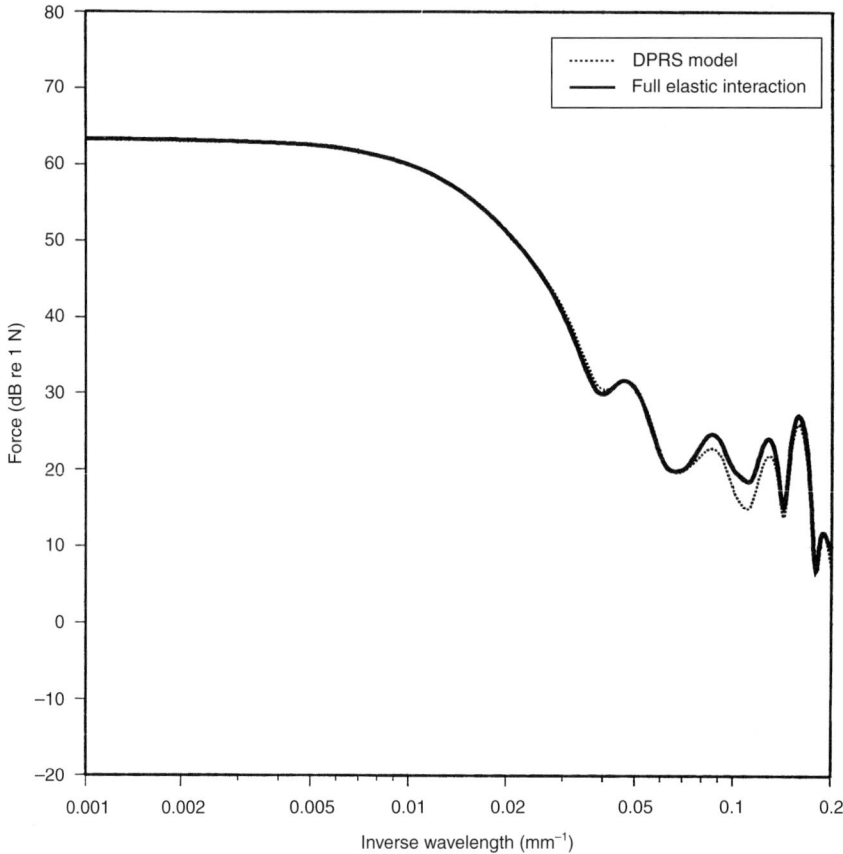

Fig. 2.12. Blocked force as a function of inverse wavelength for the DPRS and full elastic-interaction models

Table 2.2. Parameters for wheel/rail system and roughness measurement

Parameter	Value
Wheel radius	447 mm
Wheel transverse radius	Infinite
Rail radius	208 mm
Cut-off wavelength	400 mm
Loss factor	1.0
Track spacing	6 mm
Point spacing	2 mm
Static load	40 000 N
Contact area transverse width	6.8 mm
Contact area length (rolling direction)	11.4 mm

stress distribution is desired for roughness wavelengths short compared with the contact region dimensions.

2.3. Roughness measurement

Over the past 30 years a number of devices have been designed for the measurement of wheel and rail roughness. The earliest devices were designed to obtain the roughness along a single line on a wheel or rail [2.19, 2.26]. Devices employed in modern-day research measure the roughness along a number of parallel lines, allowing for better determination of the forces of interaction between wheel and rail [2.12, 2.13]. A good review of the many devices developed in Europe in the early years can be found in [2.27] and discussion of alternative sensing techniques can be found in [2.11]. Here we shall focus on two different approaches. One uses accelerometers as the sensing element, and the other uses linear variable displacement transducers (LVDTs), a displacement-sensing device. The former was the first type of device used for measuring roughness and was used only for measurement along a single line. The latter device has been used successfully in more recent years for measuring the roughness along multiple parallel lines on both wheels and rails.

2.3.1. Accelerometer-based devices

Figure 2.13 shows a sketch of one of the first devices developed for the measurement of roughness on rails [2.9]. It consists of an accelerometer attached to a lightweight, hardened probe that is pressed against the rail using a spring steel arm to ensure maintenance of good contact as the probe is pulled over the rail. The probe is attached to a six-wheeled cart, each wheel of which is made of rubber. An earlier version of the device [2.27] used a cart that slid on Teflon bearings along an aluminium I-beam. That approach was abandoned in favour of the one shown here because transport and set-up of the I-beam proved to be too cumbersome.

The cart shown in the figure was pulled along the rail by means of a variable-speed motor, and the speed was monitored using an optical probe that provided a fixed

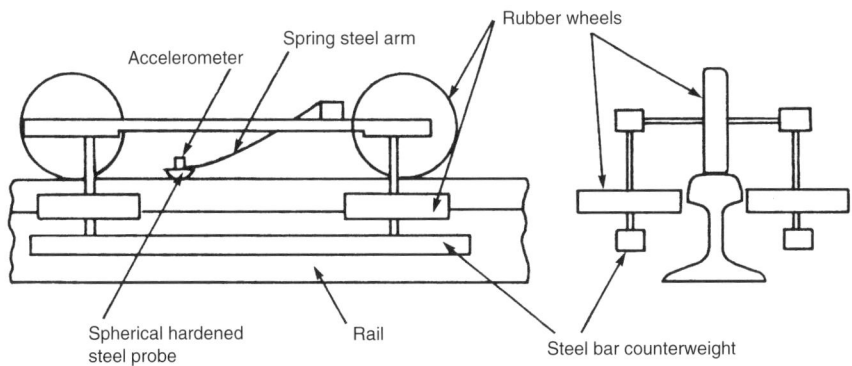

Fig. 2.13. One of the original accelerometer-based devices for measuring roughness

number of pulses per rotation of the front wheel. During measurement an accelerometer on the cart monitored cart vibration, and laboratory-measured transfer functions were used to determine the degree to which cart vibration contaminated the probe signal, if at all.

The device was also used for measuring wheel roughness. By removing the probe and the spring steel arm from the cart and mounting it on a wheel-truing machine, the probe could be made to ride on the wheel running surface as the wheel was turned slowly by the machine. Given the speed of the cart or the rotation speed of the wheel, the roughness spectrum S_R could be easily calculated from the measured acceleration spectrum S_a by means of the formula

$$\frac{S_a(\omega)}{\omega^4} \Delta\omega = S_R(k)\Delta k \qquad (2.29)$$

where Δk and $\Delta \omega$ are the wavenumber and frequency bandwidths, respectively, of the measurement. The two bandwidths are related by $\Delta k = \Delta\omega/V$, where V is the speed of the cart.

The device has been used to provide $\frac{1}{3}$ octave roughness spectral data with amplitudes in the range of 0.1 to 100 μm over the wavelength range of 20 to 2 mm. The device is best suited for the measurement of high-wavenumber (short-wavelength) roughness because, for constant roughness amplitude, the acceleration signal increases with decreasing roughness wavelength. This is advantageous because, as we shall see in the next section, the roughness spectrum actually decreases rapidly with decreasing wavelength. On the other hand, the device is hampered by noise at low frequencies. The noise at low frequencies from the instrumentation and from extraneous vibration sources interferes with the measurement of low-wavenumber (long-wavelength) roughness. In addition, the measurements can also be hampered at high frequencies by a contact resonance of the probe on the wheel or rail surface.

To some extent these limitations on the range of roughness wavenumber can be overcome by making the measurement at higher or lower speed, as necessary, because the frequencies associated with a given wavenumber depend on the speed of the measurement, i.e. $k = \omega/V$. On the other hand, with the modern-day emphasis on high-speed trains, the interest has shifted to low-wavenumber roughness, where displacement-based devices have a number of advantages.

2.3.2. Displacement-based devices

A number of researchers have used displacement-based devices in recent years to obtain roughness data on multiple parallel lines on wheels and rails [2.12, 2.13]. Figure 2.14 shows an example of one such device, developed by Mueller BBM for the German railway company (Deutsche Bahn AG) [2.13]. The principle of operation can be seen from the figure. The running surface is in contact with a measuring probe made of hard alloy with a radius of 7 mm. The vertical movement is fed via a 90° lever system to a horizontally mounted LVDT. The transducer and probe are attached to a sled, which is moved by means of a stepping motor along a

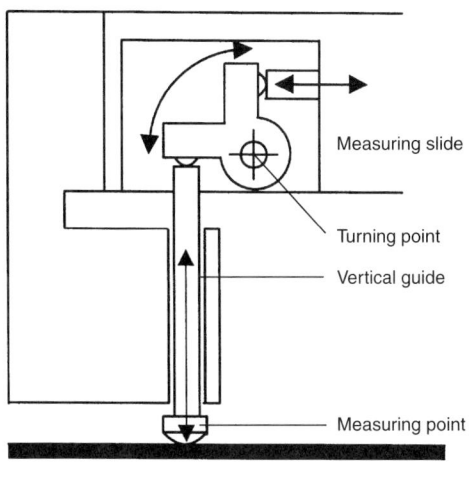

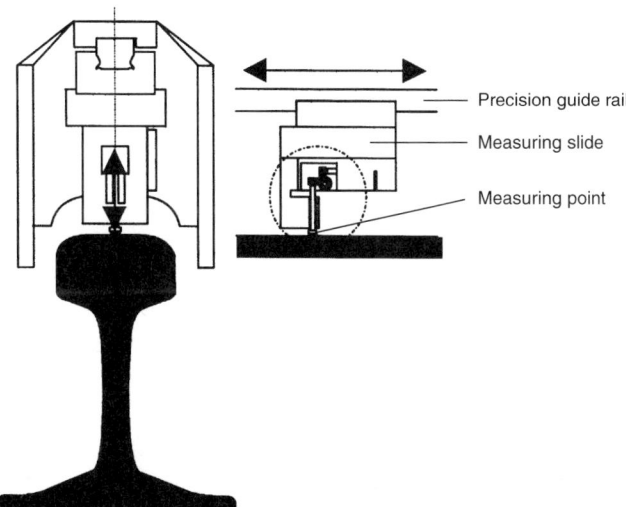

Fig. 2.14. A displacement-based device for the measurement of roughness (figure courtesy Mueller BBM, Planegg, Germany)

high-precision ball-bearing slide. Data samples are taken every 0.5 mm and stored digitally. The system is fully battery operated and provides a measuring length of 1200 mm. The resolution is 1 μm for the overall roughness signal and about 0.1 μm in any $\frac{1}{3}$ octave band of the wavenumber spectrum. The high accuracy is achieved by calibrating the device by means of a polished stone. The measurement of a trace on the surface of the stone (1500 × 70 × 200 mm, quality 00) is stored in memory and is used as a straight reference line for all succeeding measurements. A device based on the same principle has been developed for wheel roughness measurement. Both devices take advantage of modern digital technology for storing, processing and displaying the data.

Displacement-based roughness-measuring devices like this are now used almost exclusively in modern-day rail noise research. They offer significant advantages over acceleration-based devices in the measurement of low-wavenumber (long-wavelength) roughness. In addition, they are able to measure high-wavenumber roughness with sufficient accuracy so as to span the wavenumbers of interest in the generation of noise from high-speed trains. The roughness data discussed in the next section were obtained using displacement-based techniques.

2.4. Wheel and rail roughness characteristics

Displacement-based roughness-measuring devices have been used in recent years to acquire densely sampled roughness data on wheels and rails. These data have typically covered inverse wavelengths from ~0.001 mm^{-1} to ~1 mm^{-1}. For the most part these data have shown very similar characteristics, i.e. a dynamic range on the order of 60 dB with the amplitudes decreasing rapidly with increasing inverse wavelength. As mentioned earlier, roughness amplitudes on wheels depend to some degree on the type of braking system. In addition, as we shall see, the correlation lengths in the transverse direction seem also to be related to the type of braking system, although the available data are sparse enough to make any conclusions based on them tentative.

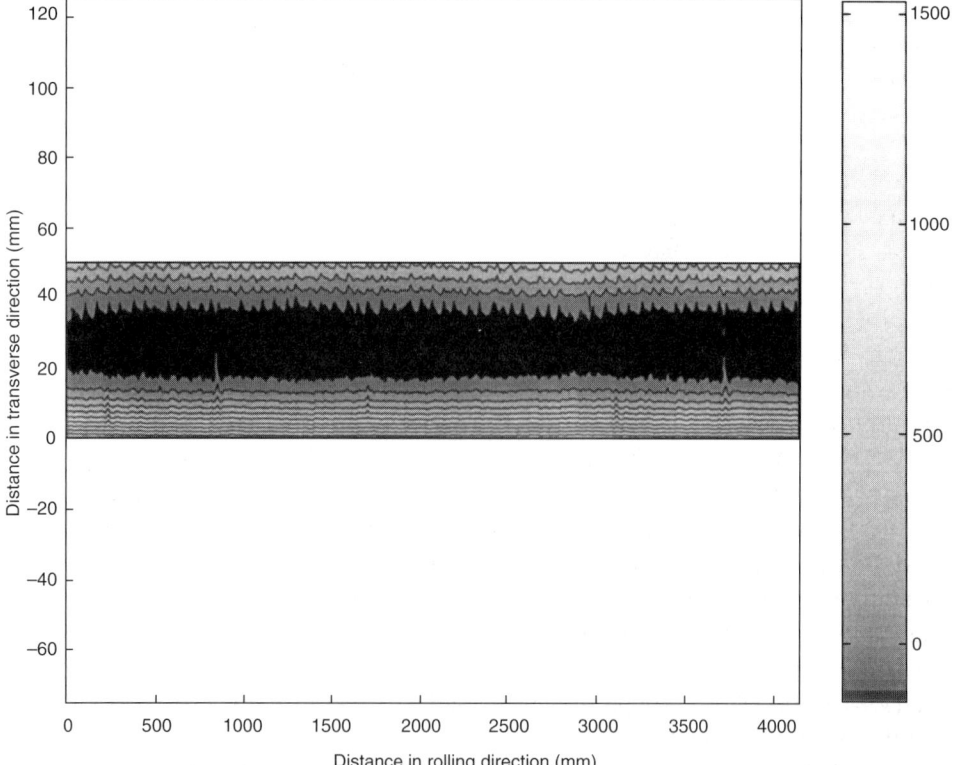

Fig. 2.15. Roughness measured on a wheel from a rail vehicle with cast-iron tread brakes

The roughness distribution on the running surfaces of rails is similar in character to that for wheels. Both have similar amplitudes and spectral content. While wheels and rails differ in that rails have a convex running surface and wheels, even after a short period of service, develop a concave running surface, the roughness that develops on those surfaces is similar in most important aspects. However, densely sampled data for rails are even sparser than for wheels. As a result, any conclusions concerning the relationship between rail roughness and type of traffic, speed, etc. can at best be conjecture.

As a typical example of roughness, Fig. 2.15 shows a contour plot of the two-dimensional roughness on the running surface of a wheel from a rail vehicle with cast-iron tread brakes [2.12]. The roughness data in Fig. 2.15 were measured on the running surface of the wheel on 26 parallel lines around the circumference. The measurement lines were each separated from each other by 2 mm and the roughness amplitude on each line was acquired every 0.5 mm. The contour plot in the figure shows the hollow worn profile typical of a wheel from normal service. There is also evidence of irregularities or roughening of the running surface, which is typical of cast-iron tread-braked wheels [2.14]. Note also that the roughness data are for more than a single rotation of the wheel. The wheel's circumference is 2887 mm. After 2887 mm the data repeat themselves, as is evident in the figure.

Figure 2.16 shows two profiles of the roughness along a single track in the middle of the tread of Fig. 2.15. Figure 2.16(a) is the roughness profile with no filtering. Figure 2.16(b) shows a profile of the same roughness with all components with inverse wavelengths less than 0.02 mm^{-1} suppressed. The first profile shows the long-wavelength irregularities in the wheel, on which are superimposed the shorter-wavelength irregularities of primary interest in the generation of rolling noise. The second profile shows the roughness primarily responsible for rolling noise. Aside from a groove at ~800 mm, the roughness is less than about 20 μm in amplitude but with occasional higher spikes.

Figure 2.17 shows the amplitude of the Fourier transform of the profile shown in Fig. 2.16(a), taken over the full range of the data and normalized by the number of points in the profile. The figure shows a very large dynamic range, with the data decreasing by about 30 dB per decade above 0.02 mm^{-1} inverse wavelength. The commonly observed dependence of wheel/rail noise on speed of $\sim N \log_{10}$(velocity) [2.28], where N is typically about 30, but may be between 20 and 40, is a consequence of this decrease of spectral amplitude with increasing inverse wavelength. Around 0.02 mm^{-1} inverse wavelength (50 mm wavelength) there is a broad hump in the spectrum corresponding to periodic irregularities of that wavelength observable in Figs 2.15 and 2.16(a). These corrugation-like irregularities are commonly observed with cast-iron tread-braked wheels and can lead to increased noise and vibration. Below 0.02 mm^{-1} inverse wavelength the spectral content of the roughness is irregular, reflecting out-of-roundness and irregular wear of the wheel.

Figure 2.18 shows a contour plot of the two-dimensional spectrum of the roughness data shown in Fig. 2.15. The spectral levels in decibels are shown versus inverse wavelength in the rolling and transverse directions. The spectrum was obtained by dividing the data into 32 records. Each record consisted of 26 lines 128 mm long in the circumferential direction, and each record followed the previous one sequentially without overlap. The two-dimensional Fourier transform of each

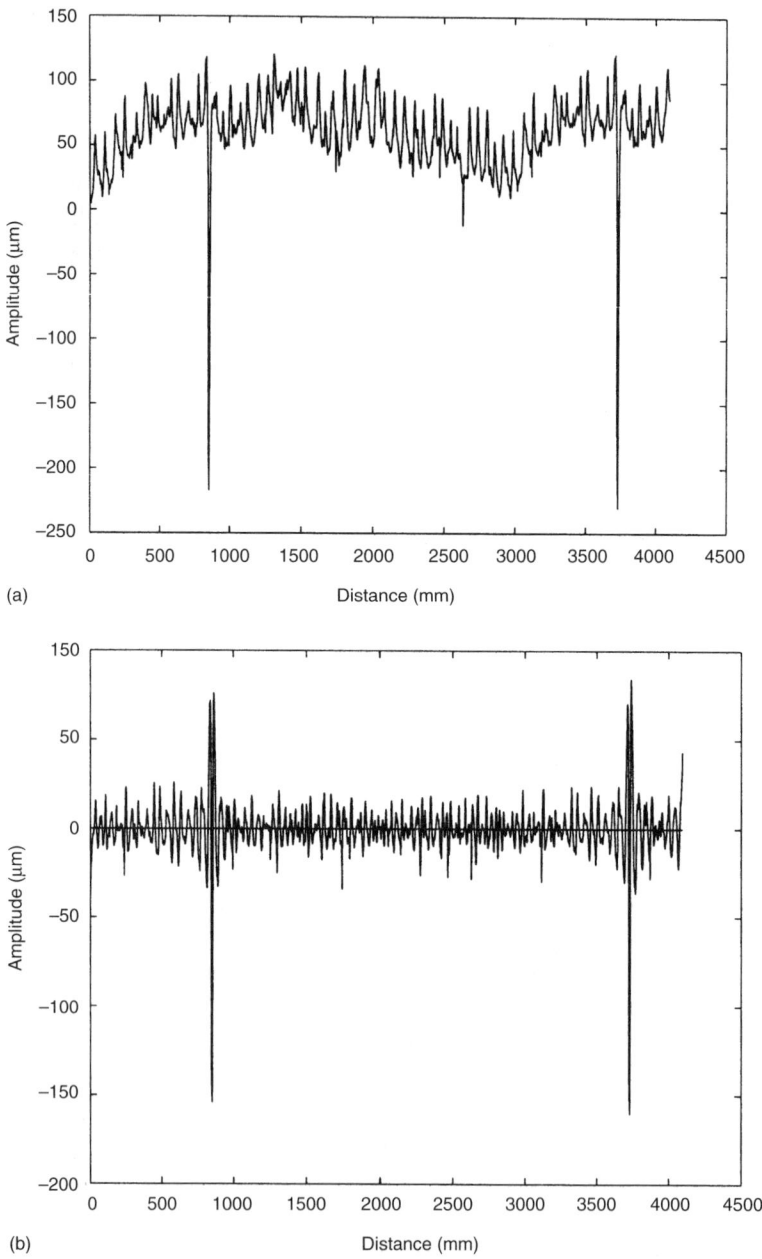

Fig. 2.16. Roughness profile along a path down the centre of the wheel tread of Fig. 2.15: (a) broadband profile; (b) high-pass profile

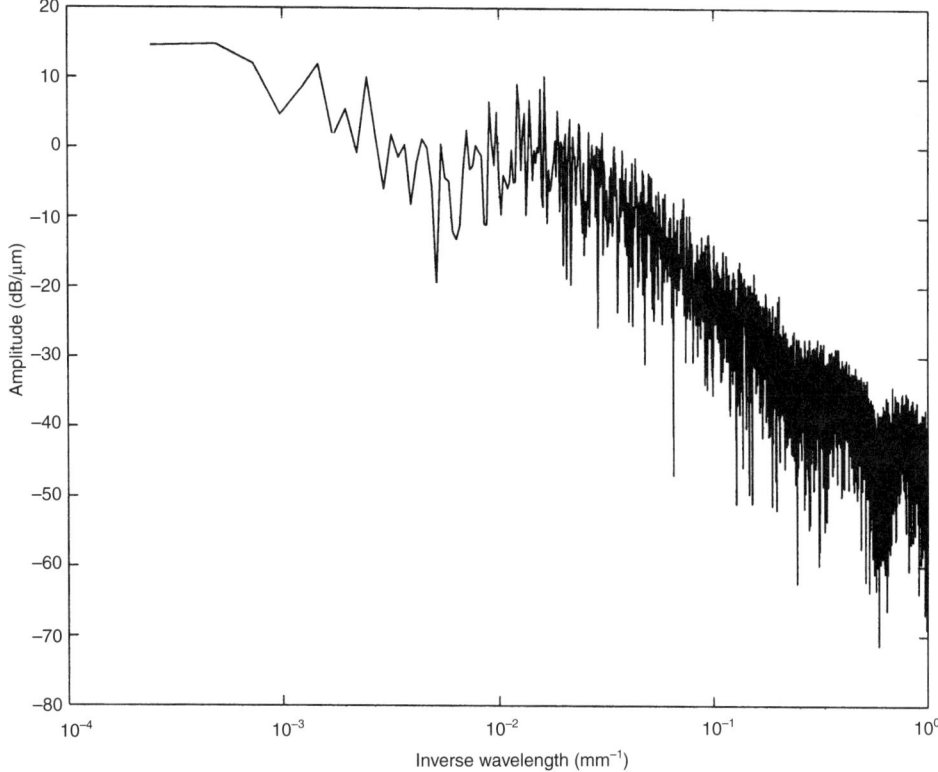

Fig. 2.17. Magnitude of the Fourier transform of the roughness profile shown in Fig. 2.16(a)

record was taken, and the average of the absolute values squared of all the records was taken to obtain the spectrum. The data were Hanning windowed to minimize spillover from spectral components of high energy into the lower-energy components, and the spectrum has been scaled such that the sum of all spectral components yields the mean square roughness.

The spectrum in the figure shows over 60 dB of dynamic range and an increased broadening of the spectrum in the transverse direction with increasing inverse wavelength in the rolling direction, as mentioned above. This can be more clearly seen in Figs 2.19 and 2.20. In Fig. 2.19 the spectrum as a function of inverse wavelength in the rolling direction is plotted for a variety of inverse wavelengths in the transverse direction. The spectrum is smoother and more broadband in nature than that in Fig. 2.17 owing to the shorter record lengths (256 points versus 8192 in Fig. 2.17). The shorter record lengths reduce the resolution in the spectrum.

Figure 2.20 shows the spectrum in the transverse direction for a number of different inverse wavelengths in the rolling direction. The figure clearly shows that the bandwidth in transverse inverse wavelength increases with increasing inverse wavelength in the rolling direction. This implies that the roughness in the transverse direction becomes less well correlated with increasing inverse wavelength in the rolling direction, or, in other words, the correlation lengths in the transverse direction decrease with decreasing wavelength. In fact, the data show that the simple two-dimensional roughness spectrum model proposed in equation (2.7) of Section 2 is in fact representative of actual measured roughness spectra.

This is more evident in Fig. 2.21. There we have modified Fig. 2.18 by normalizing the two-dimensional spectrum by the values along the crest of the distribution, where the transverse inverse wavelength is zero. Data from two wheels are shown: one with cast-iron tread brakes and one with disc brakes. In mathematical terms, the normalized spectrum S_N is obtained from the unnormalized spectrum S_0 by

$$S_N(k_x/2\pi, k_y/2\pi) = \frac{S_0(k_x/2\pi, k_y/2\pi)}{S_0(k_x/2\pi, 0)} \qquad (2.30)$$

where $k_x/2\pi$ and $k_y/2\pi$ are the inverse wavelengths in the rolling and transverse directions, respectively. The contour lines in both parts of the figure present a distribution that looks very much like that described by equation (2.7), where the value of α is the tangent of the angle made by the contour lines with the axis for the inverse wavelength in the rolling direction $k_x/2\pi$. For the cast-iron tread-braked

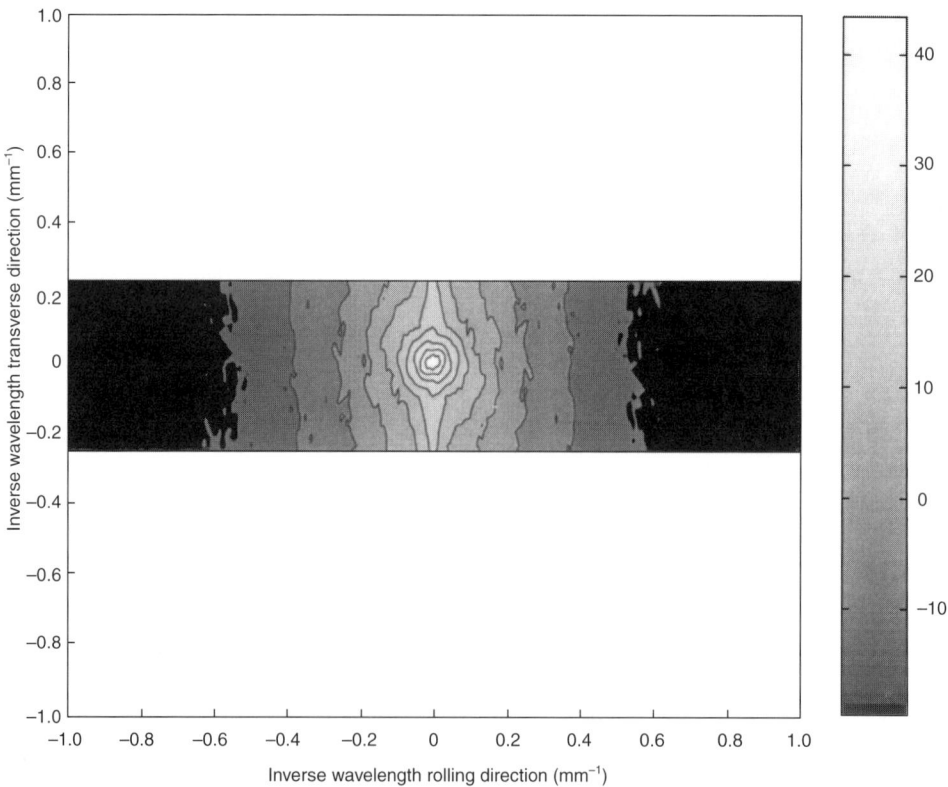

Fig. 2.18. Two-dimensional wavenumber spectrum of roughness on the wheel of Fig. 2.15

WHEEL AND RAIL EXCITATION FROM ROUGHNESS

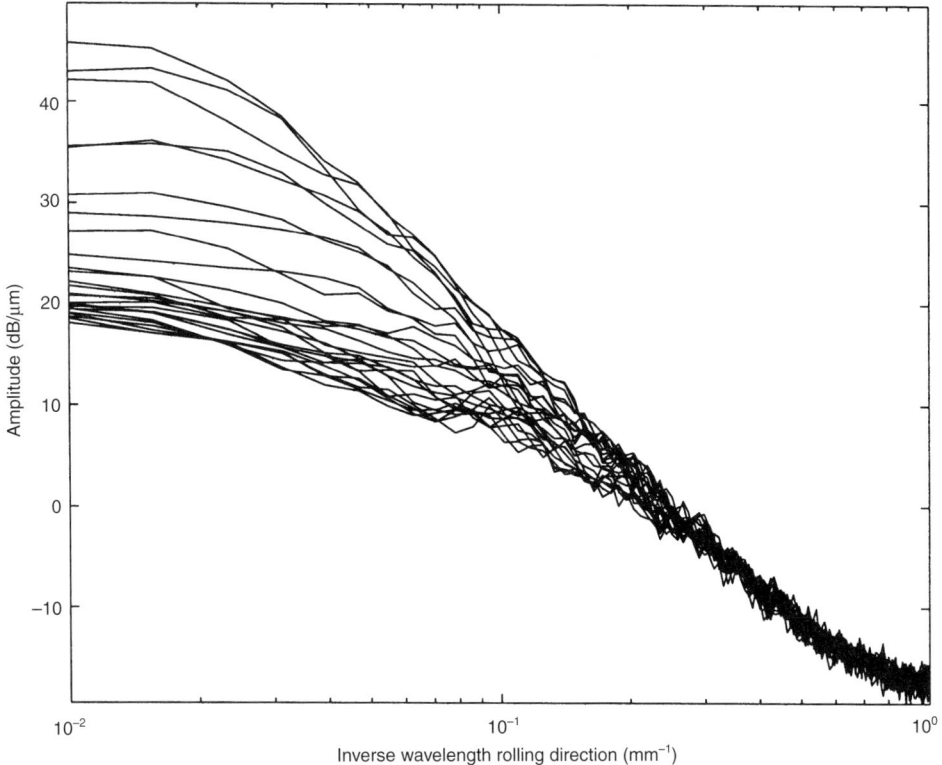

Fig. 2.19. Roughness spectrum in the rolling direction for a range of transverse wavenumbers

wheel the appropriate value of α lies somewhere between 1 and 4. For the disc-braked wheel the spectrum becomes broader much more quickly than for the tread-braked wheel. Although the simple model of equation (2.7) appears less applicable for this wheel, we might estimate, to first order for comparison purposes, an α more in the range of 4 to 10. These results suggest that the contact filter would be expected to be somewhat more effective in reducing the excitation forces from disc-braked wheels than for those from tread-braked wheels.

2.5. Controlling wheel/rail noise at the source

In this section we shall examine a number of techniques for controlling wheel/rail noise that involve directly reducing the excitation force in the contact region as opposed to modifying the wheels and rails to reduce response and sound radiation. In general, there are three approaches to controlling wheel/rail noise at the source that can be exploited:

(1) roughness amplitude reduction
(2) reduction of wheel/rail contact stiffness
(3) increase of the dimensions of the zone of contact.

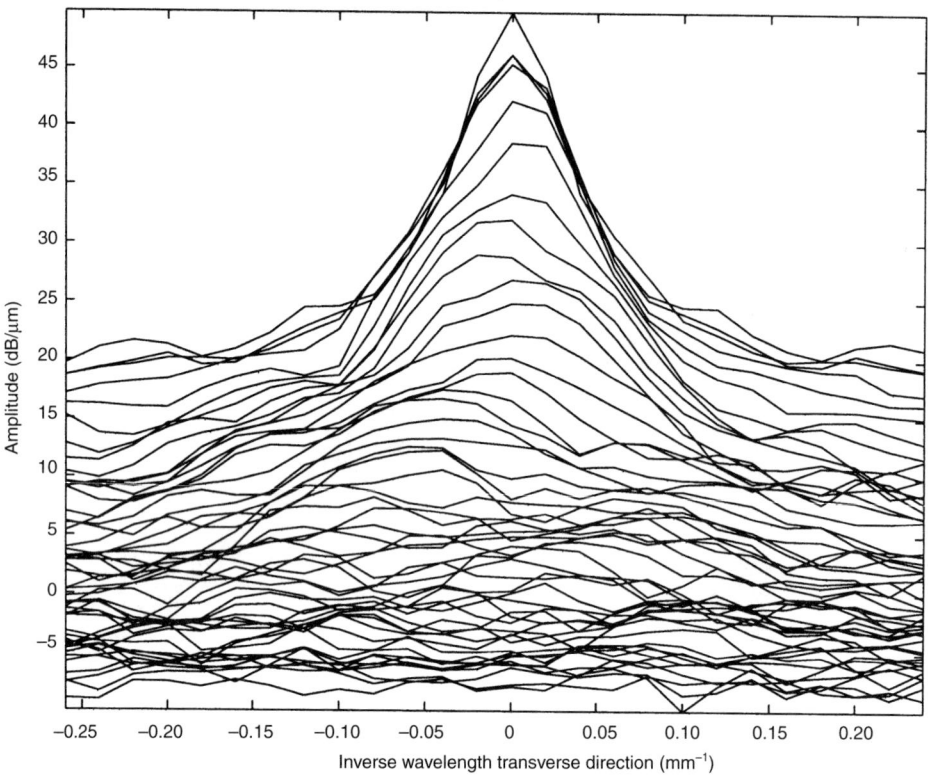

Fig. 2.20. Roughness spectrum in the transverse direction for a range of rolling-direction wavenumbers

2.5.1. Roughness amplitude reduction

The simplest way to reduce the excitation from roughness is to smoothe the running surfaces of the wheel and rail. In a study of rapid transit systems in the US [2.19] a number of alternative wheel- and rail-smoothing techniques were examined to determine their relative merit for reducing wheel/rail noise. The roughness was measured on operating transit vehicles and in-service track before and after applying each of the treatments shown in Fig. 2.22. Validated analytical models of wheel rail/noise [2.19] were then used with parameters typical of US transit systems to estimate the resulting noise reduction. Figure 2.22 shows the average pass-by noise in dBA at 7.6 m during the passage of a two-car train at 48 km/h. For the wheel-smoothing treatments, rail roughnesses representative of very smooth but readily achievable machined surfaces were used in combination with the measured wheel roughness to estimate the wayside noise. For the rail-smoothing treatments, the wheel surface was assumed to be a similar very smooth, machined surface. The intent was to show the benefits achievable with rail smoothing when the wheels are also smooth, and vice versa. The figure also shows the resulting noise if *both* wheels *and* rails are machined to a very smooth surface, not necessarily achievable with current wheel- and rail-smoothing techniques. The results in the figure show that about 8 dBA of noise reduction is achievable through wheel and rail smoothing alone.

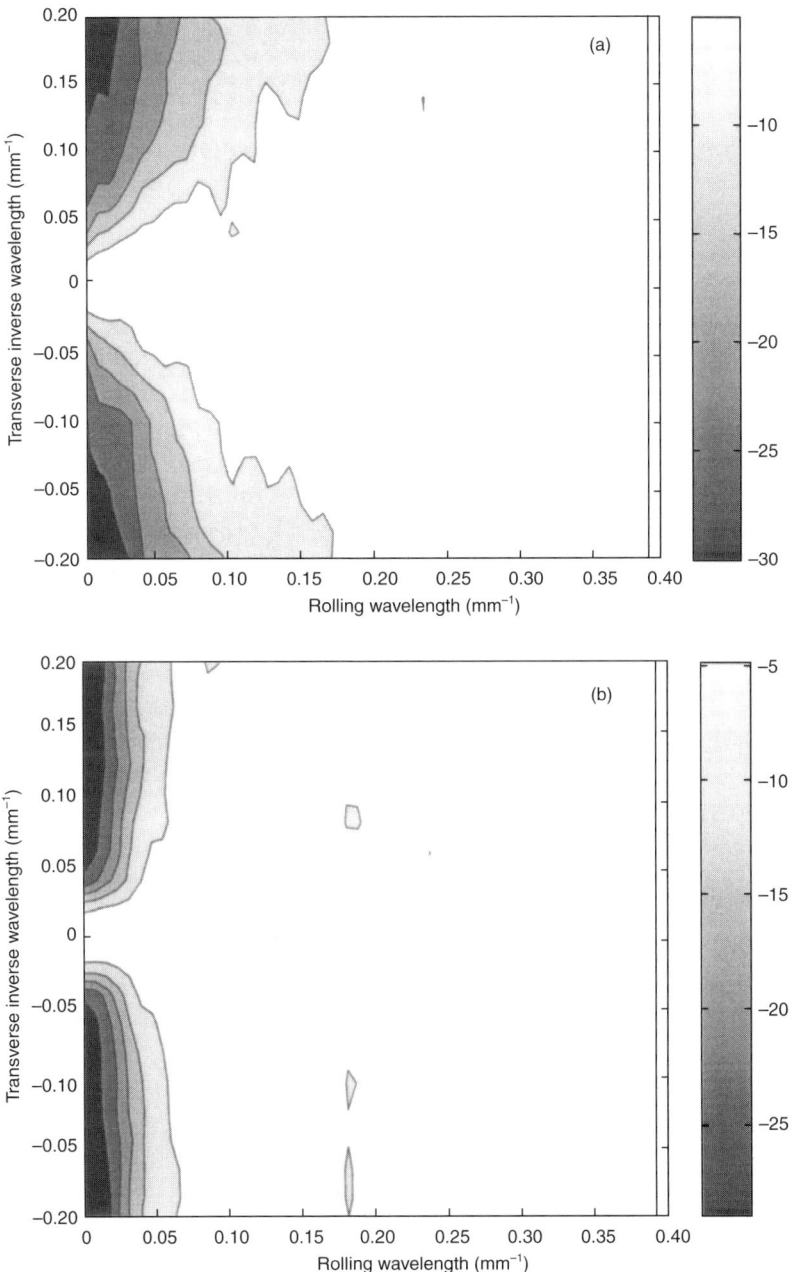

Fig. 2.21. Two-dimensional spatial spectrum normalized by the amplitude at $k_y/2\pi = 0$: (a) cast-iron tread brakes; (b) disc brakes

The smoothest surfaces of those for which results are shown in Fig. 2.22 were obtained with the rail-grinding block for the rail and with the belt grinder for the wheel. The rail-grinding block is a large grinding block mounted under a railcar. The block is contoured to match the desired profile on the railhead. It is pressed against the rail and dragged along by the railcar at near revenue speeds. When these measurements were made, rail-grinding block cars were in use by the Chicago Transit Authority (CTA) and the Toronto Transit Commission (TTC). The wheel belt grinder is similar to other under-car wheel-truing machines, except that here an abrasive belt backed by a hard rubber roller is run against the wheels as the machine turns them. The wheel belt grinder measured in this study was operated by the TTC.

While the rail-grinding block and wheel belt grinder are marginally better than the other techniques, virtually any of the modern smoothing techniques will improve the surface condition of the wheels and rails. To be effective, however, both wheels and rails must be smoothed on a periodic basis. The length of time required between smoothing operations is not known at this time, but no doubt depends in a complicated way on the condition of both wheel and rail and on the type of braking system [2.14].

2.5.2. Contact stiffness reduction and contact area increase

Figure 2.23 illustrates the change in average A-weighted wayside noise at 7.6 m during the pass-by of a two-car train at 48 km/h when the contact stiffness between wheel and rail is reduced and when the contact area is increased [2.26]. Dramatic changes in both are required to effect significant reductions in noise. A factor of ~20

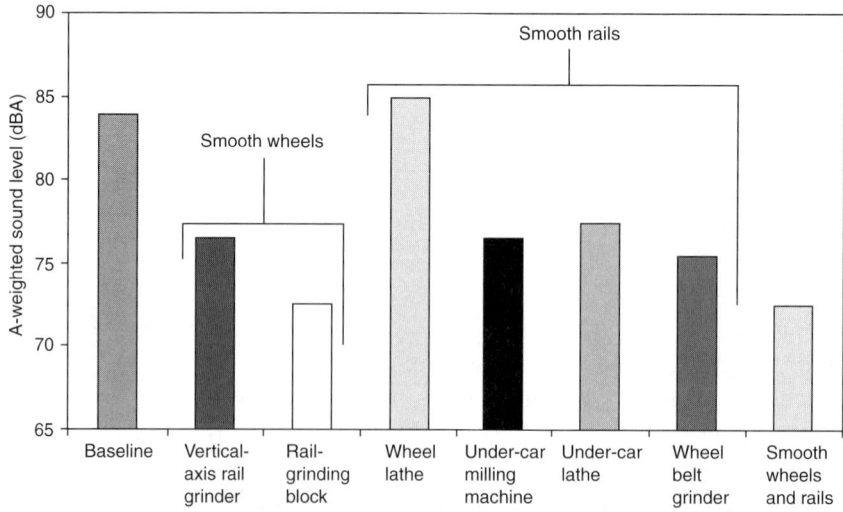

Fig. 2.22. Effect of wheel and rail smoothing on rolling noise (note that the data shown for the lathe were for a very old machine in which the wheel set had to be removed from the transit car and installed in the machine; it was apparently not functioning properly)

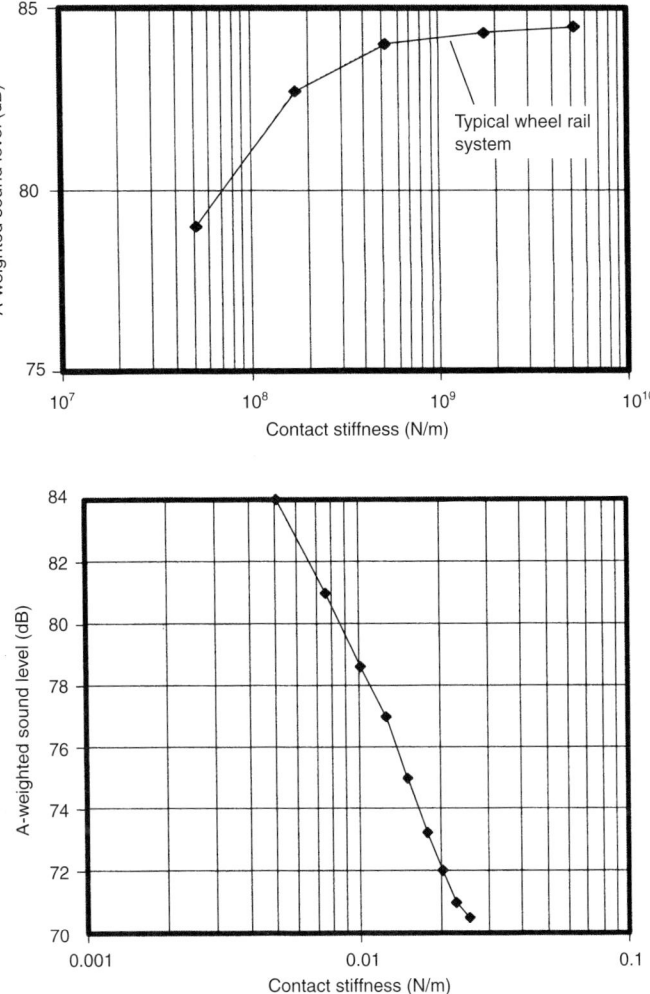

Fig. 2.23. Effect of contact stiffness and contact area on rolling noise

change in contact stiffness or an increase in the radius of the area of contact by a factor of two will each provide ~6 dBA of wayside noise reduction.

Such large changes in contact stiffness or contact area cannot be easily achieved. Simple material changes, for example, such as titanium treads on the wheel, will not be sufficient. Figure 2.24 illustrates a concept, called the *resiliently treaded wheel*, that was constructed at one-third scale and tested on a laboratory roller rig. The tread ring shown in the figure provides resiliency that reduces the contact stiffness by approximately a factor of six. Contact area dimensions were nearly a factor of two greater than the Hertzian predicted dimensions for a solid wheel. Measurements of noise from the resilient wheel and from a standard reference wheel on the roller rig showed the resiliently treaded wheel to be considerably quieter even though both had nominally the same roughness. Figure 2.25 shows the noise levels from the two

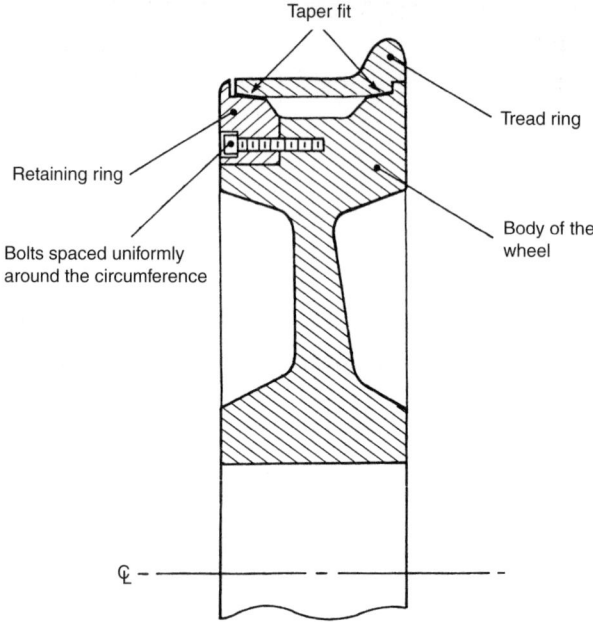

Fig. 2.24. The resiliently-treaded-wheel concept

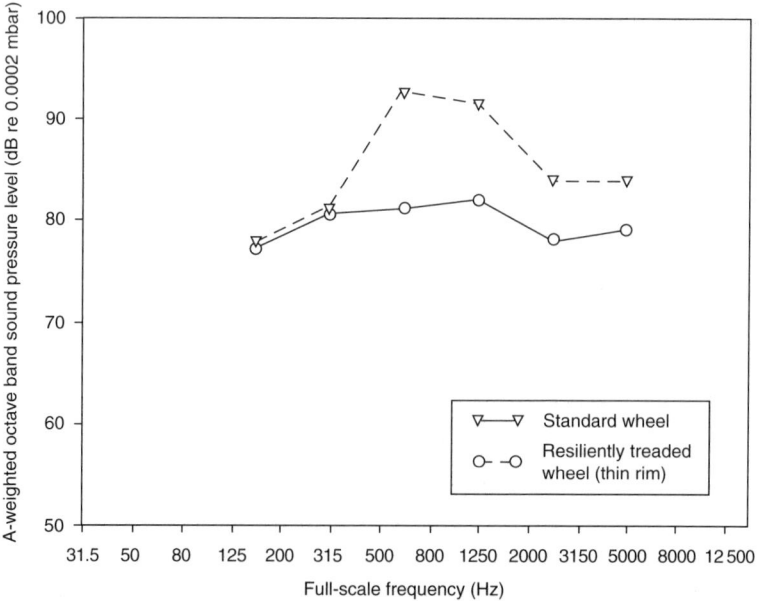

Fig. 2.25. Noise reduction achieved with a scale model test of the resiliently treaded wheel

wheels versus the equivalent full-scale frequency. At full scale, the resiliently treaded wheel would be expected to be nearly 8 dBA quieter than the standard reference wheel and would have a tread ring with a thickness of approximately 12 mm. Under normal full-scale wheel loads the maximum stresses in the tread would be a manageable 239 MPa.

The resiliently treaded wheel appears to be a promising approach to controlling rolling noise. To date only a scale model has been tested, and considerable development work would be required before a practical device would be available for use on railroads and transit properties.

It is well known that the contact stiffness and contact area geometry are affected by the wheel and rail profiles. Unfortunately, as illustrated above, the changes that are required in both contact stiffness and contact area to effect significant reductions in wheel/rail noise are very large. Analytical studies of wheel/rail contact [2.29] have shown that wheel and rail profile changes alone can change neither the contact stiffness nor the contact region dimension in the rolling direction enough to affect wheel/rail noise significantly. On the other hand, the dimensions of the contact area in the direction transverse to rolling can be changed dramatically by modifying the transverse radius of the wheel to conform to the rail, as illustrated in Fig. 2.26. The figure shows that the transverse dimension of the zone of contact can be increased almost without bound by making the transverse radius of curvature of the wheel concave so that it conforms more and more closely to the rail.

It was initially hoped that a contact area that was very long in the direction transverse to rolling could lead to reduced wheel/rail interaction forces. If the roughness on the wheel and rail were poorly correlated in the transverse direction,

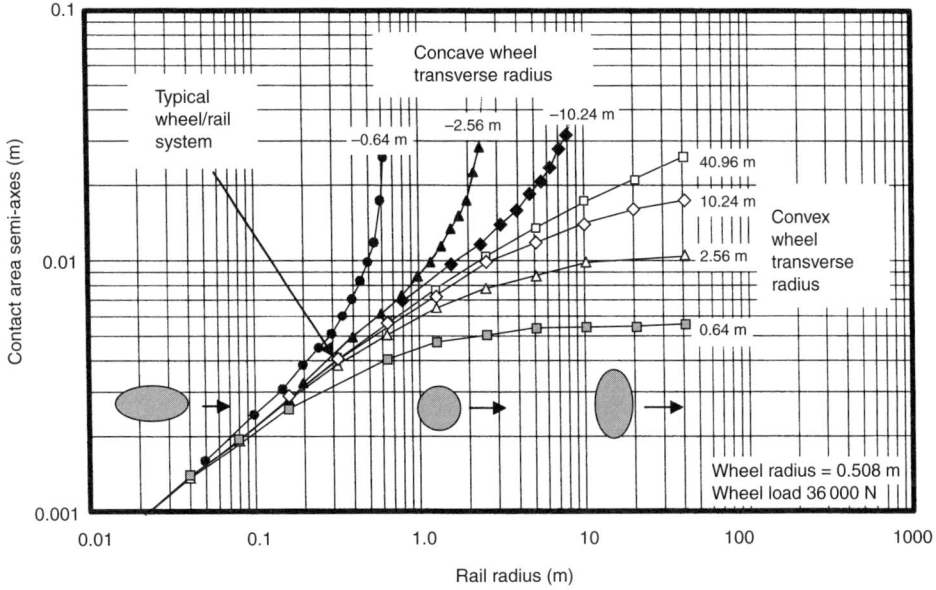

Fig. 2.26. Contact region dimensions in the direction transverse to rolling

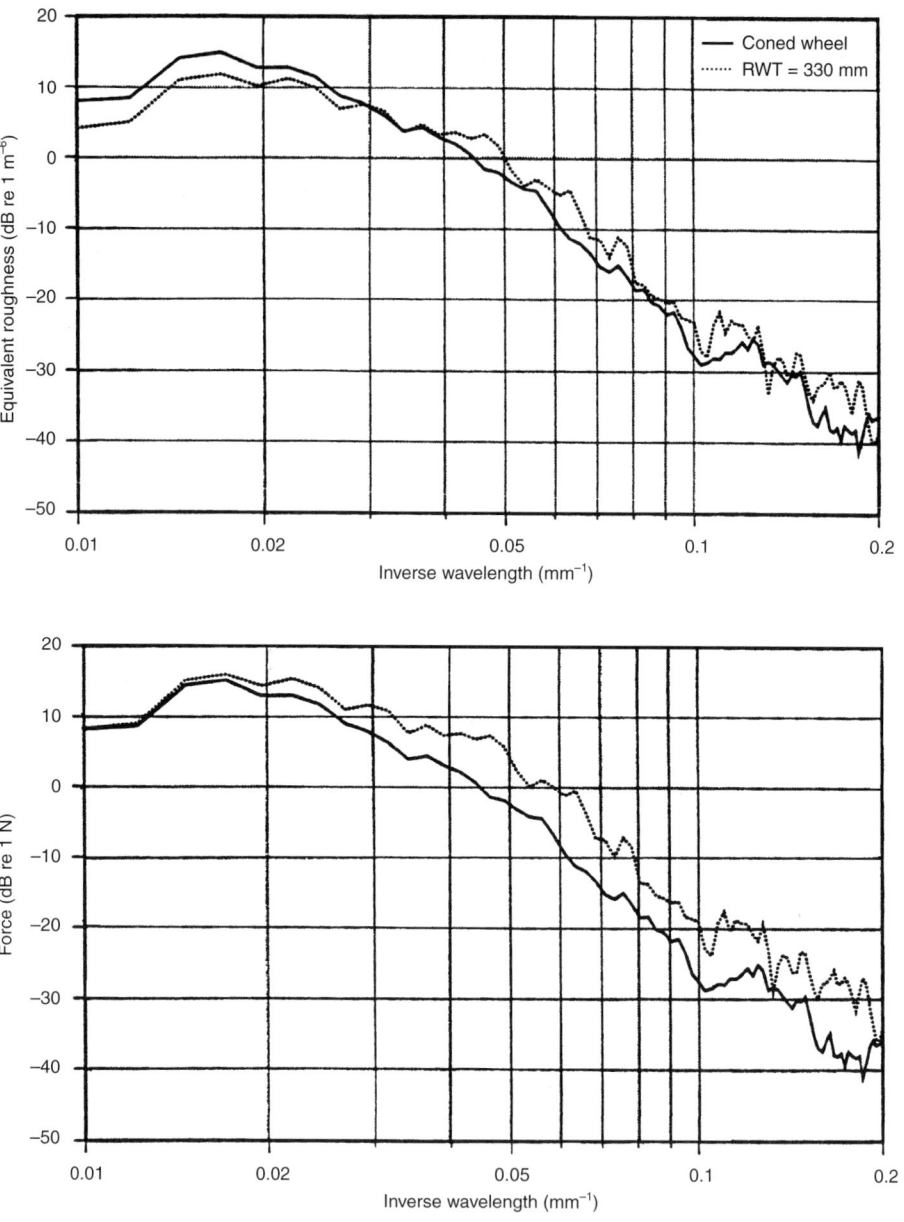

Fig. 2.27. Change in blocked force and equivalent roughness through the use of conforming wheel and rail profiles

the excitation would be essentially averaged out. Unfortunately, recent estimates using roughness data from four different rail sections and nine different wheels [2.31] have not realized that hope. Figure 2.27 shows the spectrum of the blocked interaction force between the wheel and rail computed by the DPRS model using rail roughness data [2.12] measured using a procedure similar to that described in

Section 4. The figure also shows the spectrum of the *equivalent roughness*. The equivalent roughness can be thought of as the roughness averaged over the contact area. It is computed by dividing the blocked force by the contact stiffness. In the calculation the wheel had a transverse radius of 330 mm, just slightly larger than the rail radius of 300 mm.

The results indicate that the blocked force increases owing to an increase in the contact stiffness by a factor of ~1.6 and the equivalent roughness remains essentially unchanged. There is now a general consensus that the anticipated reduction in blocked force and equivalent roughness was not achieved because the roughness was too highly correlated in the direction transverse to rolling, and that the shortening of the contact zone in the rolling direction, also a consequence of the conforming geometry of the wheel and rail, had an adverse effect on the averaging of the roughness.

2.6. Summary and conclusions

The excitation forces generated in the contact region between the wheel and rail by the roughness on their running surfaces is the primary mechanism for the generation of rolling noise. Three analytical techniques have been developed for estimating the interaction force, given measured data on the two-dimensional roughness distribution. These three techniques, called the average-roughness model, the distributed point-reacting spring model and the full elastic-interaction model, provide similar predictions of the interaction forces over most of the range of roughness wavelengths of interest for rolling noise. The distributed point-reacting spring model seems to be the best compromise between accuracy and computational efficiency for routine use and is the approach currently used in the European Rail Research Institute's TWINS computer program for predicting wheel/rail noise [2.30] (see also Chapter 1).

The earliest devices for the measurement of the roughness distribution on wheels and rails utilized accelerometers as the sensing devices and were designed for measuring roughness along a single line. Most devices in use today use displacement sensors, such as LVDTs, for measuring roughness and are capable of obtaining data on multiple parallel lines on both wheels and rails. Measuring the two-dimensional roughness distribution has allowed for more accurate predictions of rolling noise.

To date, there is very little densely sampled roughness data available. Consequently, any conclusions about the characteristics of the roughness distribution on wheels for different types of braking systems, for example, or on rails for different wheel loads, speeds, type of traffic, etc. must be somewhat tentative. The spatial spectra of roughness show a rapid decrease in amplitude with decreasing wavelength. It is common for roughness amplitudes to decrease by 60 dB from the longest to the shortest wavelengths of interest. In addition, as the wavelengths decrease, the correlation lengths in the direction transverse to rolling decrease as well. This decrease in correlation lengths implies that the contact region provides additional attenuation of the roughness excitation.

A number of wheel- and rail-smoothing techniques have been shown to provide significant reductions in roughness and, by implication, wheel/rail noise. Most of these techniques are already in active use by many railroads and rapid transit systems. However, it is currently unknown how long the surfaces will remain smooth and how often the smoothing needs to be repeated to maintain a desired noise level. Modifying wheels and rails in order to control rolling noise at the source, reducing the contact stiffness for example, is fraught with difficulty because of the severe environment in the contact region that any treatment must survive. Wheel designs for reducing the contact stiffness and increasing the contact area have been demonstrated at model scale in the laboratory. The laboratory tests showed significant reduction of rolling noise and demonstrated the durability of the concepts; however, considerable design and testing efforts lie ahead before these concepts can become viable commercial treatments. Attempts significantly to increase the filtering of the contact region through wheel profile modification have so far been unsuccessful.

Research over the past 30 years has been extremely fruitful in enhancing our understanding of the mechanisms of generation of wheel/rail noise and in clarifying the role that roughness plays in those mechanisms. New concepts in low-noise wheel and rail design have begun to appear and, as the designs mature, will no doubt lead to significant reductions in the noise impact from high-speed rail systems.

2.7. References

2.1. THOMPSON, D. On the relationship between wheel and rail surface roughness and rolling noise. *Journal of Sound and Vibration*, 1996, **193**(1), 148–160.
2.2. THOMPSON, D., HEMSWORTH, B. and VINCENT, N. Experimental validation of the TWINS prediction program for rolling noise, part 1: description of the model and method. *Journal of Sound and Vibration*, 1996, **193**(1), 123–136.
2.3. THOMPSON, D., FODIMAN, P. and MAHE, H. Experimental validation of the TWINS prediction program for rolling noise, part 2: results. *Journal of Sound and Vibration*, 1996, **193**(1), 137–148.
2.4. THOMPSON, D. Wheel–rail noise: theoretical modelling of the generation of vibrations. PhD thesis, University of Southampton, 1990.
2.5. THOMPSON, D. Theoretical modeling of wheel/rail noise generation. *Workshop on Rolling Noise Generation*. Institute for Technical Acoustics, Technical University Berlin, 1989, pp. 149–158.
2.6. REMINGTON, P. Wheel/rail noise, part I: characterization of the wheel/rail dynamic systems. *Journal of Sound and Vibration*, 1976, **46**, 359–379.
2.7. REMINGTON, P. Wheel/rail noise, part IV: rolling noise. *Journal of Sound and Vibration*, 1976, **46**, 419–436.
2.8. REMINGTON, P. Wheel/rail rolling noise, part 1: theoretical analysis. *Journal of the Acoustical Society of America*, 1987, **81**(6), 1805–1823.
2.9. REMINGTON, P. Wheel/rail rolling noise: part 2, validation of the theory. *Journal of the Acoustical Society of America*, 1987, **81**(6), 1824–1832.
2.10. REMINGTON, P. Wheel/rail rolling noise: what do we know? What don't we know? Where do we go from here? *Journal of Sound and Vibration*, 1988, **120**(2), 203–226.
2.11. VAN TOL, F. H. and VAN VLIET, W. J. *Study on Rolling Noise Task IV-A-1/1, Roughness Measuring Devices, Preliminary Report*. TNO, Delft, 1990, Report TPD-HAG-RPT-90-0013.

2.12. VAN HAAREN, R., VAN KEULEN, G. and WIERSMA, P. *Measured Wheel Noise for Various Profiles*. NS Technische Onderzoek, Utrecht, 1997, Project 9610266.
2.13. HOLM, P. *Roughness Measuring Devices*. Mueller-BBM, Planegg, Germany, 1999.
2.14. KURZWEIL, L. G. Wheel/rail noise means for control. *Journal of Sound and Vibration*, 1983, **87**(22), 197–220.
2.15. DINGS, P. C. and DITTRICH, M. G. Roughness on Dutch railway wheels and rails. *Journal of Sound and Vibration*, 1996, **193**(1), 103–112.
2.16. VINCENT, N., BOUVET, P., THOMPSON, D. and GAUTIER, P. Theoretical optimization of track components for reduced noise. *Journal of Sound and Vibration*, 1996, **193**(1), 161–172.
2.17. REMINGTON, P. and WEBB, J. Estimation of wheel/rail interaction forces in the contact area due to roughness. *Journal of Sound and Vibration*, 1996, **193**(1), 83–102.
2.18. HERTZ, H. *On the Contact of Rigid Elastic Bodies and on Hardness, Miscellaneous Papers*. Macmillan, London, 1886.
2.19. REMINGTON, P. J., RUDD, M. J. and VER, I. L. *Wheel Rail Noise and Vibration, Volume 1: Mechanics of Wheel Rail Noise Generation*. US Department of Transportation, Washington, DC, 1974, Report UMTA-MA-06–0025-75-10.
2.20. HECKL, M. Proposal for a railway simulation program. *Workshop on Rolling Noise Generation*. Institute for Technical Acoustics, Technical University Berlin, 1989, pp. 128–148.
2.21. HARRIS, T. *Roller Bearing Analysis*. Wiley, New York, 1984.
2.22. PAUL, B. and HASHEMI, J. *Fundamental Studies Related to Wheel–Rail Contact Stress*. Federal Railroad Administration, Washington, DC, 1981, Report FRA/ORD-81/05.
2.23. PAUL, B. and HASHEMI, J. An improved numerical method and computer program for counterformal contact stress problems. In: *Computational Techniques for Interface Problems* (eds K. C. Park and D. K. Gartlung), Applied Mechanics Division of the ASME, New York, 1978, vol. 30, pp. 165–180.
2.24. TIMOSHENKO, S. and GOODIER, J. N. *Theory of Elasticity*. McGraw-Hill, New York, 1951.
2.25. REMINGTON, P. J. *The Estimation of Wheel/Rail Interaction Forces due to Roughness*. BBN Corporation, Cambridge, MA, 1992, Report 7793, prepared for the European Rail Research Institute, Utrecht.
2.26. REMINGTON, P. J., DIXON, N. R., KURZWCIL, L. G., MENGE, C. W., STAHR, J. D. and WITTIG, L. E. *Control of Wheel/Rail Noise And Vibration*. US Department of Transportation, Washington, DC, Report UMTA-MA-06–0099-82-5.
2.27. OFFICE FOR RESEARCH AND EXPERIMENTS OF THE INTERNATIONAL UNION OF RAILWAYS. *Wheel/Rail Contact Noise – An Experimental Comparison of Various Systems for Measuring Rail Roughness*. Office for Research and Experiments of the International Union of Railways, Utrecht, 1988, Report 9, Question C163.
2.28. REMINGTON, P. et al. *Wheel/Rail Noise and Vibration Control*. US Department of Transportation, Washington, DC, 1974, Interim Report UMTA-MA-06-0025-74-10.
2.29. REMINGTON, P. and WEBB, J. *The Estimation and Control of Wheel/Rail Rolling Noise Due to Roughness*. BBN Corporation, Cambridge, MA, 1994, Report 8013.
2.30. THOMPSON, D. and JANSSENS, M. H. A. *TWINS Track–Wheel Interaction Noise Software User's Manual*. TNO, Delft, 1997, Report TPD-HAG-RPT-93-0213.
2.31. THOMPSON, D. and REMINGTON, P. The effect of transverse profile on the excitation of wheel rail noise. *Proceedings of the 6th International Workshop on Railway Noise*. Ile des Embiez, France, 1998, pp. 50–58.

3. High-speed train noise barrier tests at reduced scale

J. D. van der Toorn
TNO Institute of Applied Physics, Stieltjesweg 1, PO Box 155, NL-2600 AD Delft, The Netherlands

3.1. Modelling outdoor sound propagation

Acoustical scale models form a useful means of general, experimental studies of outdoor sound propagation. Scale models are very suitable for investigating trends, solutions and measurement methods and have advantages with respect to mathematical tools and models. The steady increase of available computer power eases the application of diffraction theory, ray-tracing techniques and numerical methods for the calculation of barrier effects. More intricate barrier shapes, however, cannot always be reliably modelled with conventional diffraction theory or ray-tracing techniques. Numerical methods – like the boundary element method – in principle permit one to model complex shapes, but are very demanding with respect to computer power if the source to receiver distances are not small with respect to the usual distances between railways and dwellings. Scale modelling allows one to model intricate, three-dimensional situations, over distances that are relevant for practical situations.

3.2. Scale modelling

3.2.1. Similarity

The validity of the use of a scale model is based on its similarity to the prototype. Similarity is a powerful concept, which was used and explained for structure-borne noise problems by Lord Rayleigh [3.1], who referred to it as the *method of dimensions*. A scale model is similar to the prototype in the sense that results of measurements on the scale model can be converted into prototype properties by simple rules [3.2]. A general introduction to similarity and a firm formal, theoretical basis for the use of dimensionless products and for scale model testing has been given by Langhaar [3.3].

Scale modelling starts with recognizing the important parameters of the problem under study. The important parameters for propagation of high-speed train noise are

- the geometry of the situation
- the characteristics of the high-speed train as a sound source and as a sound-reflecting object
- sound transmission loss, which is determined by the characteristics of the sound transmission paths, such as the specific acoustic impedances of the ground surface, atmospheric absorption and the sound-absorbing properties of barriers
- the directivity of the receiver.

The sound source, receiver and boundary conditions should be modelled appropriately in order to obtain reliable results from a scale model.

The acoustical scale models that are described below are geometrical models. The scale factor is 1:32, which means that all geometrical measures L – the relative positions and dimensions of the source, barrier, etc. – are 32 times smaller in the scale model than at full scale in the prototype situation. Also, the sound wavelengths λ are scaled down, according to the scaling rule that the Helmholtz number L/λ should be invariant under scaling. Usually, the measurements were performed in air. In that case the scaling factor for frequency equals 32 and the sound source should produce frequencies 32 times higher than those of the train that is modelled. Full-scale measurements in $\frac{1}{3}$ octave bands with full-scale centre frequencies of 50–5000 Hz were simulated in a 1:32 scale model in $\frac{1}{3}$ octave bands with centre frequencies of 1.5–160 kHz. The results of measurements and simulations described below are scaled back and presented with respect to the corresponding full-scale frequencies or full-scale measurements.

3.2.2. Measurable quantities

The main measurable quantity of interest in traffic noise studies is the equivalent continuous A-weighted sound pressure level L_{Aeq}, in decibels [3.4]. If different barrier designs are compared or effects of changes in the situation or in a sound transmission path are determined, the difference ΔL_{Aeq} between two levels is the measurable quantity. The accuracy of such a difference and of differences like insertion loss is usually – but not necessarily – better than that of the absolute sound pressure levels if the modelling inaccuracies are the same and cancel when two measurement results are subtracted.

3.2.3. Sound sources

3.2.3.1. Train characteristics and source type

The effect of a noise barrier depends on the characteristics of the sound source, such as the directivity and source strength distribution as a function of frequency and height. For conventional trains rolling noise is dominant; for high-speed trains higher sources, like the pantograph, can also be important [3.5]; Fig. 3.1 illustrates

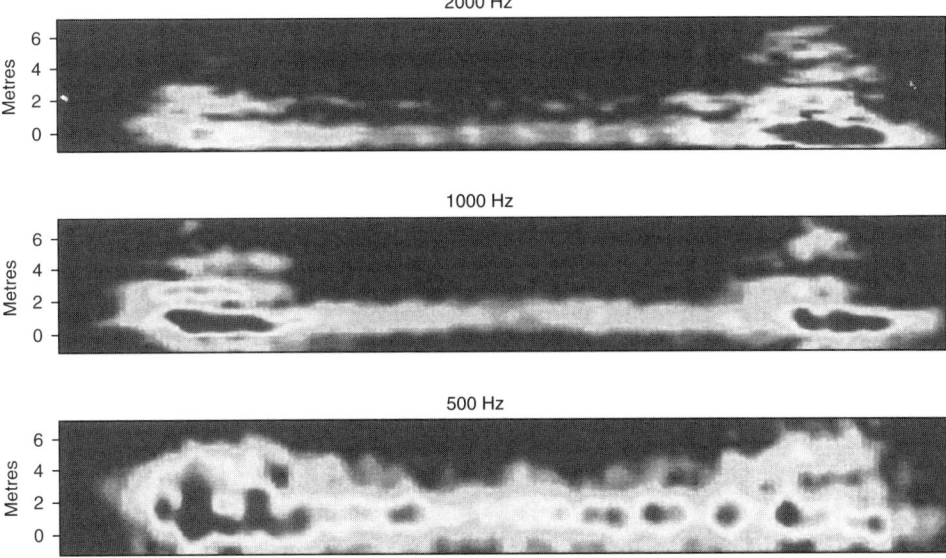

Fig. 3.1. Acoustic image of a TGV-Atlantique, travelling at 300 km/h, measured with an acoustic antenna [3.5], in 1/1 octave bands with centre frequencies of 500, 1000 and 2000 Hz. The dynamic range is 15 dB. In the higher frequency bands the leading and the rear power car are clearly visible. Bogies are more prominent as hot spots at lower frequencies. Also, the pantograph attracts attention

this. These properties of a high-speed train should be modelled properly. If noise barriers and trains are modelled, the wall of the train is also important, because the net effect of a noise barrier is affected by the sound reflections between the train wall and barrier.

We distinguish two types of trains as follows.

- A conventional train, for which rolling noise is dominant.
- A high-speed train, for which sources as high as the roof and the pantograph can also be of importance. The noise barrier experiments for which the train model has been used were not for a special type of high-speed train. Therefore a more or less generic model of a high-speed train has been developed. Measurements with a synthetic acoustic antenna of sound from the French TGV-Atlantique [3.5] driving at 300 km/h were used as the reference for the source strength distribution as a function of frequency and height.

In theoretical studies, road traffic noise is usually described by monopoles, but railway noise has often been described by using dipole elementary sound sources [3.6–3.9]. A choice to describe railway noise with dipoles was usually not justified on grounds of principle or mandatory, but was the result of a practical approach or a purely academic exercise. In this study it is assumed that trains can be described by distributions of incoherent monopoles. In the scale model that is described below, combinations of omnidirectional sound sources are simulated, with source strengths that depend on the position on the train.

3.2.3.2. A railway track in the light of L_{Aeq}

Measuring an equivalent continuous A-weighted sound pressure level L_{Aeq} involves integration over time. In the cases when L_{Aeq} of moving vehicles is measured, integration over time is equivalent to integration over the trajectory of the vehicles. Owing to the integration procedure, a railway track with passing trains (railway line) behaves as an incoherent line source with respect to L_{Aeq} [3.10]. If the track is homogeneous and trains are passing at constant speed, the track can be described as an incoherent line source. The line source strength of the track is homogeneous in the direction of motion of the trains; however, it can be a function of the height h:

$$L'_W(h) = L_W(h) + 10 \lg \left(\frac{Q d_0}{v} \right) \quad (3.1)$$

Here $L'_W(h)$ is the line source strength of the track as a function of height (dB re 1 pW/km); $L_W(h)$ is the monopole source strength of the train as a function of height (dB re 1 pW); it is the integral over the length of a train:

$$L_W(h) = 10 \lg \left(\int_{\text{train length}} 10^{L'_W(l,h)/10} \, dl \right)$$

Q is the source flow rate, i.e. the number of trains passing per unit time (h^{-1}); v is the speed of the train (km/h); and d_0 is 1 km.

Equation (3.1) shows that the source strength distribution of the railway line in the vertical direction is similar to that of the train. The vertical source strength distribution is relevant for sound transmission effects and has to be modelled. The source strength distribution on a train in the lateral direction does not play a role in L_{Aeq} and can be simplified in a model that serves to determine L_{Aeq} for trains passing at constant speed on a long, homogeneous railway track. Therefore, the train source strength distribution can be summed in the lateral directions. In the train model that will be described in Section 3.3.1, the source strength has been concentrated in the centre of the train.

3.2.3.3. Pneumatic sound source

A powerful sound source is very beneficial for avoiding signal-to-noise ratio problems, which are inherent for measurements behind barriers, sometimes at long measurement distances, with an atmospheric absorption that is relatively high at scale model frequencies (see Section 3.2.5) and with a small microphone with a relatively low sensitivity.

Nozzles, driven as pneumatic sound sources, were used in the scale model as elementary point sources to build up more complex sound sources. The sources are relatively powerful in the scaled frequency range for railway noise (see Section 3.2.1). The design was basically borrowed from Delany et al. [3.11] and adapted to a 1:32 scale factor.

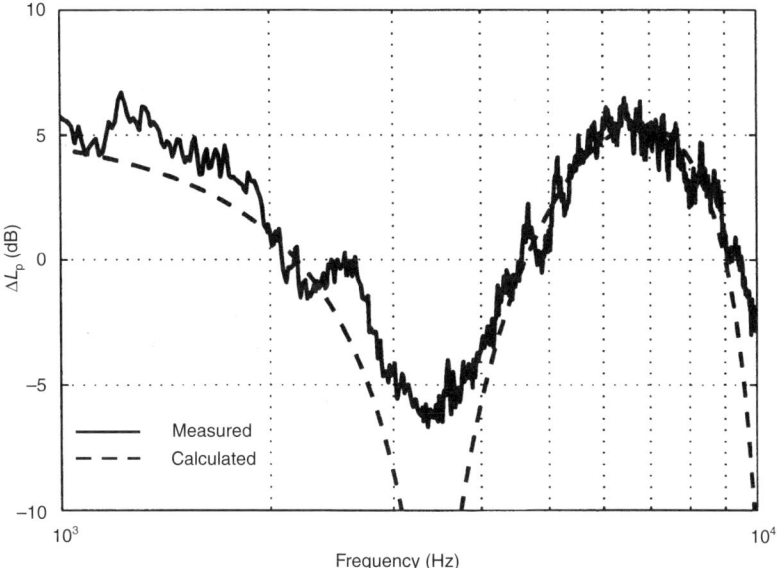

Fig. 3.2. Measured and calculated sound pressure levels of the pneumatic sound source above a hard plane, relative to free field (ΔL_p), in narrow frequency bands. In the calculations the centre of the front of the pneumatic sound source is taken as the position of the monopole source. The source height and the receiver height are 90 mm and the measurement distance is 300 mm. The first minimum is expected and measured at 3.4 kHz. The comparison demonstrates that the acoustic centre of the pneumatic sound source is in the centre of the front of the pneumatic source

A single pneumatic source of this kind – which has a diameter of 28 mm – has a clear acoustic centre, located at the centre of the front of the source. This was demonstrated by determining the sound pressure level ΔL_p relative to the free field. The sound pressure level relative to the free field ΔL_p, in decibels, measured over a ground surface, equals the negative of the excess attenuation EA and is defined as

$$\Delta L_p \equiv -EA = L_{p,\text{ground}} - L_{p,\text{free field}} \tag{3.2}$$

Here $L_{p,\text{ground}}$ is the sound pressure level above the ground at the receiver position (dB re 20 µPa), and $L_{p,\text{free field}}$ is the sound pressure level in the acoustically free field, for the same distance between the sound source and receiver (dB re 20 µPa).

The frequency at which the first interference dip in ΔL_p occurs is determined by the position of the acoustic centre of the source and by the well-defined position of the microphone. Figure 3.2 shows that the minimum in ΔL_p that was measured for the pneumatic sound source corresponds to the minimum in ΔL_p that has been calculated for a monopole source, located at the centre of the front of the pneumatic source.

In the scale model of a high-speed train, sets of two pneumatic sources were applied at different heights (see Section 3.3.1). An omnidirectional source was simulated by choosing the appropriate orientation of the two sources. The omnidirectional directivity characteristic of the model has been checked by measurements in an anechoic room.

3.2.3.4. Distribution of moving point sources

The source strength distribution of a high-speed train as a function of height was simulated by dividing the train model into four horizontal source areas, representing rails and wheels, openings in the wall – radiating engine noise and aerodynamic noise – and sources on the roof. Each area was represented with a single combination of two pneumatic sound sources. The source strength as function of frequency was simulated by calibrating the pneumatic sound source spectrum for each height. Further details are presented in Section 3.3.1.

A railway line was simulated as an incoherent line source by integrating the sound of the moving-train model for each of the four source heights and summing the results. For cases with more than one track, the contributions of the different tracks can also be measured separately. By measuring the contributions of the four source areas separately, it is easy to simulate in the postprocessing of the measurement data the effects of different source strength distributions and to determine the effect of silencing a particular source, like the pantograph.

An advantage of the use of a moving point source is that the position of the source is known during a measurement and that the sound of the point source can be integrated over small intervals of the track. This possibility was used to determine the contributions of different parts of the track and to apply corrections for microphone directivity and excess atmospheric absorption in the scale model (see Sections 3.2.4 and 3.2.5).

3.2.4. Receiver

For noise measurements it is conventional to simulate an omnidirectional observer, using an omnidirectional microphone. With a 1:32 scale factor, it is not possible to scale a conventional microphone. If a $\frac{1}{8}$ inch condenser microphone is used, the smallest commercially available microphone, it is relatively large in the scale model and is not omnidirectional for scale model frequencies. The sensitivity as a function of direction of sound incidence of such a microphone is well documented, however. This allows one to simulate an omnidirectional receiver by correcting the sound spectra that are measured during small translations of the sound source by the difference between the sensitivity of the microphone in the main direction of sound incidence and its sensitivity for frontal incidence. Usually, the main direction of sound incidence corresponds to the shortest sound transmission path. If the line of sight between the sound source and microphone is obstructed by a barrier, the direction of the sound path via the barrier rim is used.

3.2.5. Atmospheric absorption

Atmospheric absorption does not scale – it is relatively high for scale model frequencies – but it has been well documented. The atmospheric absorption in decibels per metre of sound transmission path can be calculated from the ambient temperature, the relative humidity of the air and the static air pressure by applying ISO standard 9613-1 [3.12]. The atmospheric absorption during a test was calculated for the conditions of the air in the test room and also for representative full-scale outdoor conditions. The length of the shortest sound transmission path between source and microphone was calculated for small steps of displacement of the source (if necessary via the rim of a barrier). The difference in atmospheric absorption that was calculated for this path was applied to the measured spectrum as a correction.

Compensations for atmospheric absorption and microphone directivity could be performed automatically by the computer that controlled the measurements, which 'knew' the actual positions and orientations of the source, microphone and barrier.

3.2.6. Ground plane

3.2.6.1. Characteristics
Inserting a noise barrier along a railway track also alters the effect of the ground on sound transmission [3.13]. This means that the net effect of a barrier depends on the properties of the ground between the sound source and receiver. Therefore, in order to measure realistic net sound pressure levels and barrier effects, the ground plane should have appropriate acoustic properties in the scale model. In general studies on barrier effects, the ground between the barrier and observer is usually modelled as grassland.

The frequency characteristics of the sound pressure levels relative to the free field (ΔL_p), measured above a flat surface – for appropriate point source and point receiver positions – characterize the acoustical properties of the surface [3.13, 3.14]. In particular, the first minimum in the frequency characteristic is important in this respect. The shape of the minimum and the frequency where it occurs reflect the specific acoustic impedance of the surface and its sound-absorbing properties.

Acoustical 1:32 scale models of sound-absorbing surfaces were developed by performing measurements over surfaces made of different materials and selecting the surface with which the prototype frequency characteristics of ΔL_p could be reproduced or approximated sufficiently well at scale model frequencies.

Three parts of the ground plane were distinguished in this study:

- the surface of the track: ballast bed with sleepers and rails
- the ground between the ballast bed and the barrier foot
- the ground between the barrier and the observer.

3.2.6.2. Ballast bed
If sound is measured from trains on a remote track, the ballast bed can be an important part of the sound transmission path. In that case the acoustical properties of the ballast bed have to be modelled properly. The main problem is therefore to model the bulk material of the ballast bed and to choose the appropriate thickness of

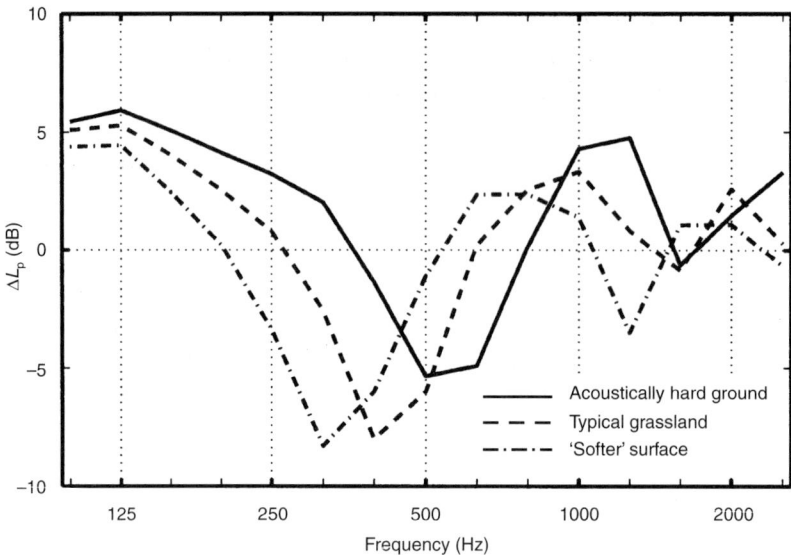

Fig. 3.3. Sound pressure levels relative to free field (ΔL_p) in $\frac{1}{3}$ octave bands measured above flat, homogeneous ground planes in order to characterize the surfaces. The source height was 2 m and the microphone height was 0.75 m above ground level; the measurement distance was 10 m, full scale. The curve for the typical grassland was used in the model between barrier and receiver. In a calculational model it can be characterized at full scale by a 35 mm thick sound-absorbing layer with a nominal flow resistivity of 200 kN s/m^4, on top of an acoustically hard surface. The 'softer' surface can be characterized in a calculational model as a sound-absorbing layer with a nominal flow resistivity of 80 kN s/m^4 and a thickness of 70 mm, on top of an acoustically hard surface. This surface was used in the experiments on the effect of the ground between the railway track and the barrier (Section 3.2.6.4) to model the surface between track and barrier and to model the sound-absorbing barrier

the bed. Acoustically hard structures like rails and sleepers can be made at scale and placed in the ballast bed models. Modelling the ballast bed is discussed in Section 3.3.2.

The ballast bed under the train from which the sound was measured was not modelled separately. The effect of the ballast bed under the train is included in the train source strength that was measured with a synthetic acoustic antenna (this apparent source strength was modelled; see Section 3.3.1).

3.2.6.3. Grassland

Studies of the acoustical properties of grassland have revealed that the acoustical impedance of such ground surfaces can be simulated – in a mathematical model and in a reduced-scale model – by an appropriate sound-absorbing layer on an acoustically hard backing [3.13].

The dashed curve in Fig. 3.3 characterizes grassland and was applied in the scale model as a typical 'average' Dutch grassland. In a calculational model [3.14] it can be described by a 35 mm thick sound-absorbing layer with a nominal flow resistivity of 200 kN s/m^4, on top of an acoustically hard plane.

3.2.6.4. Between barrier and track

The ground between the track and barrier represents a sound-absorbing or (partly) reflecting surface in the sound transmission path close to the source. Some pilot measurements of L_{Aeq} of a moving point source have been performed in a scale model to investigate the sensitivity of measured barrier effects to the properties of this part of the sound transmission path. The measurement distance was 50 m. Parameters were

- *ground between track and barrier*: hard and absorbing like a soft type of grassland (see Fig. 3.3)
- *source height*: 0.5 m and 2 m above railhead, at the nearest track
- *measurement height*: 1.5 m, 5 m and 10 m above ground level (for this test, the top railhead was at ground level)
- *barrier surface*: 2.4 m high; thin *hard* and *sound-absorbing* barriers were applied at 5 m from the sound source.

Variations in L_{Aeq} due to effects of the properties of the ground between track and barrier alone were smaller than 1 dB(A). The 'largest' differences were found for the lower source height and the situation with the hard barrier. With the source at a height of 2 m the effect was smaller than 0.5 dB. Sound-absorbing barriers that are applied in practice have higher absorption coefficients than the absorbing barrier of this test. Therefore, in practice, the effects will be even smaller.

It is concluded that the effects of variations in the properties of this part of the ground are negligible. For the tests discussed in Section 3.5, this part of the ground had the same properties as grassland.

3.2.7. Barriers

Acoustically hard barriers with special shapes, such as T-shaped barriers and overhanging barriers, can be easily scaled, using aluminium plate for example. If sound-absorbing barriers are modelled, the absorption in the scale model should be the same as for the prototype. The development of sound-absorbing panels at a scale of 1:32 is treated in Section 3.4.

3.3. Scale modelling of railway noise

3.3.1. An acoustical 1:32 scale model of a high-speed train

The more or less generic model of a high-speed train that has been adopted is constituted by the wall of the train and four sound sources. The shape of the wall, as well as its characteristic dimensions, and the sound power level of the train, as a function of height and frequency band, have been borrowed from the TGV-Atlantique [3.5] (see Fig. 3.4 and Table 3.1). Owing to the dimensions of the pneumatic sound sources, each of the four sources represents a 1 m high source area.

The four source heights represent the following:

- Noise of wheels, rails, sleepers and ballast bed (source centre at 0.5 m above railhead). The ballast bed under the train is not modelled separately; its effect is included in the source strength that was measured with the antenna, and that source is modelled.
- Noise of equipment and aerodynamic noise generated at the side of the train. The wall is represented by two source areas, with centres at 2 m and 3 m above railhead.

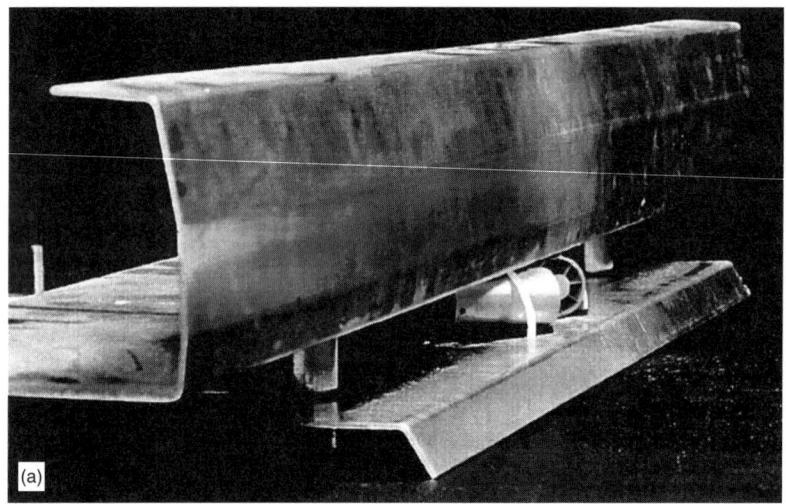

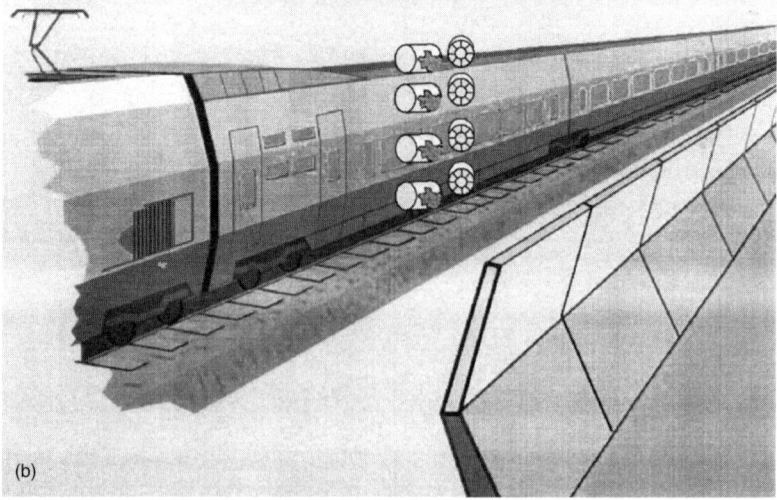

Fig. 3.4. The scale model of a high-speed train, including the wall of the train and pairs of pneumatic sources applied simultaneously at the four source heights: (a) model used to represent rolling noise; (b) artist's impression of the scale model of the complete high-speed train, with the four sets of two pneumatic sound sources

Table 3.1. Source strengths of the high-speed train. A-weighted sound power level L_W in decibels re 1 pW, in 1/1 octave bands, for the four sources that describe the high-speed train. Source height is presented with respect to the top of the railhead

Source height (m)	Octave band centre frequency, f (Hz)					
	125	250	500	1000	2000	4000
0.5	116	120	126	130	133	129
2	116	120	123	127	128	124
3	116	120	118	124	–	–
4.5	–	–	121	123	122	118

- Noise that is generated by equipment and irregularities on the roof and by the pantograph (source centre at 4.5 m above railhead).

The wall of the train in the 1:32 scale model was made of 2 mm thick aluminium plate. A pair of pneumatic sound sources was applied sequentially at each of the four source heights. The directivity patterns of each pair of sources with the train wall have been measured in an anechoic room. They are omnidirectional within typically 2 dB in the relevant measuring directions.

Figure 3.4(a) illustrates the modelling of rolling noise with a source set on the floor, under the wall. Two sets of sources were positioned behind holes in the wall to model the two source areas at 2 m and 3 m above railhead. A pair of sources was placed on the roof to represent the highest source area. The correction for the spectrum of the pneumatic sound sources for each source height was obtained by measuring the L_{Aeq} of a train model pass-by in free field, making the ground surface in the scale model fully sound-absorbing, calculating L_{Aeq} from the source strengths presented in Table 3.1 and taking the difference between the measured and the calculated L_{Aeq} for each frequency band and source height.

3.3.2. An acoustical 1:32 scale model of a railway track

Sound pressure levels relative to the free field (ΔL_p) of a single source have been measured *in situ* over ballast beds, in order to characterize the acoustical properties of this special type of ground surface. Comparable measurements have been performed over flat homogeneous areas of ballast bed, both *in situ* on a representative modern railway track and on different scale models of the ballast bed. The goal was to find a scale model with which the *in situ* measurements could be reproduced. The parameters for the scale model measurements were the thickness of the ballast bed, the chipping size and the structure of the ballast bed (layered or not).

The *in situ* measurements were performed over level ballast bed, with a chipping size of 30 to 63 mm. Measurements were performed mainly *between* two tracks, and additionally *on* a track and *across* the tracks in order to illustrate the effect of the sleepers. The sound source consisted of 12 loudspeakers, which were arranged as the faces of a dodecahedron. Its directivity characteristic was spherically symmetric within ±1 dB up to 1 kHz. The measurements were performed in $\frac{1}{3}$ octave bands with centre frequencies from 100 to 2500 Hz.

For the scale model measurements, a single pneumatic sound source and a $\frac{1}{8}$ in microphone were used with their axes of symmetry parallel (see Fig. 3.5). In this set-up the source and receiver possess approximately circular symmetry in the vertical measurement plane, as in the prototype experiment and mathematical models [3.14].

The thickness of the scale model ballast bed appeared to be an important parameter. Layered models did not perform better than a homogeneous bed. The best agreement between the results of the *in situ* measurements and scale model experiments was obtained with a model consisting of a homogeneous layer of sand with scaled chipping size, on a hard subsoil (see Fig. 3.6).

The values of ΔL_p measured in the scale model have also been reproduced by calculations, treating the top of the ballast bed as a finite-impedance plane [3.14]. The impedance was calculated describing the ballast bed as a sound-absorbing layer on an acoustically hard backing, using flow resistivity, porosity and tortuosity as parameters. The similarity that Fig. 3.7 shows between the measured and calculated curves allows one to apply, for a ballast bed, the same boundary conditions in a mathematical model and in a scale model. This offers the possibility to produce comparable and complementary results.

3.4. Design of sound-absorbing barriers at a scale of 1:32

3.4.1. Reference absorption curve

In a state-of-the-art study De Vos and Lammers [3.15] presented average sound absorption coefficients [3.16] for three types of sound-absorbing barriers: barriers

Fig. 3.5. Photograph of measurement set-up during test of scale model of ballast bed

made of perforated metal panels, concrete barriers and barriers made of wooden panels (see Fig. 3.8). The differences between the average values for different materials are small (the scatter around the average in an octave band value was typically ±0.3).

In the lower frequency bands, the perforated metal panels perform, on average, slightly better than panels made of concrete or wood. The curve for perforated metal

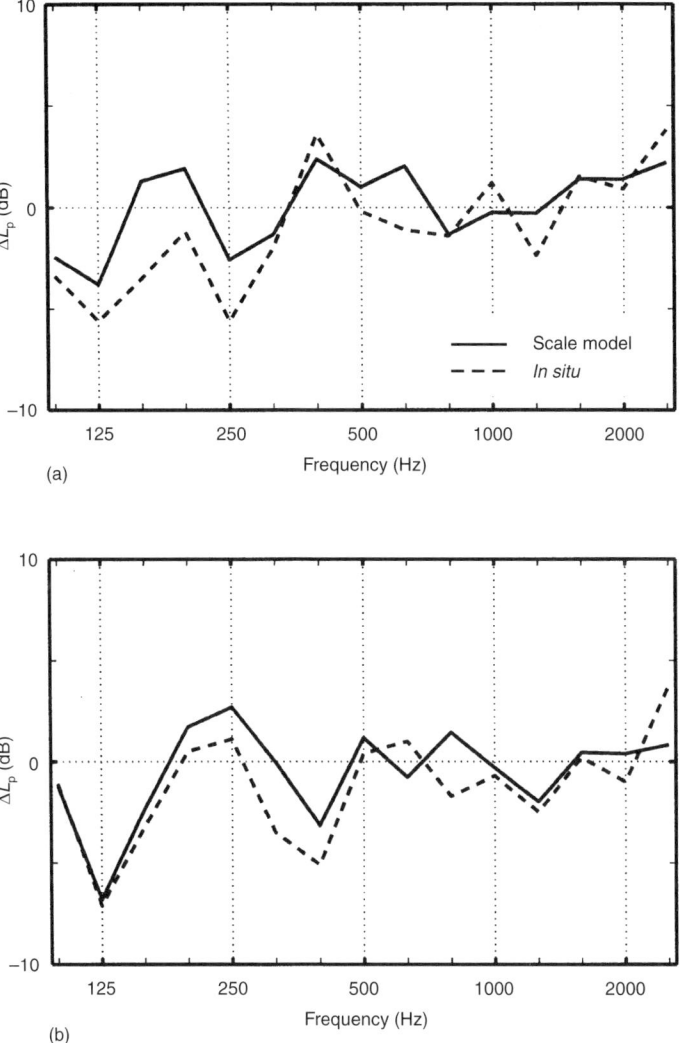

Fig. 3.6. Sound pressure levels relative to free field (ΔL_p), in $\frac{1}{3}$ octave bands, measured in situ above a railway track and at scale 1:32. The source height was 67 mm above the top of the ballast bed (2.15 m full scale); the microphone was at 28 mm above the top of the ballast bed (0.90 m full scale). Results are presented at full-scale frequencies. The scale model measurements are performed at frequencies 32 times as high. In the full-scale measurements heights were measured with respect to the railhead top, which is 0.15 m above the ballast bed. (a) Measurement distance 5 m; (b) measurement distance 10 m full scale

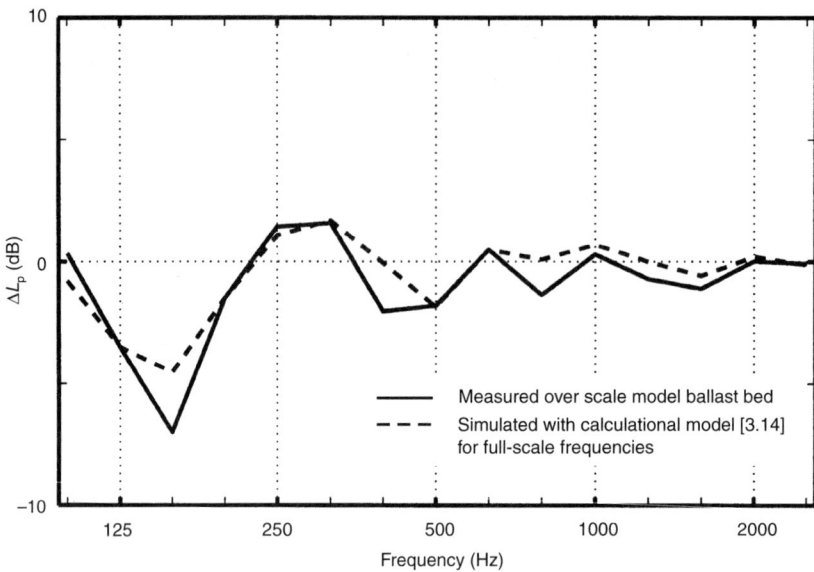

Fig. 3.7. Sound pressure levels relative to free field (ΔL_p), in $\frac{1}{3}$ octave bands

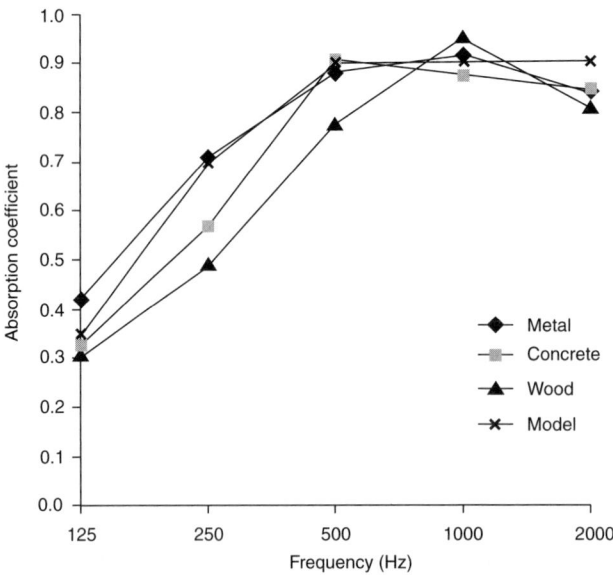

Fig. 3.8. Absorption curves for three types of full-scale noise barriers (averages of absorption curves measured in reverberation rooms) and the curve applied in the scale model: metal, barriers made of perforated metal boxes filled with sound-absorbing material; concrete, barriers made with porous concrete; wood, wooden barriers; model, calculated absorption curve, for random incidence, for a surface consisting of a layer of sound-absorbing material with an airflow resistivity $\sigma = 30\ kN\ s/m^4$ and a thickness of 0.1 m, backed by an acoustically hard surface

panels has been adopted as a reference, to be used during the development of a generic 1:32 scale model of a realistic, well-performing sound-absorbing barrier.

3.4.2. Absorption extracted from excess attenuation

Measuring sound absorption coefficients at scale model frequencies, in a scaled reverberation room, would be a clear method to develop a scale model of sound-absorbing barrier panels. The experimental conditions that are required for such measurements have been described in ISO standard 354 [3.16]. These conditions are hard to realize and to maintain at scale model frequencies, however. For example, if measurements are performed in air, the absorption of the empty 'reverberation' room will be so high – owing to the relatively high air absorption at scale model frequencies – that a reverberant sound field will not be built up. Therefore a different approach was chosen.

The absorption coefficients of a sound-absorbing surface can be calculated from the specific acoustic impedance of the surface [3.17]. This impedance can also be used to calculate excess attenuations (EA; equation (3.2)) for sound propagation over the surface [3.14], as has been demonstrated when scaling grassland and ballast bed. This relationship between absorption coefficient and excess attenuation has been used to find materials with a desired absorption coefficient by measuring excess attenuations.

Some exercises with the calculational model showed that a surface with a suitable absorption curve can be simulated with a layer of sound-absorbing material with an airflow resistivity $\sigma = 30$ kN s/m^4 and a thickness of 0.1 m, backed by an acoustically hard surface. The curve 'model' in Fig. 3.8 represents the absorption for such a surface.

The dashed curve in Fig. 3.9 represents the excess attenuation that has been *calculated* for sound propagation over such a surface, for a source height of 0.75 m, receiver height of 2 m and measurement distance of 10 m at full scale. Frequency characteristics of the excess attenuation have been measured for the same source and receiver positions, over flat 'ground' surfaces made of candidate materials for scaled sound-absorbing barrier panels. The full line in Fig. 3.9 represents the excess attenuation that has been *measured* over a surface of the material that has been adopted as a generic 1:32 scale model of a realistic, well-performing, sound-absorbing barrier surface.

3.5. Barrier tests

In the reduced-scale model, A-weighted equivalent sound pressure levels (L_{Aeq}) have been measured with the above-described model of a high-speed train, for 12 different barrier shapes and combinations of barrier types. Figure 3.10 shows four examples. The quantity determined was ΔEA_{Aeq}: the *extra excess attenuation* for L_{Aeq} with respect to the excess attenuation of the thin, vertical, sound-absorbing reference barrier, which is shown in Fig. 3.10(a). The measurement distance was 25 m and the receiver heights were 2 m, 3.5 m and 5 m.

ΔEA_{Aeq} is very appropriate for ranking barriers. The advantage of measuring the *improvement* of the barrier effect with respect to a reference barrier is that the measurement result is more specific to the barrier's performance and less sensitive to ground impedance and other properties of the test site and of the sound source.

The effects of barrier type, absorption and barrier tilt have been measured for three source configurations:

- a single source at 0.5 m above railhead, representing rolling noise
- a complete state-of-the-art high-speed train described by the four source heights listed in Table 3.1
- a high-speed train without the noise from the roof and the pantograph (omitting the highest sound source in the model).

Once the measurements with different source heights had been performed, effects of barriers could be studied for different source combinations in the postprocessing phase. Freight trains that are dominated by rolling noise, complete conventional high-speed trains, and high-speed trains with negligible noise of the pantograph and other sources on the roof can be simulated by calibrating the source spectra and taking into account appropriate sources.

The barrier with sound-absorbing panels that is shown in Fig. 3.10(d), for example, was relatively effective for a complete high-speed train, but not for trains for which rolling noise is dominant. For the complete train this barrier performed

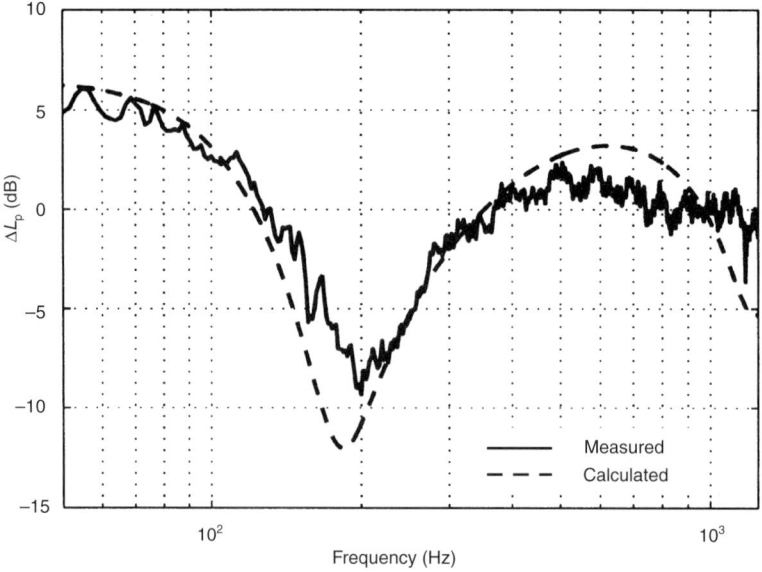

Fig. 3.9. Excess attenuation with source height 0.75 m, receiver height 2 m and measuring distance 10 m (full scale) calculated for a ground surface consisting of a 0.1 m thick layer with airflow resistivity $\sigma = 30$ kN s/m^4 on a hard planar backing, and measured at scale over a plane with the same surface as a sound-absorbing barrier used in the 1:32 scale model. Results are presented at full-scale frequencies. The scale model measurements were performed at frequencies 32 times higher

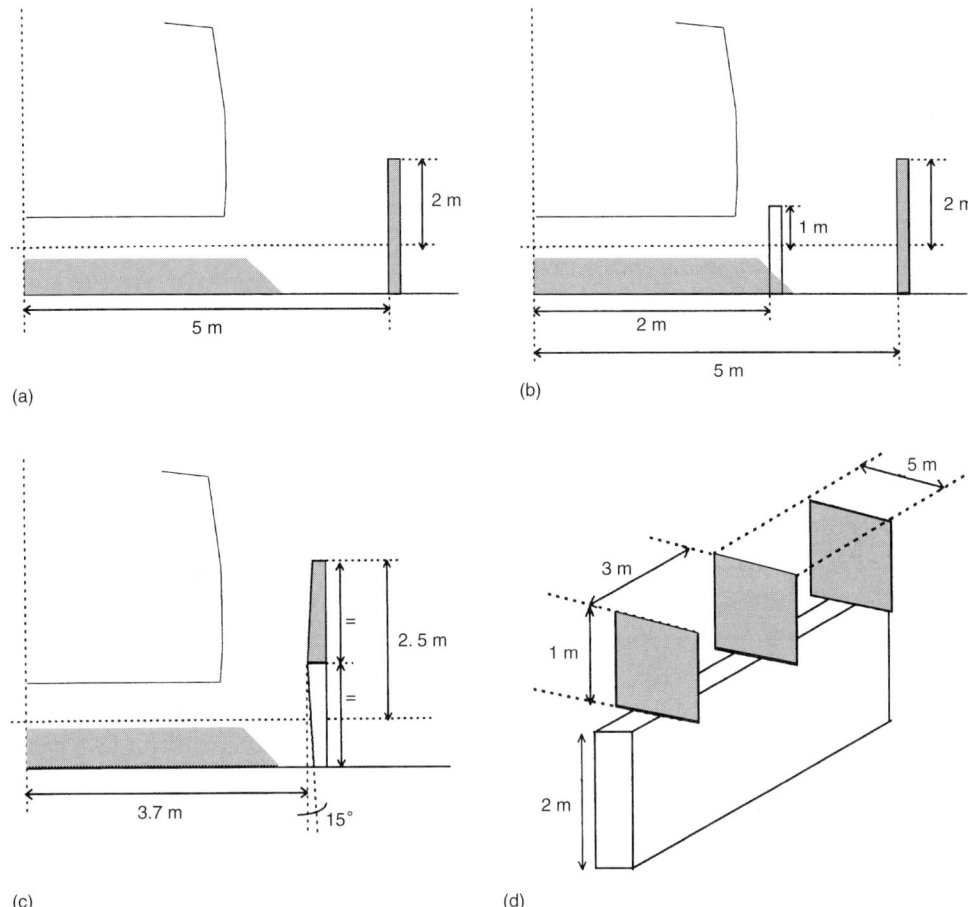

Fig. 3.10. Examples of barrier types that were compared in the scale model: (a) the reference barrier; (b) double barrier; (c) geniculated hard/absorbing barrier; (d) barrier with perpendicular cross panels. Parts (a)–(c) show sketches of the cross-section of the train wall, the railway track and the barriers (shaded barrier parts are sound-absorbing); (d) shows a barrier with sound-absorbing panels that allows train passengers a view

2–4 dB(A) better than the reference barrier. For rolling noise a low barrier close to the track (see Fig. 3.10(b)) is relatively effective, but it should be sound-absorbing. The geniculated hard/absorbing barrier shown in Fig. 3.10(c) performed 2–7 dB(A) better than the reference barrier, depending on the source distribution and the measuring position.

3.6. Concluding remarks

An acoustical scale model of a high-speed train has been developed. The scaling of the different boundary conditions for grassland, ballast bed and sound-absorbing barriers has been demonstrated, as well as the handling of anomalous air absorption and microphone directivity. Generic models have been used for the train, for the grassland and for sound-absorbing faces of barriers, with which the trends and the quality of the concepts of different solutions were studied.

The possibility to determine in a scale model the values of the parameters that should be used in calculational models to describe the boundary conditions allows one to make very closely comparable mathematical models and scale models. If such models are applied to different types of objects or situations – according to the specific possibilities and advantages of the modelling techniques – the results that are produced are complementary and closely comparable.

Scale model measurements are an efficient and effective means in a selection process such as ranking different barrier types. The barriers that are most promising, according to the scale model experiment that was presented as an example, have been selected for full-scale tests.

3.7. Acknowledgements

The work for railway applications described here was performed within the framework of the Brite-Euram II project 'Euroécran', contract number BRE2-CT94-0978. E. W. Boontje, M. N. S. Bouwhuis, E. van Doornik and W. F. A. Klein Zeggelink contributed significantly to the experiments described. A. C. Geerlings simulated measured excess attenuations by calculation and determined the parameters of the absorbing boundaries by inversion of the measurement results. T. C. van den Dool generated the two-dimensional acoustic image of a passing high-speed train shown in Fig. 3.1.

3.8. References

3.1. STRUTT, J. W., BARON RAYLEIGH. *The Theory of Sound*. Dover, New York, 1945, pp. 429–430.
3.2. VAN DER TOORN, J. D. Schaalmodelonderzoek wegverkeersgeluid [Scale model investigations on road traffic noise]. *Journal of the Acoustical Society of the Netherlands*, 1987, **87**, 91–103.
3.3. LANGHAAR, H. L. *Dimensional Analysis and Theory of Models*. Wiley, New York, 1965.
3.4. INTERNATIONAL ORGANIZATION FOR STANDARDIZATION. *Acoustics – Description of Measurements of Environmental Noise – Part 1: Basic Quantities and Procedures*. ISO, Geneva, 1982, ISO 1996/1–1982(E).
3.5. VAN DER TOORN, J. D., HENDRIKS, H. and VAN DEN DOOL, T. C. Measuring TGV source strength with Syntacan. *Journal of Sound and Vibration*, 1996, **193**(1), 113–121.
3.6. MINISTRY VROM. *Reken- en Meetvoorschrift Railverkeerslawaai '96*. [Calculation and Measuring Rules Railway Noise '96.] Ministry VROM, The Hague, 1996, distribution code 12468/164.
3.7. CATO, D. H. Prediction of environmental noise from fast electric trains. *Journal of Sound and Vibration*, 1976, **46**(4), 483–500.
3.8. LOUDEN, M. Equivalent mean energy level from relatively short parts of railway lines. *Journal of Sound and Vibration*, 1979, **66**(1), 69–73.
3.9. PETERS, S. The prediction of railway noise profiles. *Journal of Sound and Vibration*, 1974, **32**(1), 87–99.
3.10. VAN DER TOORN, J. D. Measurements of sound emission by single vehicles. *Noise Control Engineering*, 1978, **11**(3), 110–115.

3.11. DELANY, M. E., RENNIE, A. J. and COLLINS, K. M. *Scale Model Investigations of Traffic Noise Propagation*. National Physical Laboratory, Teddington, 1972, Acoustics Report Ac 58.
3.12. INTERNATIONAL ORGANIZATION FOR STANDARDIZATION. *Acoustics – Attenuation of Sound During Propagation Outdoors – Part 1: Calculation of the Absorption of Sound by the Atmosphere*. ISO, Geneva, 1993, ISO 9613–1.
3.13. DE JONG, B. A., MOERKERKEN, A. and VAN DER TOORN, J. D. Propagation of sound over grassland and over an earth barrier. *Journal of Sound and Vibration*, 1983, **86**(1), 23–46.
3.14. SALOMONS, E. M. Sound propagation in complex outdoor situations with a non-refracting atmosphere: model based on analytical solutions for diffraction and reflection. Section 2.2.3. *Acustica–Acta Acustica*, 1997, **83**, 436–454.
3.15. DE VOS, P. H. and LAMMERS, I. S. *Noise Barriers: State of the Art*. NS Technical Research, Utrecht, 1995.
3.16. INTERNATIONAL ORGANIZATION FOR STANDARDIZATION. *Acoustics – Measurements of Sound Absorption in a Reverberation Room*. ISO, Geneva, 1985, ISO 354-1985 (E).
3.17. ZWIKKER, C. and KOSTEN, C. W. *Sound Absorbing Materials*. Elsevier, New York, 1949.

4. Generic prediction models for environmental railway noise

J. J. A. van Leeuwen
DGMR Consulting Engineers BV, PO Box 82223, NL-2508 EE The Hague, The Netherlands

4.1. Introduction
The prediction of noise around a railway line is of extreme importance when a new line is under debate or when an existing line is to be reconstructed or renewed. One of the first reactions of people who live in the vicinity is about the noise and about the quality of the environment around their home. Also, the impact on other noise-sensitive buildings, such as hospitals, schools, homes for the handicapped and homes for elderly people, has to be concerned. Another aspect is noise pollution in nature. Avoiding annoyance to human beings should not lead to an unrestricted development of infrastructure in a wildlife area.

4.2. Noise indicators
A noise index must be related to the annoyance perceived. In practice the noise index must be considered before any calculation is started. Examples of such an index are the equivalent sound level during a period of the day or night, and the maximum sound level during the passage of a train. Another fact that has to be considered is whether we want to calculate the sound level for the average weather situation or for the downwind situation. In this section, a summary is given of the different indicators [4.1].

4.2.1. Annoyance
'Annoyance' is a term used to describe the negative feelings associated with noise. Since annoyance can mean a lot of different things to different people, it is not useful to define annoyance any more precisely. In general, annoyance is defined as the mean annoyance for a large group of people exposed to noise. There is no international scale for annoyance. Most of the time, the annoyance is taken as the

percentage of a group of people who assess a noise as causing light, medium or high annoyance.

Annoyance can also imply sleep disturbance. Objectively, sleep disturbance ranges from the smallest detectable physiological response to an external stimulus while asleep, to complete awakening. Sleep disturbance can also be described subjectively, using some appropriate scale, after the event. If there are any effects on mood, attitudes or performance at some task the next day, such variables could also be measured, both objectively and subjectively.

Written or spoken negative observations which are recorded by authorities can be defined as complaints. It is useful to record these complaints in such a way that statistics can be kept. Informal complaints that are not recorded are of limited value in environmental-noise assessment.

4.2.2. The noise level and the A-frequency-weighted noise level

The decibel is the unit that is most associated with noise, hearing and acoustics. It is used to describe sound pressures and intensities, and is based on a logarithmic scale. The main reason it is used so much as the fundamental unit of measurement in acoustics is that the ear itself 'hears' logarithmically and human beings judge the relative loudness of two sounds by the ratio of their intensities. The ear also has a gigantic dynamic range, and using a logarithmic decibel scale helps to keep the enormous numbers under control.

The decibel level is quoted as an absolute level, relative to a certain reference level. This reference level is the threshold of hearing, that is a sound pressure level of 20 µPa. The decibel level is represented mathematically as 20 lg (measured pressure level/reference pressure level).

The most commonly used acoustic parameter is the A-weighted noise level, quoted in dB(A). Roughly speaking, the human ear works over a frequency range from 20 Hz to 20 kHz, though this range tends to shorten with age. Over this range, the ear has a 'frequency response' and is more sensitive at some frequencies than others. The A-frequency weighting is a frequency filter used to present a noise level in such a way as to represent the human ear's sensitivity to noise. Sound-level meters have an A-frequency-weighting network built in so as automatically to measure the A-weighted noise level. Other weighting systems exist, such as B, C and D frequency weighting, representing the ear's differing response at louder levels, but A-weighting is by far the most common.

4.2.3. Root mean square average

The average sound pressure over time is close to zero, irrespective of the amplitude of the sound wave, because the positive and negative excursions of the instantaneous pressure above and below the static atmospheric pressure tend to cancel out. Therefore, the average sound pressure would be useless as an indicator of environmental noise. The root mean square average does not suffer from this disadvantage, because the squared sound pressure is always positive as a result of the squaring process. The precise method of averaging and the averaging time period must be defined. The short-time exponential average sound pressure uses a time constant of 125 ms. In a sound-level meter this time constant is defined as 'fast'. For long-time averaging over a time period, the long-time root mean square sound pressure is used, with a linear averaging time extending throughout the defined time period of observation.

4.2.4. The maximum sound level $L_{A, max}$

The maximum sound level is the maximum value of the short-time exponential average root mean square sound pressure obtained using the 'fast' detector indicator characteristic, expressed in decibels. This is not the same as the peak value, which is the maximum instantaneous sound pressure without any frequency weighting and without any averaging of any kind. The maximum sound level is usually more closely correlated with the short-term subjective loudness of the sound event than the peak sound pressure is.

4.2.5. The long-time average sound level and the equivalent sound level

For any fluctuating sound, the linear average root mean square sound pressure level gives the same numerical value as the sound level for a steady sound with the same amount of physical sound energy over a defined observation time period. The average sound level is the long-time linear average root mean square sound pressure, expressed in decibels. The average sound level over a defined time period is not the same as the average of the separate maximum sound levels of a sequence of events occurring within the same time period.

The A-weighted average sound level forms the basis of the system of the harmonized European environmental noise indicator. It is defined as the long-time linear average root mean square A-weighted sound pressure, expressed in dB(A).

4.2.6. Statistical indicators

The statistical indicator $L_{A,n}$ is defined as the sound level which is just exceeded for n per cent of the defined observation time period. The sound level used as the input for this statistical analysis is the level determined with the short-time exponential average of the root mean square sound pressure, obtained using the 'fast' detector indicator characteristic, expressed in dB(A). Most of the time, the statistical level $L_{A, 95}$ is used to define the background level. For indicating the annoyance caused by trains, this statistical level is unusable.

4.2.7. The basic indicators: $L_{A, day}$, $L_{A, evening}$, $L_{A, night}$ and $L_{A, 24h}$

The basic indicator used for the strategic assessment of policies, plans and procedures is the A-weighted equivalent sound level for a period of the day. This period can be the daytime, the evening or the night. A point of discussion can be the time periods for these indicators. In most cases, the daytime period will be from 07.00 till 19.00, the evening will be from 19.00 till 23.00 and the night will be from

23.00 till 07.00. The $L_{A, 24h}$ basic indicator is simply the A-weighted equivalent sound level for the complete 24 h period.

4.2.8. The composite indicator L_{den}

With the basic indicators being the A-weighted equivalent sound level during the daytime ($L_{A, day}$), the A-weighted equivalent sound level during the evening ($L_{A, evening}$) and the A-weighted equivalent sound level during the night-time ($L_{A, night}$), it is possible to define a composite noise indicator which is relevant to the complete 24 h. The indicator L_{den} can be determined by calculating the long-term average $L_{A, eq}$ over the day, the evening and the night. A penalty for the evening of 5 dB(A) has to be incorporated, and 10 dB(A) for the night-time. A result of this penalty is that one train per hour during the night-time results in the same equivalent sound level as ten trains per hour during the daytime.

4.3. Background to environmental-noise predictions

Before making any calculations, some important questions about the purpose of noise predictions have to be answered. These questions are why, where, what, when and how.

4.3.1. Why noise predictions?

The objective of a noise prediction is to quantify noise levels for an existing railway line and for new lines. For human beings, noise can be very subjective. The only way to get a clear picture of the extent to which anyone is bothered by a noisy train is by giving a figure in dB(A). When noise levels are quantified, something can be said about the effect or impact on the environment. A certain noise level can be checked against laws, guidelines or recommendations. In some countries this is a legal commitment. Sometimes a prescribed generic prediction model is a statutory requirement.

A generic prediction model can be used to calculate the effect of noise reduction measures on the noise source. A measure applied to only a part of the source or a measure which causes a change in the emitted noise spectrum can result in higher or, most of the time, disappointing results at a larger distance from the railway line. A component of any prediction model is the sound propagation model. A screen or a barrier will affect sound propagation. With an accurate prediction model, calculations can be performed for noise reduction measures in the sound propagation path.

4.3.2. Noise predictions for where?

The calculation of a noise level can be performed for a single receiver point. And, of course, more calculations should be performed for other points. These separate calculations are almost completely independent from each other. After all, another receptor has a different propagation effect, which will result in a higher or lower noise level. In principle, a receiver point can be located anywhere. Most prediction

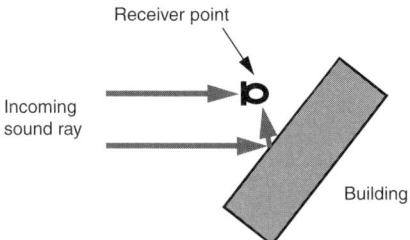

Fig. 4.1. Effect of a reflection on sound propagation

models calculate the sound propagation outdoors, so that the receiver points are in the open air. To predict noise levels inside buildings, a prediction model can be combined with calculational models for determining the facade isolation. The combination of the frequency-dependent noise isolation of the complete facade (and, if relevant, the roof) and the sound intensity from the railway line gives the inside noise level. These calculational models are not discussed in this chapter.

In most cases noise levels are calculated for houses or dwellings. However, the noise level in other noise-sensitive buildings/areas is also of interest. These buildings and areas include places where people live and sleep. Examples of such buildings/areas are hospitals, institutions for the elderly or disabled, caravan parks, etc. Less noise-sensitive, but also places where people live, stay or work, are offices, hotels, recreation areas, etc. Special receptors include nature reserves, cemeteries, etc. These places also have to be protected from noise pollution.

The height of the receiver is also a point of consideration. The average house has one or two floors. This means that receiver heights of 1.5 m and 5 m above ground level will do. Apartment buildings need receiver heights for some relevant floors. Most of the time a height of 1.5–2 m above floor level is required. Nature reserves and other areas require a height of 1.5 m above ground level.

Noise levels at receiver points in front of a reflecting facade will be higher in relation to the free-field noise level (Fig. 4.1). To calculate the noise intensity into the facade, the reflection from the facade must be left out. Thus, for sound isolation purposes, the required noise level at the facade (in the direction of the sound source) is the same as the free-field noise level. The calculated noise level in front of a facade is in most situations 2 to 3 dB higher than the free-field noise level. In noise regulation acts, these noise levels in front of a facade and the free-field noise levels can easily be confused.

4.3.3. What do we want to calculate?

To get a clear view of noise pollution, the annoyance due to noise must be calculated. For railway traffic noise the annoyance is directly related to the equivalent noise level ($L_{A, eq}$) at the facades of dwellings and other noise-sensitive buildings. For road traffic noise and for most industrial sites, the annoyance is also directly related to the equivalent noise level. Only at certain industrial sites is the annoyance related to other noise indexes such as the maximum noise level or the impulse level.

The annoyance is very limited if there is a temporarily higher noise level for only one or two days. This means that we have to calculate the $L_{A,eq}$ which is representative of a specific facade for a very long period. This period is at least a couple of months and in most cases is several years. The calculated equivalent noise level must be reproducible. Almost all prediction models calculate the equivalent noise level during one or several hours a day, in the evening and night periods.

'Representative' and 'reproducible' also mean that the average situation must be considered. Noise propagation outdoors is very much affected by weather conditions. At large distances the noise level can vary by up to 5 to 10 dB(A), and at distances over 1000 m the variation can be as large as 20–30 dB(A) (see Fig. 4.2).

Generic prediction models work with a defined meteorological situation. This could be a situation with an average weather condition, or a situation when there is average propagation. The average propagation is the average of situations with downwind, side wind and upwind. Some prediction models calculate only the downwind situation. Some other models calculate the downwind situation but have an extra correction for the average weather situation. Another possibility is to calculate the noise propagation for several different wind directions and then to calculate the time-related average noise level. The most accurate approximation to the noise propagation can be determined for the downwind situation. Side wind and upwind situations give a much larger variation in noise propagation.

Other parameters that affect the noise propagation are air temperature, humidity and other circumstances, such as the presence of water, ice or snow. Water and frozen water can give extra reflections. Snow can give extra sound absorption. When noise predictions are made, a decision must be made to include or to exclude these very important weather conditions.

Also, weather situations and other conditions and circumstances can affect the noise emission at the source. For rail traffic noise this means that calculations will be done for the average traffic flow, the average noise emission of the train or of different types of trains, the average speed of the traffic, and the average condition

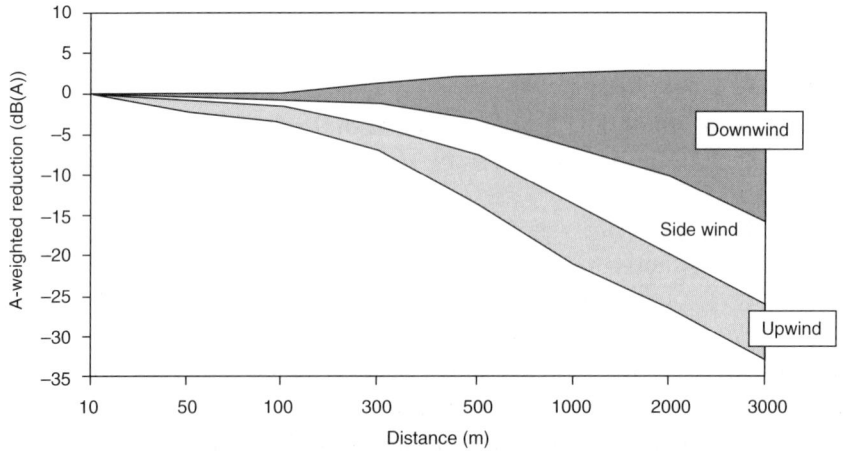

Fig. 4.2. Influence of meteorological effects on sound propagation [4.4]

PREDICTION OF ENVIRONMENTAL RAILWAY NOISE

Fig. 4.3. An example of several railway lines

of the maintenance of the track and of the wheels. The noise of trains is related to the roughness of the rails and wheels. This means that prediction of the noise level implies knowledge of the roughness of the wheels and rails. The best that can be done is to determine an average roughness level that is representative of a particular railway company or railway line.

4.3.4. When to use prediction models

Prediction models are useful in various situations. The first and most important situation is related to the fact that the results of an acoustic investigation for an existing railway line are reproducible and representative. Predicting noise levels by using a calculational model is much more representative and reproducible than measurements are. The previous section gives an explanation of the reasons that a calculation will always represent an average situation. Measuring sound levels over a very long period offers the only possibility of approaching the representative equivalent noise level. From the theoretical point of view, this very long period is at least a year.

The identification of prominent sources is another use of generic prediction models. In many situations, mostly in urban and suburban areas, there is a large number of noise sources. These sources include, for example, local road traffic, highway traffic, light rail vehicles, conventional trains, industrial installations, aeroplanes and, of course, high-speed trains. By identification of the prominent sources, it is possible to produce a ranking of these noise sources. Ranking is extremely useful for a location with complicated problems of noise pollution (such as that shown in Fig. 4.3).

Generic prediction models can be used to propose noise reduction solutions. Two types of noise reduction measures must be distinguished: measures at the source and measures in the propagation path. Calculating with the same simulation model and using a new sound power level can evaluate the acoustic effect of a measure at the source. The propagation will not be affected, so the same propagation attenuation can be used. If a noise barrier is used, the propagation will change. The propagation path must be calculated again, taking the extra diffraction into account. The sound power level of the source will not change in this case. The prediction model thus can be used to calculate the effect of noise measures prior to implementation.

There is a strong tendency for more and more authorities and owners of industrial sites to use noise management. With financial management, it is common that managers are precisely aware of their potential capacity in this respect. This is important for railway companies also. To bring more trains into action will result in higher equivalent noise levels, which is not allowed without limit. Using a prediction model for noise management provides awareness of the acoustical capacity of a railway line. A study of an area's development over a period of time or a study of the development of a railway line can be made as well.

To identify noise pollution in urban and suburban areas, and also in rural areas, a prediction model can be used to make a noise map. Noise mapping concerns the presentation of data about an existing situation and a future situation in terms of a noise indicator. Most of the time, this will be the equivalent sound level for a defined period of the 24 h (e.g. day or night) or a compound level calculated from this period. The compilation of a noise map needs a prediction model. This prediction model will calculate noise levels at a large number of points (for example, on a grid). With interpolation techniques, it is possible to draw lines between points with equal noise levels. The next step is to colour the areas between these lines, so that coloured zones are created. It must be realized that the accuracy of these coloured zones is very dependent on the degree of detail of the calculations. A grid with points for every 100 m cannot result in an accuracy of 10 m, especially if there are buildings between the grid points.

A powerful tool for making studies over a period of time is a geographical information system (GIS). A map with coloured zones can be produced in a professional way, thanks to the fact that the GIS has been developed for this purpose. With a GIS it is possible to examine several data sets. Together with data on the inhabitants per unit area, the number of people annoyed by a certain noise level can be determined. It is also possible to calculate the surface area in square metres of a nature reserve. For policy purposes, this kind of data is extremely important.

4.3.5. How do you provide your input?

For making a calculation, the acoustically relevant information is needed. For such acoustical input data, only a few items are important. The most common software packages use the following items.

PREDICTION OF ENVIRONMENTAL RAILWAY NOISE

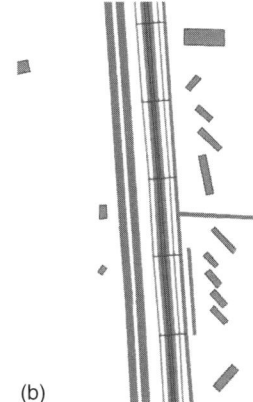

Fig. 4.4. (a) An example of a real situation (copyright Aeroview Dick Sellenraad); (b) computer simulation model for this situation

- Line source, for the characterization of a railway line.
- Buildings and other objects that give screening of the sound field or produce a reflection.
- Barriers or noise screens. These objects will give a screening attenuation, but also can give an extra contribution due to reflections.
- Height lines or surface contours. This information is needed to define the average height of the sound path above the ground.
- Characterization of ground areas (acoustically reflective or absorbing ground).
- Receiver points or a grid of points.

The combination of these items on a map characterizes a specific acoustic situation. Depending on the purpose of the study, the scale level must be defined. For a study of a complete city, a small shed is irrelevant. For the calculation of the sound level in a single house, a shed might be of importance. An example of a real situation is shown in Fig. 4.4(a). Figure 4.4(b) shows the relevant acoustical information needed for noise prediction.

4.3.6. Sequence of noise predictions

It is important that a rough picture can be given of the environmental pollution due to noise. This can be done in a very simple way, for example, by taking into account only the distance from the railway line. This calculation should be pessimistic – in other words, always determining a safe border outside of which one cannot expect annoyance. Using measurement data and incorporating traffic flow can improve these rough pictures. Then, for critical or interesting situations, it may be useful to perform a more detailed calculation, for example by using more sophisticated prediction models. Such a model can be based on theoretical and empirical formulae. Finally, in very critical situations, where prediction models might have reached their limits of application, it should be possible to proceed to more advanced techniques, such as boundary element, finite-element or parabolic equations. These techniques are only applicable to very small areas and usually require an enormous calculation time on a computer.

4.4. What is a noise prediction model?

The definition of a prediction model is not so clear. First of all, we have to distinguish two types of models. These are meta-models and specific models.

A meta-model, or framework, gives a description of how to make the calculations and which expressions, algorithms and equations have to be used. This meta-model is very general and conceptual. A meta-model is described on paper and authorized in a law, a standard, a norm or a guideline. A meta-model is a representation of the physics of a general situation, which means that a selection of the most relevant aspects is made. This model is an abstraction of reality and is highly idealized.

The meta-model on paper is the basic input for the development of computer software. The algorithms of the meta-model, in combination with a detailed explanation of the principles of the calculation, will lead to reliable software. Without this detailed explanation, the software developer must make some more assumptions. Special software developed without an acoustical background can lead to results that are not correct. Making manual calculations can validate the software. In reality there are always situations that are not covered by the meta-model.

A specific model is a representation of a particular situation. It is applicable to that situation only. This model is made as a result of observations of this situation. In a specific model, the number of passing trains and their running speed are specified. Again, there is a conversion or selection of the most relevant aspects. This model is also an abstraction of reality and is idealized.

A second distinction can be made in relation to the application of a prediction model. This distinction gives two categories of prediction methods, which can be termed survey/engineering methods and laboratory methods. Such a description is commonly used in the measurement equipment environment; the applicability here is more or less comparable.

Laboratory methods calculate the sound level of a very typical situation. This is not the average situation. This means, for example, that this method will perform a calculation for one particular train with a well-defined wheel and a well-defined rail. The roughness conditions should be measured. Also, these models use a well-defined ground surface and facade surface in the propagation path. The acoustic impedance of the ground surface and facade should be measured. The calculation can be done for only one particular weather situation, so specific weather information should be measured. Most of the time the laboratory method uses more comprehensive and more advanced numerical techniques.

The complementary method is a model comparable to the survey/engineering method. This is a model that fits better with common reality. This type of model calculates the equivalent sound level for an average situation. Most algorithms in these models are empirical formulae.

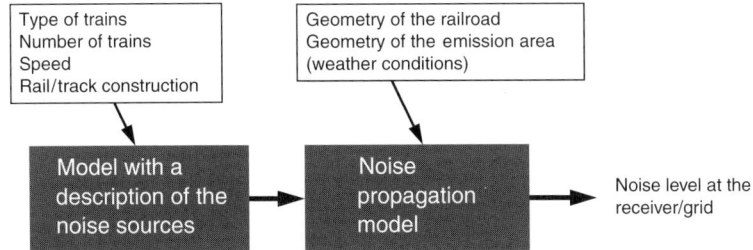

Fig. 4.5. General set-up of a prediction model

4.5. Noise prediction methodology

In general, prediction models are based on the pure theory and a number of empirical formulae. With regard to the contents of a prediction model, two main parts must be considered:

- a model with a description of the noise source
- the propagation model.

Figure 4.5 shows the general set-up of a generic prediction model. It will be clear that there are separate factors that influence the noise source and influence the acoustic propagation.

The first part of a generic prediction model is the description of a rail vehicle as a noise-generating object, which may be described as one noise source or as a combination of different noise sources. Each source has a sound power level and has to be given with its position and height. Also, the relation between the speed and the various types of track construction has to be described. Of course, the description of the sources will be dependent on the type of train.

Acoustic propagation models are (or are supposed to be) very general. A propagation model gives a description of the attenuation of noise from a source to a recipient. In the end, the noise of a vehicle will not propagate in a way different from any other kind of noise. There are several propagation models. The best known are

- the international ISO 9613 model, *Attenuation of Sound during Propagation Outdoors* [4.2, 4.3]
- the German VDI 2714/2720 model, *Schall Ausbreitung im Freien/Schallschutz durch Abschirmung im Freien* and *Schall 03* [4.4–4.6]
- the Dutch rail traffic model SRMII [4.7]
- the Nordic models from Scandinavia [4.8, 4.9]
- the French models from the *Guide du Bruit* and Mithra-Fer [4.10, 4.11]
- the *Calculation of Railway Noise* from the UK [4.12, 4.13].

In order to predict the sound level at a particular receiver point due to a known noise source, one must depart from the models for acoustic propagation in the open air [4.14–4.17]. The generic prediction formula for the calculation of the sound level is

$$L_{A, eq} = L_E - A_{propagation} \qquad (4.1)$$

in dB/octave or dB(A). Here $L_{A,eq}$ is the equivalent sound pressure level at a recipient point, L_E is the sound emission of the source and $A_{propagation}$ is the attenuation due to noise propagation. The description of the propagation of sound from source to recipient is present in the various propagation models. In these models, the various ways to calculate the attenuation and correction terms are also described. The calculation can be made directly in dB(A) or per octave frequency band.

4.6. Source description model

The description of the noise source concerns the position of the source (or sources) and the related sound power level. The most relevant is the acoustic radiated energy of the wheel (with bogie) and the rail (including the complete track structure). Other noise sources are auxiliary equipment and aerodynamic sources. The aerodynamic sources become significant at running speeds above approximately 250 km/h.

It should be clear that knowledge of the sound power level of the noise sources that can be distinguished is essential. The physical height of a noise source above the track and above ground level is essential for calculation of the acoustic propagation from source to receiver. A higher noise source, for example, will give less attenuation due to the acoustic absorption of the ground. Also, after the erecting of a noise barrier, the higher noise sources will not be screened, or will be less screened, than lower noise sources.

The positions of the noise sources are shown in Fig. 4.6.

A typical component of a generic prediction model is the sound power level of a train. This noise emission characteristic is a fixed value. As was mentioned before one of the principles of a prediction model is that the noise emission level is based on the average noise emission of the train. This noise emission level is dependent on the type of train and also on the speed. The equivalent noise emission level is also dependent on the number of vehicles passing by.

The sound power level of a train, or the sound power level of a distinguished noise source of the train can be described by the following empirical formula [4.7]:

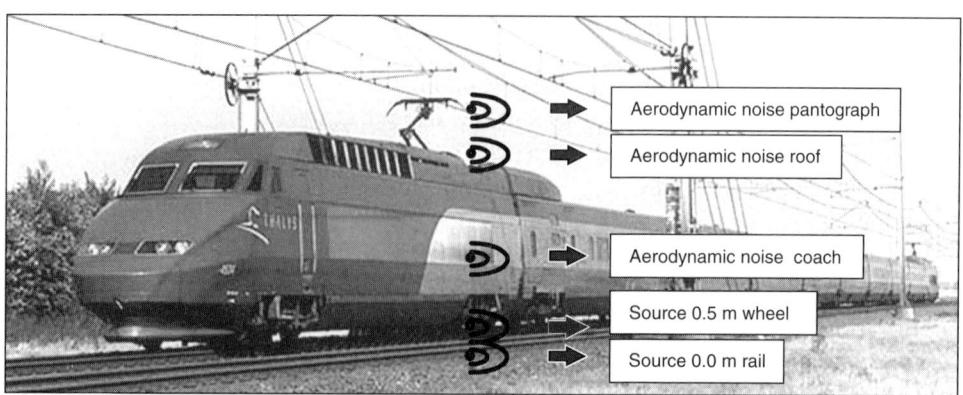

Fig. 4.6. Summary of all relevant noise sources of a high-speed train

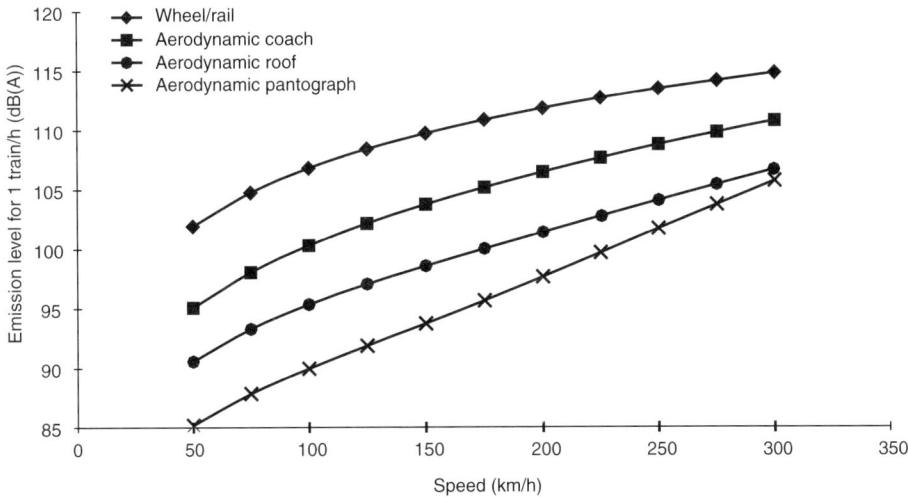

Fig. 4.7. Emission levels for a Thalys train consisting of two power cars and ten passenger cars at 250 km/h [4.7]. (The Thalys trains are TGV trains running on the Paris–Amsterdam line)

$$L_{W,n} = \alpha_n + \beta_n \lg\left(\frac{v}{v_0}\right) + C_{r,n} \tag{4.2}$$

where $L_{W,n}$ is the sound power level of the train (or of the nth source in the train) in dB per octave or in dB(A), α_n is the sound emission of one vehicle, dependent on the source height and on the frequency, β_n is the speed-dependent value of the sound emission for every source, also dependent on frequency, v is the vehicle speed in km/h, v_0 is the reference speed in km/h and $C_{r,n}$ is the correction for different types of rail construction in dB per octave or in dB(A). In generic prediction models the correction for different types of rail construction is related to a reference track construction. The reference track is, most of the time, a ballast track with concrete sleepers.

The equivalent emission level of a railway line can be seen as the emission of a complete line from which the sound power is radiated. Here the equivalent noise level is, of course, the result of the sound power level of the train and the number of trains passing by. Another relevant aspect is that when trains are running at a higher speed, the equivalent emission level may be lower owing to the fact that the exposure time is shorter.

The equivalent emission level of a railway line or the emission level of a distinguished noise source of the line is equal to [4.7]

$$L_{E,n} = L_{W,n} + 10\lg\left(\frac{Q}{v}\right) + C_{r,n} + 30 \tag{4.3}$$

where $L_{E,n}$ is the equivalent emission power level of the train (or of the nth source of the train) in dB/m per octave band or in dB(A)/m and Q is the number of trains or the traffic flow in numbers/h.

In Fig. 4.7 an example is given of the equivalent noise emission level of a high-speed railway line operating TGV trains at 250 km/h. This emission level has been calculated for the passage of one train per hour. A train consists of ten coaches and two engine cars. From the figure it is clear that, if a barrier with a height of, for example, 2 m is erected, the lower noise sources will be shielded. The reduction of these sources will be around 12 dB(A). The higher noise sources will become predominant owing to the fact that the barrier does not affect them.

4.6.1. Sound radiation characteristics

Most models use an attenuation to describe the sound radiation characteristics. Most generic models, such as the models from Austria [4.18, 4.19], Germany [4.6], Switzerland [4.20] and the Netherlands [4.7], and the UK Department of Transport model [4.13] use a horizontal and a vertical radiation characteristic. Except for the French Mithra model [4.11], all models have only a total directivity index in dB(A). Mithra has a directivity index which is different for every octave band frequency. Figures 4.8 and 4.9 give some examples of the horizontal and vertical radiation characteristics [4.17].

4.7. Propagation models

In order to predict the sound level at a receiver point due to a known noise source, models of acoustic propagation in the open air can be used. In these propagation models the general principle used to predict a sound level is based on a combination of theoretical and empirical estimations.

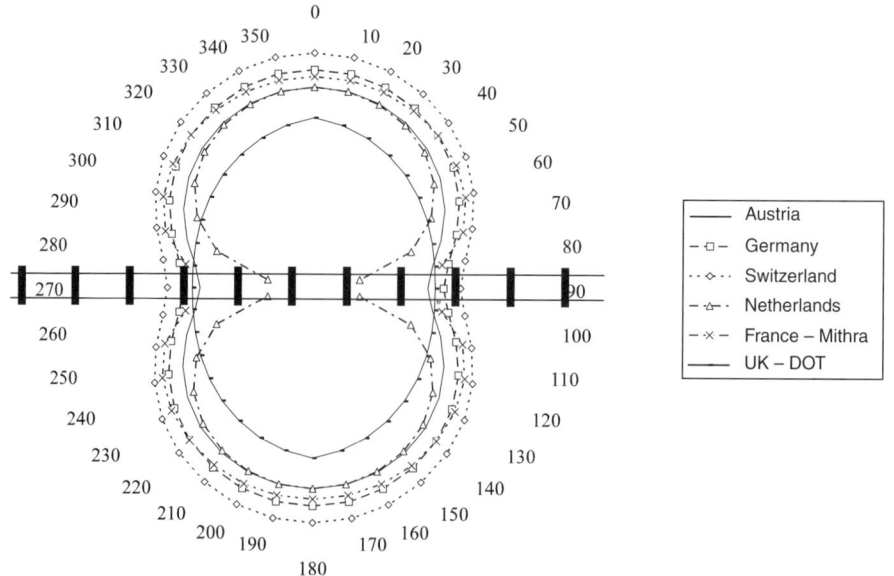

Fig. 4.8. Horizontal directivity index

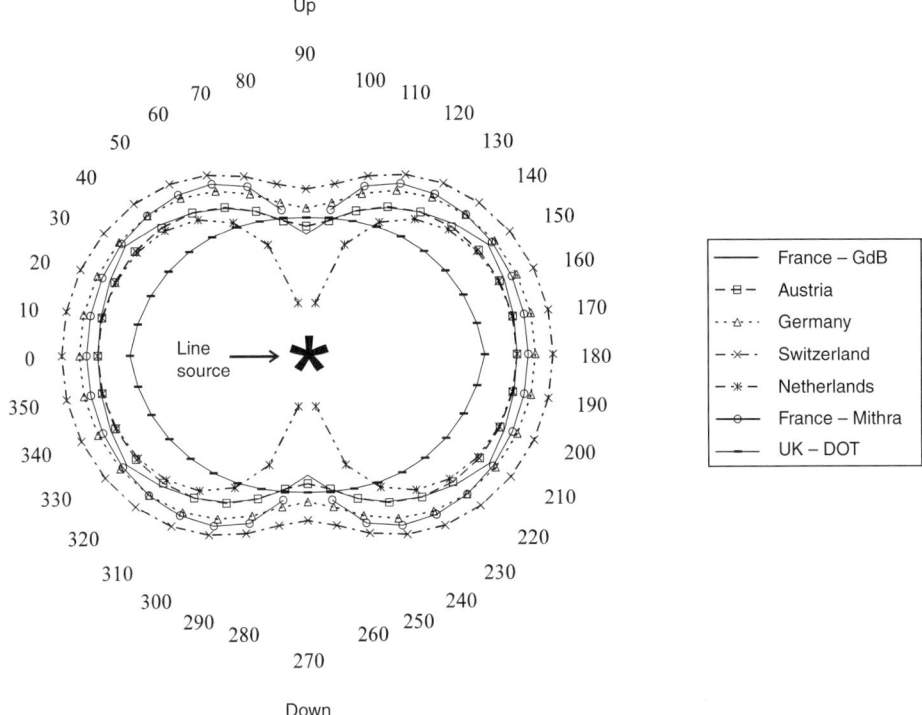

Fig. 4.9. Vertical directivity index

Propagation models calculate the attenuation for various sound propagation paths. First of all, these propagation paths must be defined. The direct propagation path is a straight line from the source point to the receiver point. This is the path without reflections from vertical obstacles. Of course, other effects that affect the propagation, such as ground reflection and attenuation or screening, can be taken into account. Besides the direct path there are also sound paths which incorporate reflection from vertical objects. The same propagation effects are taken into account, but also the attenuation due to this reflection, which depends on the absorption coefficient. Finally, the sound levels of all different propagation paths are aggregated to obtain the total sound level at the specific receiver point. The identification of the propagation paths is discussed in Section 4.9.

The calculation of the propagation is done by using a combination of theoretical and empirical formulae for downwind propagation. As described in the previous sections, the downwind situation gives much less variation of the noise compared with the upwind levels. The downwind prediction is much more accurate. On the other hand, as was also described in the previous sections, it is necessary to calculate the long-term average of the equivalent sound level. Some models do not make any correction for this. The final result for a calculation with such models is the downwind sound level. Other models have a meteorological correction to correct the downwind sound level to a long-term equivalent sound level. Simply described, if a dwelling is situated downwind from the source for a period of half a year and for the other half of the year it is situated upwind, one can expect a sound level up to 3 dB(A) lower than the downwind sound level.

In order to perform a calculation the following formula can be used:

$$L_{A, eq} = L_E - \Delta_{div} - \Delta_{air} - \Delta_{ground} - \Delta_{barrier} - \Delta_{misc} + C_{refl} - C_{meteo} + k \tag{4.4}$$

in dB/octave or dB(A). Here $L_{A, eq}$ is the equivalent sound pressure level at the recipient point, L_E is the sound emission of a noise source per metre or the sound power level per metre defined in the source model, Δ_{div} is the attenuation due to divergence or geometrical spreading of the noise, Δ_{air} is the attenuation due to atmospheric absorption, Δ_{ground} is the attenuation due to the absorption of the ground, $\Delta_{barrier}$ is the attenuation due to a barrier or another obstacle, Δ_{misc} describes the additional types of attenuation, C_{refl} is a correction for the contribution of reflections of the noise, C_{meteo} is a correction for meteorological effects and K is a constant related to the units and to other corrections.

In some models, $K = K_1 + K_2 + K_3$, where

$$K_1 = 10 \lg 1000 \tag{4.5}$$

$$K_2 = 10 \lg \left(\frac{1}{4\pi} \right) \tag{4.6}$$

$$K_3 = 10 \lg \left(\frac{\pi}{180} \right) \tag{4.7}$$

These correction factors are dependent on the definition of the units, for example, the use of kilometres instead of metres, the definition of geometrical spreading and the use of radians or degrees.

4.7.1. Geometrical spreading

The most direct, simple and basic acoustic factor is the attenuation due to geometrical spreading. If a receiver is located at a larger distance from the source the sound level will be lower. The sound intensity is reduced as the square of the distance between source and receiver for a point source or as the distance between source and receiver for a line source. For a point source the sound power is spread over a spherical surface, and for a line source the sound power is spread over a cylindrical surface. This approach implies a simple geometrical calculation of the exact distance between source and receiver for point sources or between a finite section of the line source and the receiver for a line source. For a short, finite line source this approach is in principle the same as in relation to a point source. The approach is basically similar in almost all existing models; however, in some models part of the description of the source may be included in the algorithms describing the geometrical spreading. In other models part of the ground effect may be included (e.g. hemispherical spreading for a reflecting ground).

Table 4.1. Attenuation as a result of air absorption at 10°C and 70% relative humidity [4.2]

Frequency (Hz)	63	125	250	500	1000	2000	4000	8000
Attenuation (dB/km)	0.1	0.4	1.0	1.9	3.7	9.7	32.8	117.0

The calculation of the geometrical spreading from a point source to a receiver point is performed using

$$\Delta_{div} = 10\lg\left(\frac{1}{4\pi R^2}\right) \tag{4.8}$$

or, in relation to K_2,

$$\Delta_{div} = 20\lg\left(\frac{1}{R}\right) \tag{4.9}$$

where R is the distance from the source to the receiver in metres.

4.7.2. Atmospheric absorption

The attenuation as a result of air absorption is significant only at rather large distances and at high frequencies. The influence of atmospheric absorption is dependent on the distance, frequency, temperature and relative humidity:

$$\Delta_{air} = R\alpha \tag{4.10}$$

where R is the distance from the source to the receiver in metres and α is the air absorption in dB/m for each octave band of frequency. For example, the attenuation as a result of air absorption at 10°C and 70% relative humidity (according to ISO 9613-1 [4.2]) is given in Table 4.1.

4.7.3. Absorption by the ground

For the calculation of the ground attenuation, different generic prediction models use different methods. First of all, it has to be realized that the attenuation of the ground can be either negative or positive. A negative attenuation (amplification) is a result of the reflection of the sound. This reflection from a fully reflective ground can lead to very strong interference effects and is strongly frequency dependent. Reflection from a more or less absorbing ground will lead to reduction of the sound level and a shift of phase, with strong frequency-dependent interference.

In a statistical energy approach, a full reflection will result in a sound level 3 dB higher. For the lower frequencies, as a result of the doubling of the sound pressure, the result will be a sound level 6 dB higher.

In most generic models the algorithms taking these effects into account are empirical. It is generally not possible to use complex acoustic-impedance data in propagation models. In practice, there is no information on the acoustic impedance of a patch of grass, a lawn or a meadow. For a curved sound path, multiple reflections from the ground are taken into account. The different sound paths in relation to the ground reflection or absorption are shown in Fig. 4.10.

In some prediction models the ground attenuation is based on the empirical findings of Parkin and Scholes [4.21]. The calculation of the ground absorption is dependent on the ground condition, the heights of the source and the receiver, the distance and the average height of the sound path above the ground level. According to the theory of Parkin and Scholes, one has to distinguish between three regions of the propagation path. These regions are the source region, the receiver region and the middle region. The ground effect for each region is based on the ground factor, which can be defined to be between 0 and 1, the ground may range from fully reflective (water or concrete) to fully absorptive (porous surface, such as grassland). The resulting ground attenuation is the combination of the ground attenuation of the source region, a receiver region and the middle region.

In other models the calculation of the ground absorption is based on the results of Chien and Soroka [4.22, 4.23] This uses a more theoretical and scientific approach to the ground effect. However, it should be considered whether the collection of the relevant data is practicable. For these calculations it is necessary to know the acoustic impedances of all relevant areas. Most of the time, this information is not available.

4.7.4. Attenuation due to a barrier or another obstacle

The attenuation as a result of the effect of a barrier or a screen is, in almost every generic prediction model, based on the experimental results of Maekawa [4.24], who has done a lot of experimental research based on measurements with a long, thin screen.

The approach of Maekawa is based on the calculation of the path difference (Fig. 4.11). The path difference is defined as the difference between the distance from source to receiver without the barrier and the distance from source to receiver over the diffraction edge of the barrier. This difference is simply the path difference in metres. This means that the barrier attenuation is dependent on height and distance. This vertical path difference δ is

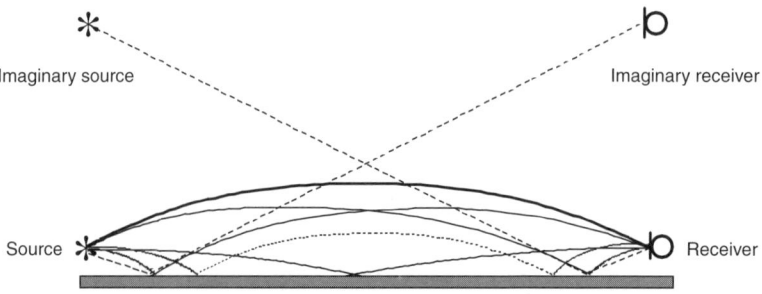

Fig. 4.10. The sound paths that can be distinguished in relation to absorption and reflection from the ground

PREDICTION OF ENVIRONMENTAL RAILWAY NOISE

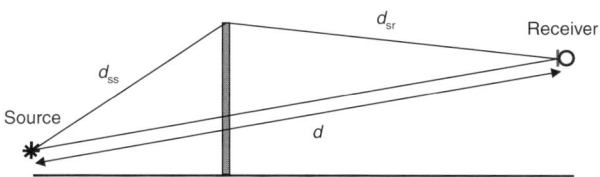

Fig. 4.11. Determination of the path difference

$$\delta = d_{ss} + d_{sr} - d \quad (4.11)$$

where δ is the path length difference in metres, d_{ss} is the path length from the source to the top of the screen/barrier in metres, d_{sr} is the path length from the top of the screen/barrier to the receiver in metres and d is the path length from the source to the receiver without the screen/barrier, in metres.

In principle one should distinguish between a vertical path difference over the top of the barrier and a horizontal path difference for a left or right detour around the barrier. For railway noise the latter is not so important because we are dealing with line sources.

In many propagation models the calculation of the path difference is based on a curved sound beam. Owing to downwind propagation, the sound rays will bear off towards the earth's surface. For short distances this does not have much effect on the result, but for larger distances, over 100 m, this effect is important. It may be found that the path difference is smaller than the path difference calculated for a straight sound ray.

Using optics terminology, we can say that the barrier attenuation should be calculated in the shadow zone and in the illuminated zone. It is clear that in the shadow zone there is a significant effect of the screen. But in the illuminated zone there is also a screen effect. Of course, at sufficiently high receiver point there is no effect at all. There is a remarkable situation when the source, the diffraction line of the screen and the receiver are exactly in line. According to the research of Maekawa, there is a 4.7 dB reduction in this case.

The barrier attenuation is dependent on frequency. At higher frequencies the effect of a barrier is, in most situations, higher than that at lower frequencies. For the calculation of the barrier attenuation the Fresnel number

$$N_f = \frac{\delta}{\lambda/2} \quad (4.12)$$

is used, where δ is the path length difference in metres, λ is the wavelength in metres, equal to $\lambda = c_{sound}/f$, c_{sound} is the velocity of sound in air (in the open air and under normal conditions, this is around 340 m/s) and f is the frequency in Hz.

A typical result for barrier attenuation is given in Fig. 4.12. In this figure there are two distinct lines shown. It is found that different propagation models show slight differences between calculations of this effect. Most of the models have a limitation on the barrier effect of 20–25 dB. This is due to the experience that at such high attenuation values of a barrier, other effects become significant.

A lot of generic models take into account complex effects in the cases of multiple barriers, of adjacent barriers and of earth walls or other barriers with a smooth diffraction edge. Almost every existing model is unable to predict accurately the efficiency of curved or inclined barrier shapes.

The use of an equivalent barrier in the model requires revision of the ground effects because the effective sound path is changed. Another very important point to which attention must be paid is that reflections at the source side of the barrier may affect its efficiency. This effect is illustrated in Fig. 4.13.

A further point to take into account is that the screen must have a superficial mass of at least 10 kg/m^2 and must not contain large gaps or cracks. Also, the dimensions of the screen normal to the source–receiver line must be larger than the wavelength in the frequency range for which the calculation is performed.

4.7.5. Additional types of attenuation

In some propagation models it is possible to calculate additional mechanisms of attenuation. Depending on the objective of the calculations, it can be useful to take these additional types of attenuation into account. The following mechanisms of attenuation can be interesting.

- The attenuation during propagation through foliage. This extra attenuation is approximately 2–3 dB per 50 m. This may be more relevant for the summer season, depending of the type of vegetation.
- The attenuation of an industrial site. This is relevant only for industrial areas, but can be important when there is an industrial area between a road or railway line and dwellings.
- The attenuation of a built-up area of residential houses. This is relevant if there is a large number of houses. A built-up area of houses can be important for impact studies. With the average attenuation, it is possible to get an impression of the number of people who would be annoyed by a high-speed train by performing a calculation of the sound level in this residential area.

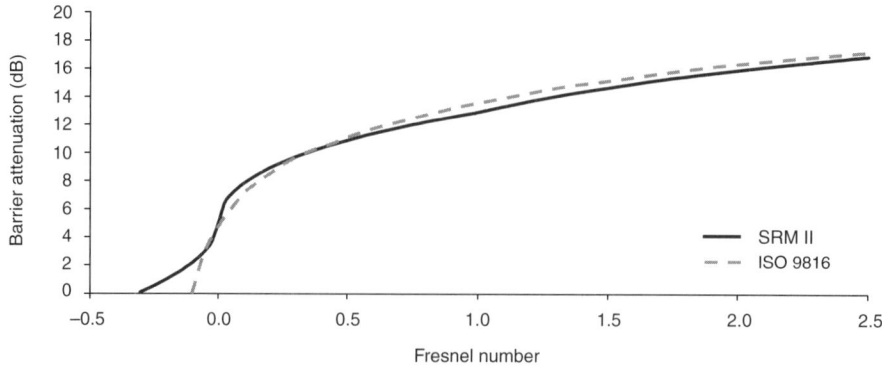

Fig. 4.12. Relation between the Fresnel number and the barrier attenuation for two propagation models [4.3, 4.7]

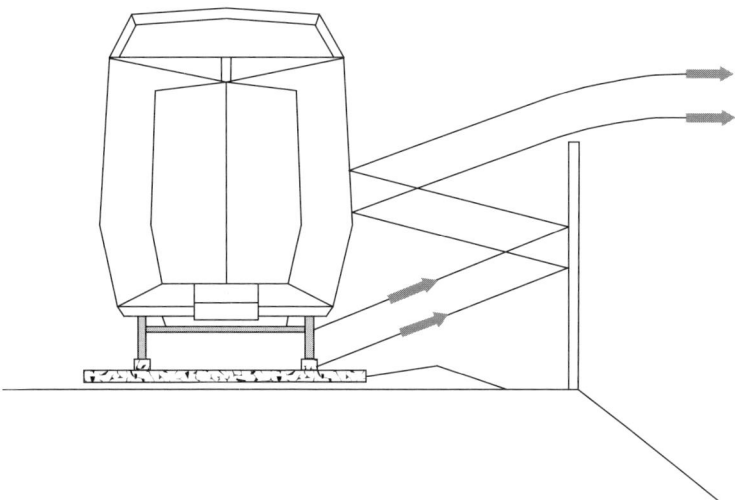

Fig. 4.13. Reduced effect of a barrier due to reflections between barrier and coach

4.7.6. Reflections

The method of calculating reflections is based on using mirror sources or mirror receivers. The requirement for a possible reflection from a surface is the first important factor. Of course, reflection can be possible only on the source–surface–receiver line. The surface must have a significant height above the ground surface (at least 1–2 m) and must be large enough compared with the wavelength.

The extra sound path is calculated as a straight line (in the x–y plane) or as a curved sound path. The effects of spreading, air absorption, ground attenuation and, if necessary, the effect of a barrier are taken into account. The calculation of the reflection contribution to the total sound level is done with the restriction that for every reflection there is a decrease of the acoustic energy. This reduction is taken into account without a phase shift and ignores the diffusing effect of the reflecting plane. For uneven facades with windows, balconies, etc., this can be an important factor. In some prediction models it is possible to incorporate a separate reflection factor for each octave band of frequency. Obviously, the number of reflections taken into account can strongly affect the overall accuracy. Including too many reflections in combination with a high reflection coefficient of reflective planes will substantially change the overall sound level. Therefore the effects of multiple reflections must be treated carefully.

4.7.7. Meteorological correction

As discussed at the beginning of this chapter, the sound propagation in the open air is highly dependent on wind and temperature differences in the atmosphere. In a downwind situation the sound path is curved towards the ground surface. This results in more or less stable sound transmission. In a prediction model, the calculation of the noise propagation is performed for the downwind situation. This is extremely important for considering the attenuation due to ground absorption and the effect of a barrier.

The downwind sound level can be corrected for the average weather conditions by using

$$C_{meteo} = C_0 \left[1 - \left(\frac{10(h_{source} + h_{receiver})}{d} \right) \right] \quad (4.13)$$

for $d > 10(h_{source} + h_{receiver})$, where C_{meteo} is the correction for the average weather conditions in dB(A), C_0 is the maximum level of the meteorological correction in dB(A), h_{source} is the height of the source above the local ground level in metres, $h_{receiver}$ is the height of the receiver above the local ground level in metres and d is the distance from the source to the receiver. The correction for the average weather condition (C_{meteo}) is always positive. In the downwind situation the sound level is always higher than in the average weather situation. C_{meteo} is equal to 0 dB(A) for short distances compared with the heights of the source and the receiver ($d = 10(h_{source} + h_{receiver})$). In some prediction models it is assumed that the maximum level for the meteorological correction C_0 is equal to 3–5 dB(A).

4.8. Calculation of the noise level

The equivalent noise level over a long time interval (e.g. several hours) calculated using a prediction model can be compared with measurements on a number of moving vehicles. The equivalent sound level is defined as:

$$L_{eq} = 10 \lg \left[\frac{1}{T} \int_0^T \left(\frac{p(t)}{p_0} \right)^2 dt \right] \quad (4.14)$$

in dB/octave or dB(A), where p is the sound pressure, p_0 is the reference pressure of 20 μPa and T is the time interval. This time interval is equal to the evaluation period, which may be the daytime, evening or night-time. This evaluation period can be equal to 4, 8 or 12 h. In fact this time interval is not relevant, because the equivalent sound level is the average over this time interval. Of course this is only valid when the same number of trains pass every hour and when these trains have the same noise emission.

The best way of determining the complete radiated sound energy of a train is to measure the sound exposure level (SEL or $L_{A,X}$) which is defined as a constant noise level for a time period of 1 s with the same energy as the actual received acoustic energy. So the SEL is equal to

$$SEL = 10 \lg \left[\int_{-\infty}^{+\infty} \left(\frac{p(t)}{p_0} \right)^2 dt \right] \quad (4.15)$$

in dB/octave or dB(A). For calculating the equivalent sound level from the sound exposure level, the following formula can be used:

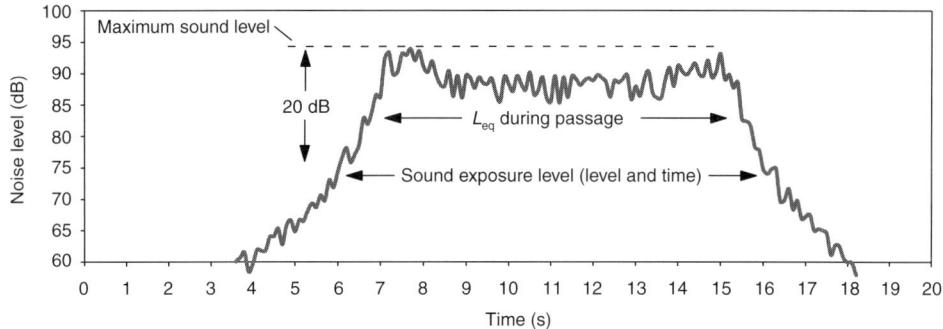

Fig. 4.14. *An example of the sound level as a function of time during the passage of a train*

$$L_{eq} = \text{SEL} + 10\lg\left(\frac{Q}{T}\right) \qquad (4.16)$$

in dB/octave or dB(A), where Q is the number of units passing by during the period, expressed as number per second or per hour, and T is the length of the period, in seconds or hours. This period can be, for example, the measurement period or the period of evaluation. The sound exposure level is equal to the equivalent sound level when Q/T is equal to 1.

In Fig. 4.14 an example of the passage of a train, measured with one microphone is given. It is obvious that the signal fluctuates strongly. Because of this the maximum sound level (defined according to ISO) will not occur when the train is passing right in front of the microphone. The determination of L_{eq} during the passage is also open for discussion because the start and stop times are not unequivocally defined. The sound exposure level is defined for start and stop times at points at least 20 dB(A) lower than the maximum sound level.

The (equivalent) sound emission level of the source can be determined by using the inverse propagation attenuation, as described in Section 4.7. This will result in the sound power level of one line source. When using several line sources one above another, measurements with and without barriers can give an indication of the sound power distribution between the line sources. Most of the time, measurements of the vertical radiation pattern are necessary to define the various sound sources. Measurement to determine the radiation pattern can be done with the use of an acoustic antenna.

4.8.1. *Calculating the noise level with monopole or dipole noise sources*

It is clear that, when making calculations, the definition of the sound power level and the directivity patterns are important. From a purely theoretical point of view, the following assumptions and definitions must be considered [4.7, 4.25, 4.26].

When a monopole source passes, the sound power is radiated in all directions in the horizontal plane with the same sound intensity. For a noise source with a dipole characteristic and with the same source power, the following equation is applicable:

$$W_{di} = 2W\cos^2\varphi \qquad (4.17)$$

where W_{di} is the radiated sound power (in watts) per unit angle in the direction of φ, W is the total radiated sound power (in watts) given by integration over 360° around the source and φ is the angle between a line from the source (or the receiver) and a line perpendicular to the track. In decibels, this relation is converted to

$$L_{w,di} = L_w + 10\lg(2\cos^2\varphi) \qquad (4.18)$$

where $L_{w,di}$ is the radiated sound power level (in dB) in the direction of φ and L_w is the total radiated sound power level (in dB). The difference in the radiation patterns is shown in Fig. 4.15, and Fig. 4.16 shows the theoretical time signature for the passage of a single monopole or dipole source.

It is clear that the maximum sound level for a dipole source is 3 dB higher than the maximum level for a monopole source. Starting from measurements of the maximum sound level, the difference between a monopole and a dipole definition will give a 3 dB variation in sound power level. This will lead to a 3 dB variation in the predicted sound level at the receptor point. Starting the measurements with the SEL or $L_{A,x}$ will not lead to this confusion.

A last remark is that the relation between the sound power level of a train and the equivalent sound level measured at a receptor point is dependent on the speed of the train. The sound power level itself will be higher when the train is running at a higher

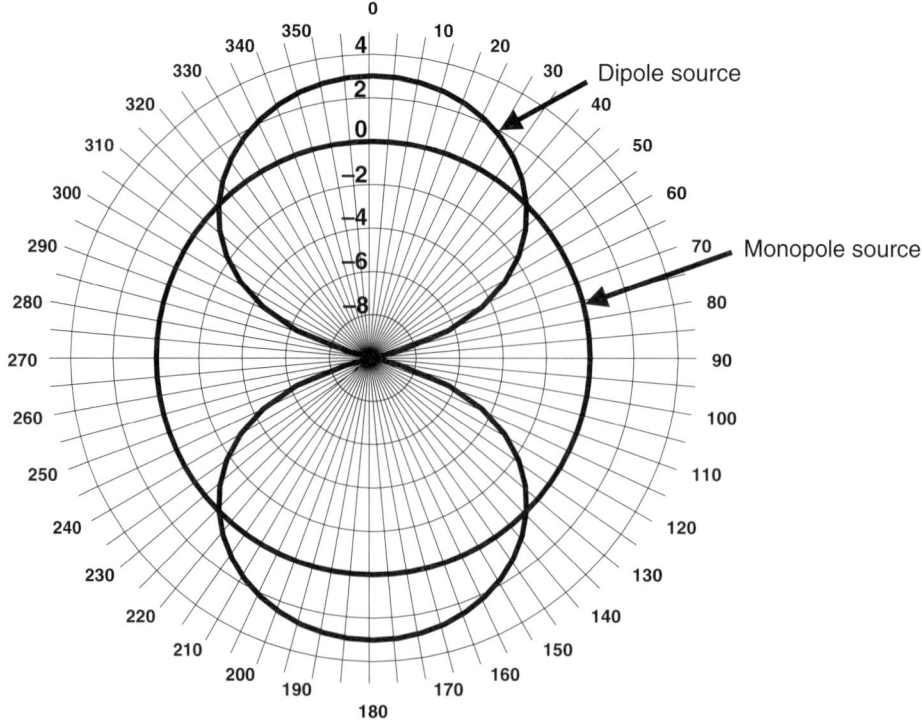

Fig. 4.15. Radiation characteristics of monopole and dipole sources

PREDICTION OF ENVIRONMENTAL RAILWAY NOISE

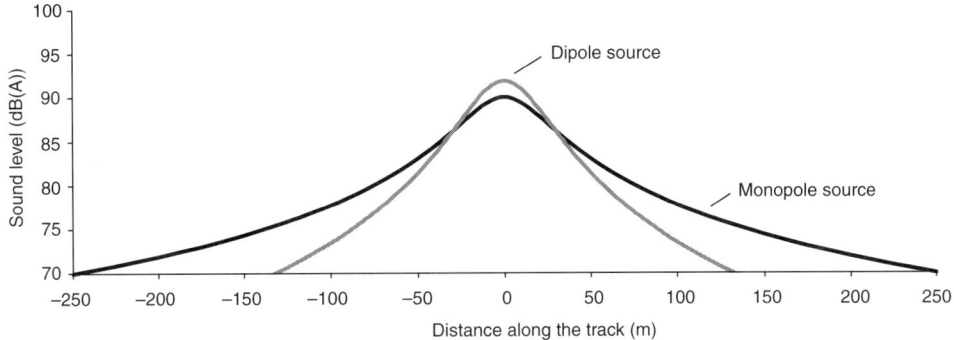

Fig. 4.16. The pass-by characteristics of a monopole and a dipole source

speed. But when this train is running at a higher speed, the exposure time will be shorter. For both a monopole and a dipole source,

$$L_w - L_{eq} \approx 10 \lg v \qquad (4.19)$$

where L_w is the total radiated sound power level (in dB) and L_{eq} is the equivalent sound level (in dB). The same applies for the relationship between the maximum sound level and the equivalent sound level. After all, the maximum sound level (when measured correctly) is directly related to the sound power level and has no exposure time corrections.

4.9. The determination of the sound propagation paths

General noise propagation models usually give only the propagation attenuation between a point source and a point receiver. A typical feature of railway noise is that a train is a moving object. This means that a calculation has to be carried out for a large number of source positions. Alternatively, the track can be seen as a line source. So, the track can be modelled in two different ways. The first is to consider the track as a line source and evaluate the propagation effects starting from the perpendicular distance between the line source and the receiver. Since the propagation from the line source to the receiver has to be the same over the whole area, this method is not commonly used. In almost every model the field of view from the receiver is partly shielded or there are other factors that affect the propagation. The second method is to split up the line source into segments. These track segments are represented acoustically by point sources, for each of which the propagation to the receiver is evaluated separately. Every segment is treated in the calculation as if it has a noise source, and all the relevant propagation factors are taken into account. All contributions from the track segments are aggregated to give the total sound pressure level of one track. For each track segment, an angle of sight Φ can be defined. An example is shown in Fig. 4.17, where $L_{segment}$ is the length of the track segment, R is the distance from the equivalent point source to the receiver and φ is the angle between R and R_p.

When the calculations are performed with the assumption that the track is divided into a large number of point sources, there is one other point to discuss. This is the method of dividing the track into sources. The two most relevant methods are shown in Fig. 4.18.

The first method uses point sources at equal distances from each other. The important issue is of course the spacing between the points. At first sight, a choice of a small spacing seems to be appropriate, but it has to be borne in mind that the calculation has to cover the complete angle of the field of view. This angle needs to be quite large for an accurate calculation (see Section 4.10). A large number of point sources must be used. Especially for receiver points at large distances from the track, one requires a very large number of point sources at the far ends of the track.

The second method is based on equal angles as seen from the receiver. It is also possible to divide the track into a number of point sources in this way. Owing to the use of an angle, the number of point sources close to the receiver is much larger than that of the sources at the start and finish of the track. The same number of noise sources is used for a receiver near to the track and for a receiver further away, so the same amount of calculation (with the same theoretical accuracy) is carried out.

The method of calculating reflections (Fig. 4.19) is very straightforward. All the models use an imaginary receiver point (or an imaginary point source, which is basically equivalent). This imaginary receiver point is determined by supposing a mirror to be at the plane of the reflecting surface. The prediction model defines the conditions of reflection. For example, the reflecting plane has to have a certain size, in relation to the wavelength, and a certain height. For the height condition, some models incorporate a curved sound ray. There might be also conditions on the vertical angle in relation to an object. In the calculation of the reflected sound ray,

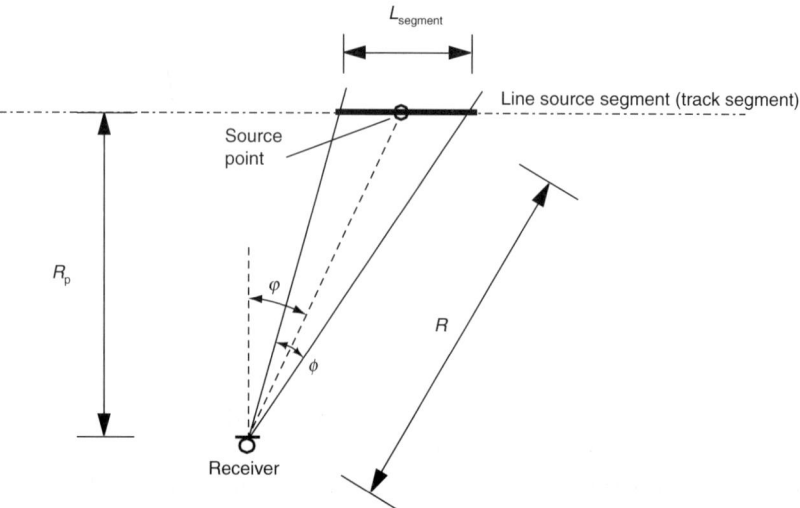

Fig. 4.17. Division of the track into segments

PREDICTION OF ENVIRONMENTAL RAILWAY NOISE

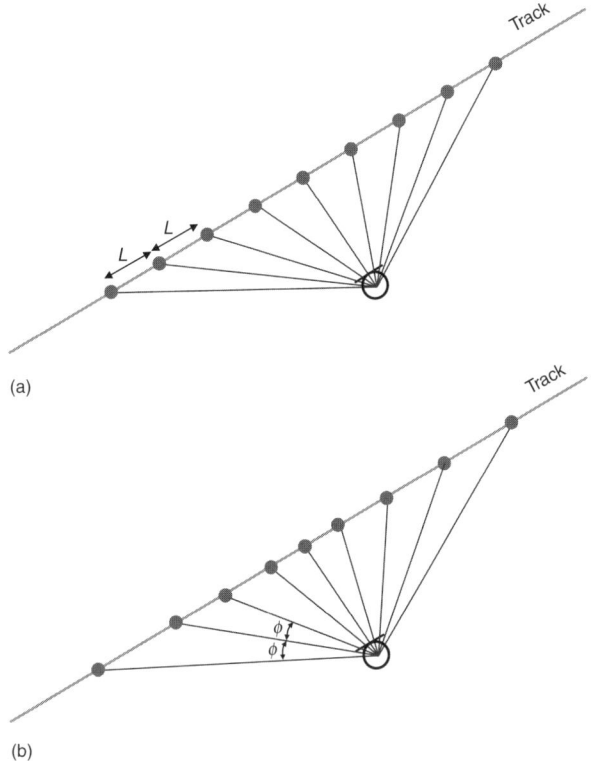

Fig. 4.18. The two methods of division of the track into point sources: (a) equal distance between source points; (b) equal angles between source points

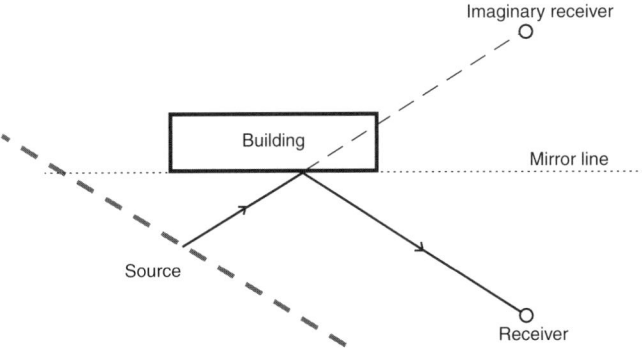

Fig. 4.19. Method for calculating reflections

the same attenuation factors as for the direct ray should be taken into account. In particular, the barrier attenuation factor is important.

For multiple reflections the same method is used sequentially. For every reflection the decrease of the sound level must be taken into account. This decrease of the sound level is dependent on the reflection coefficient of the reflecting facade. Calculating many multiple reflections will not always lead to a more accurate result.

4.10. Accuracy of a generic prediction model

Two types of accuracy have to be considered. The first is the difference between the model prediction and the actual average situation. The second is the calculation accuracy.

Measurement of the difference between the model prediction and the actual average situation is the same as the validation of the model. The objective of a prediction model is to calculate the actual situation as accurately as possible. As discussed in Section 4.2, there are a lot of factors that affect the actual situation. For the calculation of the equivalent sound level, all the relevant parameters have to be averaged. This means that when measurements are done, the results must be determined for the real situation. Only these results can be compared with the calculation results. So the first consideration about measurements is whether these measurements are valid for comparison with the calculation results. The next point is the measurement accuracy itself. For measurements an accuracy of 1 dB(A) is applicable. The calculation accuracy is very difficult to determine. From experience with a very large number of situations, it can be said that for dwellings and buildings which have a direct view of the railway line, the accuracy is 1–2 dB(A). For dwellings and buildings at a larger distance (of about 400 m) and for situations with no direct view of the railway line, an accuracy of 2–3 dB(A) must be expected. In practice these values are considered as acceptable.

The calculation accuracy must be higher when comparisons are made between different calculation models. This is important when the present situation must be compared with a new situation or when comparing two or more new lines. An important point is the correction for the view angle.

The calculation accuracy is dependent on the view angle (see Fig. 4.20). For a point (monopole) noise source this accuracy can be calculated using the differences in the distances between the closest point source and the furthest noise source. The contribution of the furthest sources is smaller than that of the closest source. The

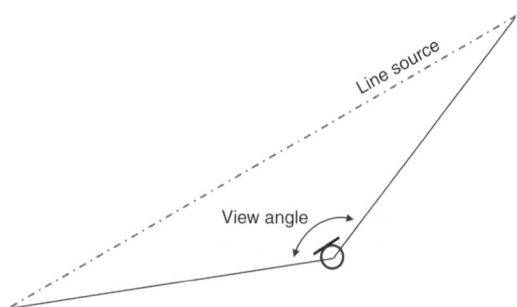

Fig. 4.20. Definition of the view angle

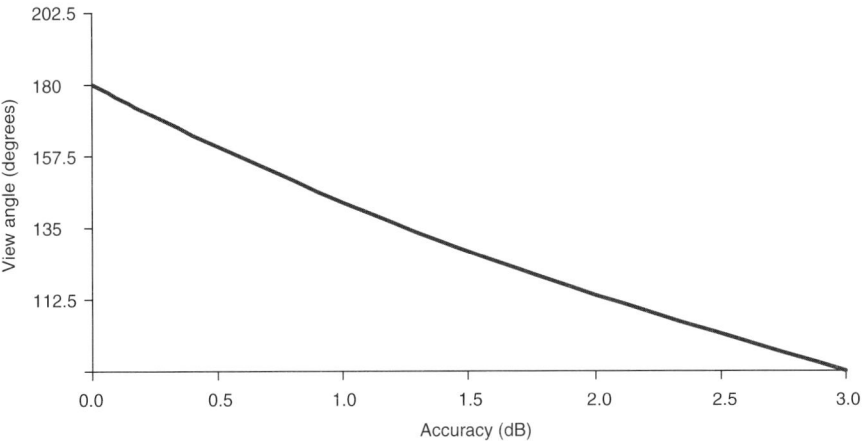

Fig. 4.21. Calculation accuracy for different view angles

difference between the maximum noise level and the lowest noise level for a given view angle Φ is a measure of the calculation accuracy.

It can be shown that the relation between the view angle and the calculation accuracy is

$$\Phi = \frac{180}{10^{L_a/10}} \tag{4.20}$$

where L_a is the desired calculation accuracy in dB.

The required view angle from the receiver to the railway line in relation to the desired calculation accuracy is given in Fig. 4.21. For a calculation accuracy of 1.5 dB the view angle must be at least 127°; for an accuracy of 0.25 dB the view angle is greater than 169°. Theoretically, this angle must be as large as possible, as a large view angle will give the best calculation accuracy.

Beside the theoretical calculation accuracy based on variation of the view angle, it is possible to calculate the sound level contributions of different parts of the line source. This is a different approach to the theoretical calculation accuracy. For small parts of the line source furthest away, the sound level contribution is low, but can result in a less accurate result of the calculation. These noise sources are seen from the receiver at very large view angles. The result of this approach gives a more or less similar accuracy to the other method discussed above.

4.11. Conclusions

The prediction of noise from high-speed railway lines is possible using generic prediction models designed for this purpose. Such prediction models are based on pure theory and on a large number of empirical formulae. In generic models, a prediction of the noise level is always made for the average situation. A measurement will always determine the actual level for a specific situation. A calculated noise level is always representative and is reproducible.

The two main parts of all generic models are the description of the noise source and the propagation model. In the latter the attenuation due to divergence, atmospheric absorption, the absorption of the ground and screening is important. Also, the correction for the contribution of reflections must be taken into account.

4.12. References

4.1. EU WORKING GROUP ON NOISE INDICATORS. *EU Noise Indicators*. Commission of the European Communities, Directorate-General XI, Brussels, 1998, position paper.

4.2. INTERNATIONAL ORGANIZATION FOR STANDARDIZATION. *Acoustics – Attenuation of Sound During Propagation Outdoors – Part 1: Calculation of the Absorption of Sound by the Atmosphere*. ISO, Geneva, 1996, ISO 9613-1.

4.3. INTERNATIONAL ORGANIZATION FOR STANDARDIZATION. *Acoustics – Attenuation of Sound During Propagation Outdoors – Part 2: General Method of Calculation*. ISO, Geneva, 1996, ISO 9613-2.

4.4. VEREIN DEUTSCHER INGENIEURE. *Schall Ausbreitung im Freien*. VDI, Düsseldorf, 1991, VDI 2714

4.5. VEREIN DEUTSCHER INGENIEURE. *Schallschutz durch Abschirmung im Freien*. VDI, Düsseldorf, 1988, VDI 2720.

4.6. DEUTSCHE BUNDESBAHN. *Richtlinie zur Berechnung der Schallimmissionen von Schienenwegen*. Deutsche Bundesbahn Zentralamt, Munich, 1990, Schall 03/Akustik 03.

4.7. MINISTERIE VOLKSHUISVESTING, RUIMTELIJKE ORDENING EN MILIEUBEHEER. *Reken- en meetvoorschriften railverkeerslawaai '96*. Ministerie volkshuisvesting, ruimtelijke ordening en milieubeheer, Directie Geluid en Verkeer, The Hague, 1996.

4.8. *Beregning af støj fra jernbaner, Fælles Nordisk beregningsmethode*. [Calculation of train noise, Nordic calculation.] Copenhagen, 1985, Environment Organization No. 5.

4.9. KILDE AKUSTIKK AS and THE NORDIC COUNCIL OF MINISTERS. *The Nordic Prediction Method for Train Noise (NMT) (Draft)*. Traffic Group (NTG), Voss, Denmark, 1995, Report R790.

4.10. MINISTERE DE L'ENVIRONNEMENT. *Guide du bruit des transports terrestres, prevision des niveaux sonores*. Ministere de l'environnement et du cadre de vie and Ministere des transports, Direction générale des transports intérieurs, Bagneux, 1980.

4.11. CENTRE SCIENTIFIQUE ET TECHNIQUE DU BATIMENT. *Logiciel Mithra-Fer*, version 2.1. Centre scientifique et technique du batiment, Grenoble, 1994, technical manual.

4.12. ASHDOWN ENVIRONMENTAL LTD and UNION RAILWAYS LTD. *Review of AEL Train Noise Prediction Methodology*. Ashdown Environmental Ltd/Union Railways Ltd, Upper Hartfield/London, 1995.

4.13. DEPARTMENT OF TRANSPORTATION. *Calculation of Railway Noise*. HMSO, London, 1995.

4.14. VAN LEEUWEN, J. J. A. Noise prediction models to determine the effect of barriers alongside railway lines. *Journal of Sound and Vibration*, 1996, **193**(1), 269–276.

4.15. VAN LEEUWEN, J. J. A. Some prediction models for the calculation of traffic noise in the environment. *InterNoise '96*, Liverpool, 1996.

4.16. VAN LEEUWEN, J. J. A. Some propagation models for the calculation of traffic noise in the environment. *InterNoise '97*, Budapest, 1997.

4.17. VAN LEEUWEN, J. J. A. Comparison of some prediction models for railway noise used in Europe. *Proceedings of the 6th International Workshop on Railway Noise*. Ile des Embiez, France, 1998.

4.18. ÖSTERREICHISCHES NORMUNGSINSTITUT. *Berechnung der Schallimmission durch Schienenverkehr, Zugverkehr, Verschub- und Umschlagbetrieb*. Österreichisches Normungsinstitut, Vienna, 1995.

4.19. ÖSTERREICHISCHER ARBEITSRING FÜR LÄRMBEKÄMPFUNG. *Schallabstrahlung und Schallausbreitung.* Österreichischer Arbeitsring für Lärmbekämpfung, Vienna, 1987, ÖAL 28.

4.20. BUNDESAMTES FÜR UMWELT, WALT UND LANDSCHAFT and BUNDESAMTES FÜR VERKEHR. *Semibel, Schweizerisches Emissions- und Immissionsmodell für die Berechnung von Eisenbahnlärm*, version 1. Bundesamtes für umwelt, walt und landschaft and Bundesamtes für Verkehr, Berne, 1990, Schriftenreihe Umweltschutz No.116.

4.21. PARKIN, P. H. and SCHOLES, W. E. The horizontal propagation of sound from a jet engine close to the ground, at Hatfield. *Journal of Sound and Vibration*, 1965, **2**, 353–374.

4.22. CHIEN, C. F. and SOROKA, W. W. Sound propagation along an impedance surface. *Journal of Sound and Vibration*, 1975, **43**, 9–20.

4.23. CHIEN, C. F. and SOROKA, W. W. A note on the calculation of sound propagation along an impedance surface. *Journal of Sound and Vibration*, 1980, **69**, 340–343.

4.24. MAEKAWA, Z. Noise reduction by screens. *Applied Acoustics*, 1968, **1**, 157–173.

4.25. REDFEARN, S. W. In: *Handbook of Noise Control* (ed. C. M. Harris), McGraw-Hill, New York, 1957.

4.26. VAN RUITEN, C. J. N. *Studie van het rolgeluid van NS reizigersmaterieel.* Distributie Overheidpublicaties, Den Haag, 1982, Rl-HR-01-01.

Part 2
Measurements and control of railway noise

5. Measurements of railway noise

M. T. Kalivoda
PsiA-Kalivoda Consult, Wiener Gasse 146/3, A-2380 Perchtoldsdorf, Austria

5.1. Introduction

Noise control is a major issue confronting European railways in terms of infrastructure management, as well as train operators from the point of view of environmental legislation. It is a restrictive factor in railway construction and operation and is important in terms of interior noise, which is the most relevant aspect for passenger comfort. With the arrival of high-speed trains and increased freight traffic at night, also at increased speeds, environmental noise is a point of general concern. Much work has already been done to reduce railway noise and further work is under way, but many questions are still unresolved [5.1–5.26].

One major issue of noise control is the question of measurement methodology. An understanding of noise generation mechanisms and a sound knowledge of the capability of measurement techniques are the foundation for the development of noise-reduced railway products. The MetaRail project [5.11], which was funded by the European Commission under the Transport RTD programme of the Fourth Framework programme, focused on the improvement of measurement techniques. When this project started, noise emission type testing was not sufficiently well defined to allow reliable independent comparisons between data on rolling stock in different countries on different types of track. At the root of this problem was a lack of understanding of the variability of the parameters influencing noise, but at the end of the project the reliability and reproducibility of measurements had increased. The most recent results from the MetaRail project are used and presented here.

There are different purposes for exterior and interior rail noise measurements, which try to give answers to a range of various questions (Table 5.1).

Diagnostics primarily support industry and suppliers in the development and assessment of particular noise control measures. Type testing, on the other hand, focuses on any kind of noise-related approval of rail products, mainly rolling stock. Approval tests may range from type approval of a series of railbound vehicles to noise recertification and conformity checking after maintenance. Monitoring measurements are used for checking conformity of single vehicles during ordinary use and for permanent observation of noise impact along railway lines.

5.2. Exterior noise

Knowledge of the noise generation mechanism is essential for finding the appropriate measurement technique. The principal noise sources are (see also Chapters 1 and 2)

- the wheel–rail rolling noise as a result of the interaction between wheel and rail
- the propulsion noise of locomotives and power cars coming from the engine, gearbox and exhaust system (for a diesel engine) or traction control (for an electric engine)
- the noise from auxiliary equipment such as engine cooling, compressors or air conditioning
- the aerodynamic noise, mainly in the case of trains travelling faster than 250 km/h.

Today the primary noise source contributing to environmental noise is wheel–rail rolling noise. Propulsion noise and noise from auxiliary equipment is generally dominant only when the train is standing or running at low speed, and aerodynamic noise is important only at very high train speeds. There are also some more specific noise sources, such as curve and braking squeal and shunting noise, which are less relevant to the overall noise impact.

Microscopic irregularities or roughness of both wheel and rail surfaces excite the system at the wheel–rail interface and induce vibrations of the wheel, bogie, car body and track. Wheel–rail rolling noise is caused by these vibrations. Each component involved has its typical vibration characteristics and thus radiates noise of a typical spectral distribution. Car bodies and sleepers contribute mainly to low-frequency noise. The spectral maximum of the car body lies in a frequency range of about 200–400 Hz. Sleepers of ballast tracks generally radiate noise with a flat maximum at frequencies of about 500 Hz. Rails contribute mainly to the mid frequency range of 750–1500 Hz. The wheel noise is the noise with the highest pitch and has a maximum level in a frequency range between 2 and 4 kHz.

The fundamental knowledge that the total noise level measured with simple methods is the sum of the partial contributions of all sources involved helps one to

Table 5.1. Purposes of rail noise measurements

Exterior noise			Interior noise	
Noise emission		Noise exposure	Diagnostics	Type testing
Diagnostics	Type testing	Monitoring		
Complexity and accuracy are high	Complexity and accuracy are low			

MEASUREMENTS OF RAILWAY NOISE

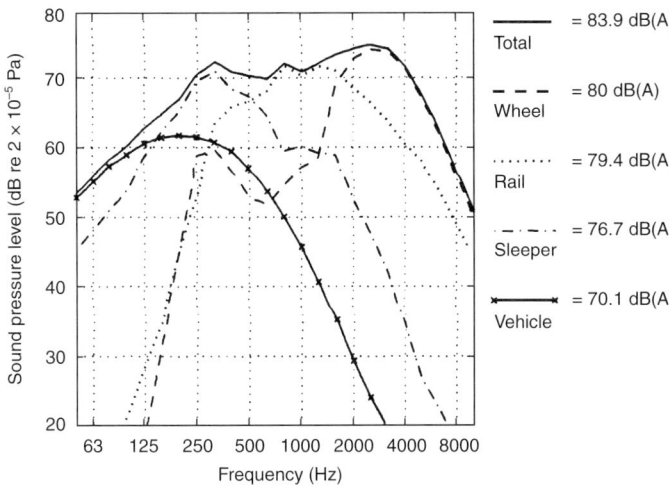

Fig. 5.1. Partial contributions of wheel, rail, sleeper and vehicle to total pass-by noise [5.25]

Table 5.2. Noise emission from different vehicle categories on ballast track with monoblock concrete sleepers and UIC60 rails

	Speed (km/h)	7.5 m/1.2 m (dB(A))	25 m/3.5 m (dB(A))
Disc-braked coach	80	83	78
	160	91	85
	200	93	87
Cast-iron-block-braked coach	80	91	84
	160	99	93
Cast-iron-block-braked wagon	80	92	85
	100	96	88
Sinter-metal-block-braked wagon	80	86	79
	100	90	83
	140	92	85
Disc-braked flat wagon	80	82	74
	100	85	76
	140	90	82

interpret results correctly. Taking the example shown in Fig. 5.1, the total A-weighted sound pressure level of the rail noise is 83.9 dB. The contribution of the wheels is predominant, at 80.0 dB, the rail contribution is 79.4 dB, the sleepers contribute 76.7 dB, and the influence of the car body is 70.1 dB and thus very low. If the noise radiation from the rails can be decreased by 3 dB to 76.4 dB, the reduction of the total noise level is less than 1 dB and much lower than *a priori* expected.

Noise emission varies with vehicle type, train speed and distance. Table 5.2 shows typical numbers for the average A-weighted pass-by level on ballast track with monoblock concrete sleepers and smooth UIC60 rails.

Variations in structure and operational conditions will influence noise emission from individual vehicles. The analysis of pass-by noise from a test train running at 100 km/h and consisting of wagons of different kinds showed the highest levels for a bogie high-sided open wagon, UIC type Eaos. Loading of hoppers, UIC type Fds, with stones reduces noise emission by about 2 dB compared with empty hoppers of the same type. Noise from an older box van, UIC type Gbs, was about 1 dB lower after overhaul than before. Finally, a modern two-axle sliding-door van, UIC type Hbils, was about 1.5 dB less noisy than an older bogie box van, UIC type Gabs. A four-axle electric locomotive, Austrian class 1044.2, hauling the whole 1300 t train was about 1.5–2.5 dB louder than the same type without traction (Fig. 5.2).

5.2.1. Diagnostics

'Diagnostics' mainly means source localization and is used by industry and research institutions to develop and assess corrective measures and to test prototypes. Diagnostics is the most accurate method, requiring advanced and expensive equipment as well as personnel with expert knowledge to handle it.

5.2.1.1. Antenna technique and source location

Acoustic-antenna systems are primarily applied for noise source localization and quantification. The main benefit of antenna systems is that they produce an image of the sound source distribution measured at a short distance of about 2–3 m from the track, overcoming the limitations of single- or two-microphone measurement methods. For railway rolling noise, antenna measurements can be particularly useful for separating the partial contributions of the vehicle, especially its wheels, and of

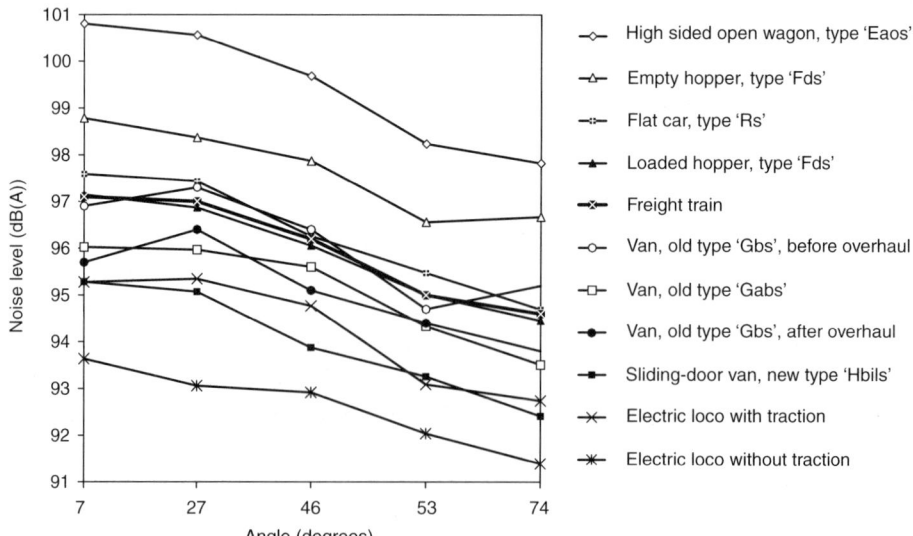

Fig. 5.2. Average pass-by levels of different wagon types at 15 m distance from the track and at 7°, 27°, 46°, 53° and 74° from horizontal

MEASUREMENTS OF RAILWAY NOISE

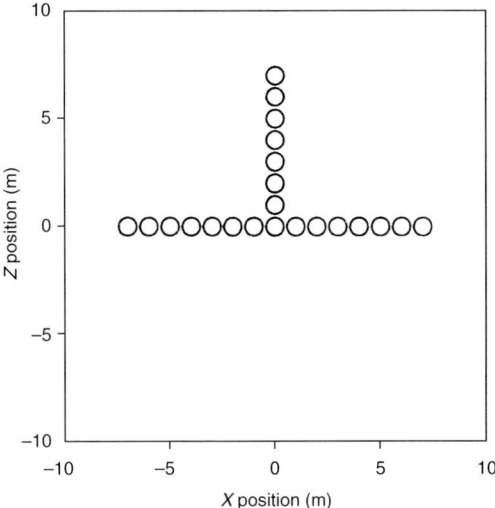

Fig. 5.3. Microphone configuration of a T-shaped array [5.4]

the track. That kind of measurement system enables visualization and quantification of the sound pressure field close to the train.

Within the MetaRail project, a swept-focus antenna measurement system was developed. This swept-focusing technique enables the antenna to 'follow' the moving source. For the antenna a two-dimensional T-array was used, which appeared to be the best configuration for localizing and quantifying source strength with the minimum number of microphones that was necessary to guarantee observing two-dimensional effects. Nevertheless, a large number of microphones was still used, which provided a good spatial resolution. Figure 5.3 shows the microphone positions of a T-shaped array.

The results of the antenna measurements are presented below together with a schematic drawing of the test train to help with the interpretation. For the same reason the total length of the train, which was almost 300 m, has been split into three pieces.

Figure 5.4 shows the perceived sound pressure level at a microphone 2 m from the nearest track for a train passing by at 80 km/h. The results have been processed for the 500 Hz, 1 kHz and 2 kHz octave bands. The 1 kHz octave band shows individual wheel locations (or pairs of wheels); however, the dynamic range around the area of the track is limited by the emission level of the track. For the 2 kHz octave band the locomotive is relatively 'silent'. Side lobe levels from the wheel–track sources are lower compared with the image at 1 kHz. In the 500 Hz octave band the locomotive shows very noisy parts near the wheels compared with the rest of the train.

Summarizing the results from the antenna measurements, one can say that the wagon superstructure radiates noise at least 15 dB(A) less than that of the wheels and track. The effect of wheel shrouds is clearly visible, as the noise is concentrated around the wheel–rail interface for an Sgs wagon. The noise radiated from

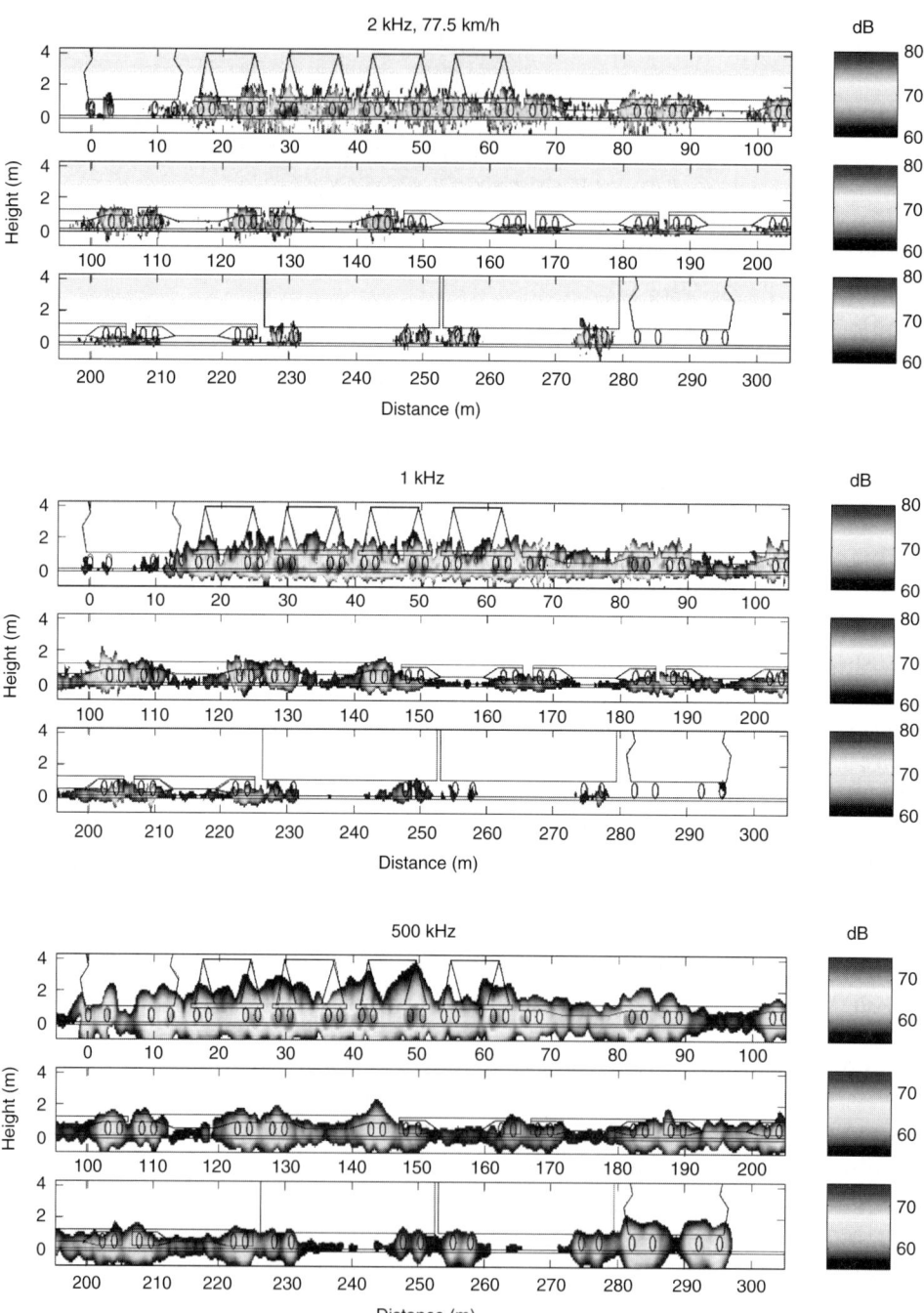

Fig. 5.4. Antenna images of the MetaRail test train at 77.5 km/h for 2 kHz, 1 kHz and 500 Hz octave bands [5.4]

wheel–rail area of Rkqss and Sgjns wagons, which have no shrouds but are comparable apart from that, extends to larger heights. It can be clearly shown that for Sgs wagons, vehicle noise is about 3–6 dB(A) lower than track noise.

5.2.1.2. Simplified approach

With regard to the high costs of diagnostic measurements, the antenna methodology and also the technique described under 'type testing' (Section 5.22) can be used frequently to get a feeling for the effects of noise-reducing measures. This sort of survey not only covers modifications of the vehicle, but also includes refinements of the track and noise barriers.

As far as the assessment of new noise-reduced rolling stock is concerned, it is clear that a sufficient number of test vehicles and a test train are needed. The need for a test train is not so obvious when the effect of track modifications or the difference in sound attenuation of two noise barriers has to be quantified. Provision and operation of a test train is certainly expensive and generally exceeds the costs of the actual acoustic part of the study. Our long experience shows that 'just a few ordinary train measurements' studies generally are not able to deliver significant results.

The objective of everyday diagnostics is to quantify acoustic effects that are of a magnitude of 3 dB or less. Measuring noise emissions from one ordinary train, even at two sites not far away from each other (e.g. 1 km apart), will lead to different results owing to different operational conditions of the train. Unfortunately, the influence of this unwanted side-effect is of the same order of or even higher than the acoustic effects actually observed, and thus significant results are not being obtained. There are two strategies to solve this problem. If there is no test train available whose operation during the test runs can be controlled, a large sample of emission measurements from ordinary trains is required to make the differences statistically significant. Generally, both approaches are equivalent; however, a test train saves a lot of measurements and makes the whole procedure more manageable.

Finally, this leads to the question of accuracy, reliability and reproducibility of results. There is always some reasonable variation in results from simple microphone noise measurements. A common guideline for repetition of measurements is that results from several measurements under the same conditions are acceptable if they are within a range of 3 dB. That is too much for the assessment of a 2 dB effect, so an additional criterion has to be used.

Confidence intervals are not yet frequently used, but they provide a very appropriate way to get an idea whether the results of an A–B comparison are significant or not. This approach uses the variance of measurement results, and the confidence interval is constructed so as to have a prescribed probability of containing the true value of the noise emission. If the confidence intervals of two very close measurement series do not overlap, the differences between these two series are statistically significant with regard to the chosen probability. If there is an overlap, it is likely that the differences between the two measurement series are random, which means the effect of noise-reducing measures could not be verified. Such a result might be unsatisfactory, but avoids misjudgement (Fig. 5.5).

5.2.2. Type testing

According to the draft European Standard *Railway Applications; Acoustics – Measurements of External Noise Emitted by Railbound Vehicles* (CEN, [5.7]), type testing means measurements performed to prove that vehicles delivered by a manufacturer comply with the noise specification.

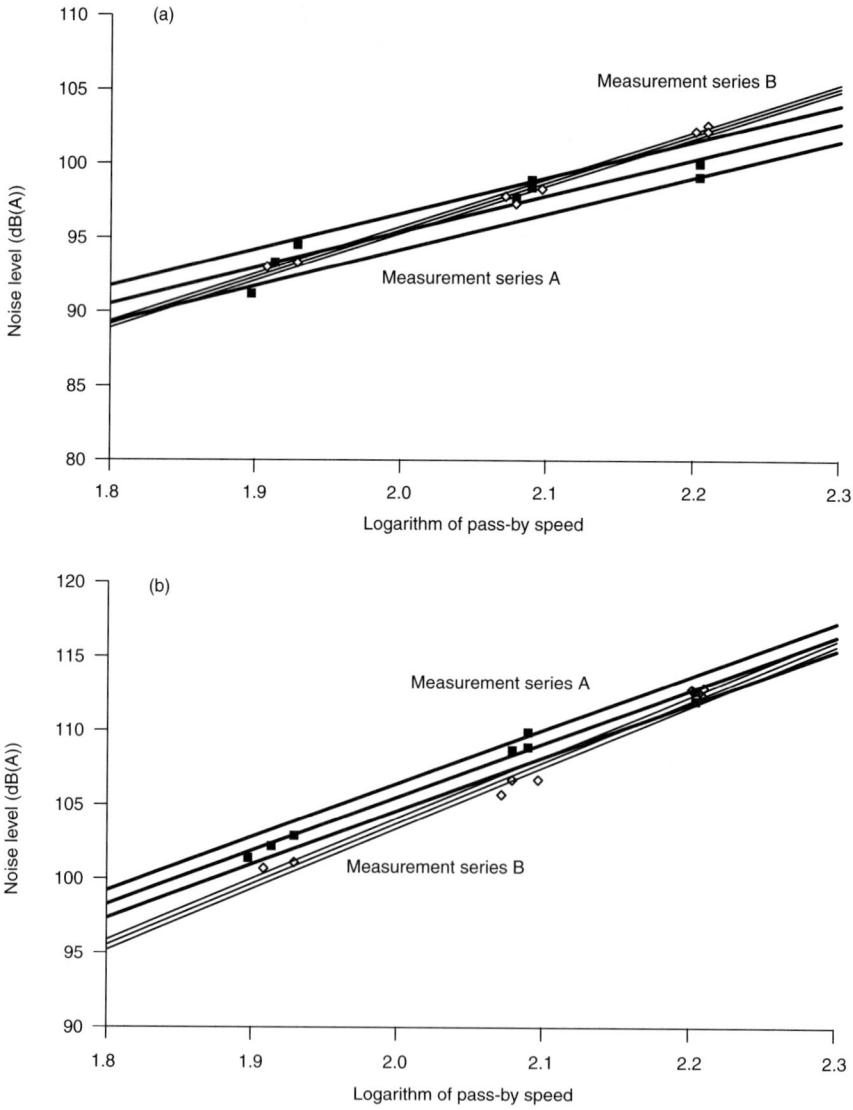

Fig. 5.5. Confidence intervals for two measurement series, A and B. In the first case (a) there is a broad area of overlap, which means no significant difference between the two series. The second case (b) shows only partial overlap, which means that in this case differences between the two series are significant

Table 5.3. Checklist for noise measurements

		Examples	Remarks
Train composition (description of each vehicle)			
!!	Vehicle type	Electric locomotive/second-class coach/hopper	
!!	International designation	Eaos/Bmpz	
!!	Vehicle number	80 81 972 0 004-7/ 70 81 20 94 047-1	
!	Vehicle manufacturer		
!	Year of construction		
!!	Maximum speed		
!!	Number of axles		
!!	Number of bogies		
!!	Overall length		
!	Bogie spacing		
!	Axle spacing		
!!	Braking system	Cast-iron block/ sinter metal block/ disc	
!!	Operational empty weight		
!!	Maximum weight		
!!	Axle load		
!!	Wheel conditions		Visual check!
Description of engine (self-propelled vehicles only)			
	Engine type	Electric asynchronous/ diesel V12	
	Manufacturer		
	Year of construction		
	Transmission		
	Rated power		
	Number, kind and rated power of all auxiliary equipment	Two compressors at 25 kW	
Operational conditions during test			
!!	Vehicles empty/loaded?		
!!	Windows, doors, hatches open/closed?		
!!	Pass-by speed		
!!	Operational conditions of main engine		
!!	Operational conditions of auxiliary equipment		Including coaches (air conditioning)

Table 5.3. Contd

		Examples	Remarks
Track description			
!!	Track type	Ballast track/slab track	
!!	Date of last maintenance works (tamping and lining)		
!!	Rail type		
!!	Rail roughness (visual inspection)		
!	Rail roughness measured		
!	Rail pad type		
!	Rail fastener type		
!!	Sleeper type	Wood B1/concrete monoblock Be19a/concrete biblock	
!!	Gradient (upline)		
!!	Radius		
Site description			
!!	Identification of railway line, location, track number		
!!	Microphone location(s)		
!	Ground absorption measured		
!!	Verbal site description		
!!	Background noise level		A-weighted and linear $\frac{1}{3}$ octave spectrum
Meteorology (for each pass-by)			
	Temperature		
	Humidity		
	Wind speed and direction		

!, recommended; !!, obligatory

However, there are some more fields of application for type-testing measurement methodologies, namely recertification and proof of conformity after overhaul. It is not only a new vehicle or vehicle series that has to be checked; also, major constructional changes of and amendments to a vehicle may change noise emission for better or worse. Therefore, a noise recertification test is obligatory according to the Austrian ordinance on permitted noise emissions for rolling stock *SchLV* [5.24] if the running gear has been rebuilt, if the braking system has been substantially modified or if the engine type, transmission or car body has been changed, and thus

MEASUREMENTS OF RAILWAY NOISE 129

the vehicle category has changed. Finally, conformity checking after maintenance work is a field of application for type-testing methodology. This has been discussed in Austria, but is not obligatory at the moment.

The measurement methodology for type testing is discussed here in detail, paying attention to the most important aspects and to the factors influencing the accuracy of results. In addition, and as guidance for all users, a checklist for type-testing measurements can be found in Table 5.3.

5.2.2.1. Sound pressure level versus sound power level

Most of the European and national standards and regulations on type-testing procedures for noise emission from rolling stock, including the Austrian ordinance on permitted noise emissions for rolling stock *SchLV* [5.24], limit the sound pressure level. However, there are some scientific papers [5.21] and an Austrian standard that propose the use of sound power level instead.

There are a lot of pros and cons for both sound pressure level and sound power level. From a theoretical acoustic point of view, the sound power level has some advantages compared with the sound pressure level. It is obvious that using sound power instead of sound pressure in characterizing noise emission from rolling stock makes the noise criterion independent of both the location of the railbound vehicle and the environmental conditions. The sound power level is independent of the distance from the point of measurement and does not change in value from one physical location to another, provided the train is operated in the same manner.

However, the sound power level is primarily used to describe noise radiation from stationary equipment. So most of the existing standards for the determination of sound power level, such as the ISO 3744 [5.18], refer to stationary sound sources. The main weakness of using sound power levels for rating the noise of mobile sources is that the measurements may not relate to actual operational situations, and give just a snapshot of the temporal variation of the noise emission. Sound power levels which are determined according to 'simple' single-microphone methods are not measured directly, but are calculated from the mean square sound pressure at a number of microphone positions and spatially averaged over an appropriate surface enclosing the source in a free sound field. For reasons of simplicity and applicability, stationary equipment is used to measure mean square sound pressures from a passing train. Increasing the speed reduces the averaging time and makes the results less reliable.

Nevertheless, there is an Austrian standard, S 5026 [5.1], that describes a method to determine the sound power level of railbound vehicles. When talking about sound power levels here, normalized sound power levels $L_{W'}$ for a train length of 1 m are always meant. A single railbound vehicle or a train – the Austrian standard S 5026 does not distinguish between them – is said to consist of a series of l incoherent point sources passing by at a height of 0.3 m above the rails where l is the length of the vehicle or train in metres (Fig. 5.6).

The measurement surface is – replacing a hemisphere by its cross-section – a semicircle centred on the theoretical emission point E. The sound pressure level is

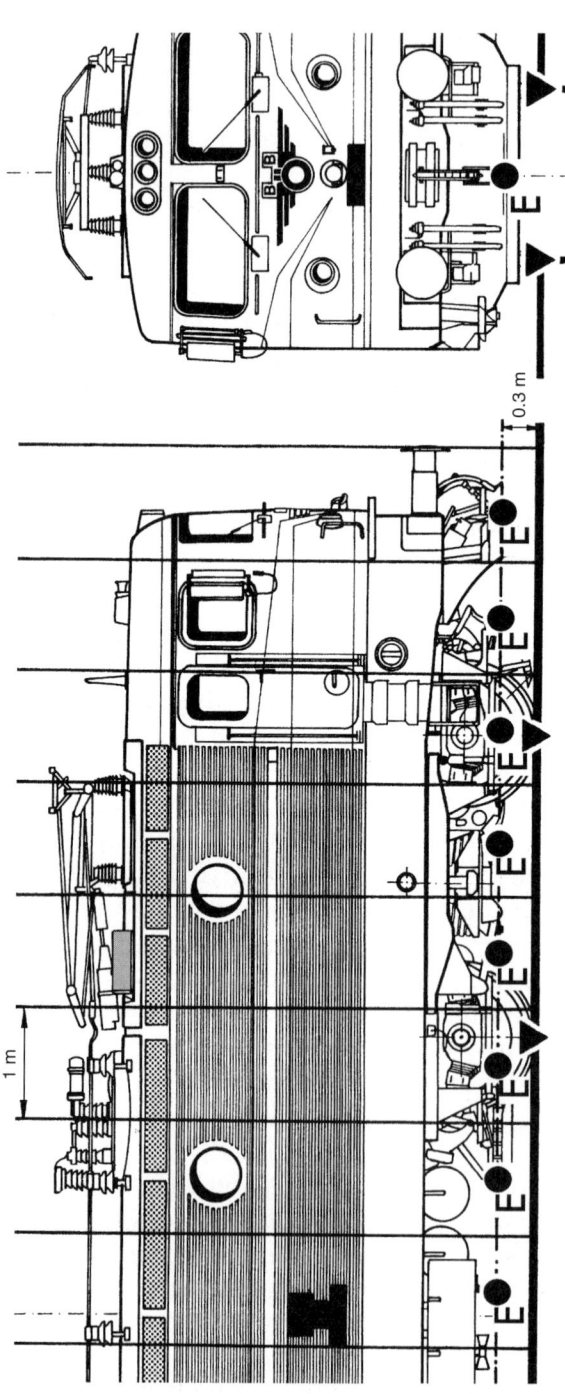

Fig. 5.6. Noise sources E at 0.3 m above rail surface, which each represent 1 m of railbound vehicle according to Austrian standard S 5026; emission points E differ from actual wheel–rail noise zone (black triangles)

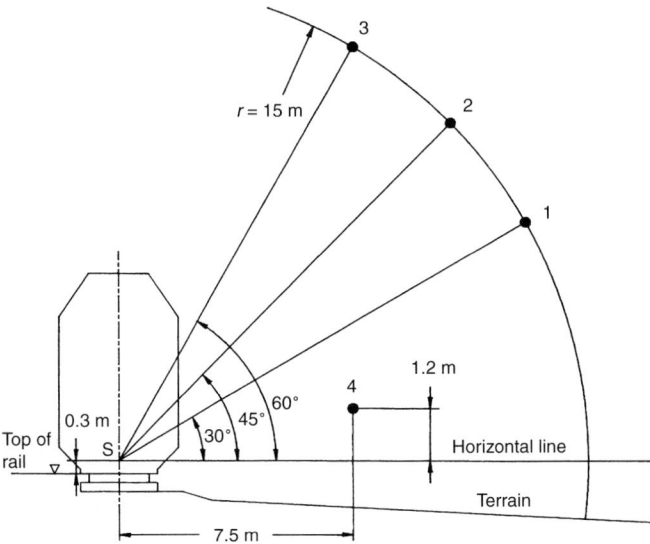

Fig. 5.7. Microphone locations for sound power level measurements of railbound vehicles according to Austrian standard S 5026 [5.1]

recorded while the whole train passes this measurement arc – we cannot call it the measurement surface any longer, for now it is in fact only an arc – and the average level, which is the equivalent level of the pass-by, is calculated.

The radius R of the semicircle measurement arc must be 15 m and three microphone positions are required. The first microphone has to be put on the arc at 30° above a horizontal line drawn through E, a second microphone is put at 45° and the third microphone is at 60° above the horizontal line (Fig. 5.7). There is also a fourth microphone required, which is located 7.5 m from the emission point E and 1.2 m above the horizontal line.

The normalized sound power level $L_{W',i}$ per metre of vehicle/train and for the $\frac{1}{3}$ octave band i is calculated from

$$L_{w',i} = 10\lg\left(\frac{1}{3}\sum_{i=1}^{3}10^{L_{W',j}/10}\right) \tag{5.1}$$

and

$$L_{w',i,j} = L_{i,j} + 19.2 - 10\lg 2.83\frac{s_j^2}{r^2}\left\{\left[\frac{z}{r^2+z^2/4} + \frac{2}{r}\arctan\left(\frac{z}{2r}\right)\right] + \frac{2}{r}\arctan\left(\frac{z}{2r}\right)\right\} \tag{5.2}$$

where L_{ij} is the $\frac{1}{3}$ octave (i) sound pressure level at the position j, corrected for the ground absorption, r is the radius (15 m), $s_j = r\cos\varphi_i$ is the horizontal distance between the noise source E and microphone j, and z is the train length in metres.

These formulae take into account a general directivity pattern of railbound vehicles consisting of 15% monopole (radial) and 85% dipole sound radiation. It is a major drawback of this methodology for determination of sound power level from sound pressure level measurements that it makes assumptions about the general directivity of rail noise. Unfortunately, directivity changes according to the type of rolling stock as well as with construction principles. Some years ago radiators and cooling fans were located on the side of the locomotive body, while newly constructed locomotives tend to have the radiators on the roof. There is also a tendency to replace multiple units with single power cars by sets with traction equipment on each car. Things like these necessarily change the directivity pattern of the rolling stock and have to be taken into account when synthesizing sound power levels from sound pressure level measurements.

Kurze and Diehl [5.21], for example, have proposed a method for calculating the sound power level from a sound pressure level measurement at just one microphone position using the general assumption of a dipole directivity of rail noise. This hypothesis of a general dipole directivity of rail noise is not valid. A dipole directivity may be accurate enough for modelling the noise perception at some distance from the track, but it cannot be used to synthesize the sound power level from the sound pressure level.

Our measurements show that it is not appropriate to replace microphone locations by the assumption of a simple dipole directivity for reasons of simplicity. Generally speaking, this will lead to an underestimation of the real sound power level by 4–6 dB. It is clear that only a reasonable number of selected microphone positions will lead to valid results.

5.2.2.2. Rail noise directivity

In the early 1990s a measurement campaign was carried out by the Federal Environmental Agency in cooperation with Austrian Federal Railways to check the methodology of the Austrian standard S 5024. In this work, a multichannel noise-recording device was used which allowed six microphones to be put in various positions on or around the semicircular measurement arc. Microphone 1 was placed at a radius of 15 m at an angle of 30° above the horizontal to make the results comparable to previous measurements. Four microphones were arranged in a quadrant at angles from 7° to 74° (Fig. 5.8). In addition, microphone 5 was placed 3.5 m above microphone 6 to form a part of a parallelepiped measurement surface, which is one of the three choices in ISO 3744 for stationary sources.

Figure 5.9 shows some typical results for one electric locomotive and three goods wagons. The diagrams show the equivalent sound pressure level as a function of the angle above the horizontal line. The open circles represent the actual microphone position, and thus the dashed line is the 7.5 m distance from E (the semi-circle reference surface) described on page 129. The solid lines and filled circles represent levels. It is clear from the graphs that there is a dipole directivity for low microphone positions, for angles φ of less than 30°, only. The higher the microphone is located on the arc, the greater the difference between the theoretical dipole directivity (solid

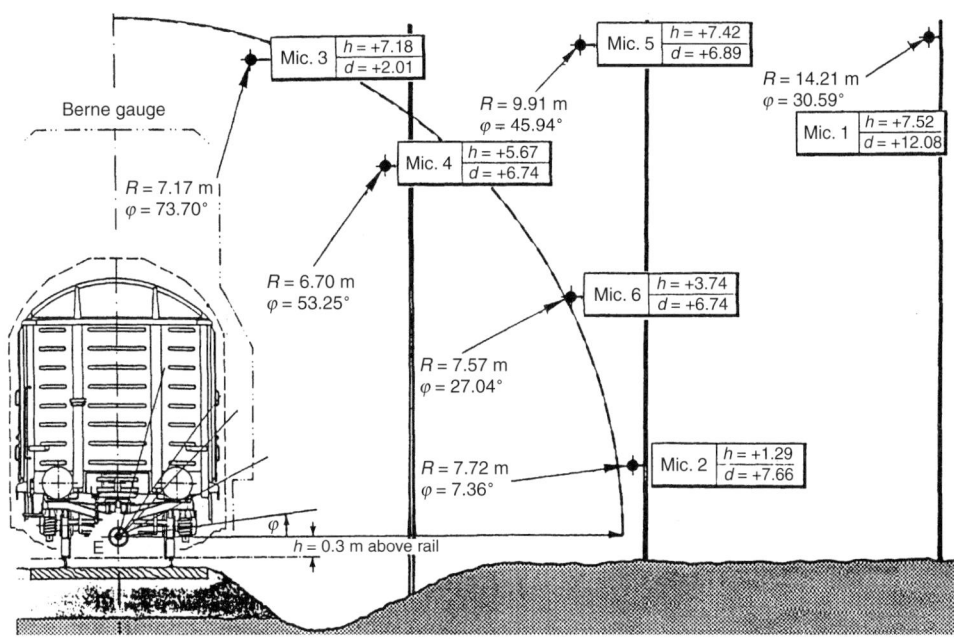

Fig. 5.8. Microphone locations of a rail noise directivity measurement (h, height above rail surface (m); d, horizontal distance from track (m))

curved line) and the measured sound pressure level (black circles). Only the conformal measurement point 5 (black square) shows negligible differences from the dipole directivity.

Another interesting aspect is the small differences in directivity between the three different wagon types. Hearing gives the impression that the car body contributes a lot to the total noise emitted from wagons, so one would expect that there would be differences in noise radiation between the flat wagons and the hoppers. There is in fact an overall difference in noise emission between the different wagon types at 15 m distance at the same speed of 100 km/h. The noisiest type was the high-sided open wagon, with noise levels between 98 and 101 dB(A), while the flat car reached 95–98.5 dB(A). But in all cases, the level reduction with increasing angle φ is almost the same.

At first sight these results contradict the literature. However, on closer examination it becomes obvious that these results cannot be used to decide whether there is a dipole rail noise directivity or not. Some of the quoted measurements, such as those of Cato [5.6] or Peters [5.22], used only low microphone positions, and other authors, such as Barsikow [5.3] and Rathé [5.23], took only the wheels as a noise source and therefore used the dipole directivity. Hohenwarter [5.15] shows results for different kinds of railbound vehicles measured with microphone positions at up to 50.5°, where there is a dipole directivity in some cases for frequencies from 630 Hz to 8 kHz.

Also, a study by the European Rail Research Institute (ORE/ERRI [5.16]) dealt with the identification of noise sources on wagons. The measurement surface used

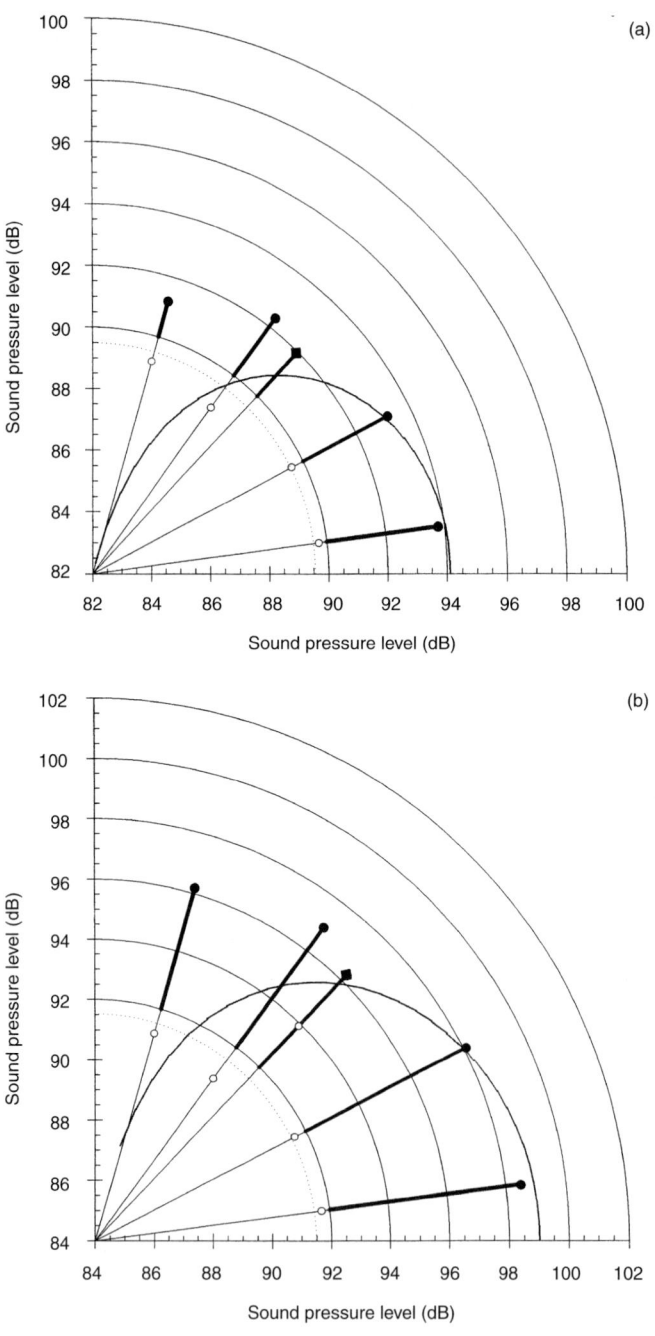

Fig. 5.9. Sound directivity diagrams. Sound pressure level measured at different angles from a horizontal line through the rail surface at 15 m distance (filled circles) from the centre line of the track and at the conformal location (filled square). The solid line represents the dipole directivity. (a) Electric locomotive at 100 km/h; (b) hopper at 100 km/h

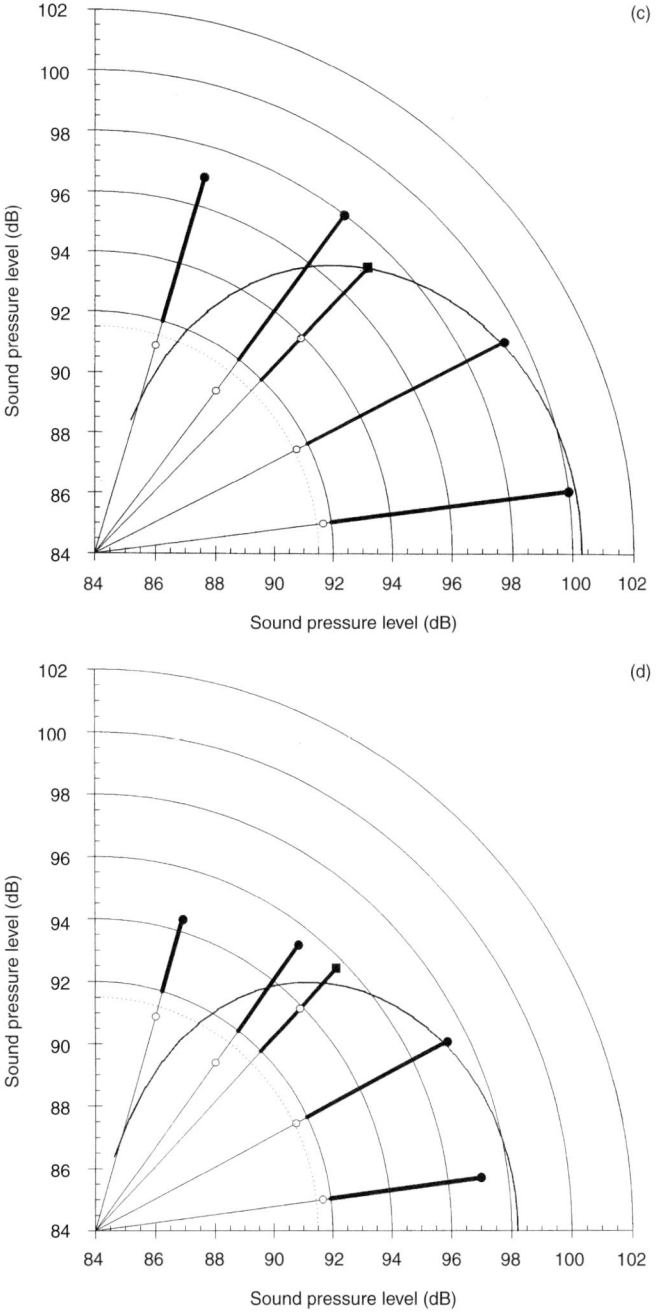

Fig. 5.9. Contd. (c) high-sided open wagon at 100 km/h; (d) flat wagon at 100 km/h

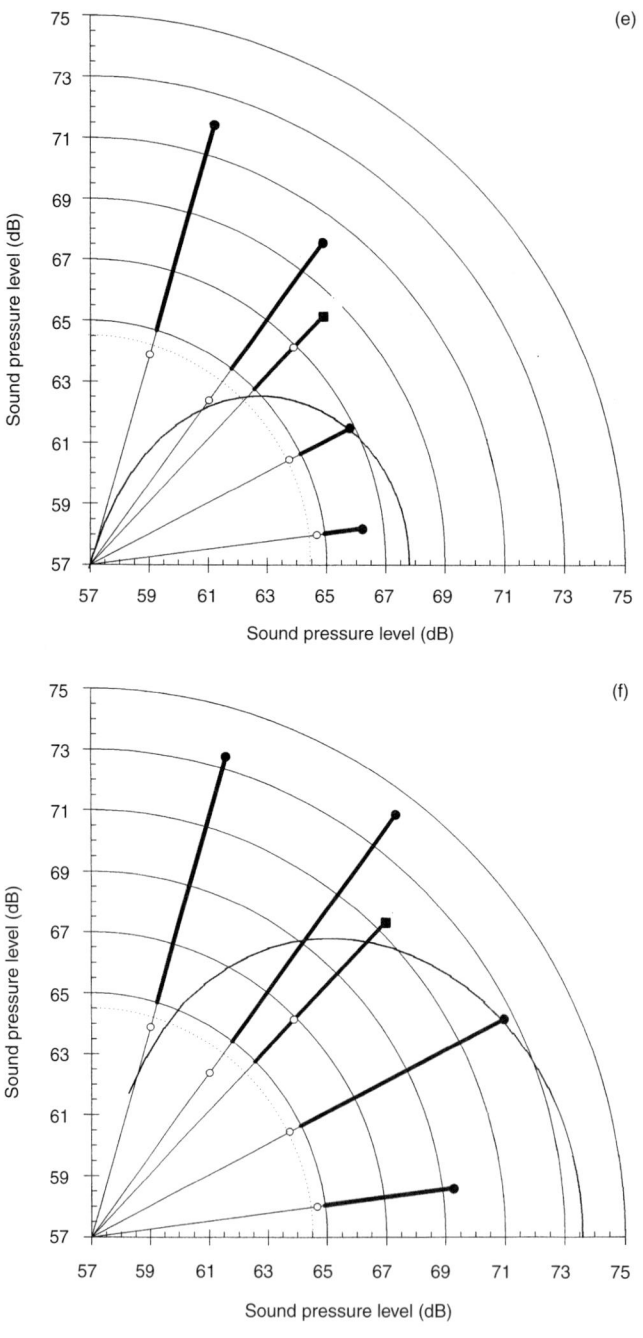

Fig. 5.9. Contd. (e) compressor of stationary electric locomotive; (f) compressor and cooling fan at low power of stationary electric locomotive

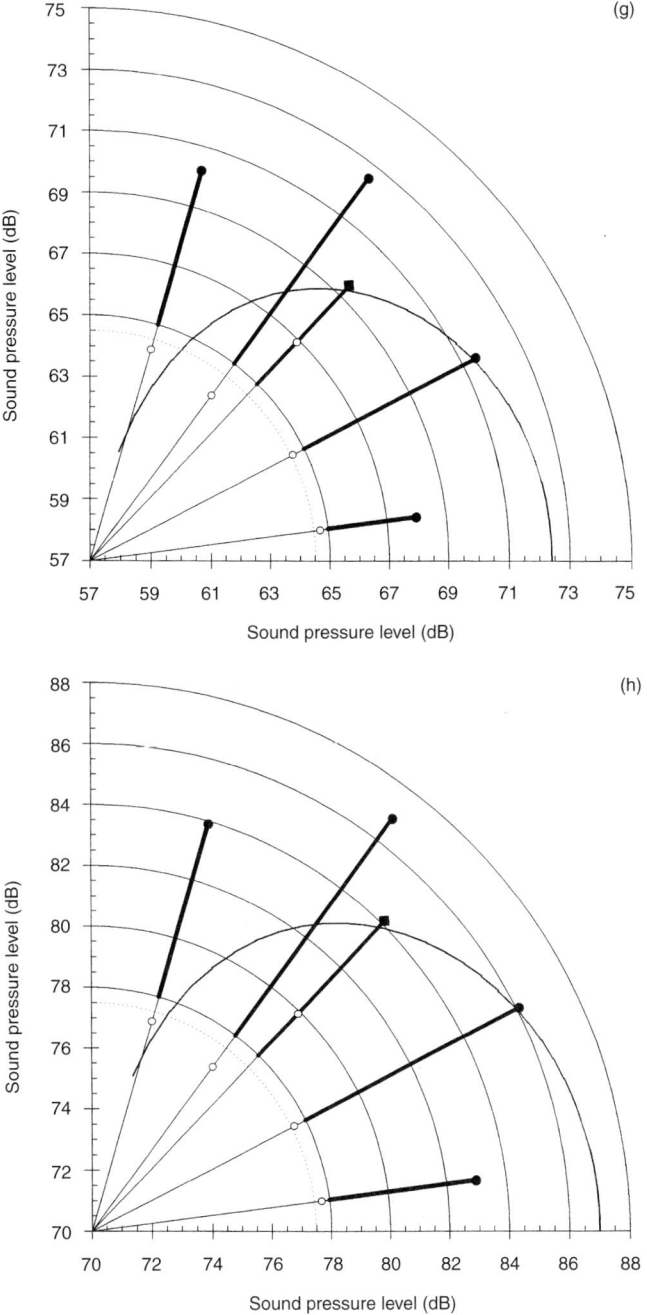

Fig. 5.9. Contd. (g) cooling fan at low power of stationary electric locomotive; (h) cooling fan at high power of stationary electric locomotive

was a conformal surface, and there were 11 microphone positions ranging from rail level up to 3.5 m. Also, for a parallel surface at a very short distance (1.6 m from the rail) there are divergences of up to 4 dB from a dipole directivity for almost all microphone positions. This is in agreement with our results that at short distances there is no dipole directivity. The car body construction has a greater influence on the total emission level than is generally assumed. This effect has to be considered when designing noise barriers since this effect, especially in the case of low barriers, strongly depends on the body construction of the railbound vehicle and its sound radiation pattern.

5.2.2.3. Time weighting

For rail noise measurements the 'fast' exponential time weighting is commonly used, as in many other areas. It is also highly recommended to use 'fast' generally for all rail noise assessments. Only in some exceptional situations might it be appropriate to choose a time constant different from 125 ms.

However, mainly in those cases when a draft value for the average noise emission from a single train is required easily and simply, without any chance to record the whole pass-by and to process the average level, it is appropriate to use the 'slow' time weighting. An integration time of 1 s leads to a smoother level history and gives a better indication of the average. Even a 'slow' weighted maximum sound pressure level is used sometimes. This parameter is called the average maximum level.

For diagnostics, on the other hand, an exponential time weighting with higher time resolution than 125 m makes it easier to detect short noisy events in the level vs time diagram. For this kind of exercise, integration times of 20–50 ms turned out to be very appropriate (Fig. 5.10).

5.2.2.4. Average level versus maximum level

The temporal character of noise along a busy railway line is very different from that close to a busy highway. Generally, rail noise can be described as noise of high intensity during a fairly short time period. Thus, there is good reason to pay attention to both the equivalent level and the maximum levels of a pass-by.

Figure 5.11 shows the average pass-by level $L_{A, av}$ and maximum pass-by level $L_{A, max}$ of three different wagon categories. Both sound pressure levels are 'fast' time weighted. The mean difference between the maximum and the average level at a 7.5 m distance from the track is 1–2 dB(A), as long as there are no prominent single noise events during pass-by, such as a wheel flat or loose parts banging against the car body. A difference greater than about 2 dB(A), on the other hand, is a good indicator that something has gone wrong.

To make sure that type-tested vehicles are in a very good condition and to detect irregularities, mainly on the wheel surfaces, the Austrian ordinance on permitted noise emissions for rolling stock SchLV limits the maximum noise level $L_{A, max}$ instead of the average pass-by level $L_{A, av}$. However, from our experience, there is no systematic difference between those criteria in general except for a level difference of about 2 dB.

MEASUREMENTS OF RAILWAY NOISE

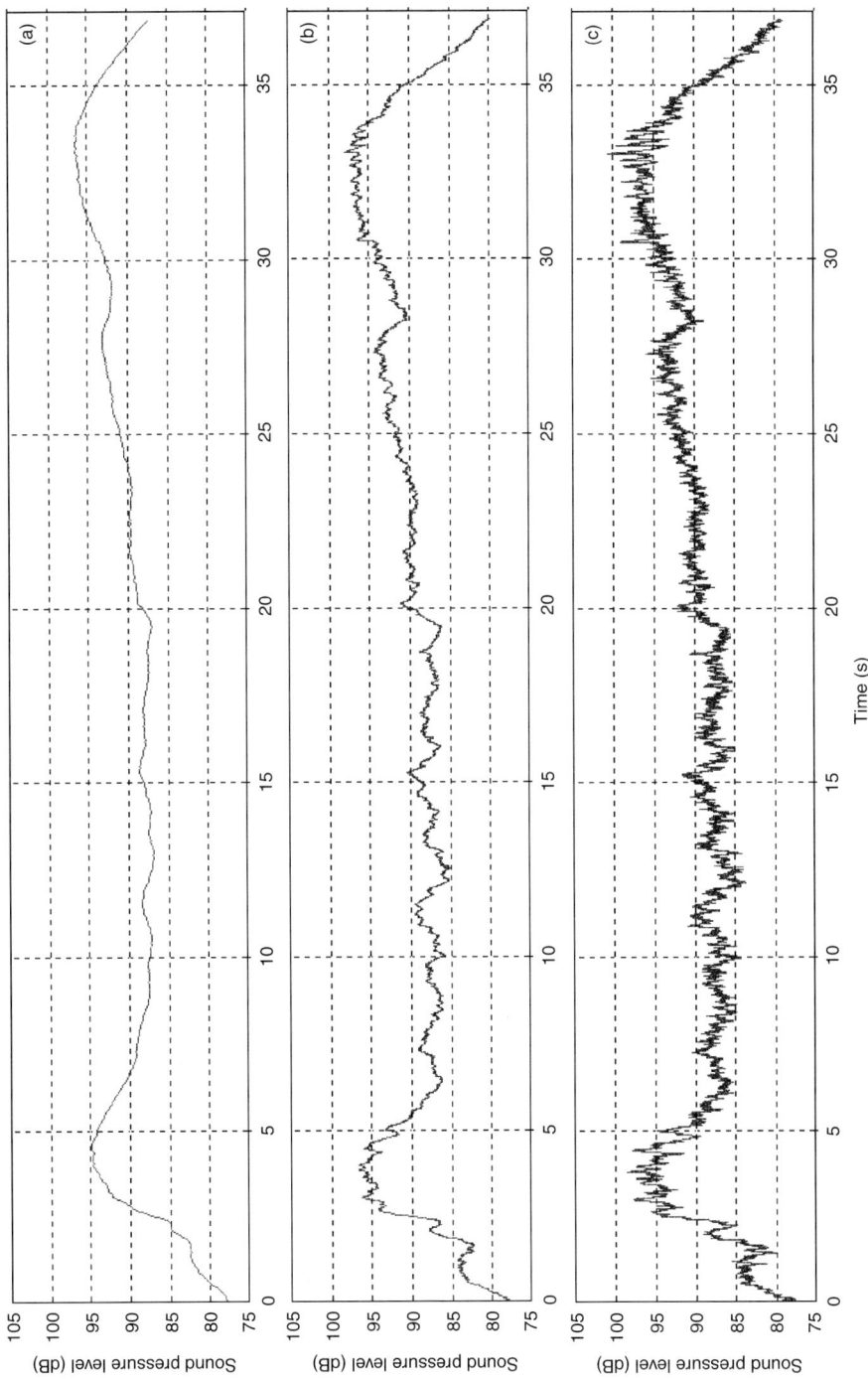

Fig. 5.10. Sound pressure level vs time diagrams of a train pass-by with (a) 'slow' and (b) 'fast' time weightings and (c) exponential time weighting with 31 ms integration time

The minimum number of vehicles required and the choice of the right test train composition are other important elements in making type-testing and monitoring results reliable. The minimum number of vehicles depends on the distance from the microphone to the track and the length of the vehicle. The Austrian regulation *SchLV* defines 70 m as the minimum length of a vehicle group tested at a measurement distance of 7.5 m. This means at least three coaches of the standard UIC length of 26.4 m, at least four wagons with bogies or five two-axle wagons. For measurement distances greater than 7.5 m, the length of vehicle groups must be increased.

For practical reasons it is very useful to have several groups of test vehicles put together in one train, measured together but analysed separately. This reduces the number of pass-bys and thus decreases the costs of the tests. Experience shows that at 7.5 m the influence of distant vehicles on the noise level of a vehicle close to the microphone is small. If the noise level difference between two vehicle groups is less than about 4 dB, the influence of louder adjacent vehicles can be ignored when the loud vehicle is at a distance of about 15 m. For noise level differences of about 10 dB, a quiet separation vehicle about 25 m in length is required.

Once all the vehicle groups, with the correct length, have been assembled and they have been arranged into a test train by following the principles above, the average pass-by level L_{av} of a vehicle group can be derived from

$$L_{av} = 10 \lg \left(\frac{1}{\tau} \int_\tau 10^{L(t)/10} \, dt \right) \qquad (5.3)$$

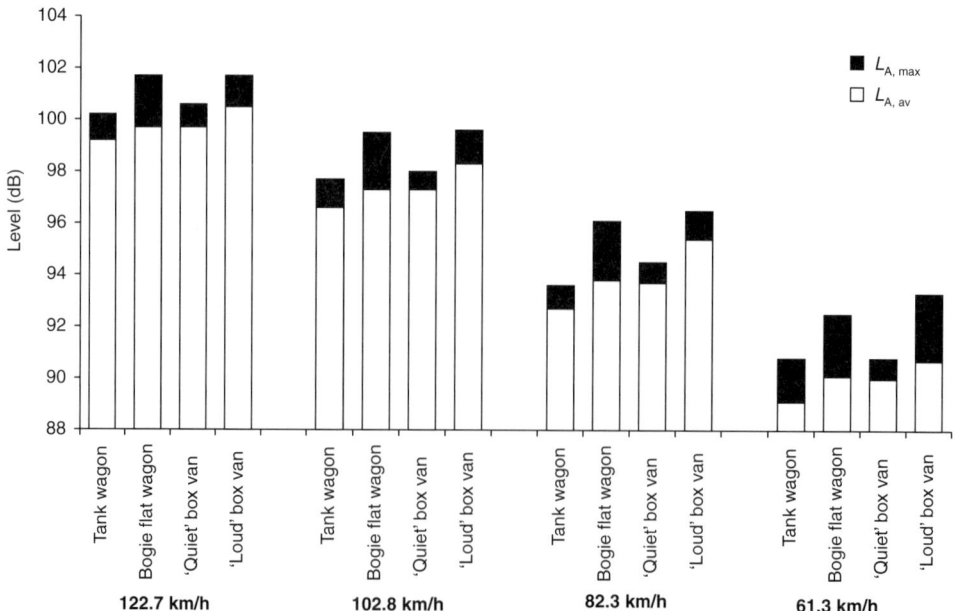

Fig. 5.11. Difference between A-weighted average pass-by level $L_{A, av}$ and maximum pass-by level $L_{A, max}$ for four different speeds and four wagon types

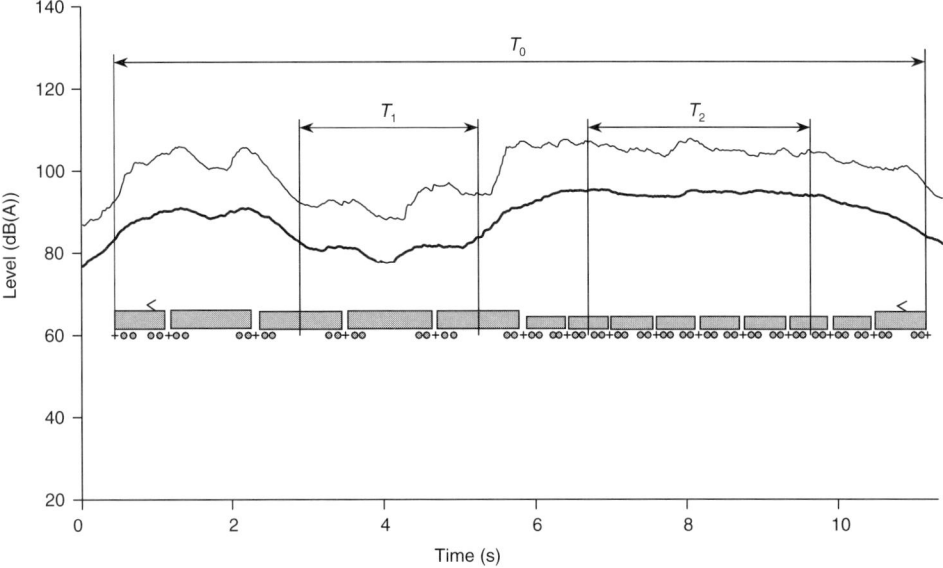

Fig. 5.12. Vehicle configuration and 'fast' weighted sound pressure level vs time: sound pressure at 0.9 m distance from track (thin line) and at 7.5 m distance (thick line). There is no influence of adjacent vehicles on the groups marked with T1 and T2

with an appropriate value for the measurement time τ of the vehicle group. Type-testing measurements in the past have shown that it is very useful to choose τ as illustrated in Fig. 5.12. To avoid any influence from the last bogie of the adjacent vehicle, it is recommended that the measurement time should start and finish in the middle of a vehicle, not at the buffers. The whole process is something like acoustic mirroring of the vehicles at their vertical axis of symmetry.

The example in Fig. 5.12 shows a test train with two groups of vehicles to be assessed. T_0 is the total measurement time for the whole train; it starts at the front of the train, top, at the front buffer, and ends at the end of the train. The measurement time τ for the group of three disc-braked coaches is marked with $T1$. This time starts in the middle of the first coach and ends in the middle of the last coach, so in total two vehicle halves and one entire vehicle are assessed. The level vs time diagram shows that the noise emission of the first bogie of the first coach is distorted by the high level of the preceding vehicle. The same happens at the end of the group, where the last bogie's noise emission is already influenced by the following wagon. The same principle applies to the group of cast-iron-block-braked wagons. Their measurement time τ is marked $T2$. Again, $T2$ starts in the middle of the first wagon and ends in the middle of the last one. The noise level vs time diagram shows that it is not too risky to start and finish the measurement time at the very beginning and end of the vehicle group. Nevertheless, as a general principle, the mid–mid rule should be applied in all cases.

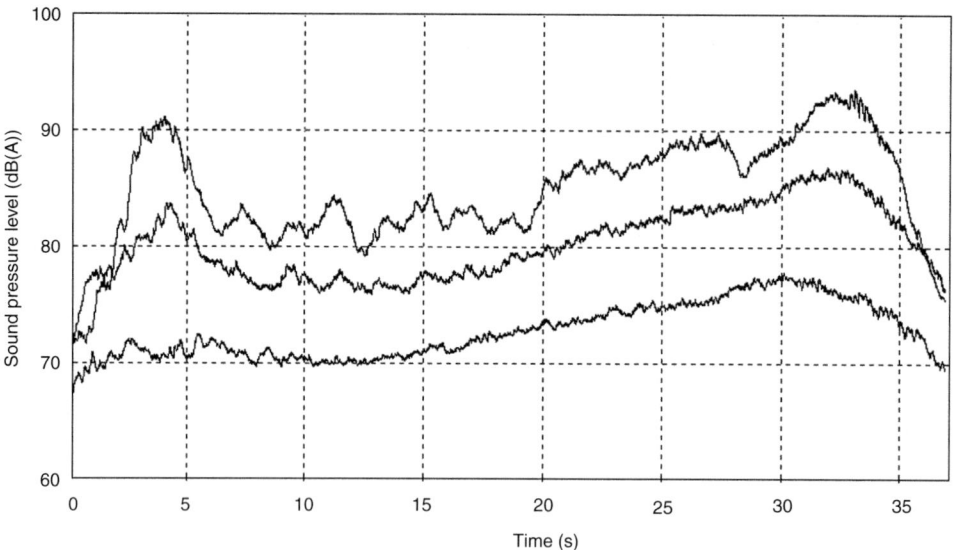

Fig. 5.13. Sound pressure level vs time diagram of a freight train at 7.5 m (top line), 25 m (middle line) and 75 m (bottom line) from the track. Increasing distance smoothes out peaks

5.2.2.5. Frequency weighting

The values to be measured and used in type testing and monitoring are preferably A-weighted sound pressure levels. For the purpose of diagnostic tests it is recommended that one records linear (non-weighted) spectra, linear total sound pressure levels and the non-weighted source signal in parallel, which adds flexibility and signal dynamics for further analysis.

5.2.2.6. Distance between track and microphone

The measurement distance for type-testing varies between 7.5 and 25 m depending on the standard or regulation used. In Germany, for example, there is a long tradition of referring noise emissions to a distance of 25 m. Legislation is also based on this figure, so rail noise emission measurements are generally carried out at this distance. Also, ISO 3095 [5.17] recommends a microphone location at a 25 m distance from the track. The Austrian standard S 5026 proposes a measurement distance of 15 m, while, according to the Austrian rail noise emission regulation *SchLV*, microphones are located at 7.5 m from the track. A distance of 7.5 m is optional according to ISO 3095.

The sensitivity of sound pressure level to measurement distance is not as simple as might be assumed. This is mainly due to ground absorption and the complexity of the source. Also, the effective length of train that contributes to the measured level increases with measuring distance and leads to a smoothing out of details (Fig. 5.13).

There is no general rule for the right measurement distance, but there are some good reasons to choose the 7.5 m position. At this distance the microphone is already far away enough from the train to be in the far field for the predominant frequency range. On the other hand, it is still close enough

- to catch fairly high levels of train noise and thus reduce the influence of background noise
- to feel differences in noise emissions of different vehicles and avoid smoothing-out effects
- to minimize the influence of variation in ground absorption and sound transmission.

5.2.2.7. General measurement conditions

The requirements and general measurement conditions for rail noise type testing are not really different from those for other sources, so it is clear that meteorological conditions such as dry, windless weather, and site conditions such as free-field conditions or no disturbing background noise are required. Additionally, there are some principles specific to rail noise measurements. The requirements for the track at the test site, apart from smoothness of the rail surface, are

- a gradient less than 1 in 3000
- a radius greater than 3500 m
- continuous welded rail for five times the measuring distance before and after the microphone position (that is, 37.5 m to the left and right for a 7.5 m measurement distance)
- the track type must not change within a section of the length described above
- no noisy elements, such as turnouts, crossings, level crossings or steel bridges, on the track for ten times the measuring distance before and after the microphone position (that is, 75 m to the left and right for a 7.5 m measurement distance).

For the vehicles to be tested, there is the requirement that they are empty when tested and all windows and doors are closed.

Each test has to be repeated at least twice to guarantee a minimum of reproducibility. If the results, which mean the A-weighted average pass-by levels L_{av} vary more than 3 dB, the test must be repeated.

5.2.2.8. Regulations and noise limits

Austria was the first European country to limit noise emission from railbound vehicles (*SchLV* [5.24]). Since 1993, the railway operator has been obliged to prove conformity of rolling stock which is newly registered in Austria. The measurement distance is 7.5 m, with two microphone positions at 1.2 and 3.5 m height above the rail surface. The maximum A-weighted sound pressure level is measured with the 'fast' time weighting for constant speed and for stationary conditions, for exterior and interior noise, and the corresponding spectra must be reported. Pass-by noise is measured at maximum speed. Noise limits are given for 80 km/h (Table 5.4), and for higher speeds V they increase as $30 \lg(V/80)$, where V is in km/h (see Section 5.2.4.2 for more detail).

On a European Union level, there was an initiative in the early 1980s to limit noise emission from railbound vehicles. There was no agreement on the measurement procedure and the limits, so the proposal for regulation failed. Up to now, no new attempt has been made.

5.2.3. Monitoring

There are two main purposes of rail noise monitoring. One is to watch for changes in noise emission from rolling stock passing a specific site, as well as for changes in rail conditions. The other purpose is to check the noise impact in a particular situation. The reliability of data is rather low compared with type testing and diagnostics. This is due to the everyday conditions under which measurements have to take place. For diagnostics and type testing, operational and environmental conditions are specified and controlled during measurements. Monitoring, on the other hand, has to be done 24 h a day and 365 days a year to serve its purpose. That means trains pass by with different speeds; they might even accelerate or brake. Also, different meteorological conditions, such as strong winds and heavy rain, as well as snow and ice, will influence results and make them less reliable.

These effects are rather easy to handle as long as averages (equivalent levels) from rather long periods, such as daytime or night-time, are measured. There is great experience with this kind of measurement, mainly from permanent, automatic aircraft noise monitoring. Modern computer technology makes it possible to install monitoring stations that are even able to 'decide' if a noise event seems to be odd and then record this sequence for a check by a human later.

Table 5.4. Noise limits of the Austrian ordinance on permitted noise emissions from rolling stock SchLV [5.24]

Vehicle type	Limit for exterior noise (dB(A))		Limit for interior noise (dB(A))	
	Stationary	Constant speed	Stationary	Constant speed
Electric locomotive	74	84	66	76
Electric unit	74	82	64	74
Diesel locomotive	80	86	66	76
Diesel unit	76	84	64	74
Shunters/special vehicles	78	86	68	78
Passenger coach Cat. 1	71	80	54	63
Passenger coach Cat. 2	71	80	57	66
Passenger coach Cat. 3	74	83	60	74
Passenger coach Cat. 4	74	83	60	72
Goods wagon Cat. 1	–	81	–	–
Goods wagon Cat. 2	–	83	–	–
Goods wagon Cat. 3	–	85	–	–

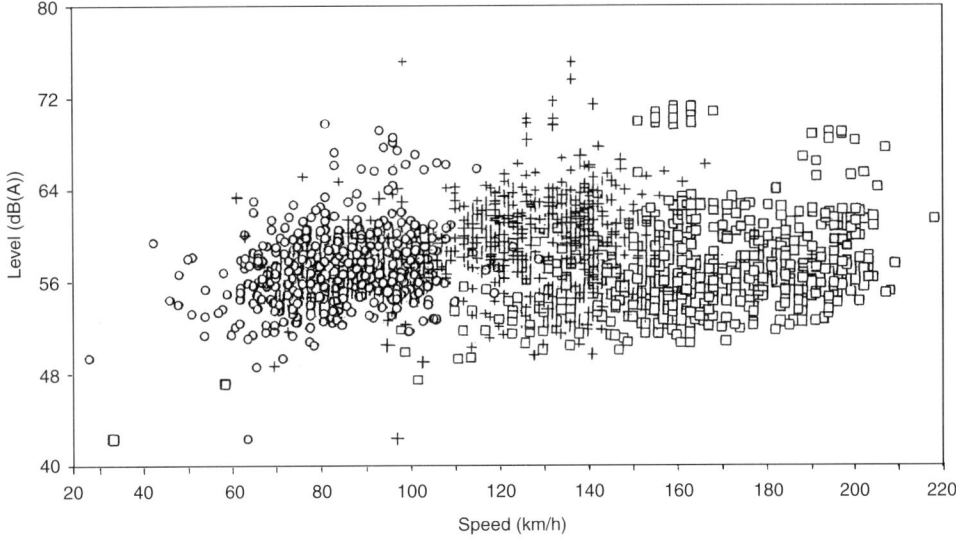

Fig. 5.14. A-weighted equivalent level $L_{A,\,eq}$ *in 25 m for 1 train/h of standardized length 100 m. Results of a measurement series of the German Umweltbundesamt [5.14]*

5.2.3.1. Permanent emission monitoring

As soon as monitoring aims to assess noise emission from a single train or a single vehicle, things get more and more difficult. Figure 5.14 shows the results of a German survey that included more than 800 trains passing by. There are some good reasons for doing this kind of monitoring. An infrastructure manager, for example, could see what were the noisiest vehicles on the network. By looking at the time series of a site he or she could also see the deterioration of the track and increasing roughness of the rails. In addition, train operators would benefit from an individual monitoring device by detecting noisy vehicles and damage such as wheel flats.

There are four principal ways (Table 5.5) of designing such an automatic monitoring device that is able to detect noise emission from single vehicles of ordinary trains.

- Sound pressure level at X m distance from the track: the sound pressure level is measured at the same position as in type testing. Theoretically this is the best method, for it allows one to compare monitored levels with results from type approval tests directly. The main drawback of this method is that varying meteorological conditions will influence the results very strongly.
- Sound pressure level in an enclosure/tunnel: measuring the sound pressure level when the train passes through a tunnel means eliminating all meteorological factors that could influence the result. This approach is described in the draft CEN standards [5.7, 5.8] and can be characterized as a reverberation room method.
- Sound pressure level in a microphone box close to the rail: the sound pressure level is measured with a microphone inside a sound-absorbing box (Fig. 5.15).

This box is airtight and is attached to the rail with elastic material. Experience with this sort of device shows that this method is able not only to quantify rail noise radiation, but also to characterize noise emission from the wheel. Meteorology does not influence the measurement: on the other hand, the results cannot be compared directly with those from type approval tests, and noise radiated from the car body is not covered.
- Rail vibration: measuring vertical and horizontal rail vibrations is a simple indirect method to quantify noise emission. This method is very similar to the microphone box method.

5.2.3.2. Permanent perception monitoring
Permanent monitoring of noise impact at sites in the vicinity of railway lines is a task not really different from that of air traffic noise monitoring. A lot of experience related to this issue is available, and there is quite a range of equipment that is commercially available and able to measure, record and automatically assess all kinds of environmental noise.

5.2.3.3. Special noise perception measurement techniques
The A-weighted equivalent sound pressure level $L_{A,eq}$ is the most frequently used noise criterion for rail noise impact assessment. However, in special situations additional criteria or assessment tools, which can be summarized as psychoacoustics, may be required. In Austria, for example, environmental impact assessment law refers neither to $L_{A,eq}$ as a noise criterion nor to noise level limits. Prevention of

Table 5.5. Assessment of monitoring methodologies

	SPL in X m distance from track	SPL in enclosure/ tunnel	SPL in microphone box close to the rail	Rail vibration
Same position as in type testing	Yes	No	No	No
Type of sound field	Free field	Diffuse field	Near field	–
SPL directly measured	Yes	Yes	Yes	No
Influence of meteorology	Yes	No	No	No
Assessment of all vehicle parts	Yes	Yes	No	No
Assessment of single wheel sets	(Yes)	(Yes)	Yes	Yes
Interference with other vehicles	High	High	Low	Low
Effort and costs (of investment)	Low	High	Low	Low
Easy to install everywhere	(Yes)	No	Yes	Yes

SPL, sound pressure level

MEASUREMENTS OF RAILWAY NOISE

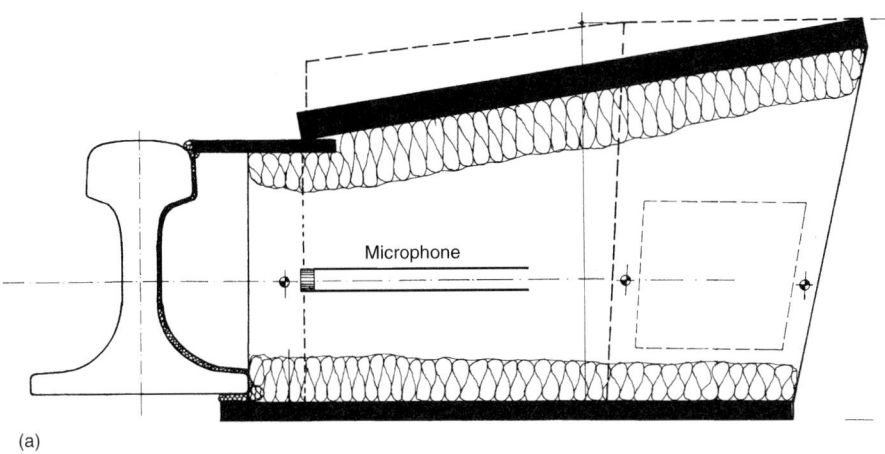

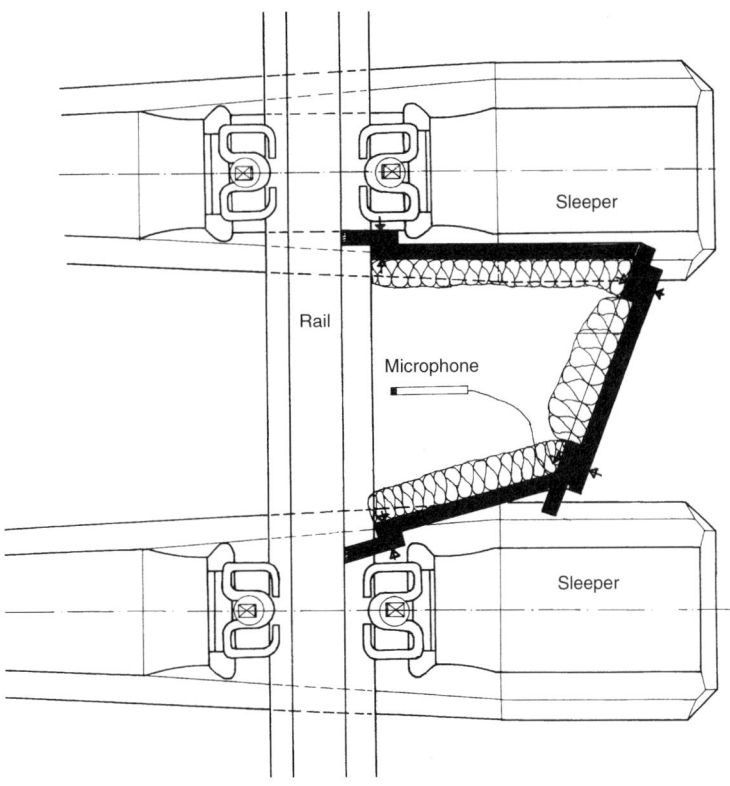

Fig. 5.15. Sound-absorbing microphone box

health risk and of annoyance is a general principle of environmental assessment. In practice this means acquisition of loudness data instead of equivalent levels.

Knowledge about alternative noise criteria of that sort is scarce and research is still in progress. Basically, a noise assessment criterion is needed which is able to describe noise perception for long time periods during which the noise intensity and loudness vary. Such a number is then called the perceived equivalent loudness. Fastl [5.12, 5.13] has shown that the N_5 loudness percentile is a valid psychoacoustic criterion for environmental-noise assessment and can be called the equivalent loudness of time-varying noise.

Apart from the above criterion, it is increasingly popular to calculate an average loudness $N_{av} = f(L_{A,eq})$ from the (non-weighted) $\frac{1}{3}$ octave spectrum of the $L_{A,eq}$ using DIN 45 631 [5.9]. Our own analysis [5.19] has shown that there is in general a good correlation between $L_{A,eq}$ and $N_{av} = f(L_{A,eq})$ and no additional information or knowledge can be obtained from this exercise.

Things are different when the N_5 loudness percentile is used to describe equivalent loudness (Fig. 5.16). There is a good correlation between N_5 and $L_{A,eq}$ for very low (squares, MP29) and very high noise impact (filled circles, MP7), but big discrepancies for medium impact. For an $L_{A,eq}$ of about 58 dB there are two different values for equivalent loudness N_5 of 5.3 and 16.8 sone GF (diamonds). On the other hand, there are situations where the equivalent loudness N_5 decreases from 10 to 8 while $L_{A,eq}$ increases from 57 to 63 dB (triangles).

Taking into account the temporal structure and the maximum levels helps to explain the loudness of these four types of noise event. During the time period represented by the diamonds labelled 2B, six trains were recorded. One train had an average maximum level of 80 dB(A); the rest had maximum levels of 70–75 dB(A).

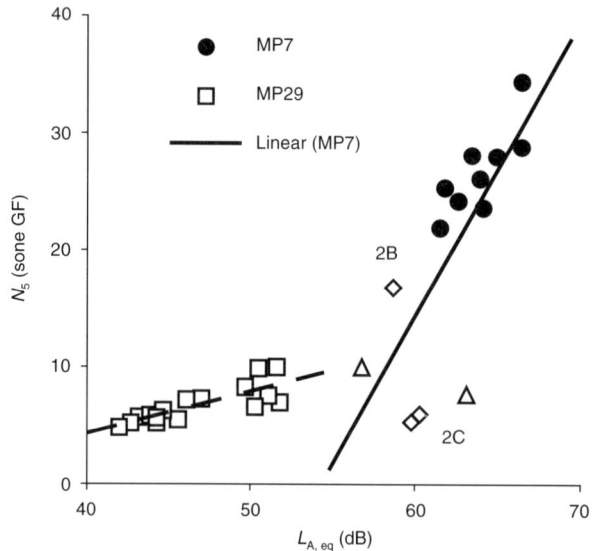

Fig. 5.16. A-weighted equivalent level $L_{A,eq}$ vs 5% percentile loudness $N_{5\%}$

Table 5.6. Indicative parameter sensitivities for total rolling noise with conventional track systems 5.10

Parameter	Parameter value for Minimum noise level	Parameter value for Maximum noise level	Level difference for minimum and maximum parameter values (dB(A))
Rail type	UIC54 E	UIC60	0.7
Pad stiffness	5×10^9 N/m	10^8 N/m	5.9
Pad loss factor	0.5	0.1	2.6
Sleeper type	Biblock concrete	Wooden	3.1
Sleeper spacing	0.4 m	0.8 m	1.2
Ballast stiffness	10^8 N/m	3×10^7 N/m	0.2
Ballast loss factor	2.0	0.5	0.2
Wheel offset	0 m	0.01 m	0.2
Rail offset	0 m	0.01 m	1.3
Wheel roughness	Smoothest	Roughest	8.5
Rail roughness	Smoothest	Roughest	0.7–3.9
Train speed	80 km/h	160 km/h	9.4
Wheel load	12 500 kg	5000 kg	1.1
Air temperature	10°C	30°C	0.2
Wind direction	Microphone–train	Train–microphone	5
Ground properties	Stable	Unstable	2

The period represented by the diamonds labelled 2C had only three trains passing by, but all of them with about 80 dB(A) maximum level. The conclusion is that $L_{A, eq}$ is strongly correlated with the peaks of noise, while N_5 takes the number of events and the quiet periods into account.

5.2.4. Non-acoustic factors influencing exterior rail noise

Wheel–rail rolling noise emission is sensitive to a lot of non-acoustic parameters, mainly related to track properties, such as track and ballast stiffness, rail pad loss factor and sleeper type. Train speed, wheel and rail roughness and rail pad behaviour have the most substantial effect on railway noise. Though other factors have less influence, their combination can cause substantial differences between measurements at different sites (Table 5.6).

5.2.4.1. Wheel roughness and rail roughness

The roughness of the wheel and rail running surfaces is one of the key parameters influencing railway rolling noise. Microscopic irregularities or roughness of both the wheel and the rail surfaces excite the system at the wheel–rail interface and induce vibrations of the wheel, bogie, car body and track (Fig. 5.17). Wheel–rail rolling noise is caused by these vibrations. Each component involved has its typical vibration characteristics and thus radiates noise of a typical spectral distribution. A knowledge of the wheel and rail roughness of a particular train passage is essential for the interpretation of measurement results and for high reproducibility (see Chapters 1 and 2 for more details).

Results of roughness measurements from wheels with different braking systems, performed during the MetaRail round-robin test, are shown in Fig. 5.18. The upper diagram in Fig. 5.18 shows the wheel roughness level vs $\frac{1}{3}$ octave spectrum for a disc-braked wheel. The main roughness amplitudes are about 1 μm at wavelengths up to 12.5 cm. The roughness decreases at lower wavelengths. The lower diagram in Fig. 5.18 shows the roughness (exaggerated) of this disc-braked wheel.

The upper diagram in Fig. 5.19 shows the roughness level vs $\frac{1}{3}$ octave spectrum for a damaged disc-braked wheel. The main roughness reaches 100 μm at a wavelength of around 30 cm. The roughness level decreases slowly at lower wavelengths. The lower diagram shows very clearly the damage to the wheel, which is not perfectly round.

The upper diagram in Fig. 5.20 shows the measured roughness spectrum for a cast-iron block-braked wheel. The roughness peak reaches about 10 μm at a wavelength of around 5 cm. The roughness level decreases at lower wavelengths. The lower diagram again shows an exaggerated image of the roughness of this wheel. Disc-braked wheels are significantly quieter than wheels from tread-braked vehicles because the running surface of disc-braked wheels is smoother.

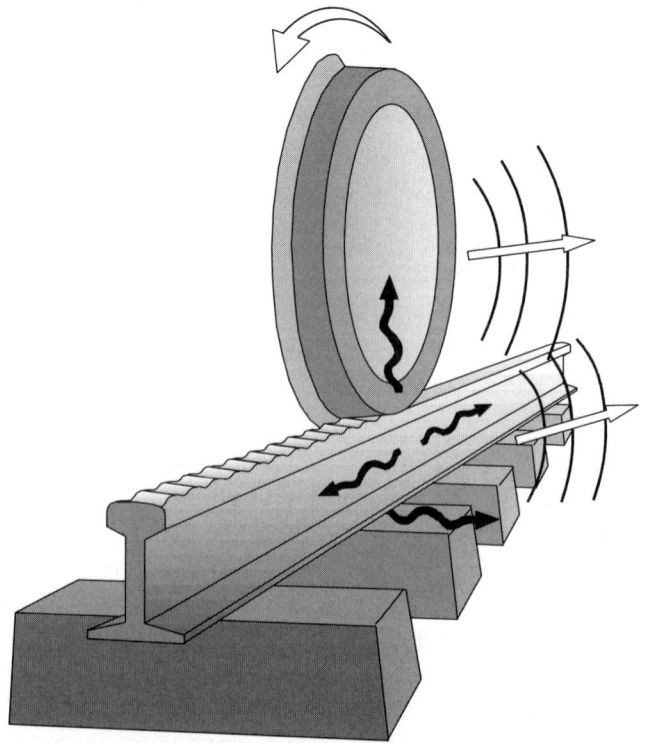

Fig. 5.17. Origin of wheel–rail rolling noise [5.25]

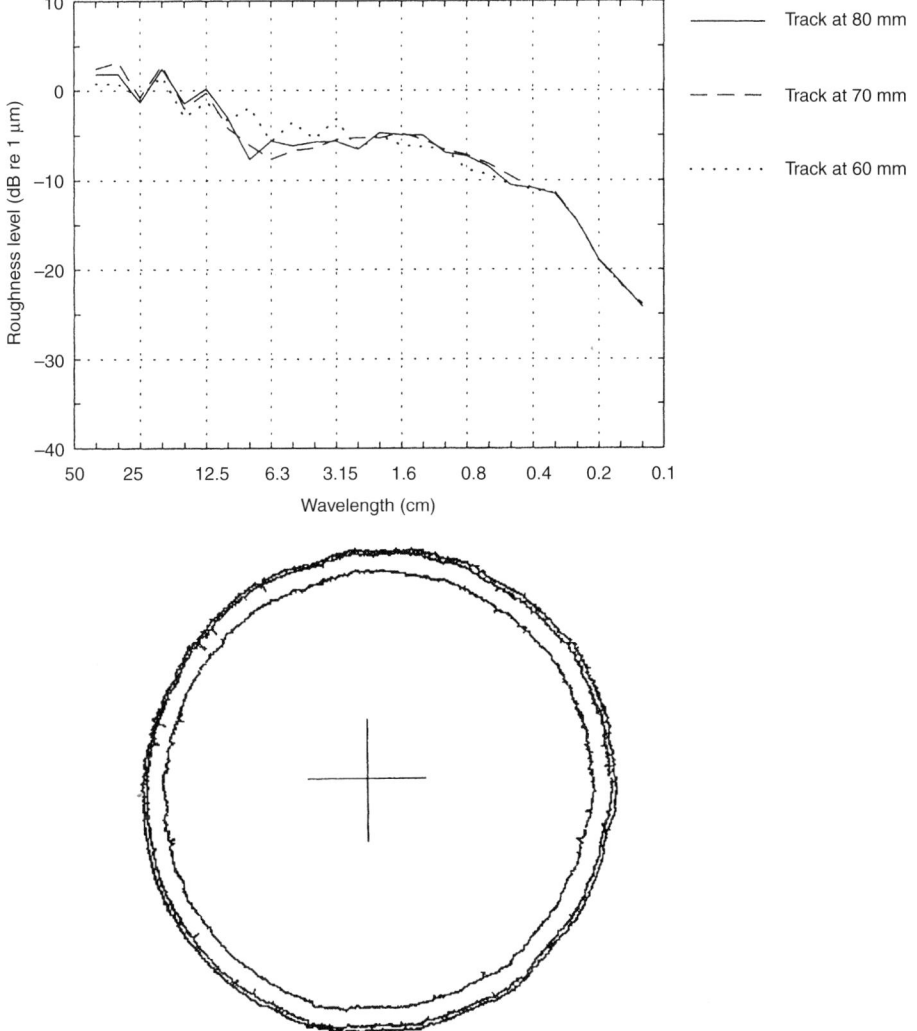

Fig. 5.18. Roughness measurement of an undamaged disc-braked wheel [5.5]

The upper diagram in Fig. 5.21 shows the roughness level vs $\frac{1}{3}$ octave spectrum for a sinter-metal-braked wheel. The main roughness has its peak at wavelengths up to 1.6 cm, with roughness amplitudes of about 1 µm. The roughness level decreases at lower wavelengths. Compared with the disc-braked system, the decrease of roughness level of the sinter-metal-braked wheel starts at a lower wavelength. The lower diagram shows an exaggerated image of the roughness of this-sinter-metal braked wheel.

Noise also increases dramatically when rails become corrugated, owing to a significant increase in rail roughness at wavelengths important for rolling-noise generation. Up to now, the surface integrity of rails has been subject only to

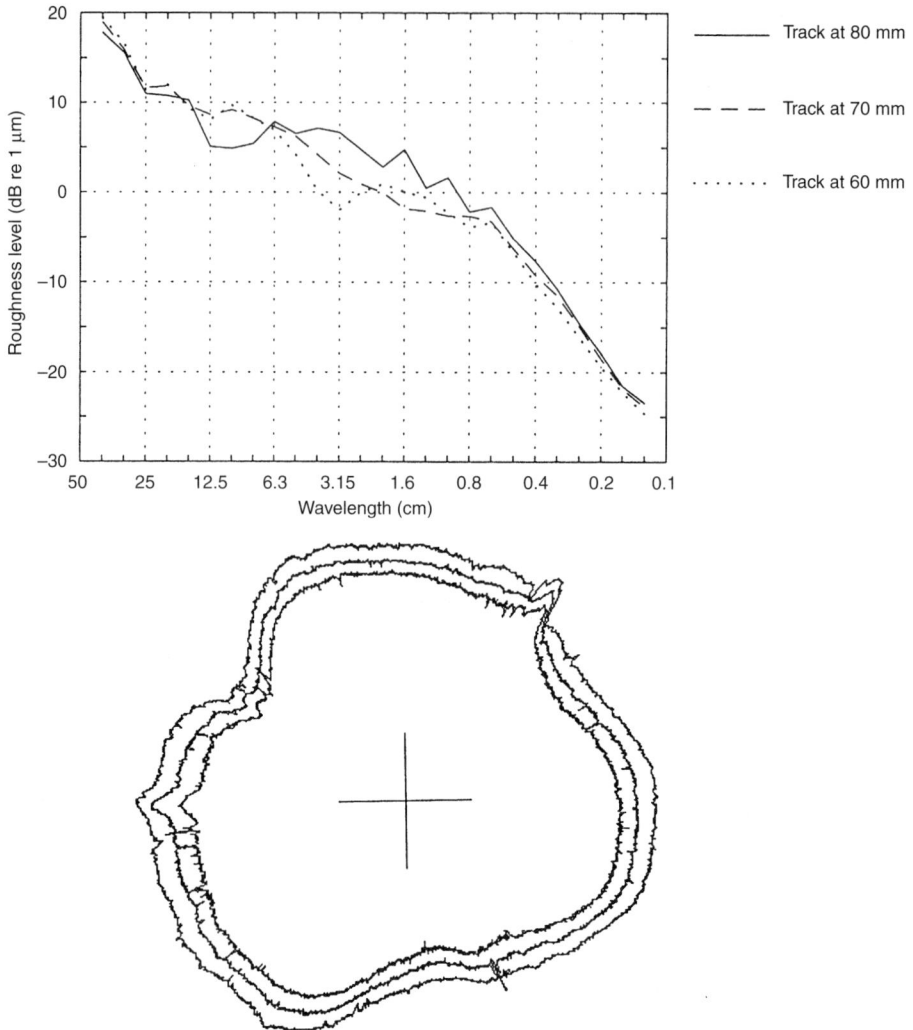

Fig. 5.19. Roughness measurement of a damaged disc-braked wheel [5.5]

general conditions. The draft standard pr EN ISO 3095 [5.17] proposes a measurement procedure and a limit curve for type-testing measurements. Keeping rail roughness below this limit should in future prevent differences of 10 dB(A) or more due to the influence of the rail from occurring. The limit proposed is based on an average of values used in different European countries as a criterion for 'smooth' rails.

Figure 5.22 shows average track roughnesses from the MetaRail measurement campaigns obtained according to pr EN ISO 3095 for all sites investigated. Although the test sites were selected so as to have roughness levels below the limit level, unfortunately some of the roughness levels exceeded the limit.

MEASUREMENTS OF RAILWAY NOISE

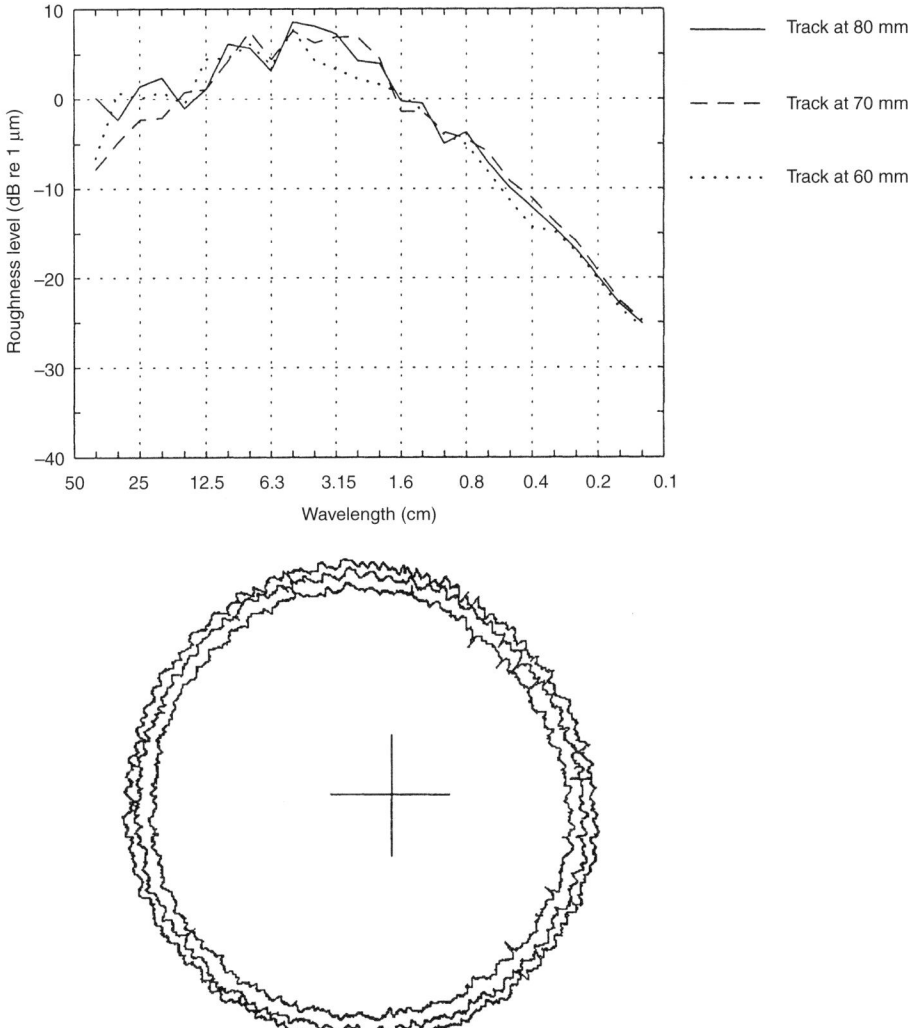

Fig. 5.20. Roughness measurement of a cast-iron-block-braked wheel [5.5]

5.2.4.2. Speed

Speed is one of the major factors influencing rail noise emission. At low speeds, up to about 250 km/h, wheel–rail rolling noise is predominant, while at speeds greater than 250 km/h aerodynamic noise becomes significant.

For wheel–rail rolling noise, a cubic dependence in speed is generally used. The Austrian standard *SchLV*, for example, defines the speed correction of noise limits for actual speeds v different from 80 km/h as follows:

$$\Delta L_{\text{Speed}} = 30 \lg(v^{(\text{km/h})}/80) \tag{5.4}$$

This formula can be used as an average for various types of rolling stock, such as

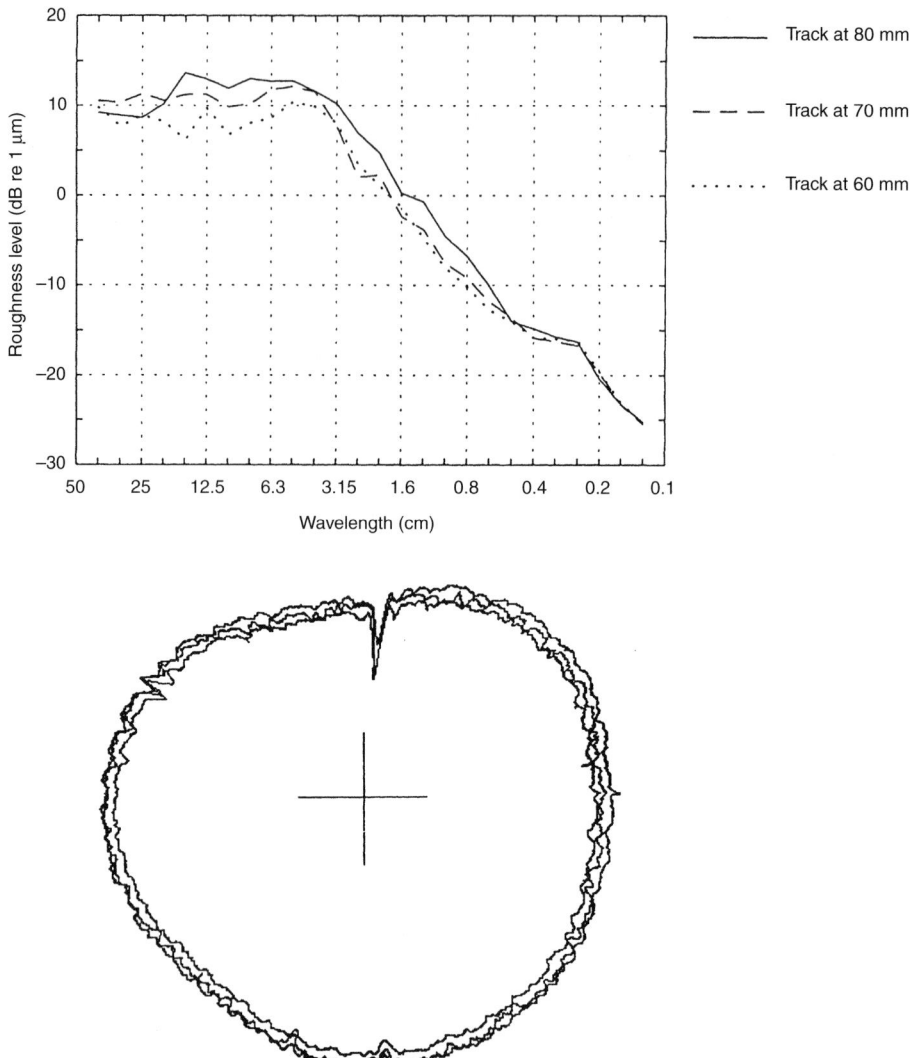

Fig. 5.21. Roughness measurement of a sinter-metal-block-braked wheel [5.5]

locomotives, coaches and wagons. The factor 30 can vary between about 25 and 35 for specific vehicle types (Table 5.7).

Aerodynamic noise increases more rapidly with train speed than does wheel–rail rolling noise. A good first estimate for the speed dependence of aerodynamic noise is

$$\Delta L_{\text{Speed, aerodyn.}} = 60 \lg(v/v_1) \tag{5.5}$$

where v_1 is a reference speed. Aerodynamic optimization of the car body and of the pantograph can reduce noise emission dramatically, so again the factor of 60 will vary depending on the properties of the single vehicle or train set (Fig. 5.23).

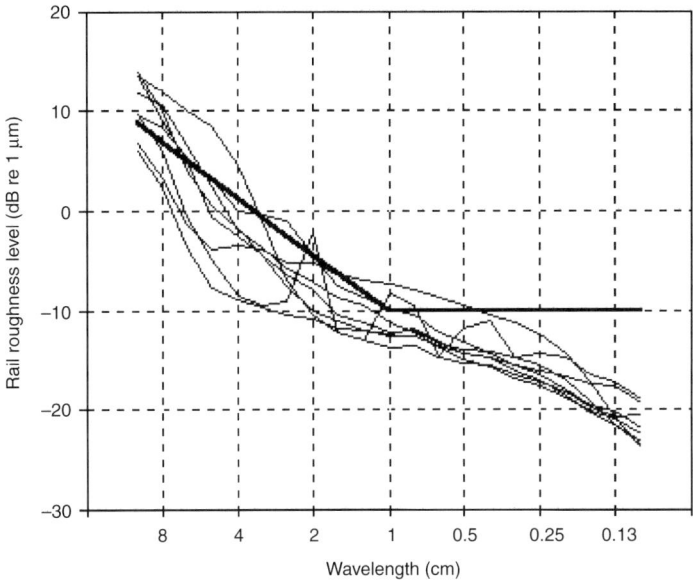

Fig. 5.22. Rail roughness measurements at the MetaRail test sites (thin lines) and proposed CEN/ISO limit for rail roughness level (thick line)

Table 5.7. Speed dependence of block- and disc-braked vehicles on ballast track with monoblock concrete sleepers and UIC60 rails

	Microphone position (m)		
	7.5/1.2	7.5/3.5	25/3.5
Cast-iron-block-braked wagons			
Linear SPL	19.8 lg ν	22.4 lg ν	23.5 lg ν
A-weighted SPL	33.6 lg ν	32.2 lg ν	34.4 lg ν
Disc-braked coaches			
Linear SPL	19.8 lg ν	20.0 lg ν	15.0 lg ν
A-weighted SPL	25.8 lg ν	26.7 lg ν	22.7 lg ν
Cast-iron-block-braked hopper			
A-weighted SPL	37.1 lg ν		
Sinter-metal-block-braked flat			
A-weighted SPL	35.9 lg ν		
Disc-braked flat			
A-weighted SPL	30.5 lg ν		

Noise measurements of one train or vehicle type are often made not only at maximum speed, but also at lower speeds, to indicate the actual speed dependence. Depending on the driver's skills, there will be some variance from the target speed. Every measurement that is within a range of ±3 km/h from the target speed is acceptable and the average noise emission level for a target speed is just the mean value of the n repetitions.

There are some situations, however, where the deviations in actual speed are greater than 3 km/h or where there is just one pass-by at each of many different speeds. As long as the pass-by speed is either below or above the region where the speed dependence of pass-by noise changes from the 30 lg v to the 60 lg v trend, a linear regression can be used to obtain an analytic function for the pass-by level $L_A(v)$ (this curve fit applies generally to the A-weighted level only):

$$L_A(v) = a_A + b_A \lg v \qquad (5.6)$$

where $L_A(v)$ is the calculated A-weighted pass-by level, v is the pass-by speed in km/h and a_A, b_A are parameters of the linear regression.

Theoretically this approach could also be applied to the frequency spectrum to obtain an analytic function $L_f(v)$ for single frequency bands:

$$L_f(v) = a_f + b_f \lg v \qquad (5.7)$$

where $L_f(v)$ is the calculated pass-by level for the frequency band f, v is the pass-by speed in km/h and a_f, b_f are parameters of the linear regression.

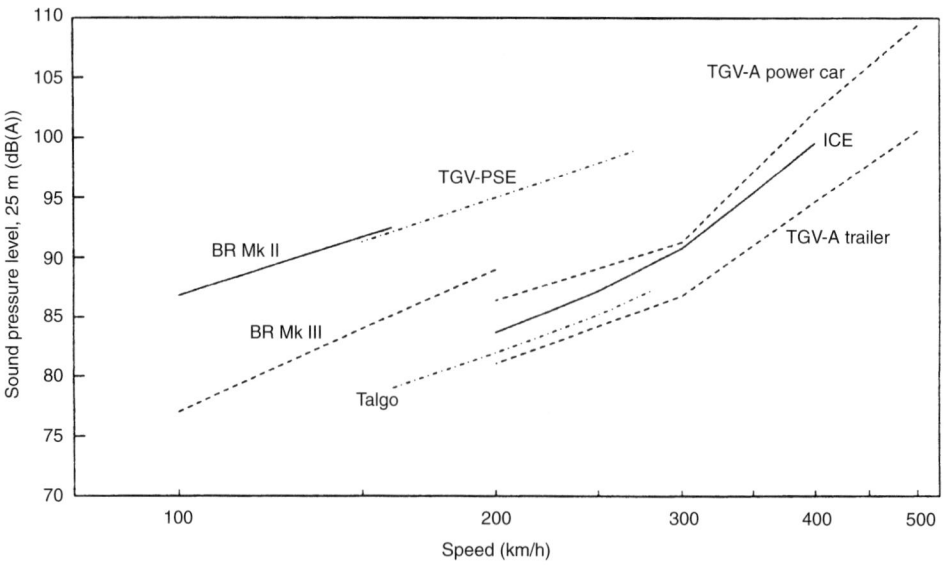

Fig. 5.23. Measured sound pressure level at 25 m (regression lines) for different high-speed train systems [5.25]

Table 5.8. Regression parameters for the logarithmic speed dependence $L_f(v) = a_f + b_f \lg v$ of the sound pressure level of wagons with different braking systems passing by

$\frac{1}{3}$ octave band f (Hz)	Cast-iron block brake		Sinter metal block brake		Disc brake	
	a_f	b_f	a_f	b_f	a_f	b_f
A-weighted total, L_A	22.16	37.06	17.97	35.85	23.93	30.50
Linear total, L_{lin}	43.60	26.52	31.16	29.38	31.16	29.38
25	−12.79	45.11	−3.41	36.34	−3.41	36.34
32	25.81	25.65	−10.53	41.05	−6.41	37.75
40	62.66	7.02	31.02	20.45	69.70	0.16
50	−6.88	42.00	6.80	32.50	6.80	32.50
63	25.33	25.53	3.13	34.11	39.77	16.23
80	125.09	−24.67	31.42	19.75	48.66	11.01
100	99.62	−9.52	41.00	14.95	41.00	14.95
125	28.39	25.74	34.25	18.13	61.49	4.28
160	44.00	17.54	45.45	12.78	40.25	15.46
200	79.29	−0.18	64.39	3.64	64.39	3.64
250	85.39	−3.31	66.33	2.67	86.21	−7.44
315	103.33	−12.71	88.20	−8.05	33.33	19.19
400	92.36	−5.79	68.65	2.62	68.65	2.62
500	71.24	5.44	64.57	4.89	47.25	13.08
630	38.49	22.88	40.55	17.60	38.62	17.68
800	58.29	13.21	32.10	22.17	32.10	22.17
1 000	38.97	23.61	37.28	20.59	10.20	32.51
1 250	20.30	32.66	24.46	29.13	2.24	36.74
1 600	−13.26	50.34	−7.05	44.59	−7.05	44.59
2 000	−20.49	55.15	−19.55	50.81	5.56	35.64
2 500	−14.95	50.07	−30.53	53.97	−7.41	40.63
3 150	25.89	28.25	−2.54	38.01	−2.54	38.01
4 000	12.91	33.38	1.36	34.61	15.70	26.62
5 000	−5.04	40.70	−28.17	48.60	0.62	33.00
6 300	−8.73	41.23	−26.29	46.50	−26.29	46.50
8 000	−15.12	42.69	−35.74	49.62	−11.71	36.70
10 000	−24.05	44.73	−36.91	48.27	−11.98	34.93

This is acceptable if only a first rough estimate is needed. However, it must be clear that there is a difference ε between the A-weighted level from the curve fit $L_A(v)$ and the A-weighted level which results from the energy sum of the A-weighted single-frequency levels $L_f(v)$. This is due to the non-linearity of energy level summation and to the fact that the speed dependence of single-frequency bands need not necessarily be linear with the logarithm of speed. For the three wagon types with different braking systems listed in Table 5.8 the error ε is +1.4 dB for the cast-iron-block-braked vehicle, +0.1 dB for the sinter-metal-braked wagon and +0.7 dB for the disc-braked wagon. One can write

$$L_A(v) = a_A + b_A \lg v = 10 \lg \sum_f 10^{[L_f(v)+A_f]/10} + \varepsilon \qquad (5.8)$$

where A_f is the A-weighting of frequency band f.

5.2.4.3. Braking

We have already seen that the type of braking system used, whether cast-iron blocks or disc brakes, will influence wheel roughness and thus the wheel–rail rolling noise emitted. Braking itself influences the sound pressure level mainly by changing the tone of noise, even if the typical braking squeal does not show up (Fig. 5.24).

5.3. Interior noise

5.3.1. Diagnostics

Comfort and quality issues are becoming more and more important for the competitiveness of passenger rail transport. Undoubtedly, interior noise is one of the major aspects of travel comfort. Thus the industry and passenger train operators will direct more attention to sound quality inside the coach.

A general definition says that noise is any kind of unwanted, disturbing or annoying sound. The human ear is not just a biological sound metre, but a selective organ able to perceive noise. This has to be taken into account when the sound quality of a coach is assessed. Sound quality and sound design are more than reduction of sound pressure level. The process is one of designing and changing the acoustic character of a technical product to increase the customer's acceptance. The methodology used has to take into account the following considerations.

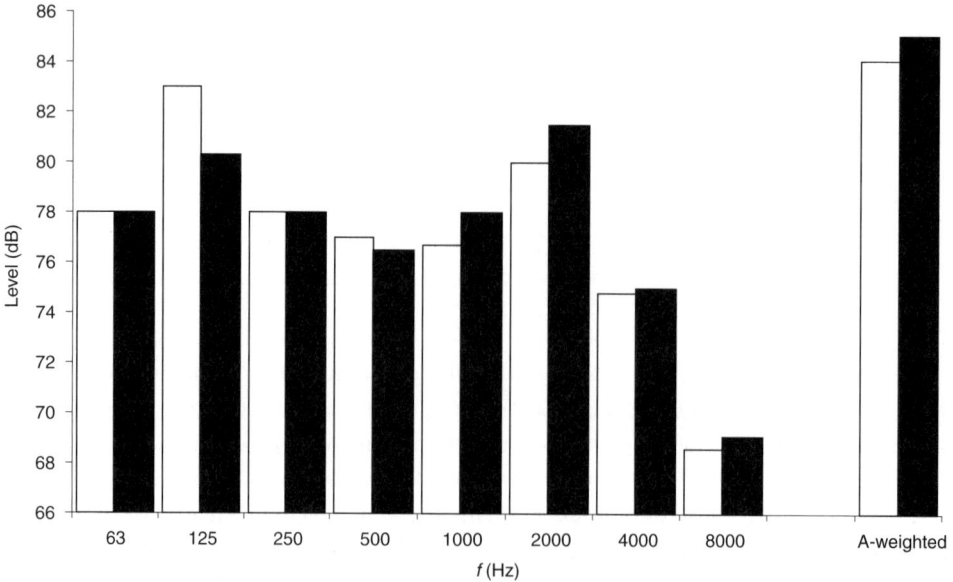

Fig. 5.24. Octave band noise level and A-weighted level of a train rolling (white bars) and braking slightly (dark bars)

Fig. 5.25. Artificial head recording noise binaurally in a first-class coach compartment

- The target group for improvements of interior noise is the travellers, so noise recordings have to be made inside the coach at those positions where travellers sit during the journey.
- The measurement methodology must try to reproduce human noise perception as accurately as possible. Reducing sound pressure level does not necessarily mean making noise more pleasant.

Conventional techniques, such as A-weighted sound pressure levels, equivalent levels or maximum levels, are not able to reflect changes in the tone or temporal character of the noise inside the coach. Only psychoacoustic methods [5.20,5.26] such as loudness, sharpness or roughness and binaural recording techniques using an artificial head (Fig. 5.25) are able to detect and quantify unpleasant sounds.

Figure 5.26 shows an example of a level vs frequency vs time diagram, which was produced from an artificial-head recording in a second-class open coach. Dark areas represent low levels and bright areas represent high levels; the levels displayed range between 25 and 65 dB. After 9 s the coach enters a tunnel, which leads to a significant level increase between 100 and 5000 Hz. The pattern does not really change for frequencies lower than 100 Hz. There is a bright line all the time at somewhat more than 200 Hz, which is the tonal noise of the air conditioning. The sealing of the windows is also bad, which can be heard in the coach as a hissing noise and appears as light-grey zones at frequencies from 1.2 to 3 kHz. Diagrams and tools like this are not able to reduce noise, but they can help to visualize critical noise.

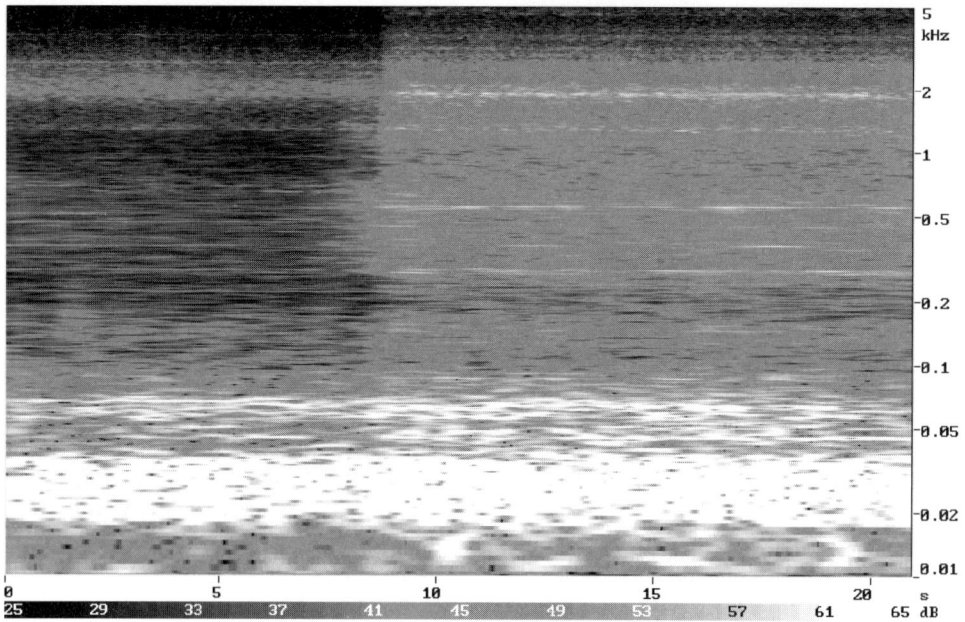

Fig. 5.26. Sound pressure level vs frequency (FFT spectrum) vs time diagram of a 21 s binaural noise recording in a second-class coach

5.3.2. Type testing

The objective of type-testing noise measurements inside railbound vehicles is, according to the CEN/TC256 WG3 draft standard [5.8], to obtain reproducible and comparable results from acoustic measurements of the noise level and noise spectrum inside all kinds of vehicles on rails or other types of fixed track. Its main application is

- to check noise levels at working positions for worker protection reasons
- to prove that vehicles delivered by a manufacturer comply with the noise specifications.

5.4. References

5.1. ÖSTERREICHISCHES NORMUNGSINSTITUT. *Measurement of Noise Emitted by Railbound Vehicles – Normalized Sound Power Level.* Österreichisches Normungsinstitut, Vienna, 1996, Önorm S 5026 [This replaced S 5024, 1991, referred to in the text.].

5.2. AUSTRIAN CIRCLE FOR NOISE CONTROL (ÖAL). *Model for Calculation of Rail Noise Exposure.* ÖAL, Vienna, 1990, Guideline 30.

5.3. BARSIKOW, B. Schallabstrahlung heutiger und zukünftiger Schienenverkehrssysteme. DAL-Fachtagung Schienenverkehrslärm Eigenverlag DAL, Düsseldorf, 1990.

5.4. BERKHOFF, A. P., KINNEGING, N. A., VAN DEN DOOL, T. C. and DITTRICH, M. G. *Antenna Systems – Intermediate report (Task II.3.1).* TNO, Delft, 1998, Deliverable D3a.

5.5. BIEGSTRAATEN, F. J. W. *Wheel Roughness of the METARAIL Test Train.* TNO, Delft, 1998, Work package 4.

5.6. CATO, D. H. Prediction of environmental noise from fast electric trains. *Journal of Sound and Vibration*, 1976, **46**, 483–500.

5.7. EUROPEAN COMMITTEE FOR STANDARDIZATION. *Railway Applications: Acoustics – Measurements of External Noise Emitted by Railbound Vehicles.* CEN, Brussels, 1995, CEN/TC256 WG3.

5.8. EUROPEAN COMMITTEE FOR STANDARDIZATION. *Railway Applications: Acoustics – Measurements of Internal Noise Emitted by Railbound Vehicles.* CEN, Brussels, 1995, CEN/TC256 WG3.

5.9. DEUTSCHES INSTITUT FÜR NORMUNG. *Berechnung des Lautstärkepegels aus dem Geräuschspektrum. Verfahren nach E. Zwicker.* DIN, Berlin, DIN 45 631.

5.10. DITTRICH, M. G. and JANSSENS, M. H. A. Improved measurement methods for railway rolling noise. *Proceedings of the International Workshop on Railway Noise IWRN.* Ile des Embiez, 1998.

5.11. DITTRICH, M. G., WIRNSBERGER, M., LUB, J., POLLONE, G., KALIVODA, M., VAN BUCHEM, P., HANREICH, W. and FODIMAN, P. *Application of Type Testing Methodologies – Round Robin Testing and Low Noise Solutions. MetaRail Deliverable D11.* TNO, Delft, 1999, Report TPD-HAG-RPT-980066.

5.12. FASTL, H. Beurteilung und Messung der wahrgenommenen äquivalenten Dauerlautheit. *Zeitschrift für Lärmbekämpfung,* 1991, **38**, 98–103.

5.13. FASTL, H., KUWANO, S. and NAMBA, S. Psychoakustische Experimente zum Schienenbonus. *Proceedings of DAGA '94.* Frankfurt, 1994, pp. 1113–1116.

5.14. GIESLER, H.-J. and NOLLE, A. Geräuschemissionen von Schienenfahrzeugen. *Zeitschrift für Lärmbekämpfung,* 1988, **35**, 103–108.

5.15. HOHENWARTER, D. Railway noise propagation models. *Journal of Sound and Vibration,* 1990, **141**, 17–41.

5.16. INTERNATIONALER EISENBAHNVERBAND (ORE). *Lärm im Eisenbahnwesen; Experimentelle Untersuchung zur Identifizierung von Lärmquellen an Güterwagen und deren Laufwerken.* ORE, Utrecht, 1985, Question C 163, Report. 3.

5.17. INTERNATIONAL ORGANIZATION FOR STANDARDIZATION. *Railway Applications/ Acoustics – Measurements of Noise Emitted by Railbound Vehicles.* ISO, Geneva, 1998, draft standard pr EN ISO 3095.

5.18. INTERNATIONAL ORGANIZATION FOR STANDARDIZATION. *Acoustics – Determination of Sound Power Levels of Noise Sources – Engineering Methods for Free-field Conditions over a Reflecting Plane,* 1st edn. ISO, Geneva, 1981, ISO 3744.

5.19. KALIVODA, M. T. Psychoacoustics and noise abatement in Austria – some practical experiences. *Proceedings of the 2nd Convention of the European Acoustic Association: Forum Acusticum.* Berlin, 1999.

5.20. KALIVODA, M. T. and STEINER, J. W. *Taschenbuch der Angewandten Psychoakustik,* 1st edn. Springer, Vienna, 1998.

5.21. KURZE, U. J. and DIEHL, R. J. Characterization of noise from railbound freight transport. *Proceedings of the 2nd International Workshop, Abatement of Railway Noise Emissions – Freight Transport.* Berlin, 1998.

5.22. PETERS, S. The prediction of railway noise profiles. *Journal of Sound and Vibration,* 1974, **32**, 87–99.

5.23. RATHÉ, E. J. Railway noise propagation. *Journal of Sound and Vibration,* 1977, **51**, 371–388.

5.24. SchLV. *Schienenfahrzeug-Lärmzulässigekitsverordnung.* Federal Ministry of Transport, Vienna, Bundesgesetzblatt der Republik Österreich BGBl. 414/1993.

5.25. THOMPSON, D. J., JANSSENS, M. H. A. and DITTRICH, M. G. An assessment of potential measures for reducing the noise emitted by railway traffic. *Proceedings of the International Conference on Noise and Vibration Engineering, ISMA 21.* Leuven, 1996.

5.26. ZWICKER, E. and FASTL, H. *Psychoacoustics – Facts and models,* 2nd edn. Springer, Berlin, 1999.

6. Means of controlling rolling noise at source

C. J. C. Jones and D. J. Thompson
Institute of Sound and Vibration Research, University of Southampton, Highfield, Southampton SO17 1BJ, UK

6.1. Introduction

As was stated in Chapter 1, rolling noise is the dominant source of noise from railway operations at conventional speeds. Aerodynamic sources increase more rapidly with speed but even at 300 km/h they are only of about equal importance to wheel/rail noise. This discussion of the reduction of noise and vibration at source, therefore, concentrates on rolling noise.

In order to reduce rolling noise at source, the theoretical model described in Chapter 1 can be used as a basis. There, it has been seen that vibrations of both the wheel and the track form significant sources of sound. It is therefore imperative, before attempting ad hoc noise reduction, to understand the relative importance of the various sources. An effective reduction in wheel noise will have little impact on the overall noise in situations where the track is the dominant source. More critical still is the need to avoid increasing the contribution of one source by measures taken to reduce another. These aspects can be studied using the theoretical model in combination with well-chosen measurements.

From the structure of the theoretical model (see Fig. 1.3 in Chapter 1), it may be seen that reduction can be sought via a number of routes. This chapter considers, first, various measures that can be applied to wheels and track in turn to reduce their radiated noise. These include added damping, structural modification and the use of vibration isolation. Consideration is then given to the reduction of the excitation due to roughness. The discussion is concluded with a section covering shielding close to the source. Traditional noise barriers and insulation of lineside properties are further measures that can be taken, but these are not considered in this chapter.

Throughout the chapter, results are drawn from research projects in which the authors have been involved. These include the research of Committee C163 of the European Rail Research Institute (ERRI) which led to the development and validation of the TWINS model [6.1, 6.2]. This was followed by the OFWHAT experiment ('Optimized Freight Wheel and Track') [6.3, 6.4] coordinated by ERRI, and then by the European-Union-funded, collaborative projects 'Silent Freight', 'Silent Track' and 'Eurosabot' [6.5–6.7]. In such a research setting it is possible to establish the effects of different measures more precisely than when they are implemented in service. It should be emphasized, however, that many of the techniques referred to, although not all, have passed the research phase and are now well on the way towards implementation in practice.

6.2. Wheel noise

6.2.1. Damping treatments

The low initial damping of wheels appears at first sight to make them very suitable for the addition of damping treatments. However, it should be remembered (see Chapter 1) that the coupling to the rail increases the initial damping considerably, so that, to be effective, additional damping should exceed this 'rolling damping'. The damping experienced by a rolling wheel is much greater than for a wheel suspended freely. Typically the damping ratios rise from around 10^{-4} for a free wheel to greater than 10^{-3} for a wheel rolling on a track, owing to the track acting as a damper to the wheel as waves are transmitted along the rail ('structural radiation damping').

Although this chapter concentrates on rolling noise, it may be pointed out that wheel-damping treatments are very effective in reducing or eliminating the squealing noise that occurs in sharp curves. This phenomenon is tonal and can, on a case by case basis, be linked to one of the wheel resonances. It is excited by a lateral instability at the contact patch caused by the fact that the relation between creep force and creepage (i.e. normalized relative velocity between the wheel and rail) has a negative slope at large values of creepage. The success in treating squeal noise, where a small amount of additional damping is often sufficient, does not indicate that a damping treatment is necessarily effective in reducing the wheel component of rolling noise.

Wheel dampers of various forms have been used in practice. Tuned resonance dampers (Fig. 6.1(a)) and laminated cover plates have both been used successfully in Germany [6.8], and reductions in overall noise of 5–8 dB(A) are reported for train speeds of 200 km/h. The relatively low levels of noise radiated by the ICE high-speed train have also been attributed to the use of such dampers. It should be pointed out that, as these results were obtained are at high speeds and with disc-braked roughness spectra, the wheel can be expected to be dominant in the overall spectrum before the wheel damping treatment is applied. Moreover, the starting point in these tests appears to be a wheel design which is noisy for other reasons. Under these circumstances, reductions in wheel noise can be effective in reducing the overall noise level.

Tests of similar dampers were performed in the Netherlands and France at speeds of 120 km/h. These were much less successful, with overall noise reductions of 1–3 dB(A) being measured [6.9, 6.10]. However, spectral results showed that there

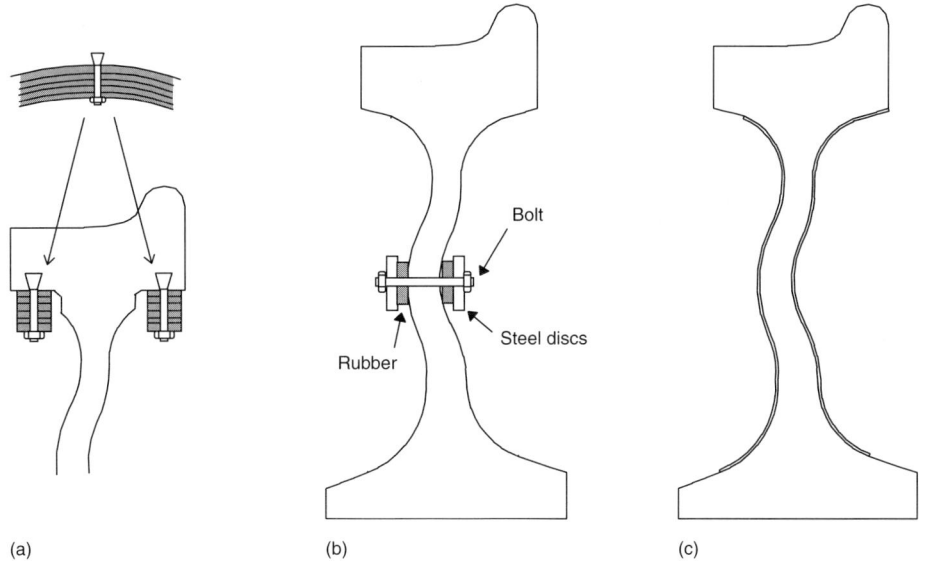

Fig. 6.1. Various wheel-damping devices: (a) tuned resonance devices, (b) tuned absorbers (OFWHAT), (c) constrained-layer damping (thin layer of viscoelastic material backed by thin, stiff constraining layer)

were greater reductions at frequencies above 1.6 kHz, where the wheel noise is expected to be significant. Thus the wheel component may have been reduced considerably, but this is masked by the track component of noise in the situations considered.

In the OFWHAT tests [6.3, 6.4], wheels fitted with 'tuned absorbers' were tested in combination with other measures. In principle such absorbers are mass–spring systems added to a structure to remove energy from particular resonances. By using a system with a high damping loss factor, such an absorber also acts as a source of added damping at frequencies above the mass–spring resonance frequency, where the mass becomes inertially 'grounded'. Two sets of absorbers were fitted to a standard UIC 920 mm freight wheel, tuned to frequencies of 1720 and 2330 Hz, composed of steel discs bolted through holes in the wheel web and supported by rubber pads. These are shown schematically in Fig. 6.1(b). It was found in running measurements that the wheel component of noise was reduced by 4 dB(A) by these absorbers. This was less than predicted using TWINS (6–9 dB), probably owing to the fact that the absorbers did not conform to the performance specification [6.3].

An alternative means of adding damping to the wheel is to use a constrained layer treatment (Fig. 6.1(c)). This system consists of a thin layer of highly damped viscoelastic material sandwiched between the wheel and a stiff constraining plate. Such treatments have been successfully used in the UK on the wheels of tread-braked multiple-unit trains for the last ten years to prevent curve squeal. For tread-braked wheels, constrained-layer damping can only be applied to the wheel web, owing to the high temperatures reached in the tread region during prolonged periods of drag braking. Under these circumstances, even on the web, the materials have to be capable of surviving temperatures of over 200°C.

Calculations of the effects of such damping treatments applied to the web of a wheel [6.11] have shown that constrained-layer treatments can reduce the wheel component of rolling noise by 3–4 dB. These analyses use parameters corresponding to a realistic choice of damping material and allow for the temperature and frequency dependence of the material properties. Predicted results are shown in Fig. 6.2 for two wheel types. These were based on a maximum thickness of the constraining plate of 1 mm steel, as the damping treatment needs to be formed to the shape of the wheel web. To improve on this result a thicker constraining plate would be needed, which would be more appropriate to disc-braked wheels with a straight web. The mass added to the wheel by such treatments is limited to a few kilograms.

6.2.2. Wheel shape optimization

The cross-sectional shape of the wheel can have a significant influence on the noise radiated. Several attempts to develop an 'optimized' wheel shape that produces the minimum noise within practical constraints have been reported. Two early attempts are presented in [6.12]. One was a 'Schalloptimiertesrad' developed by the TU Berlin on the basis of laboratory testing at a scale of 1:4. This was then tested in practice by the German Railways (DB) [6.8]. This wheel, which had no additional damping treatment, was almost as quiet as the damped wheels referred to above, also tested in [6.8]. Compared with the DB Intercity passenger wheel

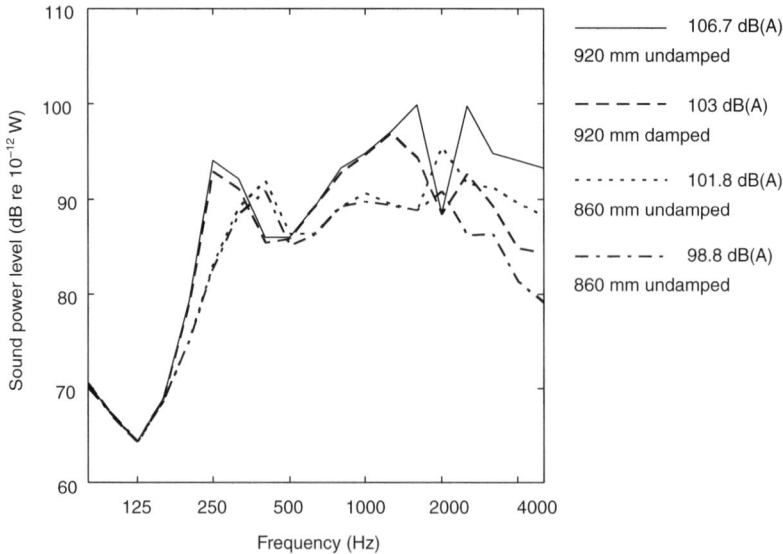

Fig. 6.2. Predicted sound power from a UIC 920 mm freight wheel at 100 km/h and from the ISVR860F wheel with and without constrained-layer damping. The roughness is typical of tread-braked stock

MEANS OF CONTROLLING ROLLING NOISE AT SOURCE

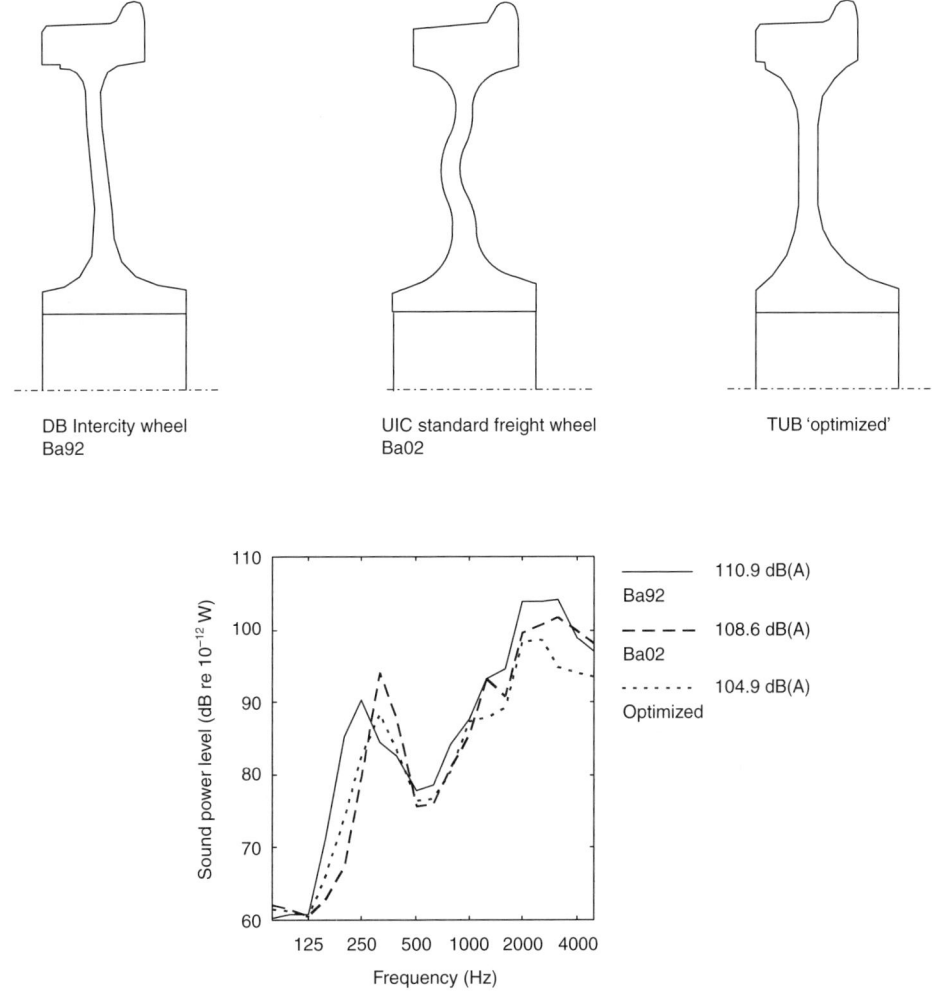

Fig. 6.3. TWINS predictions of the wheel sound power component from three types of wheel, assuming a train speed of 160 km/h and the same roughness spectrum in each case, typical of a disc-braked wheel

(Ba92), it had a thicker, straight web and larger-radius fillets. Figure 6.3 shows predicted results from the TWINS model for these two wheel designs and for a UIC standard freight wheel (Ba02) for comparison. All three wheels have a diameter of 920 mm and the roughness spectrum is assumed to be the same in each case. The predictions are given in terms of the sound power from a single wheel. A difference of up to 6 dB in the wheel component of sound can be seen, particularly at high frequencies.

A theoretical study, also reported in [6.12], used Springboard (the predecessor of TWINS) to show that the noise from a wheel could be reduced by more than 10 dB by a combination of a straight, thick web and a reduction in the wheel diameter from 920 to 740 mm. This wheel design has not been applied in practice, however.

Within the OFWHAT tests, a wheel designed to similar principles but with a diameter of 860 mm was tested. In the experiments, this was found to reduce the wheel component of noise by only 1.5 dB compared with predictions of 4 dB. Another wheel of diameter 640 mm was also tested and found to reduce the wheel component by 18 dB. However, owing to the shorter contact patch, the roughness filtering was lower and the track noise was found to increase by about 2 dB [6.3].

In [6.13], three other optimized wheel shapes are presented, designed for use on a TGV. Each increases the mass of the wheel by about 40 kg compared with the standard design. They produced a reduction of 4–5 dB in the noise above 1600 Hz and 2 dB(A) overall in tests at 300 km/h.

In practice, there are difficulties in applying such optimal wheel shapes as retrofit solutions to existing vehicles. Only a small change in wheel radius can be tolerated within an existing bogie. Moreover, for tread-braked stock, the stresses and deformations caused by thermal loading due to braking have to be kept within strict tolerances. This means that a straight web (as found in the wheels in [6.12, 6.13]) can only be used for disc-braked stock.

Nevertheless, within the Silent Freight project it was shown that it is still possible to reduce the wheel component of noise significantly whilst keeping within these limits. One such 'optimized' wheel shape, known as the ISVR860F, has a diameter of 860 mm, and a curved web of minimum thickness 27 mm. This design was predicted to produce 5 dB(A) less noise than a standard UIC 920 mm freight wheel at 100 km/h, as shown in Fig. 6.2. With constrained-layer damping, the overall benefit compared with the undamped standard wheel is predicted to rise to 8 dB(A).

The development of the 860 mm shape-optimized wheel suitable for tread braking demonstrates the conflict between the requirement for a radially stiff wheel, to obtain wheel natural frequencies as high as possible, and the requirement for a wheel with low residual stresses and low permanent deformation after onerous braking cycles. For reasons of safe operation of freight wagons internationally, the standards for thermal stresses are being set even more tightly than in the draft versions used in the shape optimization. It seems therefore that the scope for an acoustically optimized tread-braked wheel will, in future, be very limited.

6.2.3. Resilient wheels

Resilient wheels have a rubber insert between the tread and the inner part of the wheel. This has the effect of isolating the wheel web from the tread and also introducing additional damping to the wheel. In fact, to be effective in isolation the stiffness has to be quite low, whereas to be effective in damping it should be higher. Although the tragic accident of an ICE at Eschede in Germany in 1998 has cast a shadow over their use at high speed, such resilient wheels are very common in light rail and tramway applications.

In [6.11] a light rail resilient wheel is analysed theoretically for different values of stiffness of the resilient layer. Figure 6.4 shows the components of noise from the wheel and track for various values of the stiffness of the resilient element. It is found that the noise from a very stiff resilient wheel is much less than that from the track,

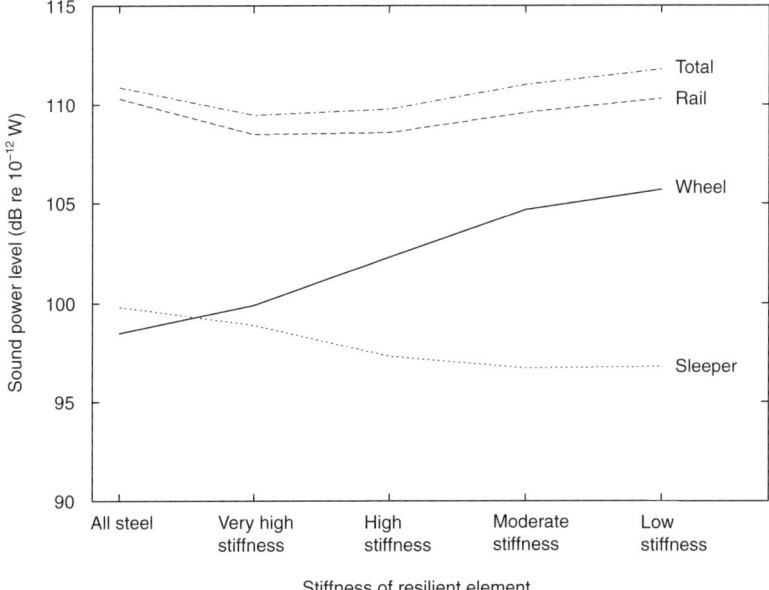

Fig. 6.4. Overall A-weighted sound power levels from resilient wheels and track for various stiffnesses of resilient element. Resilient wheel based on a 740 mm diameter light rail vehicle wheel, typical ballasted track, train speed 100 km/h, roughness from tread-braked wheels [6.11]

since the wheel has a small radius and straight web. The extreme case of this is where the resilient material is replaced by steel in the model, shown as 'all steel' in Fig. 6.4. However, at more typical stiffnesses of the resilient elements, the noise from the wheel is similar to that of the track.

A similar study is reported in [6.14] for a wheel intended for freight application. It is shown there, as in Fig. 6.4, that by appropriate selection of the stiffness, a resilient wheel can reduce the track component of noise by around 2 dB, by changing the balance of wheel and rail vibration caused by the roughness, especially around 500 to 1000 Hz. The wheel component, although increased, is still sufficiently less than the track component at these frequencies for a small overall rolling noise reduction to be predicted for a train speed of 100 km/h.

6.2.4. Reduced wheel radiation

All the above measures aim to reduce the vibration of the wheel and hence its radiated noise. A further technique that is available is to reduce the radiated sound due to a particular vibration level. In [6.15] the concept of a perforated wheel is studied theoretically. The idea is to reduce the radiation efficiency by introducing acoustic short-circuiting between the front and back of the wheel web. The effect depends on the size and spacing of the holes. Owing to the thickness of the web and practical limitations on the distance between the holes and on their size, the perforated wheel is shown to be effective at low frequencies, where a 6–9 dB reduction can be achieved, but no appreciable effect is predicted above about 1 kHz, which is where the wheel noise is usually dominant.

Another means of reducing the radiation is to mount a shield on the wheel so that the web region cannot radiate sound. It is the web that produces most of the noise from the wheel, although the tread, particularly its radial motion, is also significant. Such a shield must be mounted in such a way that it does not vibrate, i.e. it must be resiliently mounted and well damped. It must also be made of a material which gives a high enough transmission loss while still being flexible so that bending waves excited in the panel are short compared with the wavelength of sound in air and therefore have poor sound radiation. Good sealing is also required so that no sound can escape through gaps. Such a wheel was constructed and tested within the Silent Freight project [6.5].

6.3. Track noise
6.3.1. Rail pad stiffness

The noise radiated by the track is strongly related to the stiffness of the rail fastening, in particular the rail pad between the rail and the sleeper. Soft pads cause the rail to become uncoupled from the sleeper, which minimizes noise from the sleeper but allows the rail to vibrate more freely. Waves can therefore travel over a greater distance and the noise from the rail is increased. Conversely, with stiff pads the contribution from the rail is reduced but that from the sleeper is increased, as shown in Fig. 6.5. A compromise can be reached when the sleeper contribution is equal to the combined vertical and lateral rail component [6.16]. For the situation modelled here, this occurs at a pad stiffness of 2500 MN/m which is typical of a traditional stiff 5 mm thick pad but is very stiff by current standards, and unacceptable for new installations. The pad stiffnesses referred to here, and shown in Fig. 6.5, are the high-frequency, low-amplitude tangent stiffnesses, not the static or low-frequency, large-amplitude secant stiffnesses often used to evaluate rail pads.

These theoretical results have been confirmed in practice in the OFWHAT tests, where a fairly soft reference pad was compared with two 'optimum'-stiffness pads. A difference of 4–5 dB in the track component of noise was measured. Attempts to double the damping loss factor of the pad, which should produce a further 2 dB reduction in track noise, were less successful.

Tests carried out at Southampton [6.17] have shown that the value of pad stiffness obtained from a test on a single assembly is not necessarily the same as the stiffness required in the model to give a good fit to point mobility or decay rate measurements in track. The latter is usually higher. Nevertheless, Fig. 6.6 shows measured decay rates in the track for three different pads with comparisons with predictions which confirm the dependence on pad stiffness.

To reduce track forces and damage to sleepers and track components, it has become the practice to install soft rail pads. According to the prediction model, track with soft rail pads should be considerably noisier than track with stiff rail pads for the same roughness, see Fig. 6.5. For example, a track with a pad of stiffness 80 MN/m should be about 8 dB noisier than one with a stiffness of 800 MN/m.

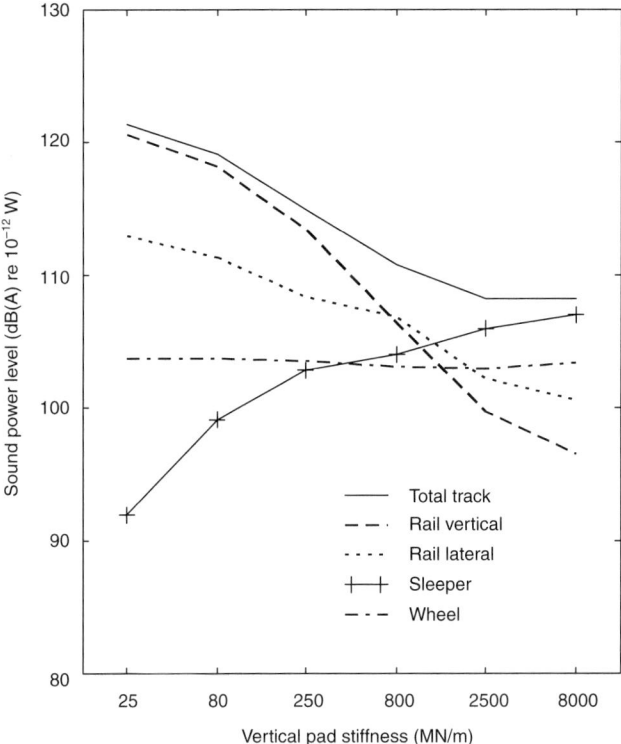

Fig. 6.5. Example of predicted sound power due to one wheel and the associated track vibration versus high-frequency rail pad stiffness. Calculations using TWINS for a standard 920 mm freight wheel at 100 km/h, with a typical tread-braked roughness

However, experimental evidence suggests that this is not the case. Part of the reason seems to be that the decay rates of the unloaded track agree with the predictions when a stiffness of 140 MN/m is used (see Fig. 6.6, solid line), even though laboratory measurements of pad stiffness indicate 80 MN/m as the correct value. Using 140 MN/m in the predictions reduces the predicted noise for this track by about 2 dB compared with a stiffness of 80 MN/m (see Fig. 6.5).

Moreover, the soft pads in question are studded rubber pads, which have a strongly load-dependent stiffness in order to prevent rail roll-over. The effect on noise of this load dependence has therefore been studied. In [6.18] an analysis is given of the effect of the preload on the point receptance of the track. This is affected significantly by the preload, but the decay rate is affected much less. However, by including the effects of the preloads exerted by additional wheels [6.17], it has been shown that the noise predicted using the unloaded stiffness is about 4 dB too great. The corresponding difference for a stiff pad is 1 dB. Altogether, therefore, this particular soft pad, which has a stiffness of 80 MN/m as measured on a single assembly, should give an increase in track noise of only about 3 dB compared with a pad of stiffness 800 MN/m.

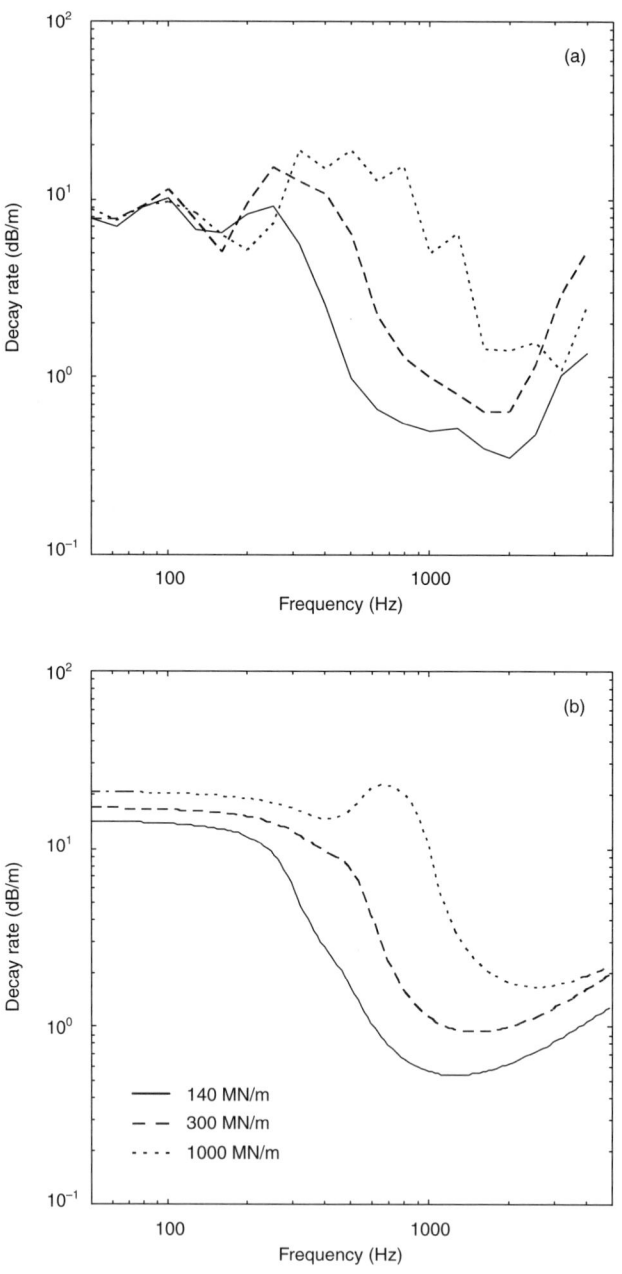

Fig. 6.6. Decay rate of vertical vibration along the track for three different rail pads: (a) measured, (b) predicted using continuously supported Timoshenko beam model for different pad stiffnesses [6.17]

6.3.2. Damping treatments

Whilst the rail pad stiffness affects the balance between the rail and sleeper components of noise, adding damping treatments to the rail has the advantage of producing high decay rates in the rail without adversely affecting the isolation of the sleeper. If used in combination with a soft pad, therefore, added rail damping can achieve significant noise reductions.

The problem in adding damping to the rail is that the rail is already highly damped. The decay rates presented in Fig. 6.6 correspond to an equivalent damping loss factor of the rail of around 0.02 at high frequencies, and greater at low frequencies. This is many times larger than the damping of the wheel (the damping loss factor is approximately twice the damping ratio), even allowing for the additional damping present during rolling.

Adding constrained-layer damping to the rail has been studied numerically and found to give only a very small benefit. A 1 mm layer of damping material and a 2 mm constraining layer were studied on either the foot or the web. Even on a track with a pad stiffness of 80 MN/m, the track noise was predicted to be reduced by only about 2 dB(A), whilst for stiffer pads, the reduction would be negligible.

Some commercial systems are available which have the form of a constrained layer treatment. However, these have much thicker layers of both viscoelastic material and constraining plate. Analysis of their performance indicates that they operate on the principle of a tuned absorber, a damped mass–spring system attached to the rail.

Indeed, tuned-absorber systems seem to be a much better alternative for application to rails. In the OFWHAT tests a prototype absorber was tested which was mounted on the rail foot. This produced a measured reduction in the track noise component of 2 dB(A) for a train speed of 100 km/h when mounted on a track with optimum-stiffness pads. Unfortunately, the combination of absorbers with soft pads was not tested, but this could be expected to produce a considerably greater benefit.

Within the Silent Track project, a novel absorber system has been developed. Decay rates measured on a 4 m length of free UIC60 rail fitted with this absorber are presented in Fig. 6.7. These decay rates are very high at high frequencies but quite small at low frequencies. This is in part a consequence of the design, since the tuning frequencies are set to correspond to the broad peak in the rail noise spectrum between 500 and 2000 Hz. However, these decay rate values should be added to the decay rates already present in the track to give an overall decay rate for a damped track. Comparing these results with Fig. 6.6, it can be seen that the damping of vertical vibration would be increased by the absorber for the soft pad above about 500 Hz, and for the intermediate pad above about 800 Hz.

For a track with a pad stiffness of 300 MN/m (i.e. equivalent to the intermediate pad of Fig. 6.6), this system is predicted to reduce the track component of noise by about 6 dB(A). The reduction will be less for stiffer pads and greater for softer pads. Because the track decay rates of a damped track are largely determined by the damping from the absorbers, the noise level for a track with absorbers is found to be almost independent of pad stiffness.

6.3.3. Rail shape optimization

A reduction in the size of the rail could lead to significant reductions in its sound radiation. This occurs by a combination of reduced radiating area and reduced radiation ratio. Figure 6.8 shows the radiation ratio for two notional modified rail sections as well as for the standard UIC60 section. One is a rail with a reduced height (107 mm instead of 172 mm) achieved by shortening the web; the other, labelled 'bullhead', has the same height as UIC60 but has a 'head' at the bottom as well as the top, so that it is 70 mm wide instead of 150 mm. The radiation efficiency curves for a standard rail peak between 500 and 1000 Hz, above which they tend to 1. At lower frequencies they fall at a rate proportional to the cube of the frequency. By reducing the size of the rail cross-section this frequency can be raised, as shown in Fig. 6.8. Reducing the height can lower the radiation from lateral motion, while reducing the width can minimize the vertical component of sound.

In practice, the scope for reducing the dimensions of the rail is limited. Nevertheless, this concept led to the development by British Steel and British Rail Research of a low-height rail known as 'hush rail', which is 110 mm high with a weight of 50 kg/m. This was tested in practice in 1991, although the tests were inconclusive. The rails were installed on wooden sleepers for practical reasons, and this led to high levels of noise from the sleepers, which unfortunately masked the effect of the low-height rail [6.19].

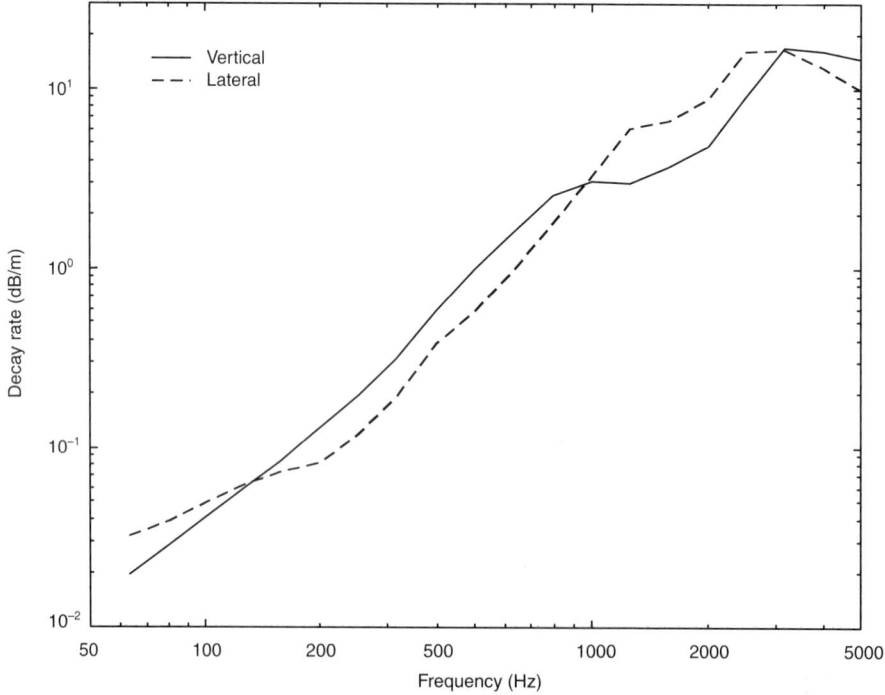

Fig. 6.7. Measured decay rates on a 4 m length of free rail fitted with the Silent Track tuned-absorber system

MEANS OF CONTROLLING ROLLING NOISE AT SOURCE

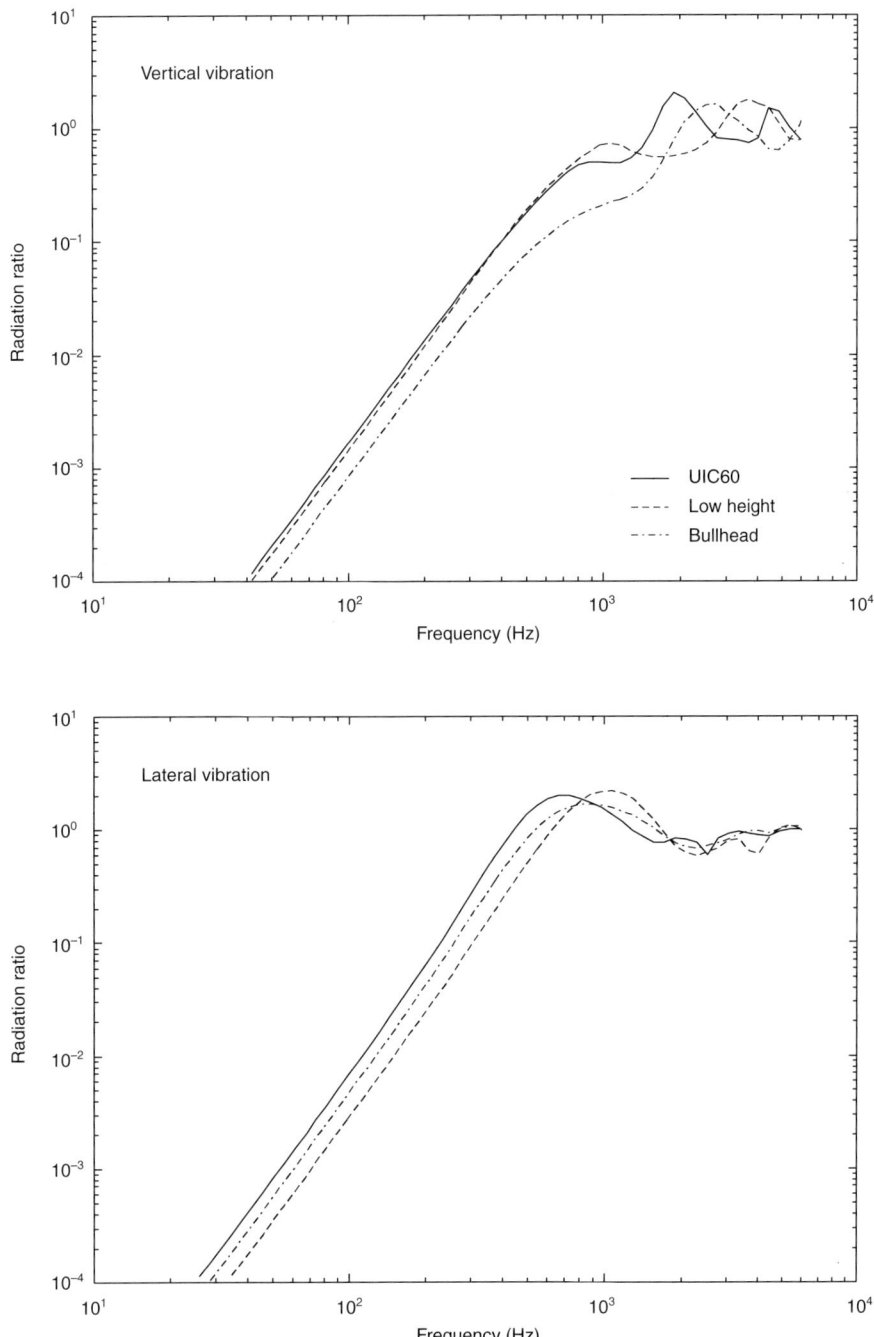

Fig. 6.8. Radiation ratio of rails of various dimensions vibrating uniformly vertically or laterally, calculated using boundary elements

Where the dominant source of noise from the rail is its vertical vibration, as is the case for low pad stiffnesses (see Fig. 6.5), the dimension which should be minimized is the width rather than the height. Gains of up to 4 dB(A) can be expected from a rail with a narrower foot [6.20], although this requires a modified rail support system. Such a solution has been tested within the Silent Track project.

6.3.4. Track mobility

The dynamic response of the track, as characterized by its mobility (velocity/force), has an effect on the interaction of the wheel and track. As seen in Chapter 1, the component with the highest mobility at a particular frequency is caused to vibrate with the amplitude of the roughness. Between about 100 and 1000 Hz the track mostly has a higher mobility than either the wheel or the contact spring, as shown in Fig. 6.9. It therefore vibrates with an amplitude close to that of the roughness. Above 1000 Hz the contact spring is more flexible than the rail and, apart from near to wheel resonances, the contact spring absorbs most of the roughness excitation.

Changing the track mobility has no effect at all in the region where it has a greater mobility than the wheel or the contact spring. Only if the mobility can be brought down well below that of those elements is there any effect. At high frequencies, where the rail mobility drops below that of the contact spring, reducing the rail mobility has a beneficial effect. However, in this region above 1 kHz, the rail

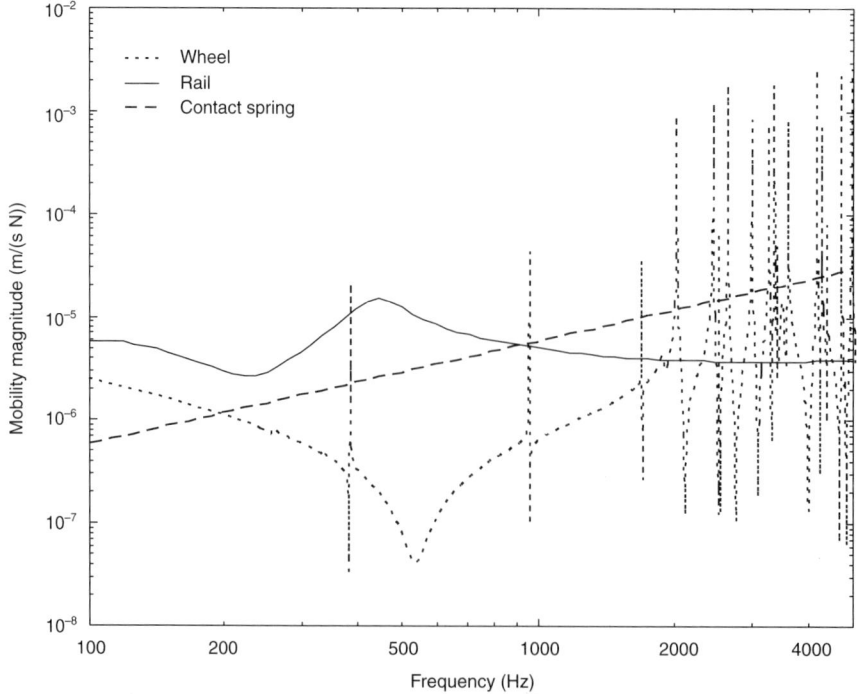

Fig. 6.9. Predicted vertical point mobilities of wheel, rail and contact spring. The wheel is from a TGV, and the track represents UIC60 rail on twin-block sleepers with moderately soft rail pads

mobility is influenced only by the mass and bending stiffness of the rail itself, not by the support structure. Increasing the size of the rail to reduce its mobility here would have negative effects in terms of the sound radiation (see Section 6.3.3). In any case, the rail is not the dominant source at these higher frequencies.

6.3.5. Ballastless track forms

Ballastless track, or 'slab track', is growing in popularity in some countries for high-speed lines. In such a track, the rail is attached by resilient fasteners to a concrete base, typically 0.5 m thick. This gives a track with lower maintenance costs, since the track vertical and lateral profiles do not need to be maintained by tamping the ballast at regular intervals.

Unfortunately, slab track has a reputation for being rather noisy. Increases in level of between 2 and 4 dB(A) are typically found. The reason for this is often given as the fact that the acoustic absorption offered by the ballast has been removed. However, this is likely to explain only about 1 dB of the difference. A more important effect lies in the fact that slab track has softer rail fasteners. The aim of these is to substitute the vertical compliance normally provided by the ballast layer with an extra compliance in the rail support. As seen above, soft rail pads can lead to higher rail noise as the rail can vibrate over a greater length.

The concrete slab has such a high impedance that its vibration is much less than that of the rail, and its noise radiation is thus also negligible. Therefore the stiffness of the rail support can be increased considerably without this component of noise becoming significant compared with the rail noise. Use of a stiffer support will increase the attenuation of vibration along the rail. An approach to noise reduction on a slab track is to support the rail continuously by embedding it in a viscoelastic material within a channel in the slab. This makes it possible to use a much smaller rail section, as the track bending stiffness can be partly provided by the slab. Using a small rail allows the radiating surface area of the rail and embedding material to be minimized, thus reducing the sound radiation. This geometry is shown in Fig. 6.10(a). According to [6.21] these two measures are found to turn the usual noise penalty of about 2 dB(A) of a slab track into a benefit of 4–5 dB(A) in track noise relative to ballasted track, see Fig. 6.10(b).

6.4. Roughness

6.4.1. Effects of braking system

It has been known since the late 1970s that if cast-iron block brakes are replaced by disc brakes this can lead to reductions in noise levels on good track of up to 10 dB [6.22]. This is related to the presence of corrugations on the running surface of tread braked wheels with cast-iron blocks. More recently, Dings and Dittrich [6.23] surveyed the roughness of many Dutch wheels and rails and confirmed this conclusion. They showed, moreover, that the roughness of wheels with disc brakes and supplementary cast-iron tread brakes could be greater than that of purely tread-braked wheels; see Fig. 1.11.

The use of disc brakes and the elimination of supplementary block brakes on high-speed trains, such as the ICE and the latest-generation TGV, is largely responsible for the fact that these trains are no noisier at 300 km/h than traditional (tread-braked) stock at conventional speeds of 140–160 km/h; see Fig. 1.1.

Alternative brake block materials also seem to offer a solution, which could be less costly than disc brakes. However, although the sinter block materials tested in [6.23] were found to produce very smooth wheel surfaces, the noise was not reduced in proportion. This is probably due to hollow wear of the wheel profiles, leading to greater noise generation. The sintered material also has the disadvantage that excessive amounts of copper are emitted into the environment. Research is continuing into new brake block materials [6.7].

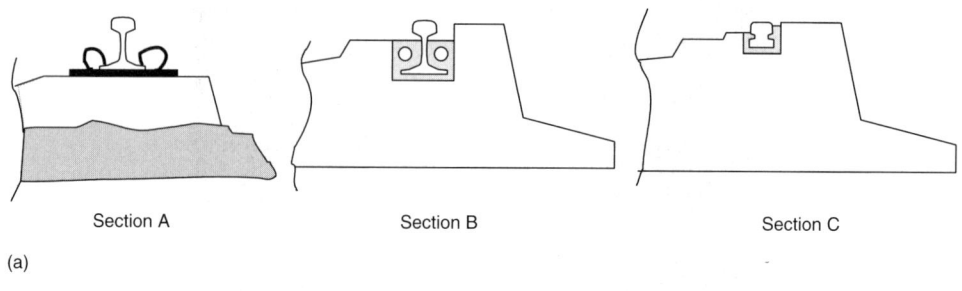

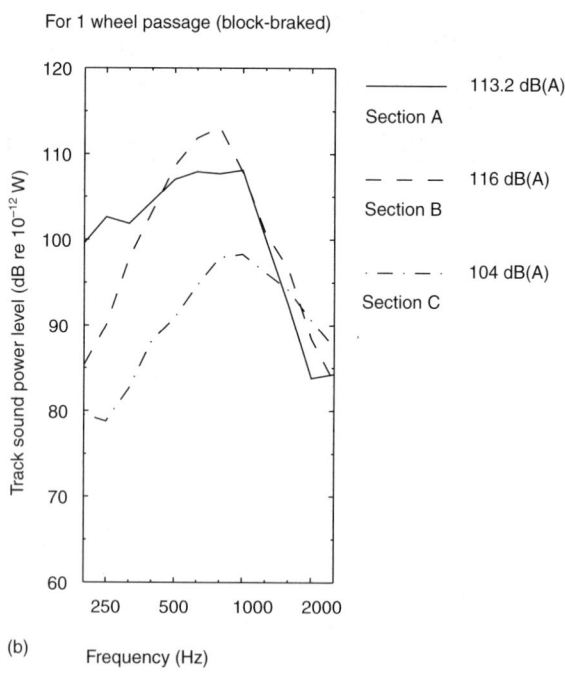

Fig. 6.10. Low-noise slab track: (a) cross-section of the novel track form; (b) total track sound powers based on measurements. Section A, ballasted track; section B, conventional slab track with embedded rail; section C, novel slab track [6.21]. Used with permission (copyright TNO)

Fig. 6.11. Photograph of rail corrugations

6.4.2. Rail corrugation

When corrugations form on the rail head, the noise from both tread-braked and disc-braked stock rises to a level that can be 10 dB higher than the normal level for tread-braked stock [6.22]. As a result of this, the phenomenon of rail corrugations is known to many as 'roaring rail'. An example of a corrugated rail is shown in Fig. 6.11. Wavelengths of around 5 cm and peak-to-trough amplitudes of up to 0.1 mm (more for longer wavelengths) are common. The normal remedy is to grind the rail, although until recently this was done only for reasons of preventing rail defects and fatigue cracks, not for acoustic reasons.

It is, of course, preferable to understand the reason for the formation and growth of corrugations and other severe roughness. In this area, much research has been carried out, e.g. [6.24–6.26], and a full discussion of rail corrugation and roughness growth in general is beyond the scope of this chapter. It is a complex phenomenon involving the metallurgical as well as the dynamic properties of the rail, but there are good indications that reducing the rail pad stiffness can reduce the chances of the formation of short-wavelength corrugation [6.24]. In particular, the antiresonance dip found at the pinned–pinned frequency around 1 kHz for excitation above a sleeper induces high dynamic forces and has been implicated in the formation of corrugations. This dip, and the corresponding peak at midspan, can be reduced by the use of a softer rail pad.

6.4.3. Changes to the contact zone

The wheel/rail contact zone can affect noise production in two ways. First, the Hertzian contact spring introduces resilience between the wheel and rail. At high frequencies this absorbs some of the roughness excitation. If the contact spring can be made softer, this effect can be enhanced and the noise from the track can be reduced. The effect on overall noise of softening the contact spring is shown in Fig. 6.12 [6.27]. Unfortunately, such changes are very difficult to achieve. Large changes to the wheel and rail profiles lead to only marginal differences in contact stiffness. Only very radical changes such as the use of different materials or a resiliently mounted tread [6.28] are likely to achieve these changes, and they probably have too many associated risks for practical application.

6.5. Shielding

The final area of noise reduction that is possible close to the source is to reduce the sound transmission to the receiver by introducing shielding in the form of vehicle-mounted 'shrouds' and track-mounted low barriers. Such a configuration is shown schematically in Fig. 6.13. This has been considered in theoretical and practical studies in a number of countries. A successful demonstration project carried out in the UK is reported in [6.29]. Reductions of 8–10 dB(A) were found for combinations of low barriers mounted close to the rail and bogie-mounted shrouds. The use of either low barriers or vehicle-mounted shrouds in isolation is expected to lead to much smaller reductions in noise [6.30].

Modelling of the performance of such systems can be achieved using a combination of statistical energy analysis and boundary element predictions [6.30, 6.31]. Not surprisingly, one of the most critical parameters of such systems is the size of the gap between the top of the barrier and the bottom of the shield. Where the design has to be able to accommodate the different structure and vehicle gauging constraints of several countries, the benefit is expected to be limited to around 5 dB(A). Another limiting factor is that, when these systems are used on tracks with a low decay rate of vibration, the contribution of the part of the rail not contained within the bogie shield is large enough to limit severely the overall noise reduction.

6.6. Measures in combination

A number of measures have been described in this chapter which can reduce one or more of the components of noise from the wheel/track system. Several solutions for wheel noise have been commercially available for over a decade; successful solutions for track noise have only more recently reached the prototype stage.

None of the noise reduction technologies mentioned in this chapter can achieve reductions of 10 dB or more in overall noise by themselves. To obtain such reductions, it is necessary to use a combination of different measures. The first step is a correct identification of the relative importance of the various sources in the initial situation. Wheel and track components are often similar in their contribution to the overall sound. To achieve a reduction of the order of 10 dB, a large reduction in both track noise and wheel noise is required, usually meaning that separate treatments of each source are required.

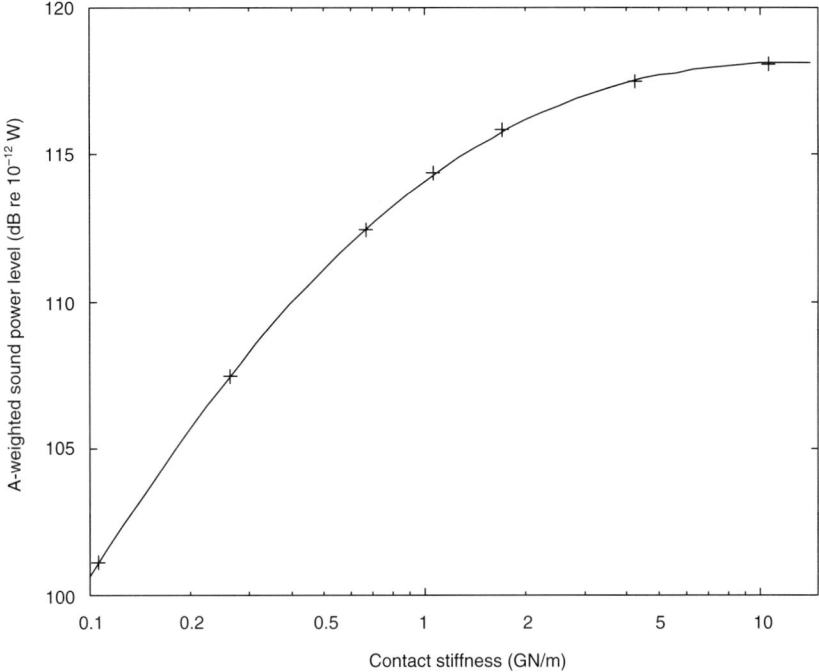

Fig. 6.12. Effect of change in contact stiffness on overall noise radiation

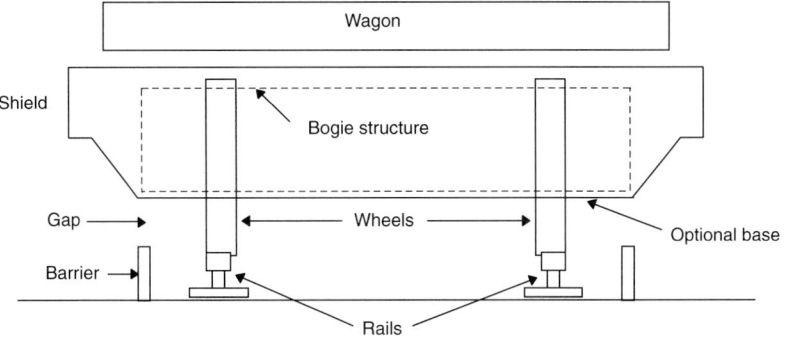

Fig. 6.13. The principle of bogie shields and low, close barriers. The gap should be as small as possible, and absorbent material should be included on the inside faces of the shield

6.7. Summary

The development of theoretical models has allowed attempts at noise reduction to change in recent years from traditional empirical methods to a more scientifically based approach. It is clear that substantial reductions in railway noise are achievable, but only if the relative contributions of all sources are properly quantified and suitable combinations of measures are developed and applied. The various measures currently being tested at the prototype stage need to be industrialized and refined. Increased understanding of the growth of roughness remains a priority for further research.

6.8. References

6.1. THOMPSON, D. J., HEMSWORTH, B. and VINCENT, N. Experimental validation of the TWINS prediction program for rolling noise, part 1: description of the model and method. *Journal of Sound and Vibration*, 1996, **193**, 123–135.

6.2. THOMPSON, D. J., FODIMAN, P. and MAHÉ, H. Experimental validation of the TWINS prediction program for rolling noise, part 2: results. *Journal of Sound and Vibration*, 1996, **193**, 137–147.

6.3. JONES, C. J. C. and EDWARDS, J. W. Development of wheels and track components for reduced rolling noise from freight trains. *Proceedings of Internoise*. Liverpool, 1996, pp. 403–408.

6.4. FODIMAN, P. Line test validation of low noise railway components. *Proceedings of World Congress on Railway Research*. Colorado Springs, Colorado, 1996, pp. 497–502.

6.5. JONES, R. R. K. The Silent Freight project. *Second International Workshop on Abatement of Railway Noise Emissions – Freight Transport*. Berlin, 1998.

6.6. CASTEL, L. and BIRD, W. The Silent Track project. *Second International Workshop on Abatement of Railway Noise Emissions – Freight Transport*. Berlin, 1998.

6.7. HEMSWORTH, B. and BIRD, W. European Union sponsored railway noise research projects. *Second International Workshop on Abatement of Railway Noise Emissions – Freight Transport*. Berlin, 1998.

6.8. HÖLZL, G. A quiet railway by noise-optimized wheels. *Zeitschrift für Eisenbahnwesen und Verkehrstechnik und Die Eisenbahntechnik Glasers Annalen*, 1994, **188**(1), 20–23 [in German].

6.9. HEMELRIJK, R. The effect of wheel dampers on the sound emission of trains [in Dutch]. *Proceedings of the Dutch Acoustical Society (NAG)*, 1983, **67**, 27–37.

6.10. ORE. *Reduction of Running Noise by Wheel-mounted Devices. Question C163 Railway Noise*. ORE, Utrecht, 1989, Report RP13.

6.11. JONES, C. J. C. and THOMPSON, D. J. Rolling noise generated by wheels with visco-elastic layers. *Journal of Sound and Vibration*, 2000, **231**(3), 779–790.

6.12. GAUTIER, P. E., VINCENT, N., THOMPSON, D. J. and HÖLZL, G. Railway wheel optimization. *Proceedings of Inter Noise*. Leuven, 1993, pp. 1455–1458.

6.13. FODIMAN, P., CASTEL, L. and GABORIT, G. Validation of a TGV-A trailing-car wheel with an acoustically optimised profile. *Proceedings of the 11th International Wheelset Congress*. Paris, 1995.

6.14. BOUVET, P., VINCENT, N., COBLENTZ, A. and DEMILLY, F. Optimisation of resilient wheel for rolling noise control. *Proceedings of the 6th International Workshop on Railway and Tracked Transit System Noise*. Ile des Embiez, France, 1998, pp. 264–273.

6.15. DANERYD, A., NIELSEN, J., LUNDBERG, E. and FRID, A. On vibro-acoustic and mechanical properties of a perforated railway wheel. *Proceedings of the 6th International Workshop on Railway and Tracked Transit System Noise*. Ile des Embiez, France, 1998, pp. 305–317.

6.16. VINCENT, N., BOUVET, P., THOMPSON, D. J. and GAUTIER, P. E. Theoretical optimization of track components to reduce rolling noise. *Journal of Sound and Vibration*, 1996, **193**, 161–171.
6.17. THOMPSON, D. J., JONES, C. J. C., WU, T. X. and DE FRANCE, G. The influence of the non-linear stiffness behaviour of rail pads on the track component of rolling noise. *Proceedings of the Institution of Mechanical Engineers, Journal of Rail and Rapid Transit*, 1999, **213F**, 233–241.
6.18. WU, T. X. and THOMPSON, D. J. The effects of local preload on the foundation stiffness and vertical vibration of railway track. *Journal of Sound and Vibration*, 1999, **219**, 881–904.
6.19. JONES, C. J. C. Reduction of noise and ground vibration from freight trains. *Institution of Mechanical Engineers International Conference on Better Journey Times – Better Business*. Birmingham, 1996, pp. 87–97.
6.20. VINCENT, N. Rolling noise control at source – state of the art survey. *Proceedings of the 6th International Workshop on Railway and Tracked Transit System Noise*. Ile des Embiez, France, 1998, pp. 372–387.
6.21. JANSSENS, M. H. A. Low noise slab-track design: acoustic development and final tests. *Proceedings of the 6th International Congress on Sound and Vibration*. Lyngby, Denmark, 1999, pp. 2643–2652.
6.22. HEMSWORTH, B. Recent developments in wheel/rail noise research. *Journal of Sound and Vibration*, 1979, **66**, 297–310.
6.23. DINGS, P. C. and DITTRICH, M. G. Roughness on Dutch railway wheels and rails. *Journal of Sound and Vibration*, 1996, **193**, 103–112.
6.24. HEMPELMANN, K. *Short Pitch Corrugation on Railway Rails – a Linear Model for Prediction*. VDI, Düsseldorf, 1994, Fortschrittberichte, Series 12, No. 231.
6.25. IGELAND, A. Dynamic train/track interaction: simulation of railhead corrugation growth under a moving bogie using mathematical models combined with full-scale measurements. Doctoral thesis, Chalmers University of Technology, Göteborg, Sweden, 1997.
6.26. GRASSIE, S. L. and KALOUSEK, J. Rail corrugation: characteristics, causes and treatments. *Proceedings of the Institution of Mechanical Engineers, Part F, Journal of Rail and Rapid Transit*, 1993, **207**, 57–68.
6.27. THOMPSON, D. J. and REMINGTON, P. J. The effects of transverse profile on the excitation of wheel/rail noise. *Journal of Sound and Vibration*, 2000, **231**(3), 537–548.
6.28. REMINGTON, P. J. The resiliently treaded wheel – a concept for control of wheel/rail rolling noise. *Proceedings of Internoise*. Avignon, France, 1988, pp. 1401–1406.
6.29. JONES, R. R. K. Bogie shrouds and low barriers could significantly reduce wheel/rail noise. *Railway Gazette International*, 1994, July, 459–462.
6.30. JONES, C. J. C. and THOMPSON, D. J. Application of numerical models to a system of train and track-mounted acoustic shields. *Proceedings of the 6th International Congress on Acoustics and Vibration*, Lyngby. Denmark, 1999, pp. 2661–2668.
6.31. JONES, C. J. C., HARDY, A. E. J., JONES, R. R. K. and WANG, A. Bogie shrouds and low trackside barriers for the control of railway vehicle rolling noise. *Journal of Sound and Vibration*, 1996, **193**, 427–431.

Part 3
Bursting noise associated with non-linear pressure waves in tunnels

7. Micropressure waves radiating from a Shinkansen tunnel portal

T. Maeda
Environmental Engineering Division, Railway Technical Research Institute, 2-8-38 Hikari-cho Kokubunji-shi, Tokyo 185-8540, Japan

7.1. Introduction

In 1975, trial runs on the newly extended Okayama–Hakata section of the San'yo Shinkansen line resulted in protests from residents living along the railway line because an explosive sound was generated and windows of houses were rattled near the portal of a long tunnel with slab track when a Shinkansen train entered the tunnel. This environmental problem had not occurred near the tunnel with ballasted track on the Tokyo–Okayama section of the Tokaido and San'yo Shinkansen lines [7.1]. This phenomenon is caused by pressure waves radiating from the tunnel portal when a compression wave, generated by the train entering the tunnel, is propagated through the tunnel at the speed of sound and reaches the tunnel portal. At the tunnel exit, the greater part of the compression wave is reflected as an expansion wave, which causes ear discomfort to passengers in the train, while a part of the compression wave radiates out of the tunnel portal as an acoustic pressure pulse, which causes an environmental problem; namely, it generates an explosive sound and strikes the windows and doors of houses with a noisy thud near the tunnel portal. This acoustic pressure pulse is often called a micropressure wave to emphasize its origin, or a tunnel sonic boom.* In Japan, micropressure waves are considered to constitute a low-frequency problem, because they are impulsive waves and include low-frequency components. Japan still has no environmental quality standards or guidelines covering these micropressure waves; however more and more people are calling for measures to protect the environment. The train speeds cannot be raised without assessing the negative impact on the environment and taking countermeasures against the environmental problems.

---

*The use of the phrase 'sonic boom' for this phenomenon does not mean that the train speed is higher than the speed of sound – *Ed*.

The phenomenon of microvpressure waves has three phases: generation of the compression wave at the tunnel portal which the train enters, propagation of the compression wave through the tunnel and radiation of the micropressure waves from the tunnel portal (Fig. 7.1). In this chapter, hereafter, the portal which the train enters will be called the tunnel entrance, and the portal out of which the micropressure waves radiate will be called the tunnel exit.

The peak value of the acoustic micropressure wave is nearly proportional to the pressure gradient of the wavefront of the compression wave arriving at the tunnel exit and is inversely proportional to the distance from the tunnel exit. The pressure gradient of the wavefront of the compression wave at the tunnel entrance is nearly proportional to the third power of the train speed, and depends on the ratio of the cross-sectional area of the train to that of the tunnel and on the shapes of the train nose and the tunnel entrance. In a short tunnel, as the wavefront scarcely changes during propagation irrespective of the type of track, the peak value of the micropressure wave is proportional to the third power of the train speed. In a long tunnel, the wavefront of the compression wave changes during propagation. The wavefront steepens in a tunnel with slab track owing to the non-linear effects accompanying the propagation of the compression wave, while the wavefront attenuates in a tunnel with ballasted track because the effect of the ballast is more dominant than the non-linear effects. Consequently, the peak value of the micropressure wave in a long tunnel with slab track has a dependence exceeding the third power of the train speed; however, the peak value of the micropressure wave in a long tunnel with ballasted track becomes smaller than it would be in a short tunnel [7.2].

The principle of the countermeasures for mitigating micropressure waves is to reduce the pressure gradient of the wavefront of the compression wave arriving at the tunnel exit. The methods applied to Shinkansen tunnels are as follows: (1) the installation of a tunnel hood at the tunnel entrance, (2) the use of side branches in the tunnel and (3) the installation of a shelter with slits between two adjacent tunnels. The methods applied to the Shinkansen train are as follows: (1) the reduction of the cross-sectional area of the train and (2) the optimization of the train nose.

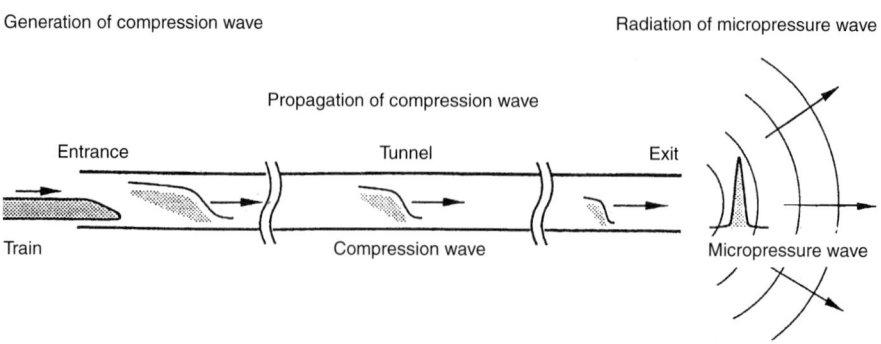

Fig. 7.1. Phenomenon of generation of micropressure waves

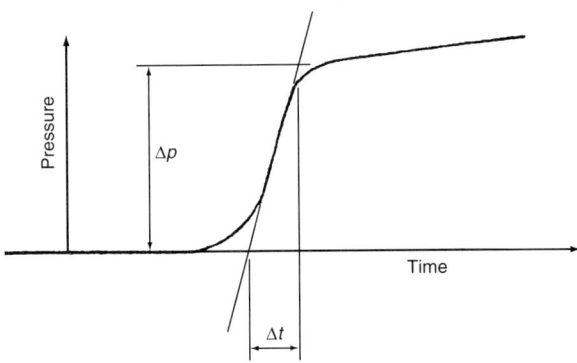

Fig. 7.2. Wavefront of a compression wave generated by a train entering a tunnel

Section 7.2 describes the generation of a compression wave by a train entering a tunnel. Section 7.3 is dedicated to the propagation of the compression wave through the tunnel. Section 7.4 discusses the radiation of the micropressure wave out of the tunnel portal. The measures to decrease the amplitude of the micropressure waves generated will be described in detail in Section 7.5.

7.2. Generation of a compression wave by a train

The waveform of the compression wave generated by a train entering a tunnel depends on the cross-sectional area of the train, the cross-sectional area of the tunnel, the shape of the train nose, the shape of the tunnel entrance and the train speed (Fig. 7.2). The pressure rise Δp at the wavefront is expressed by the following equation:

$$\Delta p = \frac{1}{2} p_0 V^2 \frac{1-(1-R)^2}{(1-V/c_0)[V/c_0 + (1-R)^2]} \tag{7.1}$$

derived by Hara [7.3]. Here V is the train speed, R is the ratio of the cross-sectional area of the train to that of the tunnel, c is the sound speed in air, and ρ is the density of air. The pressure rise Δp depends on the train speed V and the ratio of the cross-sectional area of the train to that of the tunnel, R. The time interval Δt for the pressure rise is inversely proportional to the train speed V and depends on the shapes of the train and the tunnel entrance. Accordingly, the maximum pressure gradient of the wavefront of the compression wave is given by the following equation:

$$\left(\frac{\Delta p}{\Delta t}\right)_{MAX} = \frac{1}{2} \frac{p_0 V^2}{\kappa d} \frac{1-(1-R)^2}{(1-V/c_0)[V/c_0 + (1-R)^2]} \tag{7.2}$$

where d is the tunnel hydraulic diameter of the tunnel, and κ is a coefficient indicating the effect of the shapes of the train nose and the tunnel entrance on Δt. Equation (7.2) shows that the maximum pressure gradient of the wavefront $(\Delta p/\Delta t)_{MAX}$ reduces as the ratio R becomes smaller and the coefficient κ increases, namely, the nose becomes longer.

To validate the equation (7.2) and determine the value of κ, model experiments were performed. Figure 7.3 shows the equipment for the model experiments. The model train and the model tunnel are axisymmetric. The model scale is 1/60. The maximum model train speed is about 400 km/h. The parameters of the similarity principle for the generation of the compression wave are the Mach number and the Reynolds number, however, the effect of the Reynolds number can be neglected because the flow around the streamlined noses of modern high-speed trains does not separate. If the model train speed is the same as the actual train speed, the actual waveform of the compression wave can be estimated by transforming the timescale of the waveform in the model experiment by the model scale. Figure 7.4 shows the results of the model experiments for the effects of the

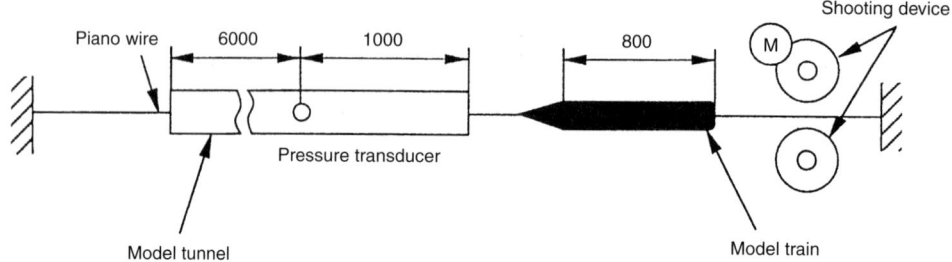

Fig. 7.3. *Equipment for model experiments at scale 1/60 (dimensions in mm)*

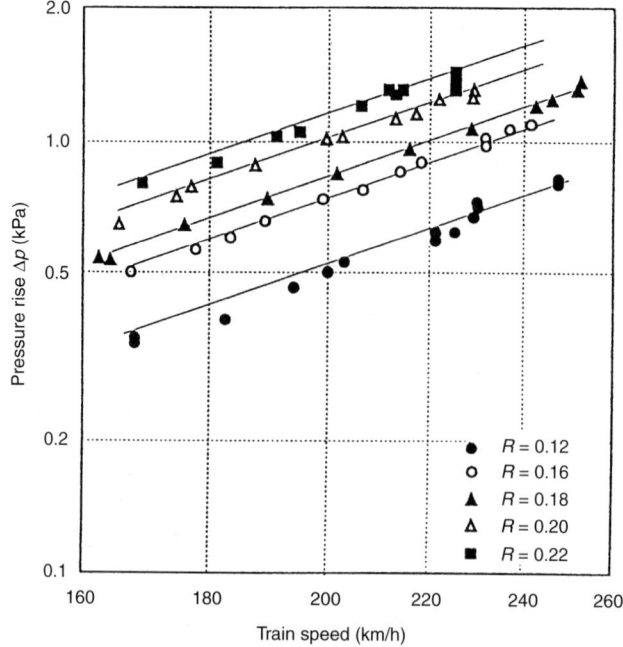

Fig. 7.4. *Pressure rise Δp of the wavefront*

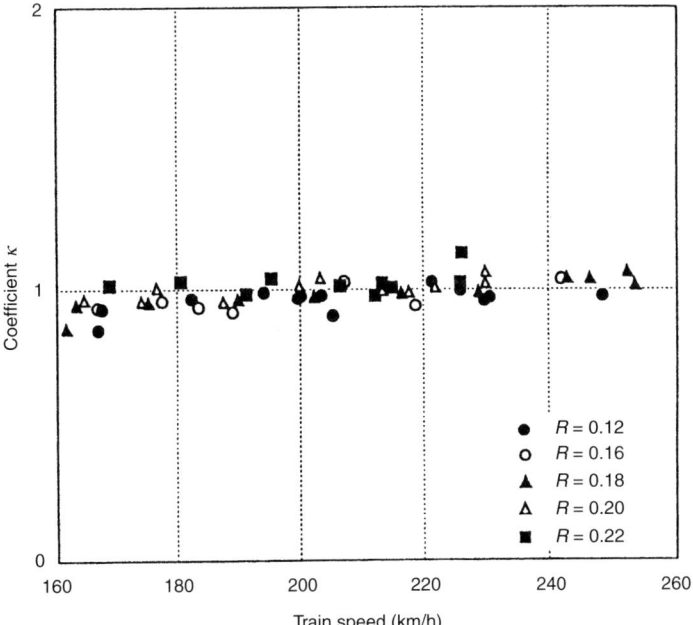

Fig. 7.5. Coefficient κ

cross-sectional area ratio of the train to the tunnel on Δp, using a shape of the train nose such that $a/b = 3$ (a is the length of the nose, and b is the radius of the nose). In Fig. 7.4, the solid lines show the results calculated from equation (7.1), which agree with the results of the model experiments. Figure 7.5 shows the values of the coefficient κ calculated from equation (7.2) using the results of the model experiments for Δp and $(\Delta p/\Delta t)_{MAX}$. In Fig. 7.5, the coefficient κ is almost constant within the range of R from 0.12 to 0.22; this confirms that κ is the parameter indicating the effect of the shape of the train nose on Δt, provided that the shape of the tunnel entrance does not change. Accordingly, the maximum pressure gradient of the wavefront of the compression wave is nearly proportional to the third power of the train speed and can for practical purposes be estimated by equation (7.2) [7.4].

The waveform of the compression wave generated by the train at the tunnel entrance has also been evaluated by numerical simulation as well as from model experiments. A numerical simulation has been performed by Iida *et al.* [7.5] and Ogawa *et al.* [7.6]. The results of the numerical simulation also agree with the results obtained in the field. The generation of the compression wave has been studied by Sugimoto and Ogawa [7.7], Howe [7.8] and Gregoire *et al.* [7.9].

The effect of the shape of the train nose on the compression wave and the optimization of the train nose as a countermeasure to reduce micropressure waves will be described in Section 7.5.

7.3. The propagation of the compression wave through the tunnel

The compression wave generated by the train at the tunnel entrance is propagated through the tunnel at the speed of sound in air. During propagation, the compression wave is affected by the structure inside the tunnel: track, branches, etc. In a long tunnel, the waveform of the compression wave changes during propagation. The wavefront of the compression wave steepens in a tunnel with slab track owing to the non-linear acoustic effects of the compression wave, while the wavefront of the compression wave attenuates in a tunnel with ballasted track because the effect of ballast is more important in this case than the non-linear effects of the compression wave. In a short tunnel, as the distortion of the waveform of the compression wave is small, the waveform of the wave arriving at the tunnel exit is nearly the same as that at the tunnel entrance. Accordingly, the pressure gradient of the wavefront of the compression wave arriving at the tunnel exit is nearly proportional to the third power of the train speed. However, the pressure gradient of

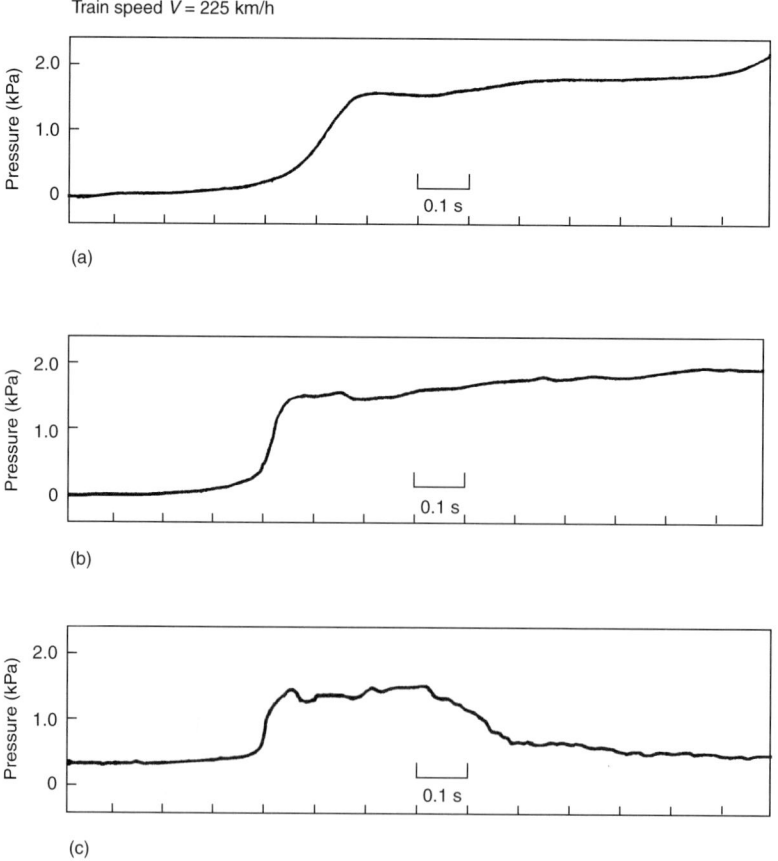

Fig. 7.6. Distortion of a compression wave in a slab-track tunnel on a Shinkansen line: compression wave at (a) 0.1 km, (b) 3.5 km and (c) 6.7 km from tunnel entrance

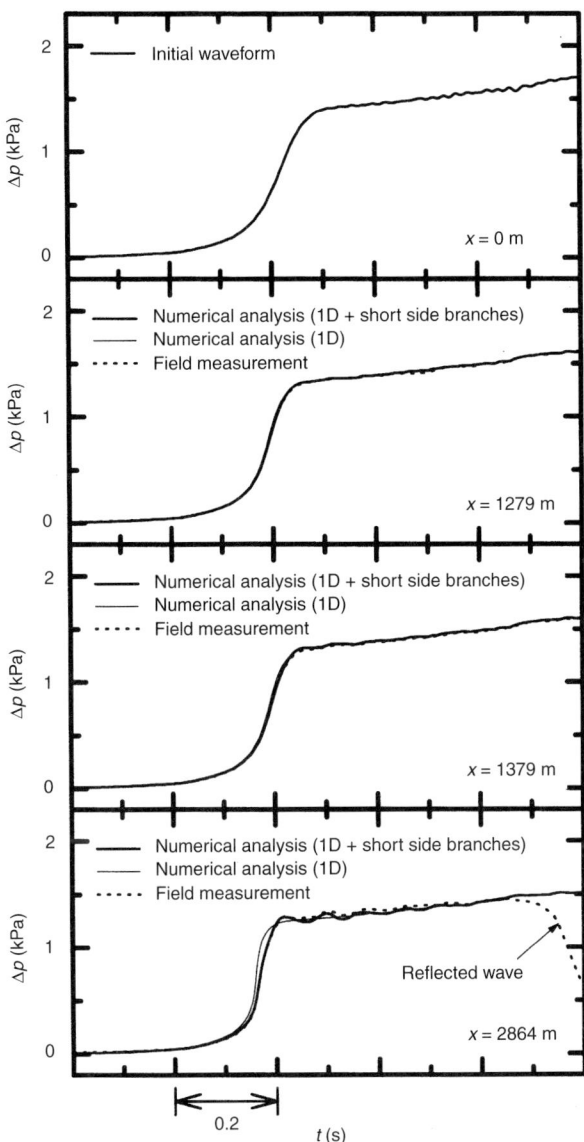

Fig. 7.7. Comparison of steepening of wavefront between numerical simulation and field measurements

a compression wave arriving at the tunnel exit in a long tunnel with slab track has a dependence exceeding the third power of the train speed, while that in a long tunnel with ballasted track becomes smaller than that in a short tunnel.

Figure 7.6 shows the steepening of the wavefront of a compression wave propagating through a slab track tunnel in field tests. Numerical simulations on the distortion of the waveform of a compression wave propagating through a slab track tunnel have been performed by Ozawa [7.1], Maeda [7.10] and Fukuda *et al.* [7.11].

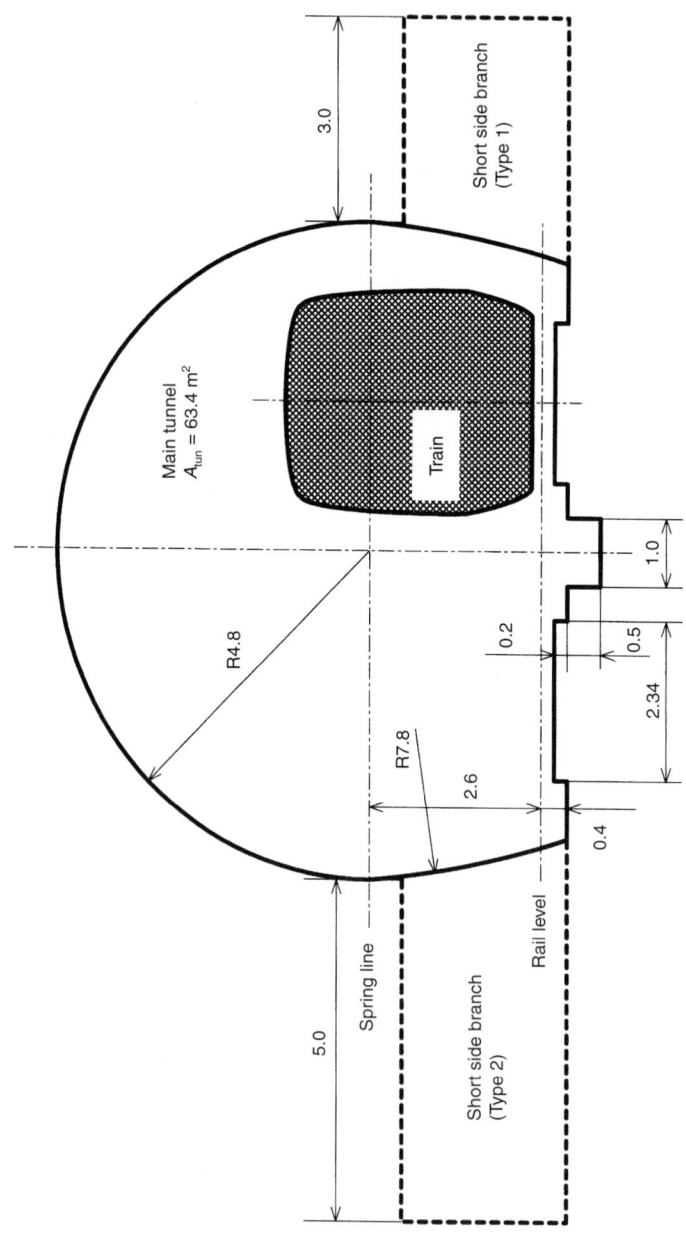

Fig. 7.8. Cross-section of a slab track tunnel on a Shinkansen line (dimensions in m)

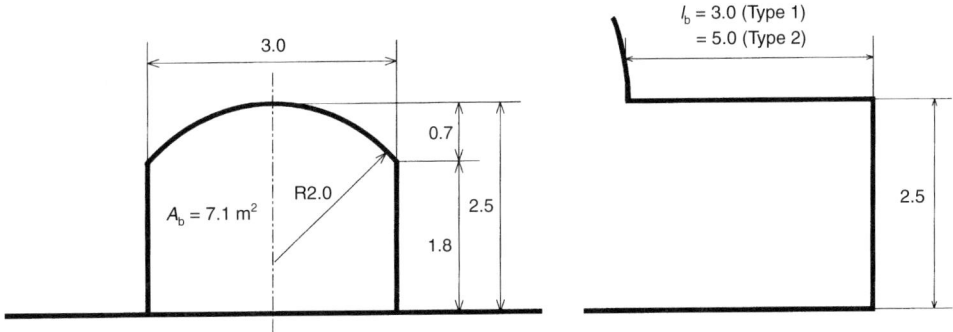

Fig. 7.9. Geometry of two types of short side branches in a tunnel (dimensions in m)

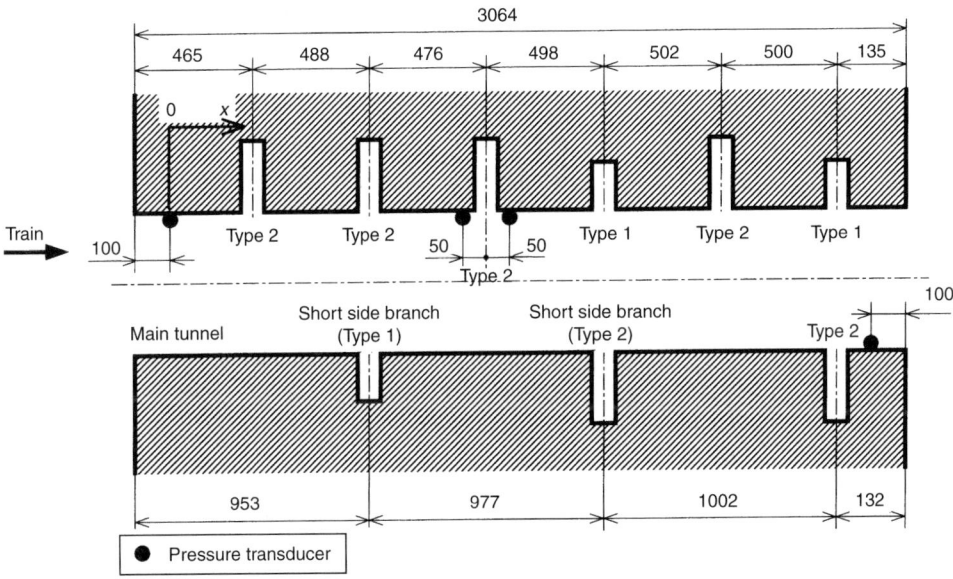

Fig. 7.10. Arrangement of short side branches in a tunnel (dimensions in m)

Model experiments have been performed by Aoki *et al.* [7.12]. The distortion of the compression wave is affected not only by the non-linear effects of the compression wave, but also by the wall friction and the effect of the short side branches arranged at an interval of 500 m in the tunnel. Figure 7.7 shows numerical results obtained by Fukuda *et al.* [7.11] using one-dimensional compressible-flow analysis with unsteady wall friction terms [7.13] and an acoustic theory of the effect of the side branches. In Fig. 7.7, the results of field tests are also shown. Figure 7.8 shows a cross-section of the slab track tunnel of the Shinkansen line. Figure 7.9 shows the geometry of the side branches. Figure 7.10 shows the arrangement of side branches in the tunnel. The results of the numerical simulation shown in Fig. 7.7 agree with the results of the field tests. Further study will be needed for slab track tunnels longer than 3 km and train speeds higher than 270 km/h.

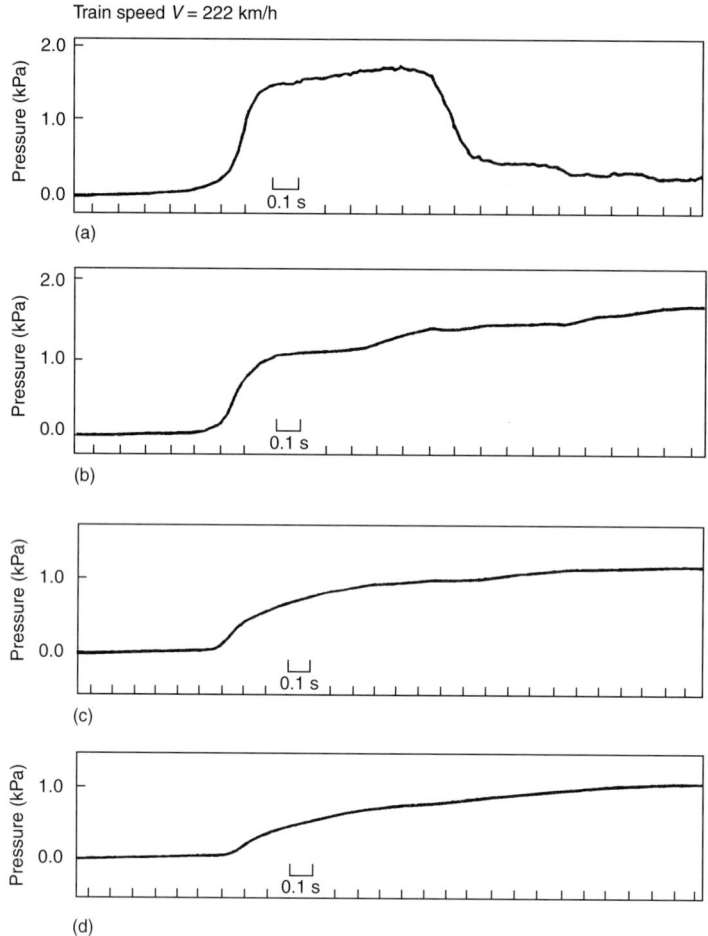

Fig. 7.11. Distortion of a compression wave in a ballasted-track tunnel on a Shinkansen line: compression wave at (a) 0.07 km, (b) 3.64 km, (c) 7.25 km and (d) 11.05 km from tunnel entrance

Figure 7.11 shows the attenuation of the wavefront of the compression wave during propagation through a ballasted-track tunnel in field tests. The effect of ballasted track on the compression wave is larger than the non-linear effects of the compression wave. Figure 7.12 shows the attenuation of the maximum pressure gradient of the wavefront for initial values less than 10 kPa/s. Each series of points connected by solid lines corresponds to data from one train. The attenuation in Fig. 7.12 is approximated well by the following formula:

$$\left(\frac{\Delta p}{\Delta t}\right)_{MAX, x} = \left(\frac{\Delta p}{\Delta t}\right)_{MAX, 0} \exp\left(-\frac{\eta x}{d}\right) \qquad (7.3)$$

where d is the hydraulic diameter of the Shinkansen tunnel (8.1 m) and x is the distance from the tunnel entrance. Figure 7.13 shows the attenuation coefficient η

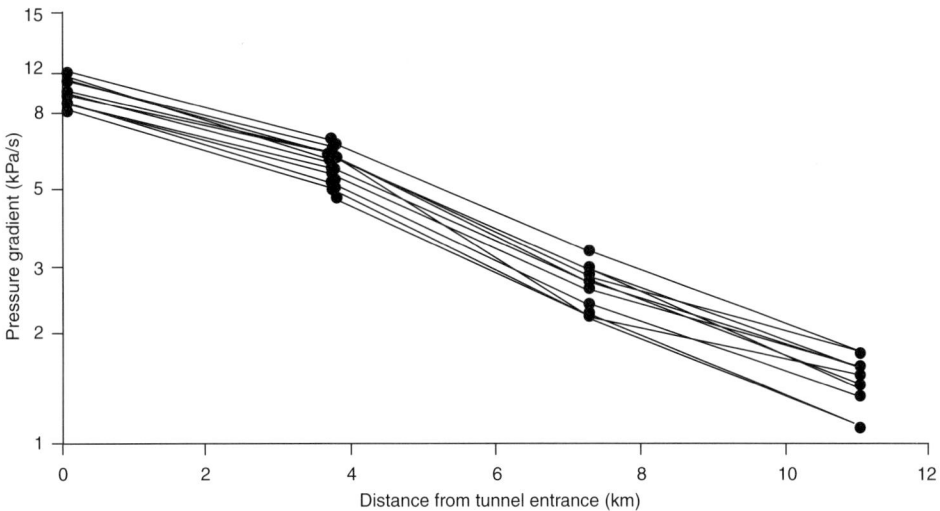

Fig. 7.12. Attenuation of wavefront in a ballasted-track tunnel

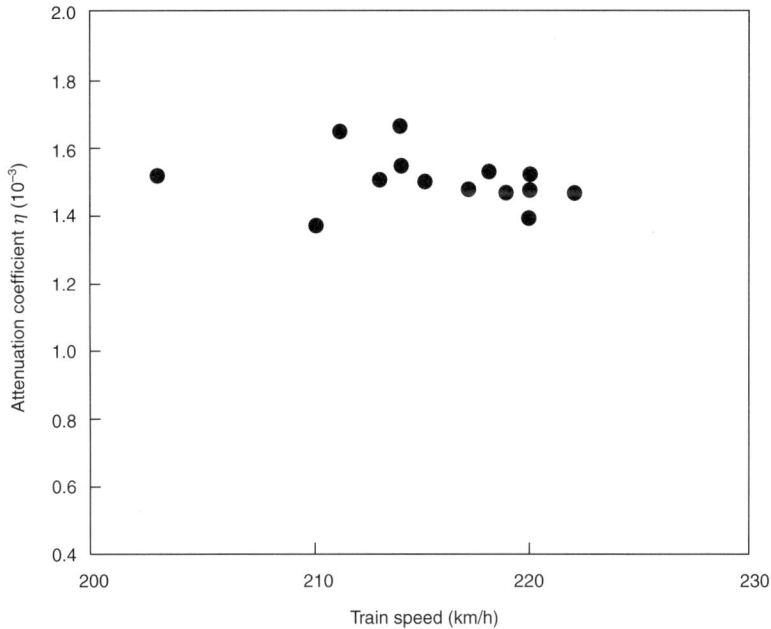

Fig. 7.13. Attenuation coefficient η of a ballasted track on a Shinkansen line

calculated from the equation (7.3). The average value of the coefficient η is 1.5×10^{-3}. Figure 7.14 shows a cross-section of the ballasted-track tunnel of the Shinkansen line. It is considered that the effect of ballasted track depends on the porosity, thickness, width and length of the ballasted track. The porosity of the ballasted track of the Shinkansen line is about 0.4. The distortion of a compression wave propagating through a ballasted-track tunnel has been studied by Ozawa *et al.* [7.14] and Aoki *et al.* [7.15]. The effect of the structure inside the tunnel on pressure wave propagating through the tunnel has been also studied by Sugimoto [7.16, 7.17] (see Chapter 8).

7.4. Radiation of the micropressure wave out of the tunnel portal

The relation between the compression wave arriving at the tunnel exit and the acoustic micropressure wave radiating from the tunnel exit has been evaluated by Ozawa using the theory of acoustic radiation [7.1]. For simplicity, the sound waves radiating from a semi-infinite circular tube (of radius a) with an infinite baffle plate at the end of the tube are considered (Fig. 7.15). The amplitudes of the incident and reflected waves propagating through the tube are $\alpha(\omega)$ and $\beta(\omega)$, respectively, where ω is the angular frequency. The pressure P and the flow velocity u of the plane wave propagating through the tube are expressed as follows:

$$P = [\alpha(\omega)\exp(-ikx) + \beta(\omega)\exp ikx)]\exp(i\omega t) \qquad (7.4)$$

$$u = [\alpha(\omega)\exp(-ikx) - \beta(\omega)\exp ikx)]\exp(i\omega t)/\rho_0 c_0 \qquad (7.5)$$

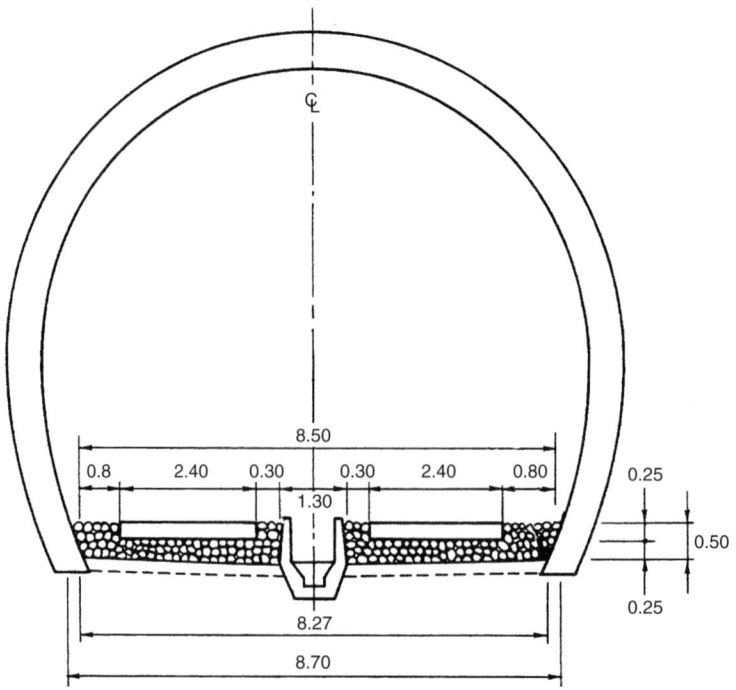

Fig. 7.14. Cross-section of a ballasted track tunnel on a Shinkansen line (dimensions in m)

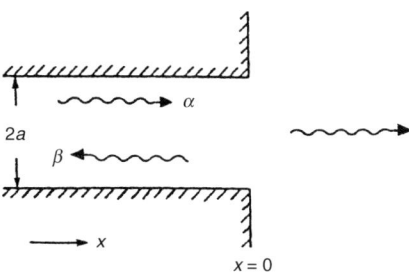

Fig. 7.15. Sound waves radiating from a semi-infinite tube with an infinite baffle plate

Here the ratio of P to $\rho_0 c_0 u$, namely the acoustic impedance at the tube end $x = 0$, is assumed to be equal to that of a vibrating circular plate (of radius a) with an infinite baffle plate. Consequently

$$\frac{\alpha(\omega)+\beta(\omega)}{\alpha(\omega)-\beta(\omega)} = R(2ka)+iX(2ka) \qquad (7.6)$$

where $R(2ka) = 1 - 2J_1(2ka)$, $X(2ka) = 2S_1(2ka)$, J_1 is a Bessel function, S_1 is a Struve function, $k\ (=\omega/c_0)$ is a wavenumber and c_0 is the sound speed of air. Using equations (7.5) and (7.6), the flow velocity u at the tube end $x = 0$ can be expressed as follows:

$$u = \frac{2\alpha(\omega)}{\rho_0 c_0 [R(2ka)+1+iX(2ka)]} \exp(i\omega t) = \xi \exp(i\omega t) \qquad (7.7)$$

Meanwhile, the acoustic pressure wave p radiating out of the tube can be expressed using the theory of radiation from a circular plate vibrating at a velocity u with an infinite baffle plate [7.18]. The corresponding solution along the axis of the circular plate is as follows:

$$p(\omega) = \rho_0 c_0 \left[2\xi \exp(i\omega t) \sin\left\{ \frac{ka}{2}\left[\sqrt{\left(\frac{r}{a}\right)^2 + 1} - \frac{r}{a} \right] \right\} \times \right.$$
$$\left. \exp\left(i\left\{ \frac{\pi}{2} - \frac{ka}{2}\left[\sqrt{\left(\frac{r}{a}\right)^2 + 1} + \frac{r}{a} \right] \right\} \right) \right] \qquad (7.8)$$

Substituting equation (7.7) into equation (7.8) gives

$$p(\omega) = \frac{4\alpha(\omega)}{R(2ka)+1+iX(2ka)} \exp(i\omega t) \sin\left\{ \frac{ka}{2}\left[\sqrt{\left(\frac{r}{a}\right)^2 + 1} - \frac{r}{a} \right] \right\} \times$$
$$\exp\left(i\left\{ \frac{\pi}{2} - \frac{ka}{2}\left[\sqrt{\left(\frac{r}{a}\right)^2 + 1} + \frac{r}{a} \right] \right\} \right) \qquad (7.9)$$

Accordingly, the micropressure wave in the time domain $p(t)$ can be expressed as follows:

$$p(t) = \frac{1}{\sqrt{2\pi}} \int_{-\infty}^{\infty} \frac{4\alpha(\omega)}{R(2ka)+1+iX(2ka)} \sin\left\{\frac{ka}{2}\left[\sqrt{\left(\frac{r}{a}\right)^2+1}-\frac{r}{a}\right]\right\} \times$$
$$\exp\left(i\left\{\frac{\pi}{2}-\frac{ka}{2}\left[\sqrt{\left(\frac{r}{a}\right)^2+1}+\frac{r}{a}\right]\right\}\right)\exp(i\omega t)d\omega \quad (7.10)$$

$$\alpha(\omega) = \frac{1}{\sqrt{2\pi}} \int_{-\infty}^{\infty} P(t)\exp(-i\omega t)dt \quad (7.11)$$

Here $\alpha(\omega)$ is the Fourier transform of the incident wave $P(t)$ arriving at the tube exit (see also equations (7.4) and (7.5)).

Equation (7.10) can be simplified if the low-frequency approximation $ka < 1$ and the far-field approximation $r/a > 1$ are satisfied. In this case,

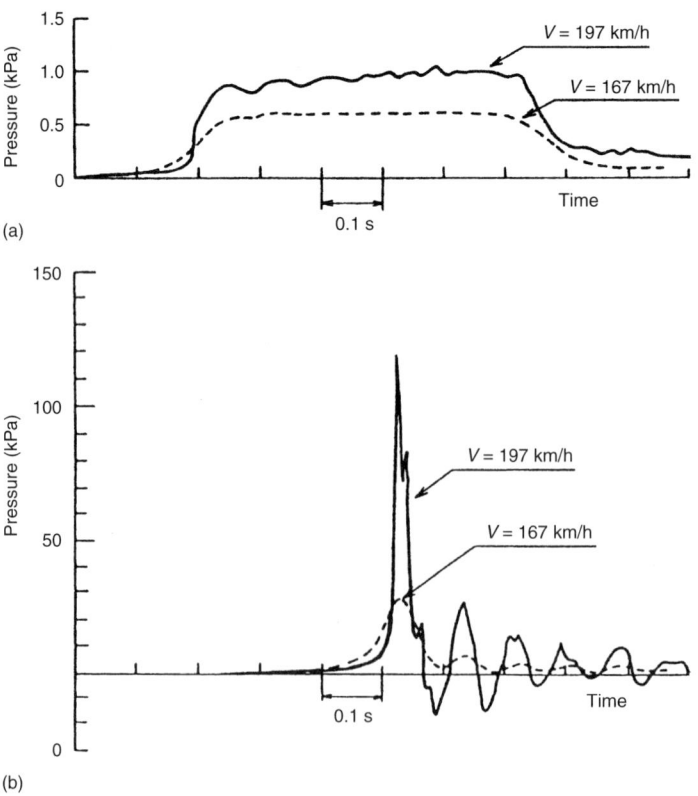

Fig. 7.16. *Micropressure waves and compression waves: (a) compression waves near tunnel exit (measured at a point 90 m inside tunnel); (b) micropressure wave (measured at 22 m inside tunnel)*

$$\sin\left\{\frac{ka}{2}\left[\sqrt{\left(\frac{r}{a}\right)^2+1}-\frac{r}{a}\right]\right\}\exp\left(i\left\{\frac{\pi}{2}-\frac{ka}{2}\left[\sqrt{\left(\frac{r}{a}\right)^2+1}+\frac{r}{a}\right]\right\}\right)$$
$$\cong \frac{ka^2}{4r}\exp\left[i\left(\frac{\pi}{2}-kr\right)\right] \quad (7.12)$$

$$R(2ka)+1+iX(2ka)\cong 1 \quad (7.13)$$

Substituting equations (7.12) and (7.13) into equation (7.10), the micropressure wave $p(t)$, in the low-frequency and far-field approximations, is as follows:

$$p(t)=\frac{1}{\sqrt{2\pi}}\int_{-\infty}^{\infty}\frac{ika^2\alpha(\omega)}{r}\exp[i(\omega t-kr)]d\omega$$

$$=\frac{1}{\sqrt{2\pi}}\int_{-\infty}^{\infty}\frac{i\omega\alpha(\omega)\pi a^2}{\pi c_0 r}\exp[i(\omega t-kr)]d\omega$$

$$=\frac{A}{\pi c_0 r}\frac{1}{\sqrt{2\pi}}\int_{-\infty}^{\infty}i\omega\alpha(\omega)\exp[i(\omega t-kr)]d\omega \quad (7.14)$$

$$=\frac{2A}{2\pi c_0 r}\left[\frac{dP}{dt}\right]_{t-r/c}$$

$$=\frac{2A}{\Omega c_0 r}\left[\frac{dP}{dt}\right]_{t-r/c}$$

where A is the cross-sectional area of the tube and Ω ($=2\pi$) is the solid angle for a semi-infinite space. Equation (14) means that the micropressure wave is proportional to the pressure gradient of the compression wave arriving at the tunnel exit and inversely proportional to the distance from the tunnel exit, provided that the low-frequency and far-field approximations are valid. Equation (7.14) is the same equation as one derived earlier by Yamamoto [7.19]. The actual solid angle Ω calculated from the results of the field tests, when the conditions for the low-frequency and far-field approximations are satisfied, is in the range from 2π to 4π, and depends on the configuration of the tunnel exit. The solid angle Ω is small when the tunnel exit is connected to a cutting section and large when the tunnel exit is connected to an elevated viaduct section.

Figure 7.16 shows the waveforms of micropressure waves and compression waves arriving at the tunnel exit. It is clear that the waveforms of the micropressure waves are nearly proportional to the pressure gradient of the waveform of the compression waves arriving at the tunnel exit. The micropressure waves should be measured with an ultra-low-frequency sound level meter or a pressure transducer because the waveform of the micropressure waves is impulsive and includes low-frequency components. Figure 7.17 shows the relation between the peak value of the micropressure waves and the train speed. In short tunnels, the peak values of the micropressure waves are nearly proportional to the third power of the train speed, irrespective of the type of track. In long slab track tunnels, the peak value of the

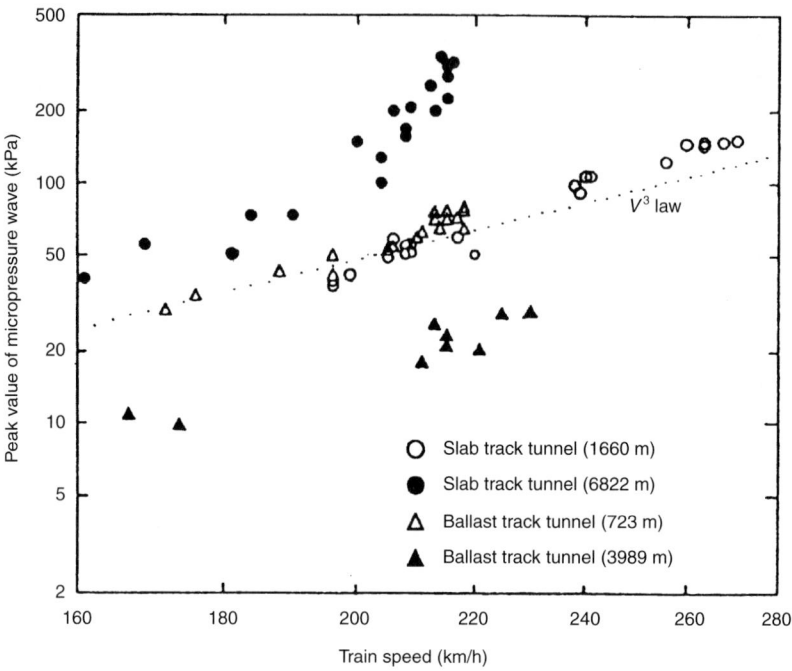

Fig. 7.17. Relation between peak value of micropressure waves and train speed (measured at a distance of 20 m)

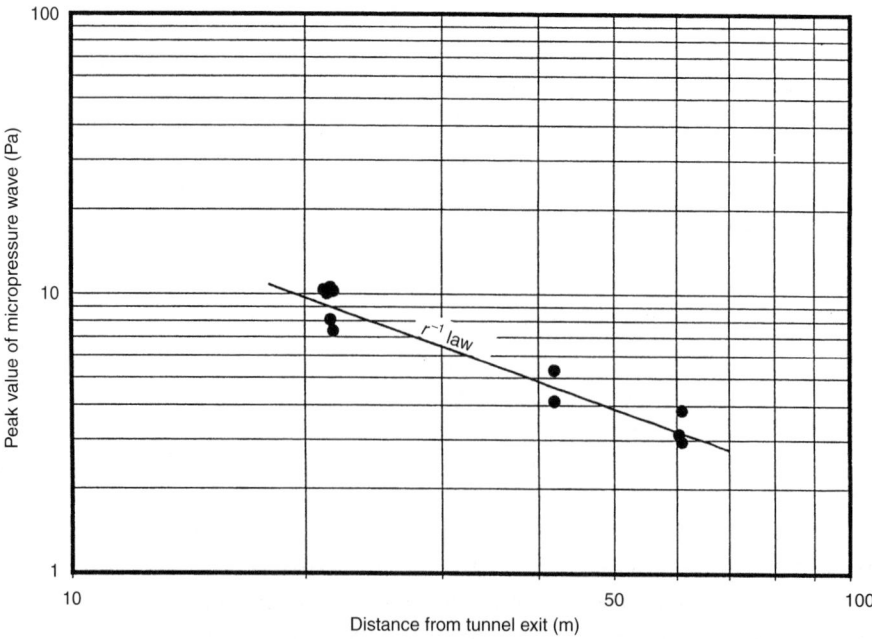

Fig. 7.18. Attenuation of micropressure waves with distance

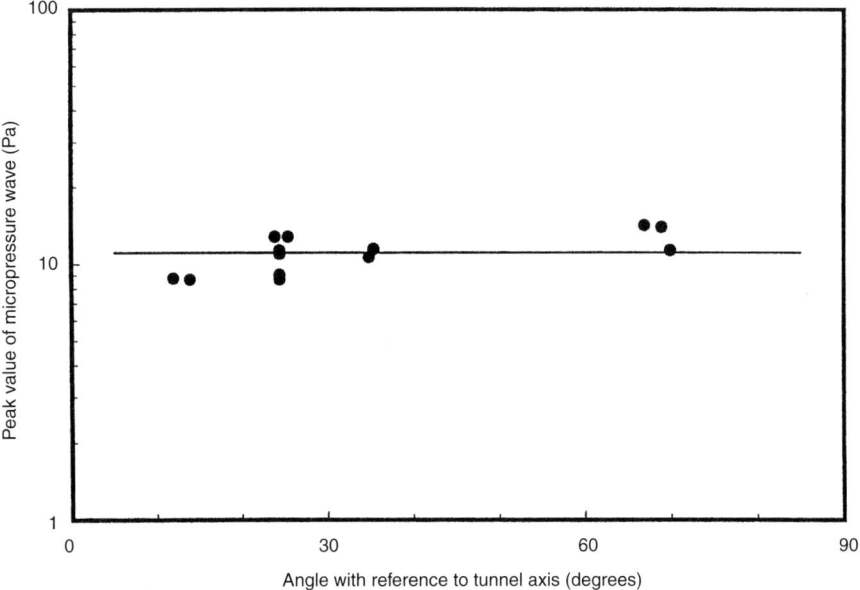

Fig. 7.19. Directivity of radial micropressure waves (r = 20 m)

micropressure waves has a dependence exceeding the third power of the train speed, while in the long ballasted-track tunnel, the peak value is smaller than that in the short tunnels. Figures 7.18 and 7.19 show that the attenuation is inversely proportional to the distance and that there is no directivity of the micropressure waves, when the low-frequency and far-field approximations are valid. As the train speed increases further and the steepening of the wavefront of the compression wave arriving at the tunnel exit becomes larger, the characteristics of the micropressure waves become increasingly different from those in the low-frequency and far-field approximations.

The radiation of micropressure waves has been studied by Matsuo *et al.* using numerical simulation [7.20].

7.5. Measures to decrease the micropressure waves

The peak value of the micropressure wave is nearly proportional to the pressure gradient of the wavefront of the compression wave arriving at the tunnel exit. Accordingly, the principle of the measures taken to decrease the micropressure waves is to reduce the pressure gradient of the wavefront of the compression wave arriving at the tunnel exit. As the micropressure waves are impulsive waves and include low-frequency components, measures taken outside the tunnel exit, such as a noise barrier, are not effective. The mitigation measures must instead be taken at the stage of generation of the compression wave or at the stage of its propagation through the tunnel. Some of these measures have been applied to Shinkansen tunnels and Shinkansen trains.

7.5.1. Measures applied to Shinkansen tunnels

The measures applied to Shinkansen tunnels are as follows: (1) the installation of a tunnel hood at the tunnel entrance, (2) the use of side branches and (3) the installation of a shelter with slits between two adjacent tunnels. Method (1) is a measure taken at the stage of generation of the compression wave at the tunnel entrance. Methods (2) and (3) are measures taken at the stage of propagation of the compression wave through the tunnel.

The tunnel hood has the effect of reducing the pressure gradient of the wavefront of the compression wave at the tunnel entrance, that is, expanding the time interval Δt shown in Fig. 7.2. Figure 7.20 shows the 49 m long tunnel entrance hood. A typical tunnel entrance hood has a cross-section about 1.4 times as large as that of the tunnel itself and has openings in the sides. The optimum area and the optimum position of the openings were decided by model experiments (Fig. 7.3). Figure 7.21 shows a comparison between the waveform for the 49 m long tunnel hood with optimum openings and the waveform obtained without the tunnel hood in the model experiments. In Fig. 7.21, the pressure does not rise so quickly with the tunnel hood, and the pressure gradient of the wavefront is only 0.2 times that obtained without the tunnel hood. The effect of the tunnel hood depends on the length of the hood, provided that it has the optimum openings. Figure 7.22 shows the relation between the effect of the hood and the length of the hood. The vertical axis denotes the reduction ratio of the pressure gradient ϕ and the equivalent reduction ratio of the train speed φ. The reduction ratio of the pressure gradient ϕ can be converted into the reduction ratio of the train speed φ by the formula $\varphi = \phi^{1/3}$, because the pressure gradient of the wavefront of the compression wave at the tunnel entrance is nearly

Fig. 7.20. Tunnel hood (49 m long)

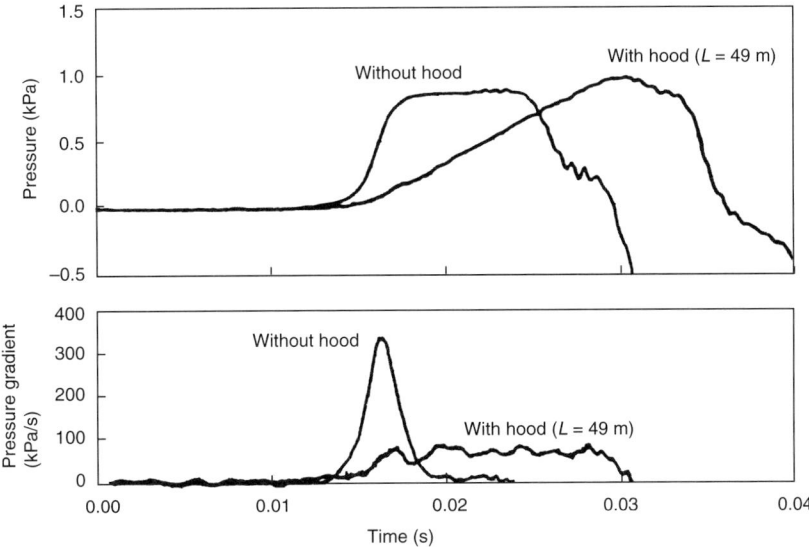

Fig. 7.21. Compression waves for a tunnel hood with optimum openings (high-speed-train nose, model train speed 223 km/h)

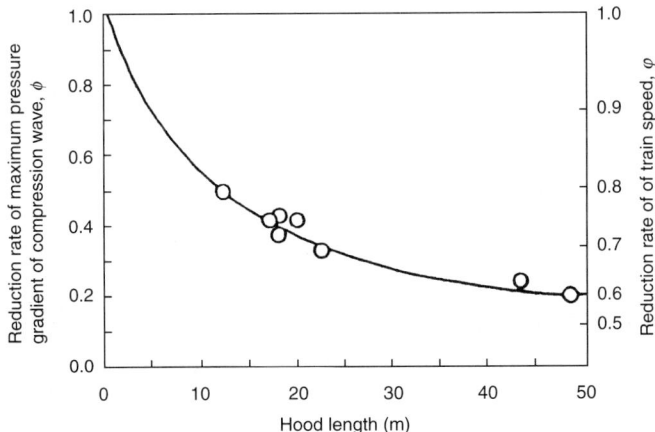

Fig. 7.22. Effect of tunnel hood

proportional to the third power of the train speed. For example, the equivalent reduction ratio of the train speed φ of the 49 m long tunnel hood is $0.6 = (0.2)^{1/3}$, using a reduction ratio of the pressure gradient $\phi = 0.2$. This means that the 49 m long tunnel hood has the effect of reducing the peak value of the micropressure waves at 300 km/h to that which occurs at 180 km/h without the hood. Figure 7.23 shows the effect of the 49 m long tunnel hood on the radiated micropressure waves. The micropressure wave decreases from about 300 Pa to 20 Pa, and the explosive sound disappears. At present, 175 tunnel hoods are installed on Shinkansen lines.

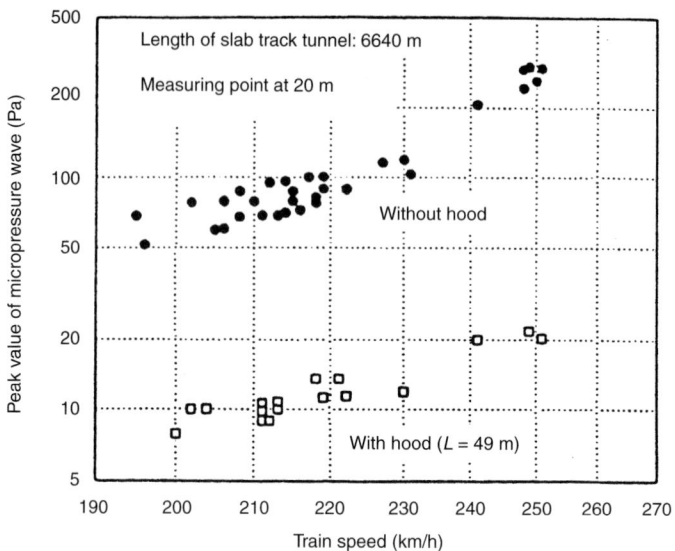

Fig. 7.23. Effect of the 49 m long tunnel hood in the field

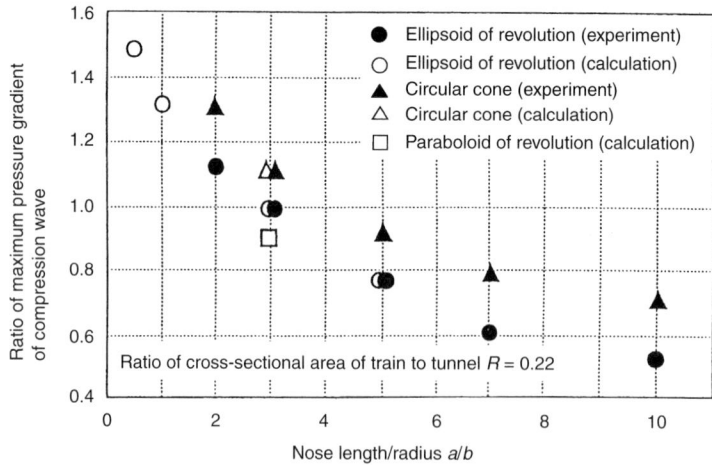

Fig. 7.24. Relationship between maximum pressure gradient of compression wave and nose length for basic nose shapes

7.5.2. Measures applied to Shinkansen trains

The methods applied to the train are as follows: (1) the reduction of the cross-sectional area of the train and (2) the optimization of the train nose. Methods (1) and (2) are measures taken at the stage of generation of the compression wave at the tunnel entrance. The effect of method (1) can be evaluated from equation (7.2) in Section 7.2. Here the effect of the shape of the train nose and its optimization will be discussed.

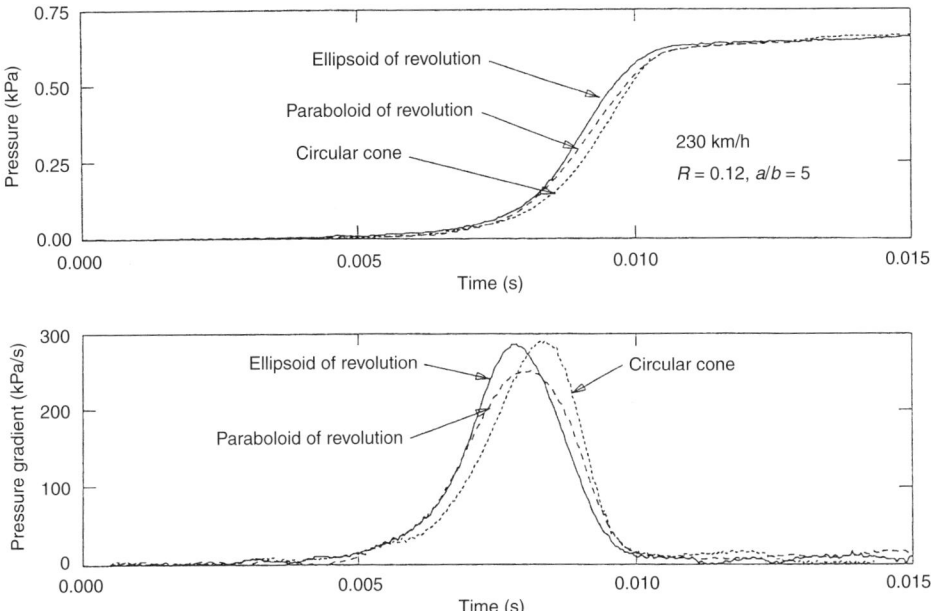

Fig. 7.25. Compression waves and pressure gradients for the basic nose shapes

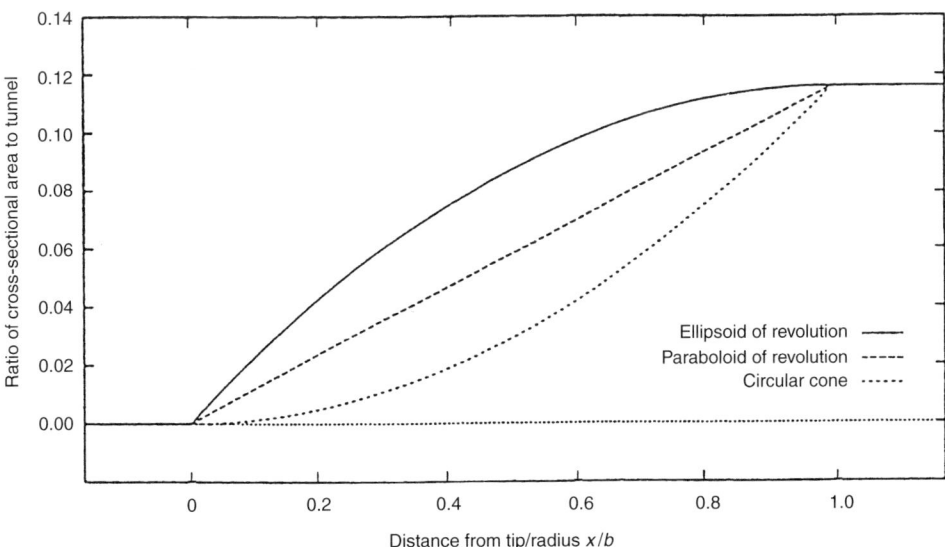

Fig. 7.26. Variation of cross-sectional area of basic nose shapes

To investigate the effect of the shape of the train nose on the waveform of the compression wave, model experiments on basic nose shapes of ellipsoids of revolution, circular cones and paraboloids of revolution were performed using the equipment shown in Fig. 7.3. Figure 7.24 shows the relation between the maximum pressure gradient of the compression wave and the length of the nose (a is the nose length and b is the radius). A longer nose is more effective in reducing the pressure gradient of the compression wave. Figure 7.25 shows a comparison of the waveforms of the compression wave for an ellipsoid of revolution, a circular cone and a paraboloid of revolution, each of which has the same length ($a/b = 5$). The paraboloid of revolution is the most effective of all basic shapes, because the variation of its cross-sectional area is linear and the smallest of all the variations shown in Fig. 7.26. Also, model experiments on shortened versions of the basic shapes (Fig. 7.27) were performed for the purpose of consideration of a practical nose design. Figure 7.28 shows the results for the shortened noses. It is clear that the tip of the nose has little effect on the maximum pressure gradient of the compression wave. Finally, the distribution of the cross-sectional area of the nose shown in Fig. 7.29 was proposed for an effective train nose for reducing micropressure waves [7.4]. Later, a numerical simulation method to find the optimum distribution of the cross-sectional area of the nose was developed by Iida et al. [7.5], using a non-linear optimization method under the constraint that the cross-sectional area and the length of the nose were given. The optimum nose shape is approximated by the following formula, based on the experimental results shown in Fig. 7.29:

$$\frac{A(x)}{\pi b^2} = (1-\alpha_2)\left[(1-\alpha_1)\frac{x}{a}+\alpha_1\sqrt{\frac{x}{a}}\right]+\alpha_2\left(\frac{x}{a}\right)^2 \tag{7.15}$$

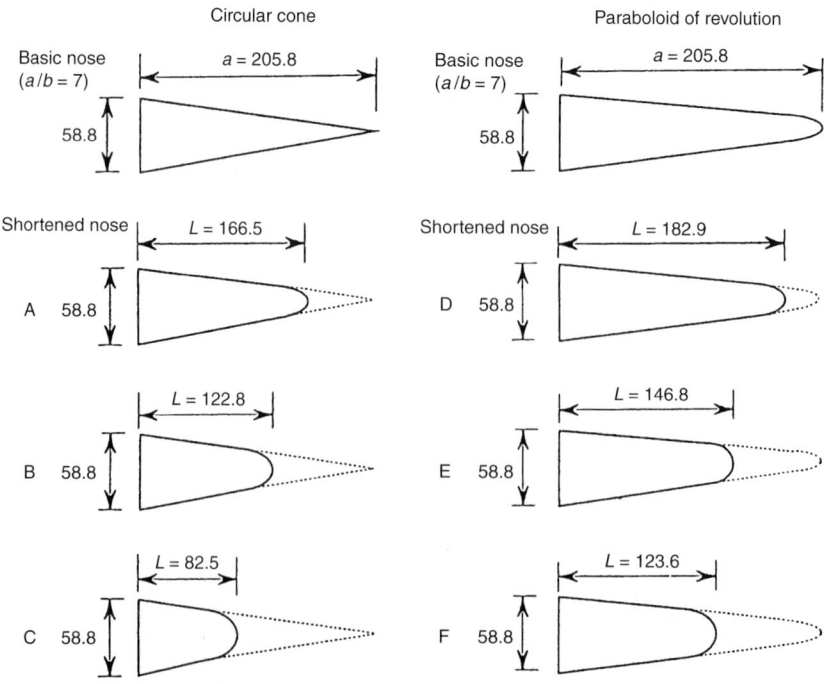

Fig. 7.27. *Basic noses and shortened noses (dimensions in mm)*

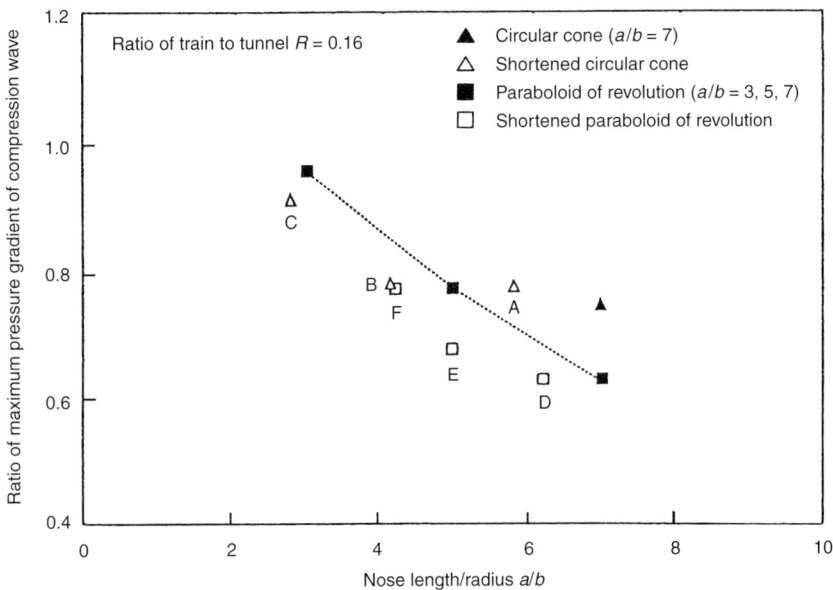

Fig. 7.28. Relation between maximum pressure gradient of compression wave and nose length for shortened nose shapes

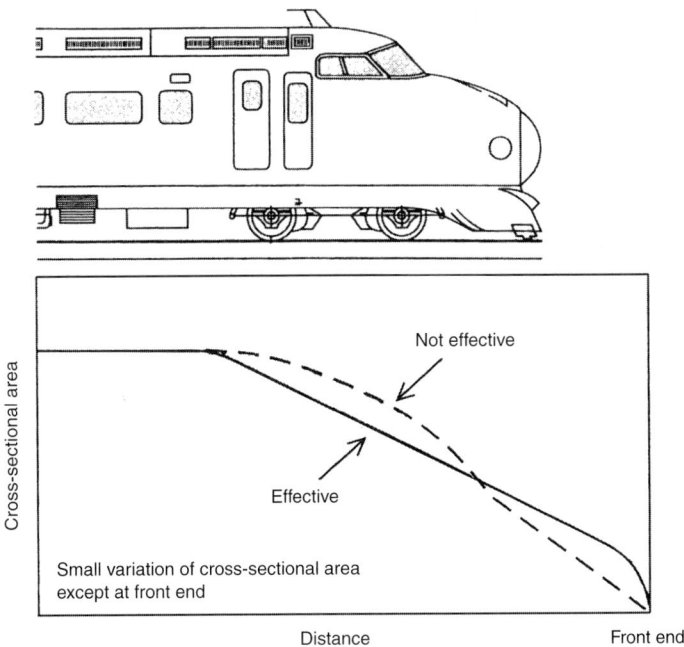

Fig. 7.29. Variation of cross-sectional area of train nose that is effective for reducing micropressure waves

Fig. 7.30. Example of a train nose that is effective for reducing micropressure waves: 500 series Shinkansen train (JR West)

where α_1 and α_2 are design variables used to control the shape of the nose for optimization, a is the nose length and b is the nose radius. The latest Shinkansen trains, series 500 of JR West, series 700 of JR Central and West, and series E1–E4 of JR East, have a nose shape which mitigates micropressure waves. Figure 7.30 shows an example of an effective train nose for reducing micropressure waves, for the JR West 500 series Shinkansen train.

7.6. References

7.1. OZAWA, S. Studies of micropressure wave radiated from a tunnel exit. *Railway Technical Research Report*, 1979, **1121**, 1–92 [in Japanese].
7.2. OZAWA, S., MAEDA, T., MATSUMURA, T., UCHIDA, K., KAJIYAMA, H. and TANEMOTO, K. Countermeasures to reduce micro-pressure waves radiated from exits on Shinkansen tunnel. *Proceedings of the 7th International Symposium on Aerodynamics and Ventilation of Vehicle Tunnels*. Brighton, 1991, pp. 253–266
7.3. HARA, T. Aerodynamic problems when a train is running into a tunnel with a large velocity. *Railway Technical Research Report*, 1960, **153**, 1–20 [in Japanese].
7.4. MAEDA, T., MATSUMURA, T., IIDA, M., NAKATANI, K. and UCHIDA, K. Effect of train nose on compression wave generated by train entering tunnel. *Proceedings of the International Conference on Speedup Technology for Railway and Maglev Vehicle*. Yokohama, 1993, vol. 2, pp. 315–319.
7.5. IIDA, M., MATSUMURA, T., NAKATANI, K., FUKUDA, T. and MAEDA, T. Effective nose shape for reducing tunnel sonic boom. *Quarterly Report of RTRI*, 1997, **38**(4), 206–211.
7.6. OGAWA, T. and FUJII, K. Numerical simulation of compressible flows induced by a train moving into a tunnel. *Computational Fluid Dynamics Journal*, 1994, **3**(1), 63–82.
7.7. SUGIMOTO, N. and OGAWA, T. Acoustic analysis of the pressure field in a tunnel, generated by entry of a train. *Proceedings of the Royal Society of London A*, 1998, **454**, 2083–2112.

7.8. HOWE, M. S. The compression wave produced by a high-speed train entering a tunnel. *Proceedings of the Royal Society of London A*, 1998, **454**, 1523–1534.

7.9. GREGOIRE, R., RETY, J. M., MASBERNAT, F., MORINIERE, V., BELLENOUE, M. and KAGEYAMA, T. Experimental study (scale1/70) and numerical simulations on the generation of pressure waves and micro-pressure waves due to high-speed train-tunnel entry. *Proceedings of the 9th International Symposium on Aerodynamics and Ventilation of Vehicle Tunnels*. Valle d'Aosta, 1997, pp. 877–903.

7.10. MAEDA, T. Reduction of micropressure wave radiated from tunnel exit by branches in tunnel. *Railway Technical Research Report*, 1981, **1167**, 1–35 [in Japanese].

7.11. FUKUDA, T., IIDA, M., MAENO, K. and HONMA, H. Distortion of compression wave propagating through slab track tunnel with short side branches of Shinkansen. *Proceedings of the 22nd International Symposium on Shock Waves, Imperial College, London, UK*, 1999, Paper No. 3040, pp. 1–6.

7.12. AOKI, T., NAKANO, S., MATSUO, K., KASHUMURA, H. and YASUNOBU, T. Attenuation and distortion of compression wave and shock waves propagating along high-speed railway model tunnels. *Proceedings of the 21st International Symposium on Shock Waves*. Great Keppel Island, Australia, 1997, Paper No. 1140, pp. 1–5.

7.13. VARDY, A. E. and BROWN, J. M. B. Transient, turbulent smooth pipe friction. *Journal of Hydraulic Research*, 1995, **33–4**, 435–455.

7.14. OZAWA, S., MURATA, K. and MAEDA, T. Effect of ballasted track on distortion of pressure wave in tunnel and emission of micropressure wave. *Proceedings of the 9th International Symposium on Aerodynamics and Ventilation of Vehicle Tunnels*. Valle d'Aosta, 1997, pp. 935–947.

7.15. AOKI, T., NAKAO, S. and MATSUO, K. Ballast effect on distortion of propagating compression wave in high-speed railway tunnel. *Proceedings of the 22nd International Symposium on Shock Waves*. London, 1999, Paper No. 2990, pp. 1–6.

7.16. SUGIMOTO, N. Propagation of nonlinear acoustic waves in a tunnel with an array of Helmholtz resonators. *Journal of Fluid Mechanics*, 1992, **244**, 55–78.

7.17. SUGIMOTO, N. Shock-free tunnel for future high-speed trains. *Proceedings of the International Conference on Speedup Technology for Railway and Maglev Vehicle*. Yokohama, 1993, vol. 2, pp. 284–292.

7.18. HAYASAKA, T. *Theory of Acoustic Vibration*. Maruzen, Tokyo, 1974, pp. 621–618 [in Japanese].

7.19. YAMAMOTO, A. Micropressure wave radiated from tunnel exit. *Spring Meeting of the Physical Society of Japan, Yamaguchi*. 1977, preprint [in Japanese].

7.20. MATSUO, K., AOKI, T., KASHIMURA, H., IWAMOTO, K., NOGUCHI, Y. and TUJIMOTO, Y. Numerical simulation of an impulsive wave emitted from an exit of high-speed railway tunnel. *Computers in Railways III*. Washington, DC, 1992, **2**, 455–463.

8. Emergence of an acoustic shock wave in a tunnel and a concept of shock-free propagation

N. Sugimoto
Department of Mechanical Science, Graduate School of Engineering Science, University of Osaka, Toyonaka, Osaka 560-8531, Japan

8.1. Introduction

When a train enters a tunnel at high speed, it is known that a bursting sound, just like a sonic boom from a supersonic flight, occurs at the tunnel exit (Fig. 8.1). This phenomenon happened first in test runs just before the Sanyo Shinkansen[*] began service in 1975 at a speed of 210 km/h. This bursting sound, called simply a 'burst' hereafter, was observed at exits of tunnels several to ten kilometres long with

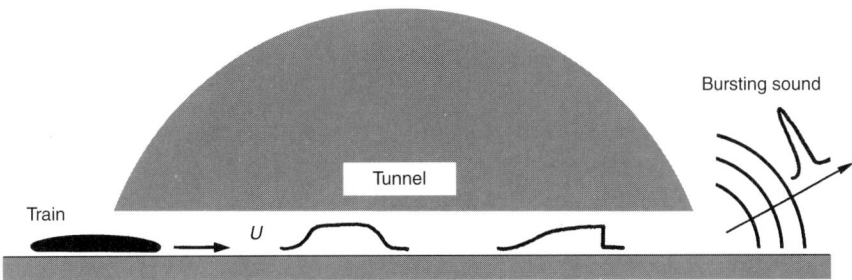

Fig. 8.1. Train entering a tunnel and the resulting radiation of a bursting sound from the tunnel exit

---

[*] The word 'Shinkansen' in Japanese originally meant 'new trunk line'. Nowadays, however, it is often used to mean high-speed passenger trains themselves or even whole systems of high-speed, ground mass transportation. The Sanyo Shinkansen links Osaka and Fukuoka.

concrete slab track. Interestingly enough, however, no bursts have been reported in tunnels with ballasted track (see also Chapter 7).

The burst results from radiation of an acoustic shock wave from the exit. Because the train speed is well in the subsonic range even in the case of the present fastest trains with speeds of 300 km/h, the shock wave is never produced from the outset on entry of a train. Instead, pressure disturbances generated in the tunnel by train entry transform their profile progressively in the course of propagation to give rise to a discontinuity in pressure, i.e. a shock. The emergence of the shock has unfavourable effects not only on the environment around the tunnel exit, but also on trains and tunnels because they are loaded repeatedly.

To alleviate the burst, at present, hoods of suitable length and with slits are installed at tunnel portals [8.1]. The hoods serve to reduce the gradient of the pressure wave at the wavefront, so that the wave may leave the tunnel before the shock is formed. In future, magnetically levitated trains are expected to travel at a speed of 500 km/h or even higher. In this case, the burst would be more severe unless suitable countermeasures were taken. This chapter is devoted to elucidation of the physical processes leading to shock formation and to the proposal of a concept for shock-free propagation of pressure waves in tunnels.

Here we briefly review the history of investigations of the aerodynamics associated with the passage of a train through a tunnel. To the author's knowledge, it was Tollmien [8.2] who first applied the then modern subject of fluid dynamics to such a practical problem. He was interested in the drag acting on a train due to travelling inside a tunnel. The train speed was then 100 km/h. After this, no progress was made until the 1950s when the Tokaido Shinkansen between Tokyo and Osaka was planned. In the course of its development, attention was paid to aerodynamic problems. Extending the analysis of Tollmien, Hara [8.3] derived in 1961 an estimate of the maximum pressure generated by entry of a train. Considering the air flow in the tunnel as a flow in a stream tube, he employed Bernoulli's theorem for the flow displaced by the train and the shock conditions at the wavefront. His formula provides a good estimate of the maximum pressure. Also, several aerodynamic analyses were performed in a different context of high-speed tube transportation [8.4, 8.5].

But the investigations were rather slow until 1975, when the burst first happened. Ozawa and his coworkers studied the propagation of these pressure waves intensively, not only in theory, but also both in experiments and in the field, and arrived at the solution of installing the hoods [8.1]. Thanks to their efforts, the bursts have now been suppressed. But much interest has been kindled from the viewpoints of speeding up the Shinkansen and also of development of magnetically levitated trains. The analysis is based on a one-dimensional model of a stream tube [8.6, 8.7]. Recently, however, there have emerged new techniques using computational fluid dynamics to tackle the problem three-dimensionally [8.8, 8.9]. These techniques can almost simulate a real flow field in a tunnel generated by a real train, though only over a short distance.

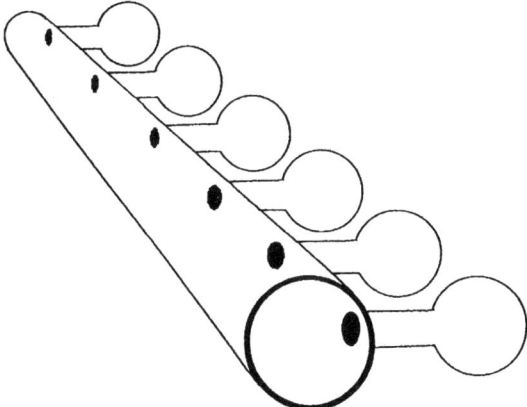

Fig. 8.2. Schematic illustration of a tunnel with an array of Helmholtz resonators

As the magnitude of the pressure disturbances increases it is physically quite natural that a shock wave tends to emerge. Rather, propagation free from a shock is unnatural, so its suppression is not an easy task. Even worse, because the pressure disturbances generated by entry of a train are of very low frequency and are propagated as infrasound, they are too pervasive to be damped or controlled by local action alone. As will be seen later, the processes leading to shock formation are very slow because the non-linear effects essential to them are small locally. But their cumulative effects during propagation give rise to a sudden emergence of a shock in the far field. The shock-free wave propagation described in this chapter is not what is achieved by delaying shock formation by installing hoods at tunnel portals or by controlling the pressure disturbances locally by the action of external agents. Instead, the concept is based on changing or 'remodelling' of the physical mechanisms involved in propagation of the pressure waves so as to counteract shock formation intrinsically. One method for realizing this concept is to exploit dispersion, which is absent in the propagation of pressure waves in the usual situation. Dispersion means that the wave propagation has properties such that the phase velocity of a sinusoidal wave varies depending on its frequency. Dispersion can be generated by connecting an array of Helmholtz resonators to a tunnel, as shown in Fig. 8.2. It has been shown that shock-free propagation is achievable by controlling the dispersion [8.10].

In what follows, an overview of the problem is first given in Section 8.2, where the near and far fields in the pressure field are introduced. An analysis of the near field is given in Section 8.3 and then an analysis of the far field is given in Section 8.4, where the emergence of the shock is explicitly demonstrated. In Section 8.5, propagation in a tunnel with an array of Helmholtz resonators is formulated, and shock-free propagation is examined. Section 8.6 is devoted to the results of experiments performed to confirm the theoretical findings.

8.2. Overview of the problem

Before going into the details of the physical processes of shock formation, it may be appropriate, first, to outline the problem by considering the pressure field generated in a tunnel by entry of a train. In this problem, there are two important parameters: one is obviously the train speed U on entry, and the other is the blockage ratio χ, defined by the ratio of the train's cross-sectional area S to that of the tunnel A, i.e., $\chi = S/A < 1$. The train speed is slower than the sound speed a_0, i.e. subsonic, so the train Mach number $M\,(= U/a_0)$ is less than unity. Even in the case of magnetically levitated trains, U will be about 150 m/s (540 km/h), and then M is about 0.44. On the other hand, the blockage ratio takes relatively small values in the case of tunnels for high-speed trains. For example, for the tunnels of the Shinkansen system, the value is 0.21, and for magnetically levitated trains the value is 0.1.

One of the structural characteristics of tunnels is that tunnels are usually extremely long relative to their lateral dimension. In fact, the tunnels in which the shock wave emerges are several to ten kilometres long, while their typical diameter is merely about 10 m. The aspect ratio of length to diameter is indeed of the order 10^3! Similarly, trains are also long relative to their typical diameter. Usually they are a few hundred metres long and a few metres wide. The aspect ratio is of the order 10^2. Such a geometrically extreme configuration may be a helpful feature in developing a theory. From the viewpoint of wave propagation, tunnels are waveguides in which pressure disturbances can be propagated freely without any geometrical spreading.

We now consider the pressure disturbances generated by entry of a train. When a train rushes into the tunnel, it displaces the air in the tunnel near to the entrance. One part of the air is expelled from the tunnel entrance, while the other part is pushed forward into the tunnel, giving rise to local compression of the air. The pressure disturbances thus produced are propagated down the tunnel with a speed close to sound speed. Figure 8.3 illustrates a pressure field in the x–t diagram, where x measures the axial distance from the entrance and t measures the time, with its origin taken when the train nose just enters the tunnel. The exit of the tunnel is not shown here. With two typical speeds U and a_0 and with a typical tunnel diameter of D as a length scale, two timescales, D/U and D/a_0, are involved in this problem. Although there is another scale of the train length $l\,(\gg D)$, the times l/U and l/a_0 are much larger. Supposing that $U = 150$ m/s, $a_0 = 340$ m/s and $D = 10$ m, D/U and D/a_0 are equal to 1/15 and 1/34 s, respectively. Thus most frequencies will be limited to below 34 Hz. Of course, other sources of sound, such as vortices, involve higher frequencies, but the frequency associated with the train's entry is so low that it may be regarded as infrasound.

Next, let us estimate the magnitude of the pressure disturbances. There may be estimated by applying Bernoulli's theorem to the tunnel, treated as a stream tube. In a frame moving with the train, the velocity of the air will be increased in an annular region between the tunnel and the train. Disregarding the compressibility of the air because of the low Mach number of the train, the continuity equation gives a velocity $UA/(A-S)$ in the region. Application of the theorem at a location just behind the wavefront and at the entrance, where the pressure is equal to the atmospheric

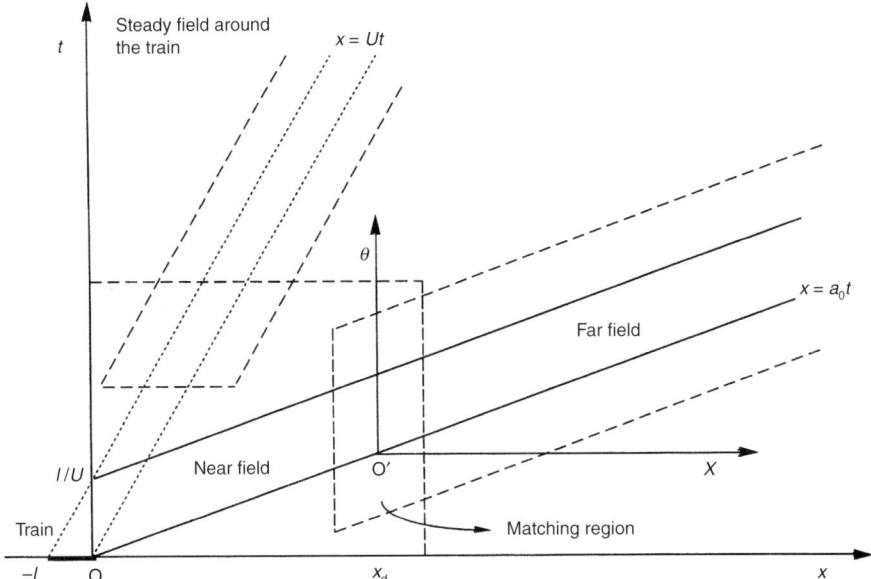

Fig. 8.3. Schematic diagram illustrating a near field, a far field and a steady field moving with a train of length l, indicated by broken lines; the point O corresponds to the entrance, and the point O' indicates a point at $x = x_d \gg R$ in the distant near field, where the matching is executed between the near and far fields; θ and X denote the retarded time $t - x/a_0$ in a frame moving with the sound speed a_0, and the far-field coordinate $\varepsilon \omega x/a_0$ associated with the smallness of the non-linearity ε, ω being a typical angular frequency

pressure p_0, indicates that $p_0 + \Delta p + \rho_0 U^2/2 = p_0 + \rho_0[UA/(A-S)]^2/2$, ρ_0 being the density of air in the atmospheric and Δp the pressure rise behind the wavefront. Retaining the first-order terms in S/A ($=\chi$), Δp is estimated to be $\rho_0 \chi U^2$. The magnitude Δp relative to the atmospheric pressure p_0 is given by $\gamma \chi M^2$, γ being the ratio of specific heats, and the sound speed a_0 is given by $a_0 = \sqrt{(\gamma p_0/\rho_0)}$. In fact, Hara [8.3] developed this stream-tube model for the entry of a long train (in which no account is taken of the tail) and used the shock conditions at the wavefront. From a detailed analysis, he derived the formula

$$\frac{\Delta p}{p_0} = \frac{\gamma M^2}{2} \frac{1-(1-\chi)^2}{(1-M)[M+(1-\chi)^2]} \tag{8.1}$$

Here it should be remarked that this formula assumes implicitly a small blockage ratio ($\chi \ll 1$). The estimate $\gamma \chi M^2$ corresponds to the first-order term of equation (8.1) expanded in terms of M and χ around $M = \chi = 0$. Even in the case of a magnetically levitated train, $\Delta p/p_0$ is estimated to be as small as 3.7% for $M = 0.44$ and $\chi = 0.1$.

With this estimate of the magnitude, the non-linearity in the gas dynamic equations is found to be sufficiently small that they may be linearized, to the lowest approximation. This is nothing but linear acoustic theory. In addition, the pressure disturbances have extremely low frequencies, so that they are propagated in the form of plane waves, as the fundamental mode of the tunnel. Although there exist an infinite number of modes, this fundamental mode is responsible for propagation along a long tunnel because other modes are dispersive and have cut-off frequencies. The lowest angular cut-off frequency of the axisymmetric mode in a waveguide of circular cross-section is given by $7.66 a_0/D$ [8.11], which corresponds to about 40 Hz for $D = 10$ m. The timescale D/a_0 is associated with higher modes. But it should be remarked that in the field around the train, the higher modes are present and the pressure field is not one-dimensional, but three-dimensional.

Such a picture of propagation of pressure disturbances will hold in short tunnels. But since the tunnels in which the shock emerges are very long, it is likely that non-linearity, even though small locally, may accumulate to give rise to pronouncedly different phenomena. In the linear theory, the fundamental mode is non-dispersive and the pressure disturbances are propagated with a constant sound speed. But when the magnitude becomes large, the pressure disturbances propagate faster where the pressure is higher. The deviation of the propagation speed from the sound speed is given by $(\gamma + 1)u/2$ for a simple wave [8.12], u being the particle velocity of the air, and u is almost proportional to the local excess pressure p' ($\approx \rho_0 a_0 u$). Although the deviation is small locally ($|u/a_0| \ll 1$), the tunnel is so long that the pressure profile is distorted slowly but progressively so as to steepen forward, and eventually there emerges a discontinuity in the pressure, i.e. a shock. This is the processes of shock formation. Because the magnitude of the shock is much smaller than that of shocks in gas dynamics, it is called an acoustic shock wave.

The above considerations suggest separating the pressure field into a *near field* (near to the entrance) and a *far field* (in which the shock emerges) (Fig. 8.3). In the near field, the linear acoustic theory holds to the lowest approximation. The near field will disappear soon when the train travels down deep into the tunnel. There then remains a three-dimensional steady field around the train. In the far field, a non-linear theory is to be employed. The magnitude of the pressure waves is moderately small, so that a quadratic theory of non-linearity should be adequate. To complete the pressure field in the tunnel, both fields need to be matched in an overlap region called the matching region, according to the method of the matched asymptotic expansion [8.13].

In reality, there is a tunnel exit at which both reflection and radiation of the pressure waves take place, and then the pressure field in the tunnel becomes complicated. Here we mention briefly radiation of the pressure waves from the tunnel exit. The pressure wave radiated from the tunnel is sometimes called a micropressure wave [8.1]. The excess pressure p'_{exit} outside of the tunnel at a large distance s from the exit is given by linear acoustic theory as (see also Chapter 7)

$$p'_{exit} = \frac{2A}{\varpi a_0 s} \frac{\partial}{\partial t}\left[p'_{exit}\left(t - \frac{s}{a_0}\right)\right] \tag{8.2}$$

where $p'_{exit}(t)$ denotes the excess pressure of the wave about to leave the exit. Here ϖ

means the solid angle of radiation, which takes a value of 2π for a semi-infinite space. Thus, if p'_{exit} contains a shock, the derivative of p'_{exit} takes a very large value, so that a burst is radiated from the exit. For example, p'_{exit}/p_0 has been measured to have a peak value of 1.5×10^{-3} and a duration of order 10^{-2} s at a point $s = 20$ m from the exit of a tunnel with concrete slab track of length 8.9 km for a train speed of 200 km/h [8.1]. In the case of a sonic boom, for reference, p'/p_0 for the pressure rise due to the N-wave is of the order 10^{-3}, while the rise time is of the order 10^{-4} to 10^{-2} s [8.14]. Unless the pressure wave contains a shock, no burst is heard. This chapter does not intend to discuss radiation of pressure waves any further; we shall pause here only to mention that the following analysis assumes a semi-infinitely long tunnel.

8.3. Analysis of the near field

8.3.1. Linear acoustic theory

Let us now examine the near field in the framework of linear acoustics. The train's motion may be taken into account through a source term in the linear wave equation [8.15–8.18]:

$$\frac{\partial^2 p'}{\partial t^2} - a_0^2 \Delta p' = a_0^2 \frac{\partial q}{\partial t} \tag{8.3}$$

where p' and q denote, respectively, the excess pressure and the source term reflecting the steady field around the train; Δ stands for the three-dimensional Laplacian. In order to determine the near field, equation (8.3) must be solved not only in the tunnel but also in a free space outside the tunnel. Therefore, the geometrical configuration of the free space must be specified. The tunnel wall and the surface bounding the air in the free space are assumed to be rigid, and the boundary condition of a vanishing normal derivative $\partial p'/\partial n = 0$ is used. Given a smooth function q, a solution p' to equation (8.3) can be expressed in integral form by using the Green's function [8.17]. But obtaining this function is a difficult task, and we attempt instead to seek solutions under simplified conditions.

When a train is travelling, air mass outflow and inflow of the same magnitude are created in front and at the rear, respectively. At the same time, the train's motion exerts a force on the surrounding air. Since the train is not acoustically compact axially (though compact laterally), its motion may be modelled by a distribution of monopole and dipole sources over finite axial extent. But the train is streamlined to avoid flow separation, so that the flow around the train may be regarded as a potential flow and the force exerted on the air may be considered negligible [8.16, 8.17]. Hence, a long train with a small lateral dimension may be modelled well by a pair of acoustic monopoles of strength $\rho_0 SU (\equiv m)$ with opposite signs. For a train of axisymmetric shape and of length l, we assume that

$$q = \frac{m}{\pi r} \delta(r)[\delta(x - Ut) - \delta(x - Ut + l)] \tag{8.4}$$

where r and x denote, respectively, the radial and axial coordinates, $\delta(x)$ is the one-dimensional delta function defined over $-\infty < x < \infty$ and $\delta(r)/\pi r$ represents the axisymmetric delta function. When the pair of monopoles of equation (8.4) is placed in an unbounded fluid, it represents a flow around a Rankine ovoid. But because there exists a tunnel wall, this model should be applied only for a thin train, i.e. for $\chi \ll 1$. In this section, we assume a tunnel of radius R and seek a solution for an axisymmetric pressure field due to the source (8.4) on the central axis. This may also model a train and a tunnel of semicircular cross-section if only a half-space is considered, by using the mirror image.

8.3.2. Evaluation of the pressure field

If we attempted to solve equation (8.3) with the source (8.4) not only within the tunnel, but also in the free space specified, the analysis would be difficult. To simplify the problem, first, the effect of the free space is ignored by imposing a boundary condition $p' = 0$ at the entrance $x = 0$. If an end correction is made for the tunnel entrance, the boundary condition is applied at $x = -x_e$, where x_e is taken as $0.82R$ for a 'flanged' tunnel and $0.61R$ for an 'unflanged' tunnel [8.11, 8.19]. The application of the boundary condition neglects radiation out of the tunnel as well as scattering as the train approaches the tunnel.

As the pressure waves are propagated down the tunnel, we can expect that the pressure field tends to become one-dimensional. This field may be called a distant near field, where multidimensional behaviour disappears, but non-linear effects have not yet been accumulated significantly. To evaluate the distant field, we average equation (8.3) over the tunnel's cross-section to use the boundary condition $\partial p'/\partial r = 0$ at $r = R$. It then follows that

$$\frac{\partial^2 \tilde{p}'}{\partial t^2} = a_0^2 \frac{\partial^2 \tilde{p}'}{\partial x^2} + \frac{a_0^2 m}{\pi R^2} \frac{\partial}{\partial t}[\delta(x-Ut) - \delta(x-Ut+l)] \tag{8.5}$$

where $\tilde{p}'$ is the averaged pressure defined by

$$\tilde{p}'(x,t) = \frac{1}{\pi R^2} \int_0^R p'(r,x,t) 2\pi r\, dr \tag{8.6}$$

This is the one-dimensional wave equation, and the solution p' is easily obtainable as follows:

$$\tilde{p}' = \frac{m a_0 M}{\pi R^2 (1-M^2)} \varphi(x,t) \tag{8.7}$$

where $\varphi(x, t)$ is defined by

$$\varphi(x,t) = \frac{1}{2}\left[-\operatorname{sgn}\left(t-\frac{x}{U}\right) + \operatorname{sgn}\left(t-\frac{x}{a_0}\right) + \operatorname{sgn}\left(t-\frac{l}{U}-\frac{x}{U}\right) - \operatorname{sgn}\left(t-\frac{l}{U}-\frac{x}{a_0}\right)\right] \tag{8.8}$$

$$= h(x-Ut) - h(x-a_0 t) - h(x-Ut+l) + h\left(x-a_0 t + \frac{l}{M}\right)$$

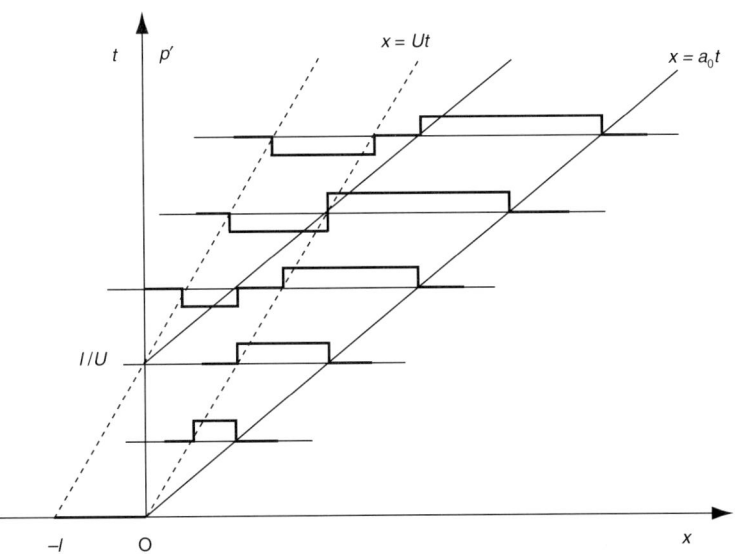

Fig. 8.4. The x–t diagram for the evolution of the pressure disturbance in the one-dimensional near field averaged over the tunnel's cross-section; the bold lines indicate the spatial profile of the excess pressure and the dotted lines represent the paths of the train's nose and tail; the magnitude of the positive and negative square humps relative to p_0 is given by $\gamma\chi M^2/(1 - M^2)$

and $\text{sgn}(x) = 2h(x) - 1$, $h(x)$ being the unit step function. The x–t diagram is depicted in Fig. 8.4, together with the spatial profiles of the excess pressure (solution (8.7)) at several typical times. It can be seen that a positive square hump is radiated forward, while a negative hump appears around the train, i.e. in the region between the two monopoles. The discontinuity in the humps is not a shock wave produced essentially by non-linear processes but is merely a discontinuous Mach wave. The shock wave will appear in the far field beyond the field concerned here.

The magnitude Δp of the positive hump radiated forward, relative to the atmospheric pressure p_0, is given from the solution (8.7) as

$$\frac{\Delta p}{p_0} = \frac{\gamma \chi M^2}{1 - M^2} \qquad (8.9)$$

where χ is the blockage ratio $S/\pi R^2$, and the factor $ma_0/\pi R^2$ divided by p_0 is equal to $\gamma\chi M$. This expression agrees with the first-order term of formula (8.1) expanded in terms of χ. If the formula is expanded up to the second order in χ, the correction to (8.9) is given by $\gamma\chi^2 M^2(3 - M)/[2(1 - M^2)(1 + M)] > 0$. Hence the leading term, i.e. expression (8.9), is found to underestimate expression (8.1). Although the higher-order terms in χ may be interpreted to provide non-linear corrections to formula (8.9) their validity is an open question, given the assumption of the one-dimensional stream tube.

Figure 8.4 shows a negative pressure of magnitude $-\Delta p$ on average over the cross-section, moving steadily with the train. But the field around the train will depart from being one-dimensional. So we look for a steady solution to formula (8.3) which depends on $x - Ut$ only. The detailed analysis shows that the steady field around the train is given by [8.16]

$$p' = \frac{ma_0 M}{2\pi R^2 (1 - M^2)} [E(r, x - Ut) - E(r, x - Ut + l)] \qquad (8.10)$$

where the function $E(r, \xi)$ is defined by the eigenfunction expansion

$$E(r, \xi) \equiv \sum_{n=0}^{\infty} \frac{1}{J_0^2(\zeta_n)} \mathrm{sgn}\,\xi \exp\left(-\zeta_n \frac{|\xi|}{\alpha R}\right) J_0\left(\zeta_n \frac{r}{R}\right) \qquad (8.11)$$

Here $\alpha = \sqrt{(1 - M^2)}$, where ζ_n ($n = 1, 2, 3, \ldots$) denote all the positive roots of $J_1(\zeta_n) = 0$ ordered as $0 < \zeta_1 < \zeta_2 < \zeta_3 < \ldots$; J_0 and J_1 stand for the zeroth- and first-order Bessel functions respectively, of the first kind. The function $E(r, \xi)$ behaves substantially as $\mathrm{sgn}\,\xi$ for $|\xi| \gg 1$. The spatial profiles of p' at some radial points are displayed in Fig. 8.5 for $l = 5\alpha R$, where one division of the ordinate represents $(ma/\pi R^2) M/[(1 - M^2) p_0] = \gamma \chi M^2/(1 - M^2)$. These profiles now replace the negative, square pressure profiles around the train shown in Fig. 8.4.

Although the wavefront of the pressure hump radiated is not a shock, its behaviour is not well treated, because the analysis has simplified the problem too much to take account of radiation and scattering at the entrance. Howe [8.17, 8.18] has succeeded in evaluating the smooth pressure profile and its temporal gradient at

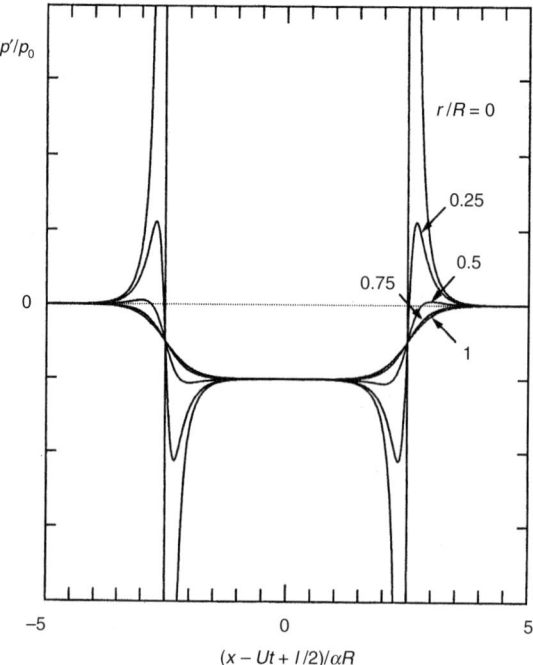

Fig. 8.5. *Spatial profiles of the excess pressure moving steadily with the train at several radial positions $r/R = 0, 0.25, 0.5, 0.75$ and 1, where the train length is chosen to be $5\alpha R$ and one division of the ordinate corresponds to $\gamma \chi M^2/(1 - M^2)$*

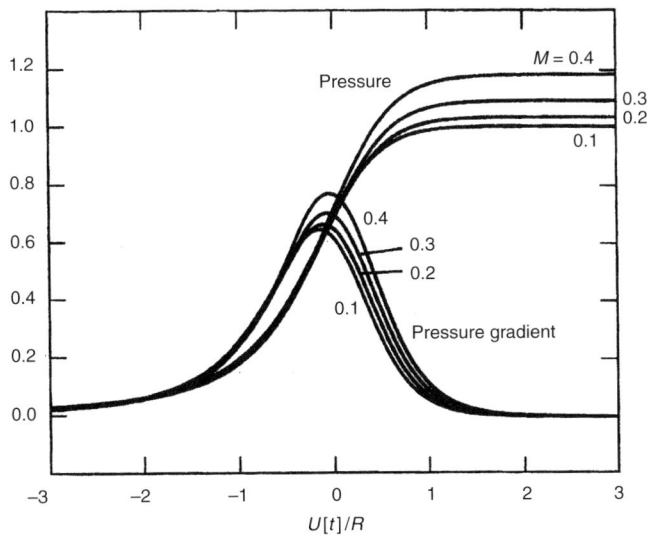

Fig. 8.6. Temporal profiles of the pressure and its gradient near the wavefront in the distant near field, where $[t] = t - x/a_0$ and the pressure and its gradient are normalized as $(\gamma \chi M^2)^{-1} p'/p_0$ and $(\gamma \chi M^2)^{-1}(R/U)\partial(p'/p_0)/\partial t$, respectively [8.17]. © Cambridge University Press

the wavefront by using the Green's function not only within the tunnel, but also in the free space outside when a single monopole in the pair in expression (8.4) enters an unflanged tunnel. Figure 8.6 reproduces his results, which show that the pressure increases so as to approach a constant value for $U[t]/R \gg 1$, where $[t]$ is equal to $t - x/a_0$ ($x \gg R$). The pressure in the figure is represented by $(\gamma \chi M^2)^{-1} p'/p_0$, and the constant value for $U[t]/R \gg 1$ seems to agree with expression (8.9). Slight differences in these values with respect to M will result from the factor $1/(1 - M^2)$. A typical time here is found to be R/U, so a typical frequency is given by U/R. The pressure gradient in the figure represents $(\gamma \chi M^2)^{-1}(R/U)\partial(p'/p_0)/\partial t$. Howe stated that the rise time is essentially independent of the Mach number (≤ 0.4), being equal to the effective transit time $2R/U$ of the nose across the entrance plane of the tunnel [8.17]. But note that $\partial p'/\partial t$ increases in proportion to M^3 because U/R is a typical frequency, which is proportional to M. If the pressure shown in the figure is radiated from the exit, as occurs in short tunnels, it is found from expression (8.2) that the pressure level increases in proportion to M^3 unless a shock is formed.

8.4. Analysis of the far field

8.4.1. Formulation

To perform an analysis of the far field, it is vitally important to take account of dissipative effects due to viscosity and heat conduction in the air. Because the pressure disturbances are of low frequency, these effects are negligibly small over a short distance. But in the long tunnels, they may be accumulated, as well as the nonlinear effects.

The dissipative effects appear through friction at the tunnel wall and through the diffusivity of sound (whose definition will be given shortly). But the former is more significant than the latter. We first introduce the acoustic Reynolds number R_e defined by $a_0^2/\omega\nu$, which refers to the sound speed, a typical wavelength a_0/ω and the kinematic viscosity ν, ω being a typical frequency. This number takes a very large value, of order 10^8 for $\omega = 10\pi$ rad/s (5 Hz), for example. For the induced air flow, the Reynolds number R_f may be defined as $a_0 u_0/\omega\nu$, u_0 being a typical speed of the flow. Defining the acoustic Mach number ε by u_0/a_0, R_f is equal to εR_e. As will be shown later in Section 8.4.2, ε is comparable with $\Delta p/p_0$ so it is of order 10^{-2} at most. Thus the Reynolds number R_f is still found to be very large, of order 10^6.

The presence of the boundary layer makes the flow field deviate from being completely one-dimensional. But the main flow outside the boundary layer may still be regarded as quasi-one-dimensional. Thus, all physical quantities in the main flow are defined as a sum of quantities averaged over the cross-section and small deviations. Of course, deviations result from non-uniformity in the thickness of the boundary layer. We start with the gas dynamic theory and average the equations over the cross-section of the tunnel except for the boundary layer. Then we retain terms up to the first order in the deviations, but neglect the deviations if they are multiplied by the viscosity or the thermal conductivity.

The equation of continuity is then given by

$$\frac{\partial \rho}{\partial t} + \frac{\partial}{\partial x}(\rho u) = \frac{1}{A}\int \rho v_n \, ds \tag{8.12}$$

where ρ and u denote, respectively, the mean values of the density and axial velocity of the air, and v_n denotes the velocity on the edge of the boundary layer within the cross-section of the tunnel normal to the axis, directed inwards, and ds is a line element along the edge. Strictly speaking, A is the cross-sectional area of the main-flow region and therefore should depend on x and t. But since the boundary layer is very thin, A may be identified as the cross-sectional area of the tunnel, and the difference becomes of higher order because v_n is of the first order.

Within the same approximation, the axial equation of motion becomes simply the Navier–Stokes equation as follows:

$$\frac{\partial u}{\partial t} + u\frac{\partial u}{\partial x} = -\frac{1}{\rho}\frac{\partial p}{\partial x} + \frac{1}{\rho}\left(\frac{4}{3}\mu + \mu_v\right)\frac{\partial^2 u}{\partial x^2} \tag{8.13}$$

where p is the mean pressure, and μ and μ_v denote, respectively, the coefficients of the shear and bulk viscosities, which are assumed to be constant, as well as the thermal conductivity k below.

The equation of energy takes a one-dimensional form as follows:

$$\rho T\left(\frac{\partial S}{\partial t} + u\frac{\partial S}{\partial x}\right) = k\frac{\partial^2 T}{\partial x^2} + \left(\frac{4}{3}\mu + \mu_v\right)\left(\frac{\partial u}{\partial x}\right)^2 \tag{8.14}$$

where T and S denote the temperature and the entropy, respectively. By assuming the equation of state of an ideal gas for the air, i.e.

$$\frac{p}{p_0} = \frac{\rho T}{\rho_0 T_0} \qquad (8.15)$$

where the subscript 0 implies the equilibrium value of the attached variable, the entropy is related to the pressure and the density as follows:

$$\frac{p}{p_0} = \left(\frac{\rho}{\rho_0}\right)^\gamma \exp\left(\frac{S-S_0}{c_v}\right) \qquad (8.16)$$

with $\gamma = c_p/c_v$, c_p and c_v being the specific heats at constant pressure and volume, respectively.

Equations (8.12)–(8.16) govern the quasi-one-dimensional propagation of the pressure waves. The equations may be simplified further by using the fact that the viscous and thermal effects are small in the main flow. Making use of the adiabatic relation given by equation (8.16), with $S = S_0$ to the lowest approximation, we evaluate a small change of the entropy. This is done by linearizing equation (8.14) as

$$\rho_0 T_0 \frac{\partial S}{\partial T} = k \frac{\partial^2 T}{\partial x^2} \qquad (8.17)$$

Using the adiabatic relation and the equation of state, $(T - T_0)/T_0$ is related to the variation of $(\rho - \rho_0)/\rho_0$. Furthermore, using the linearized relation of equation (8.13), the entropy change is related to u as [8.10]

$$\frac{\partial S}{\partial x} = -\frac{(\gamma-1)k}{\gamma p_0} \frac{\partial^2 u}{\partial x^2} \qquad (8.18)$$

This small entropy change appears in the equation of motion through the pressure gradient as follows:

$$\frac{\partial p}{\partial x} = \left.\frac{\partial p}{\partial \rho}\right|_S \frac{\partial \rho}{\partial x} + \left.\frac{\partial p}{\partial S}\right|_\rho \frac{\partial S}{\partial x} \qquad (8.19)$$

Following the standard procedure for adiabatic processes, it is convenient to express ρ in terms of the local sound speed a defined by $a = \sqrt{(\partial p/\partial \rho)_{S=S_0}} = a_0(\rho/\rho_0)^{(\gamma-1)/2}$. After doing so, we have

$$\left(\frac{\partial}{\partial t} + (u \pm a)\frac{\partial}{\partial x}\right)\left(u \pm \frac{2}{\gamma-1}a\right) = \pm\frac{a}{A}\int v_n\, ds + v_d \frac{\partial^2 u}{\partial x^2} \qquad (8.20)$$

with the plus and minus signs vertically ordered, where v_d ($= v[4/3 + \mu_v/\mu + (\gamma-1)/Pr]$) is called the diffusivity of sound and Pr ($= \mu c_p/k$) is the Prandtl number.

Equations (8.20) are closed for u and a if v_n can be related to these quantities by performing an analysis of the boundary layer. Since the full analysis has been performed by Chester [8.20], we describe only the essential results, briefly. The boundary layer consists of two layers, for the velocity and the temperature. To obtain v_n, it suffices to develop the linear theory of the boundary layer. Under the boundary layer approximation, it is well known that the pressure in the main flow prevails throughout the boundary layer. Thus, the density and the temperature are subjected to an isobaric change, i.e. $\rho'/\rho_0 = -T'/T_0$ where $\rho + \rho'$ and $T + T'$ represent, respectively, the density and the temperature in the boundary layer. Using the linearized relations of equations (8.12) and (8.13), and denoting the axial velocity in the boundary layer by $u + u'$, we derive two heat equations for u' and T':

$$\frac{\partial u'}{\partial t} = \nu \frac{\partial^2 u'}{\partial n^2} \quad \text{and} \quad \frac{\partial T'}{\partial t} = \frac{\nu}{Pr} \frac{\partial^2 T'}{\partial n^2} \qquad (8.21)$$

where n denotes the coordinate along the inward normal to the tunnel wall ($n = 0$). These equations need to be solved under non-slip and isothermal boundary conditions on the tunnel wall, $u' = -u$ and $T' = T_0 - T$ at $n = 0$, and matching conditions with the main flow, $u' \to 0$ and $T' \to 0$ as $n \to \infty$. When u' and T' are available, v_n is obtained by integrating the linearized equation of continuity for the boundary layer from the wall to infinity over n:

$$v_n = \int_0^\infty \frac{\partial v'}{\partial n} dn = -\int_0^\infty \left(\frac{\partial u'}{\partial x} + \frac{1}{\rho_0} \frac{\partial \rho'}{\partial t} \right) dn \qquad (8.22)$$

where v' is the velocity component along the n axis. As a consequence, v_n can be expressed in terms of u by using the definition of the minus half-order derivative as follows [8.21]:

$$v_n = C \sqrt{\frac{\nu}{\pi}} \int_{-\infty}^{t} \frac{1}{\sqrt{t-t'}} \frac{\partial u(x,t')}{\partial x} dt' \equiv C \sqrt{\nu} \frac{\partial^{-1/2}}{\partial t^{-1/2}} \left(\frac{\partial u}{\partial x} \right) \qquad (8.23)$$

with $C = 1 + (\gamma - 1)/\sqrt{Pr}$. As v_n is expressed in the form of a hereditary integral from the past to the present, it is found that the effect of the boundary layer, i.e. the wall friction, is accumulated.

8.4.2. Non-linear wave equation for the far field

Equations (8.20) with relation (8.23) describe bidirectional propagation of the pressure waves. But since we are concerned with unidirectional propagation because of the neglect of reflection at the tunnel exit, the equations may be simplified further. Noting that the dissipative terms on the right-hand side of equations (8.20) are much smaller than the terms on the left-hand side, $u \pm 2a/(\gamma - 1)$ may be taken as constants, to the lowest, lossless approximation, along each characteristic defined by $dx/dt = u \pm a$, with the signs vertically ordered. Using the lowest-order relation $u - 2a/(\gamma - 1) = -2a_0/(\gamma - 1)$ in equation (8.20) with the upper sign, i.e. the relation for a simple wave [8.12], we include the small dissipative effects on the right-hand side for propagation in the positive direction of x, to derive

$$\frac{\partial u}{\partial t} + a_0 \frac{\partial u}{\partial x} + \frac{\gamma+1}{2} u \frac{\partial u}{\partial x} = \frac{Ca_0 \sqrt{\nu}}{R} \frac{\partial^{-1/2}}{\partial t^{-1/2}} \left(\frac{\partial u}{\partial x} \right) + \frac{\nu_d}{2} \frac{\partial^2 u}{\partial x^2} \qquad (8.24)$$

where R may be taken to be the hydraulic radius of the tunnel if its cross-section is not circular.

To put this equation in a compact form, we normalize it. To specify the small order of non-linearity, we introduce a parameter ε ($0 \ll \varepsilon \ll 1$) by setting $[(\gamma + 1)/2]u/a_0$ equal to εf ($f = O(1)$). This parameter represents the order of the acoustic Mach number. Further, by introducing a retarded time θ ($= \omega(t - x/a_0 - \text{constant})$) measured in a frame moving with the sound speed, ω being a typical frequency, and a far-field coordinate X ($= \varepsilon\omega x/a_0$) associated with the non-linearity, we rewrite equation (8.24) in terms of these variables. It then follows that

$$\frac{\partial f}{\partial X} - f\frac{\partial f}{\partial \theta} = -\delta \frac{\partial^{1/2} f}{\partial \theta^{1/2}} + \beta \frac{\partial^2 f}{\partial \theta^2} \qquad (8.25)$$

where the half-order derivative is defined by [8.21]

$$\frac{\partial^{1/2} f}{\partial \theta^{1/2}} = \frac{1}{\sqrt{\pi}} \int_{-\infty}^{\theta} \frac{1}{\sqrt{\theta - \theta'}} \frac{\partial f(X,\theta')}{\partial \theta'} d\theta' = \frac{\partial}{\partial \theta}\left(\frac{\partial^{-1/2} f}{\partial \theta^{-1/2}}\right) \qquad (8.26)$$

and the parameters δ and β are defined as follows:

$$\delta = C\frac{\sqrt{\nu/\omega}}{\varepsilon R} \quad \text{and} \quad \beta = \frac{\nu_d \omega}{2\varepsilon a_0^2} \qquad (8.27)$$

This is the dimensionless non-linear wave equation for the far field. When the half-order derivative is absent, it is the well-known Burgers equation, expressed in the form of a spatial evolution [8.12]. Thus, equation (8.25) generalizes the Burgers equation to include the hereditary effect of the boundary layer. As will be seen shortly, however, β is so small in the present problem that the second-order term is almost negligible except for a thin shock layer, which is replaced by a discontinuity in the following analysis [8.22, 8.23].

In terms of f, p'/p_0 is given by $[2\gamma/(\gamma + 1)]\varepsilon f$ because the relation $p' \approx \rho_0 a_0 u$ holds to the lowest approximation. Let the maximum values of u and p' at $X = 0$ be Δu and Δp, respectively, and take f to be unity there. Then ε is related to Δp and Δu by

$$\varepsilon = \frac{\gamma+1}{2\gamma}\frac{\Delta p}{p_0} = \frac{\gamma+1}{2}\frac{\Delta u}{a_0} \qquad (8.28)$$

Hence ε is the acoustic Mach number defined for a typical speed of $(\gamma + 1)\Delta u/2$. In the definition of δ, $\sqrt{(\nu/\omega)}/R$ measures the thinness of the boundary layer relative to the radius, and is assumed to be much smaller than unity. Therefore, δ represents the competition between the effects due to the wall friction and to the non-linearity ε. Similarly, β represents the competition between $1/R_e$ and ε due, respectively, to the diffusivity and the non-linearity, and it is nothing but $1/R_f$ multiplied by $\nu_d/2\nu$. As the non-linearity becomes strong, the dissipative effects become relatively small. It is to be noted that since ω is proportional to M, while ε is proportional to M^2 for $M^2 \ll 1$ from expression (8.9), δ and β decrease as $M^{-5/2}$ and M^{-1}, respectively, as M increases.

For air at 15°C, we have $\gamma = 1.4$, $Pr = 0.72$, $C = 1.47$, $\nu = 1.45 \times 10^{-5}$ m²/s and $\nu_d = 2.49\nu$, if μ_v/μ is set equal to 0.6 [8.11]. Supposing a tunnel of diameter 10 m and a typical frequency $\omega = 3\pi$ rad/s (1.5 Hz), δ and β take values of $3.7 \times 10^{-4}/\varepsilon$ and $1.5 \times 10^{-9}/\varepsilon$, respectively. If ε takes a moderate value of 3.7×10^{-3} ($\Delta p/p_0 = 4.3 \times 10^{-3}$), δ takes the value 0.1, but β takes a very small value of the order of 10^{-7}. This case is close to the one in which a train travelling at a speed of 200 km/h enters a tunnel with $\chi = 0.1$.

8.4.3. Evolution of the pressure wave into a shock

In order to examine the evolution of the pressure waves by solving equation (8.25), an initial (physically a boundary) condition must be supplied at $X = 0$. This condition can be provided by an analysis of the distant near field. In fact, we now know of the emergence of a pressure hump of duration l/U in this field. With the magnitude (expression (8.9)) and the temporal gradient available, the two parameters ε and ω left unspecified can now be determined and the phase constant in θ also given. The matching is executed at an arbitrary point $x = x_d$ ($\gg R$) in the distant near field such that $\varepsilon \omega x_d/a_0 \ll 1$. This point is now regarded as $X = 0$ in the far field. Hence, we remark that the point $X = 0$ does not mean the entrance.

For the evolution of the pressure disturbances generated by the entry of a real high-speed train, the duration of the hump is too long to be implemented in numerical calculations. To reduce the amount of computation, yet demonstrate the emergence of the shock, we assume a short train and model the hump by a Gaussian-shaped pulse as follows:

$$f(\theta, X = 0) = \exp(-\theta^2) \tag{8.29}$$

where the constant in the definition of θ is chosen suitably so as to set the peak at $\theta = 0$. Figure 8.7 shows the spatial evolution of the pulse up to $X = 4$ for $\delta = 0.1$ and $\beta = 0$. The profile tends to steepen forward (leftward) to form a discontinuity at

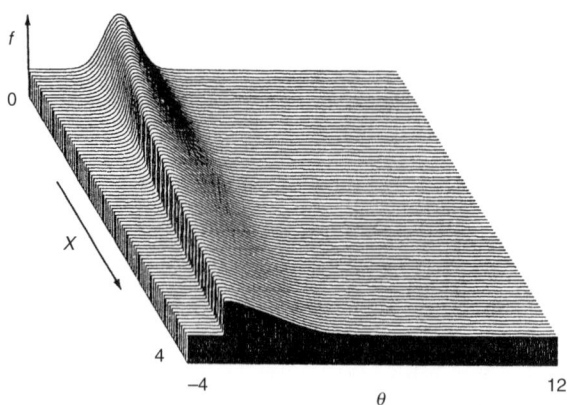

Fig. 8.7. Evolution of a Gaussian-shaped pulse at $X = 0$ into a shock wave in the far field; the shock formation point is located at $X = 1.255$ for $\delta = 0.1$ and the height of the vertical arrow corresponds to a value of unity of f

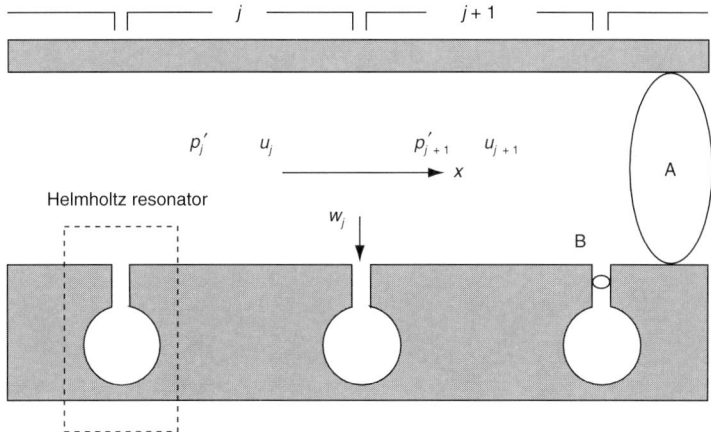

Fig. 8.8. *Periodic array of Helmholtz resonators connected to a tunnel*

$X = 1.255$. Incidentally, we remark that since β is now set equal to zero, the shock fitting can be done for further evolution of the discontinuity [8.22, 8.23]. By the definition of X, this shock formation point corresponds to the point $1.255\, a_0/\varepsilon\omega$ in terms of x. For $\varepsilon = 3.7 \times 10^{-3}$ and $\omega = 3\pi$ rad/s, the shock formation distance is about 12 km. If δ is completely ignored, the shock formation point in terms of X is given by $\sqrt{(e/2)}\ (= 1.166)$. In view of this, it is found that the wall friction delays the shock formation, but fails to suppress it. Hence, as a train speed is increased, the emergence of a shock tends to be unavoidable even in shorter tunnels.

8.5. Shock-free propagation

In order to suppress emergence of the shock wave, we have introduced the concept of shock-free propagation and proposed one method to realize it. This method is to exploit dispersion, which is realized by connecting an array of Helmholtz resonators to the tunnel (Fig. 8.8). A Helmholtz resonator is a cavity closed everywhere except for a throat with an opening. For simplicity, it is assumed that the resonators in the array are identical and that the axial spacing between neighbouring resonators is constant. Then the tunnel has spatially periodic structure, if it extends infinitely. This periodicity is expected to give rise to dispersion and also damping, owing to the geometrical configuration. In the following, we first show this by discarding all dissipative effects to focus on the effect of spatial periodicity. For an analysis taking account of dissipative effects, see [8.24].

8.5.1. Linear dispersion characteristics

With the periodic array connected, the tunnel consists of infinite number of intervals as shown in Fig. 8.8. Let each interval be numbered by consecutive values of j from minus to plus infinity ($j = \ldots, -1, 0, 1, \ldots$), and take an axial coordinate x along the tunnel with its origin at the midpoint in the interval $j = 0$. Of course, no entrance of the tunnel is considered here. It is assumed that the cavity's volume V is small compared with the tunnel's volume Ad in each interval, d being the spacing between the neighbouring resonators, and that the throat's cross-sectional area B is also much smaller than that of the tunnel A:

$$\frac{V}{Ad} \equiv \kappa \ll 1 \quad \text{and} \quad \frac{B}{A} \ll 1 \tag{8.30}$$

where the parameter κ may be called a 'size parameter' of the resonator. The assumptions (8.30) enable us to assume plane wave propagation in each interval.

The lossless linear wave equation in each interval is given by

$$\frac{\partial^2 p'_j}{\partial t^2} = a_0^2 \frac{\partial^2 p'_j}{\partial x^2} \quad (j = \ldots -1, 0, 1, \ldots) \tag{8.31}$$

where p'_j denotes the excess pressure in the interval j. Assuming a temporally sinusoidal disturbance of the form $\exp(-i\omega t)$, ω being the angular frequency, p'_j is given by a superposition of two waves propagating in the positive and negative directions of x as follows:

$$p'_j = f_j \exp[i(kx_j - \omega t)] + g_j \exp[i(-kx_j - \omega t)] \tag{8.32}$$

where k is a wavenumber defined by ω/a_0 and $x_j \equiv x - jd$ ($-d/2 < x_j < d/2$); no confusion should arise from the use of k for both the wavenumber and the thermal conductivity. Here f_j and g_j represent the complex wave amplitudes, which are to be determined from the relations among neighbouring intervals and the resonator in between. Given the excess pressure (expression (8.32)), the axial velocity u_j in the interval j can be derived immediately by using the acoustic impedance $\rho_0 a_0$ for plane waves as follows:

$$u_j = \frac{f_j}{\rho_0 a_0} \exp[i(kx_j - \omega t)] - \frac{g_j}{\rho_0 a_0} \exp[i(-kx_j - \omega t)] \tag{8.33}$$

Since the resonators may be regarded as being connected at 'points' $x = (j + 1/2)d$ because of the second assumption of equation (8.30), the continuity of mass flux and of pressure require that

$$A\rho_0(u_j - u_{j+1}) = B\rho_0 w_j \tag{8.34}$$

$$p'_j = p'_{j+1} \tag{8.35}$$

at $x = (j + 1/2)d$, where w_j denotes the velocity directed into the resonator from the tunnel.

For w_j, we need to specify the response of the resonator to pressure fluctuations at the opening of the throat. Since the cavity's volume is much greater than that of the throat, the motion of the air in the cavity is negligible. Hence it suffices to consider only the conservation of mass in the cavity as follows:

$$V\frac{\partial \rho_c}{\partial t} = B\rho_0 w_c \tag{8.36}$$

where ρ_c denotes the mean density of the air in the cavity, and w_c denotes the velocity of the flow into the cavity, averaged over the throat's cross-section. For the air in the throat, the linearized inviscid equation of motion is assumed. This equation is integrated along the throat's length L as follows:

$$\rho_0 L\frac{\partial w}{\partial t} = -p_c + p_t \tag{8.37}$$

where p_c and p_t represent the pressures at the ends of the throat on the cavity side and on the tunnel side, respectively. In deriving this equation, the mean axial velocity w has been assumed to be uniform along the throat because the characteristic wavelength is much greater than the throat's length, so that the compressibility of the air is negligible in the throat. Thus w_c is equal to w. Use of the adiabatic approximation for the air in the cavity leads to the relation $\partial \rho_c/\partial t \approx (d\rho_c/dp_c)\partial p_c/\partial t$, p_c being the mean pressure in the cavity. Furthermore, by noting that the $dp_c/d\rho_c \approx a_0^2$, equations (8.36) and (8.37) can be combined into

$$\frac{\partial^2 p'_c}{\partial t^2} + \omega_0^2 p'_c = \omega_0^2 p'_t \tag{8.38}$$

where $\omega_0 (= (Ba_0^2/LV)^{1/2})$ is the natural angular frequency of the resonator, and p'_c and p'_t denote the respective excess pressures over p_0. Here L is usually lengthened by the so-called end correction $0.82r$ at each end [8.11, 8.19]. Assuming that p'_t varies sinusoidally in the form $P \exp(-i\omega t)$, the volume flow Bw from the tunnel into the throat is induced in a similar form, $Q \exp(-i\omega t)$, where P and Q denote complex amplitudes, and the ratio P/Q defines an acoustic impedance of the resonator Z, depending on ω. By using equations (8.36) and (8.38), Z is given as follows:

$$Z = i\frac{\rho_0 L}{B}\left(\frac{\omega_0^2 - \omega^2}{\omega}\right) \tag{8.39}$$

Since p'_t must be equal to p'_j and p'_{j+1} at $x = (j + 1/2)d$, we can express w_j in terms of those pressures by using the acoustic impedance Z. The relation between (f_j, g_j) and (f_{j+1}, g_{j+1}) is then established through a transmission matrix $\mathbf{W}$ as follows:

$$X_{j+1} = \mathbf{W}X_j \tag{8.40}$$

where

$$X_j = \begin{bmatrix} f_j \\ g_j \end{bmatrix} \tag{8.41}$$

and

$$W = \begin{bmatrix} (1-1/2\mathcal{R})\exp(ikd) & -1/2\mathcal{R} \\ 1/2\mathcal{R} & (1+1/2\mathcal{R})\exp(ikd) \end{bmatrix} \tag{8.42}$$

Here $\mathcal{R} = Z/Z_A$, where Z_A ($= \rho_0 a_0/A$) is the acoustic impedance of the tunnel. Here $1/\mathcal{R}$ is also expressed by using κ, as follows:

$$\frac{1}{\mathcal{R}} = -\mathrm{i}\kappa\psi_0\left(\frac{\omega/\omega_0}{1-\omega^2/\omega_0^2}\right) \tag{8.43}$$

where ψ_0 ($= \omega_0 d/a_0$) is the ratio of the axial spacing d relative to the typical wavelength a_0/ω_0.

We consider here an elementary solution to equation (8.40) of the form $X_j = \lambda^j C$, where C is an arbitrary column vector. For this to be a solution, λ must be an eigenvalue of $\mathbf{W}$. When λ is set equal to $\exp(\mathrm{i}qd)$, q being allowed to be complex, it is found that q must satisfy the following dispersion relation:

$$\cos(qd) = \cos\left(\frac{\omega d}{a_0}\right) - \frac{\mathrm{i}}{2\mathcal{R}}\sin\left(\frac{\omega d}{a_0}\right) \tag{8.44}$$

The ratio f_j/g_j in each interval is given by q and is independent of j, so that the excess pressure p' at an arbitrary point x in the tunnel can be expressed in the form $\Phi(x)\exp[\mathrm{i}(qx - \omega t)]$ where $\Phi(x)$ [$= \Phi(x + d)$] is a periodic function of x with period d. This elementary solution for p' is a 'Bloch wave function', which is known to occur generally for propagation in a spatially periodic structure, and q is called the 'Bloch wavenumber' [8.25, 8.26].

We examine the dispersion characteristics $q = q(\omega)$ given by relation (8.44). Without the side branches, i.e. $\mathcal{R} \to \infty$, q is given by ω/a_0 ($= k$), which is nothing but the dispersion relation for the non-dispersive fundamental mode. For a finite value of $\mathcal{R}$, $\mathrm{d}\omega/\mathrm{d}q$ is no longer a constant independent of ω. This means that the pressure waves exhibit dispersion. When one solves for q with a given real frequency ω, q is usually found to be real, but there appears a frequency range in which q becomes complex. Figure 8.9 shows the real and imaginary parts of qd, as the abscissa, versus the frequency ω/ω_0, as the ordinate. Here a tunnel of diameter 10 m is assumed, to which a spherical cavity of diameter 6 m is connected through a throat of circular cross-section of diameter 2 m and length 3 m, with an axial spacing 10 m. The natural frequency of the resonator is then 5.2 Hz. In this case, κ and ψ_0 take values of 0.144 and 0.960, respectively. Only the positive branches of the real and imaginary parts obtained by using these values are shown. Note that $-q$ is also a solution.

The imaginary part gives rise to damping or amplification of pressure waves owing to the geometry of the tunnel. Obviously the amplification should not be considered, because no energy sources are present, but what does the damping imply? Since we are concerned with the lossless case, this damping has nothing to do with dissipation of energy that is transformed into heat. It is an evanescent mode that occurs owing to the reflection of pressure waves by the resonators. At a frequency very close to ω_0, the resonators resonate with the incident wave and strong damping occurs. In fact, the imaginary part of qd diverges as $\omega/\omega_0 \to 1$. But its real part remains fixed at π for $\omega/\omega_0 < 1$ and at zero for $\omega/\omega_0 > 1$. At frequencies near $\omega/\omega_0 \approx 3.3$ and 6.5, an imaginary part also appears, but it remains finite, while the real part is fixed at π and

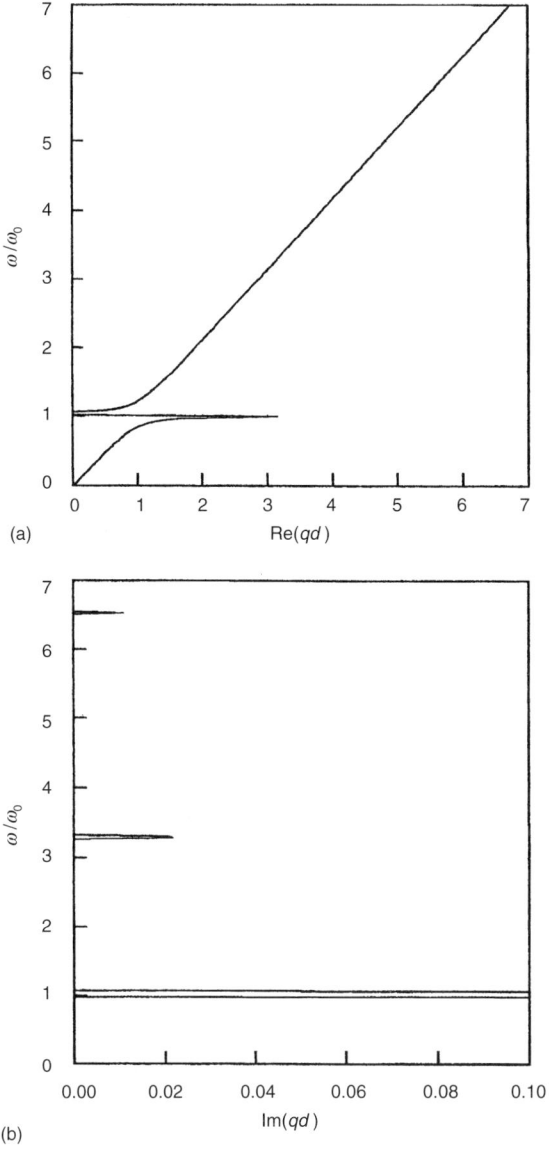

Fig. 8.9. Dispersion characteristics of pressure waves in a tunnel with an array of Helmholtz resonators for $\kappa = 0.144$ and $\psi_0 = 0.960$; (a) the real and (b) imaginary part of qd as a function of ω/ω_0

2π, respectively. This damping is brought about by Bragg reflection when the axial spacing becomes a multiple of the half wavelength $\pi a_0/\omega$, i.e. $\omega = m\pi a_0/d$ ($m = 1, 2, ...$). When an imaginary part exists, the pressure wave cannot be propagated forward. Such a frequency range defines a stopband, in contrast to the passband outside of it. For pressure waves containing many frequencies, only components in the stopband are indeed 'stopped', and the others are passed but dispersed. The stopband due to the side-branch resonance will be useful, but the ones due to the Bragg reflection are too narrow and the damping is too small to be exploited in the present array.

Here it is important to examine the dispersion characteristics at long wavelengths because the axial spacing d is taken to be much shorter than a typical wavelength. Expanding equation (8.44) in terms of ψ ($= \omega d/a_0 \ll 1$), we have

$$(qd)^2 = (1+\kappa)\psi^2 + \left(\frac{\kappa}{\psi_0^2} + \frac{\kappa^2}{12}\right)\psi^4 + O(\psi^6) \tag{8.45}$$

where $\psi_0 = \omega_0 d/a_0$, provided $\omega \ll \omega_0$. If ω_0 is small and comparable with ω, expression (8.45) is replaced by

$$(qd)^2 = \left(1 - \frac{\kappa}{(\omega/\omega_0)^2 - 1}\right)\psi^2 + O(\psi^4) \tag{8.46}$$

In any case, it is found that dispersion appears because of the periodic structure of the tunnel.

8.5.2. Effects of the array of Helmholtz resonators

By discarding the assumption of a lossless linear wave, we next formulate the problem of non-linear wave propagation in a tunnel with an array of resonators. Here we make the assumption that the axial spacing is much shorter than the wavelength. As long as the size parameter κ remains small, the effects of the resonators on the main flow are small locally and they may be considered in a way similar to that for the effects of the boundary layer. The assumption of a narrow axial spacing enables us to make a continuum approximation for the distribution of the resonators, i.e. to smear out the effects of the resonators over the unit axial length of the tunnel. Then the same governing equations (8.12)–(8.16) can still be employed. The new effects of the array appear through the integral in equation (8.12). Now v_n should take account of not only the velocity due to the boundary layer given by expression (8.23), denoted by v_b, but also the velocity w of the flow into the throat where there is an opening in the wall. Thus the integral can be approximated, per unit axial length of the tunnel, as

$$\frac{1}{A}\int v_n \, ds \approx \frac{1}{A}\left[\left(\frac{2A}{R} - NB\right)v_b - NBw\right] \tag{8.47}$$

where R is the hydraulic radius of the tunnel and NB accounts for the total cross-sectional area of the openings per unit axial length, N ($= 1/d$) being the number density of the resonators.

Proceeding in the same way as before, it follows that

$$\left(\frac{\partial}{\partial t} + (u \pm a)\frac{\partial}{\partial x}\right)\left(u \pm \frac{2}{\gamma - 1}a\right) = \\ \pm \frac{2Ca_0 \nu^{1/2}}{R^*}\frac{\partial^{-1/2}}{\partial t^{-1/2}}\left(\frac{\partial u}{\partial x}\right) + \nu_d \frac{\partial^2 u}{\partial x^2} \mp \frac{V}{\rho_0 a_0 Ad}\frac{\partial p'_c}{\partial t} \tag{8.48}$$

ACOUSTIC SHOCK WAVE IN A TUNNEL

where the signs are vertically ordered and $1/R^*$ is defined as $(1 - NRB/2A)/R$. These equations are closed when taken together with equation (8.38), if a linear and lossless response of the resonator is assumed. But if the friction at the throat's wall is taken into account, it can be included in the form of a similar hereditary integral known as the derivative of three-half order. Furthermore, if the non-linear response of the air in the cavity and the non-linear loss due to formation of jets are to be included, equation (8.38) is modified as follows [8.10]:

$$\frac{\partial^2 p'_c}{\partial t^2} + \frac{2L'\nu^{1/2}}{L_e r}\frac{\partial^{3/2} p'_c}{\partial t^{3/2}} + \omega_e^2 p'_c - \frac{\gamma-1}{2\gamma p_0}\frac{\partial^2 p'^2_c}{\partial t^2} + \frac{V}{BL_e \rho_0 a_0^2}\left|\frac{\partial p'_c}{\partial t}\right|\frac{\partial p'_c}{\partial t} = \omega_e^2 p'_t \quad (8.49)$$

where $L' = L + 2r$, r stands for the hydraulic radius of the throat and the derivative of three-half order is defined by differentiating the half-order derivative with respect to t once; p'_t is identified as p' in the tunnel. Here $\omega_e\ (= (Ba_0^2/L_e V)^{1/2})$ is defined by taking the end corrections to the throat as $L_e\ (= L + 2 \times 0.82r)$. Assuming unidirectional propagation and making the same approximations used in deriving equation (8.24), equation (8.48) with the upper sign is reduced to the following equation:

$$\frac{\partial u}{\partial t} + a_0 \frac{\partial u}{\partial x} + \frac{\gamma+1}{2} u \frac{\partial u}{\partial x} = \frac{Ca_0\sqrt{\nu}}{R^*}\frac{\partial^{-1/2}}{\partial t^{-1/2}}\left(\frac{\partial u}{\partial x}\right) + \frac{\nu_d}{2}\frac{\partial^2 u}{\partial x^2} - \frac{V}{2\rho_0 a_0 Ad}\frac{\partial p'_c}{\partial t} \quad (8.50)$$

Thus, the effects of the array of Helmholtz resonators appear through the last term on the right-hand side coupled with equation (8.49).

For normalization of equations (8.50) and (8.49), we set $[(\gamma + 1)/2]u/a_0$ and $[(\gamma + 1)/2\gamma]p'_c/p_0$ equal to εf and εg, respectively, and use θ and X defined as before. Then it follows that

$$\frac{\partial f}{\partial X} - f\frac{\partial f}{\partial \theta} = -\delta_R \frac{\partial^{1/2} f}{\partial \theta^{1/2}} + \beta \frac{\partial^2 f}{\partial \theta^2} - K\frac{\partial g}{\partial \theta} \quad (8.51)$$

and

$$\frac{\partial^2 g}{\partial \theta^2} + \delta_r \frac{\partial^{3/2} g}{\partial \theta^{3/2}} + \Omega g = \Omega f + \varepsilon\left[\left(\frac{\gamma-1}{\gamma+1}\right)\frac{\partial^2 g^2}{\partial \theta^2} - \frac{2V}{(\gamma+1)BL_e}\left|\frac{\partial g}{\partial \theta}\right|\frac{\partial g}{\partial \theta}\right] \quad (8.52)$$

where δ_R, K, δ_r and Ω are defined by

$$\delta_R = C\frac{\sqrt{\nu/\omega}}{\varepsilon R^*}, \quad K = \frac{\kappa}{2\varepsilon}, \quad \delta_r = \frac{2c_L\sqrt{\nu/\omega}}{r} \quad \text{and} \quad \Omega = \left(\frac{\omega_e}{\omega}\right) \quad (8.53)$$

with $c_L = L'/L_e$. Here the non-linear terms in equation (8.52) are small because of ε but are retained for the case in which Ω becomes small and comparable to ε. The effect of the wall friction appears through δ_R and δ_r. The effect of the array of resonators appears through the two parameters K and Ω called, respectively, the 'coupling parameter' and the 'tuning parameter'. The former measures the competition between the size of the resonators and the non-linearity, while the latter measures how far the typical frequency of the pressure waves is detuned from the natural frequency of the resonator. For the same array of resonators as that in Fig. 8.9, these parameters take the following values: $\delta_R = 2.0 \times 10^{-4}/\varepsilon$, $\delta_r = 1.4 \times 10^{-3}$, $K = 0.072/\varepsilon$ and $\Omega = 1.1$ for $\omega = 10\pi$ rad/s (5 Hz).

8.5.3. Suppression of shock formation

We now examine the evolution of the pressure wave by solving equations (8.51) and (8.52) under the same initial condition (8.29). The initial condition for g is determined from a solution to equation (8.52) with f prescribed. In following simulations, we consider a case with $\varepsilon = 0.05$, $\delta_R = 4.0 \times 10^{-3}$ and $\delta_r = 1.4 \times 10^{-3}$, for $\omega = 10\pi$ rad/s, where β is neglected. This pressure level corresponds to that generated by a train at a speed of 660 km/h in a tunnel with $\chi = 0.1$.

To see the effect of the tuning parameter Ω, we examine first the evolution for three values of Ω with K fixed. Figures 8.10–8.12 show the evolution for $\Omega = 0.1$, 1 and 10 with $K = 1$. In each figure, the upper part (a) and the lower part (b) show the evolution of f and g, respectively, where the direction of the X axis is reversed in part (b) to exhibit the initial profile of g. The height of the vertical arrow in each figure corresponds to unity for f or g, and a value of X of unity corresponds physically to

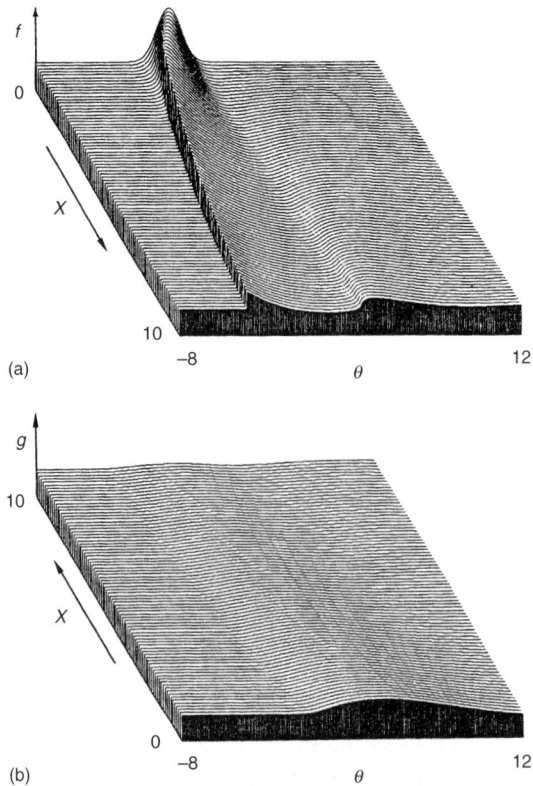

Fig. 8.10. Spatial evolution of a Gaussian-shaped pulse in the case $\Omega = 0.1$ and $K = 1$; (a) and (b) show the evolution of f and g, respectively, and the height of the vertical arrow corresponds to unity

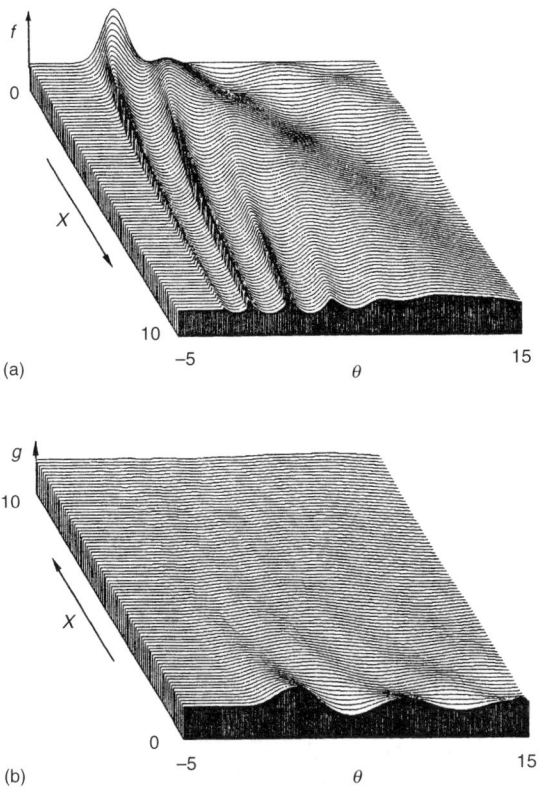

Fig. 8.11. Spatial evolution of a Gaussian-shaped pulse in the case $\Omega = 1$ and $K = 1$; (a) and (b) show the evolution of f and g, respectively, and the height of the vertical arrow corresponds to unity

about 0.2 km for the present choice of ε. Figure 8.10 shows the emergence of a shock wave, while Fig. 8.11 shows the disintegration of the initial pulse into three shock waves, up to $X = 10$. The latter situation is obviously worse than that without the array of resonators. Even for $\Omega = 1$, where the damping is highest, the shock cannot be suppressed when $K = 1$. But Fig. 8.12 shows no indication of shock formation at all, though the initial pulse is not damped.

Next we examine the effect of the coupling parameter K. As it is increased, the effects of both damping and dispersion become enhanced, so that no emergence of shock waves is expected. For $K = 10$ and $\Omega = 1$, in fact, the shock wave is inhibited as shown in Fig. 8.13, and the initial pulse has decayed significantly at $X = 10$. For $\Omega = 10$, of course, no shock waves emerge, as in the case with $K = 1$. For $\Omega = 0.1$, however, it is found that two shock waves still emerge. These results suggest several important implications. Figure 8.11 shows that the damping alone cannot compete with the non-linearity, and allows the emergence of two shock waves. For $\Omega = 0.1$ and $\Omega = 10$, of course, no substantial damping appears. Although the dispersion relation suggests that there is a stopband near $\omega/\omega_0 \approx 3.3$, i.e. $\Omega \approx 0.09$, the continuum approximation has smeared out the discrete distribution of the resonators so that the damping due to the Bragg reflection cannot be taken into account.

As K is increased, the shock wave tends to be inhibited, but only if $\Omega \geq 1$. An interesting finding is that for $\Omega = 10$, shock formation is suppressed for a moderate value of K ($K = 1$) or even for a smaller value. This may be understood by considering the extreme case $\Omega \gg 1$. Ignoring all dissipative effects in equation (8.52), g can be approximated as

$$g = f - \frac{1}{\Omega}\frac{\partial^2 g}{\partial \theta^2} = f = \frac{1}{\Omega}\frac{\partial^2 f}{\partial \theta^2} + O\left(\frac{1}{\Omega^2}\right) \tag{8.54}$$

Substituting this into equation (8.51), we derive the Korteweg–de Vries equation (called simply K–dV equation hereafter) for f, on neglecting terms of order $1/\Omega^2$:

$$\frac{\partial f}{\partial X} + K\frac{\partial f}{\partial \theta} - f\frac{\partial f}{\partial \theta} = \Gamma\frac{\partial^3 f}{\partial \theta^3} \tag{8.55}$$

with $\Gamma = K/\Omega$. The well-known properties of this equation suggest that initial

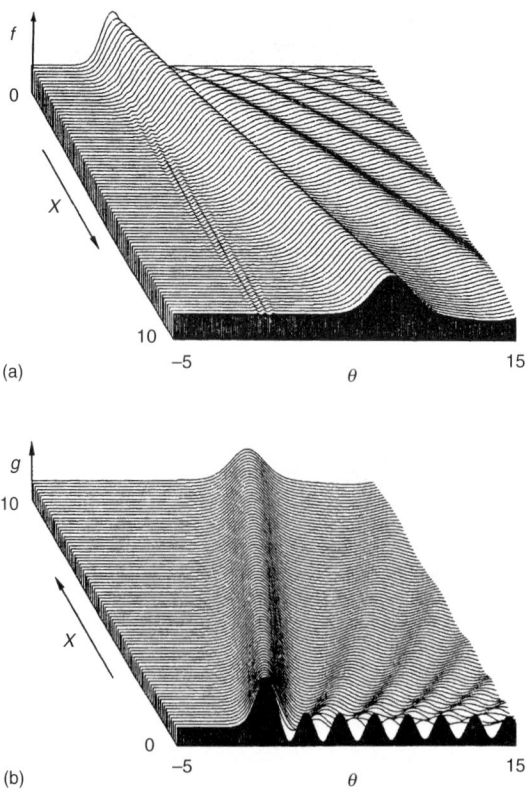

Fig. 8.12. *Spatial evolution of a Gaussian-shaped pulse in the case $\Omega = 10$ and $K = 1$; (a) and (b) show the evolution of f and g, respectively, and the height of the vertical arrow corresponds to unity*

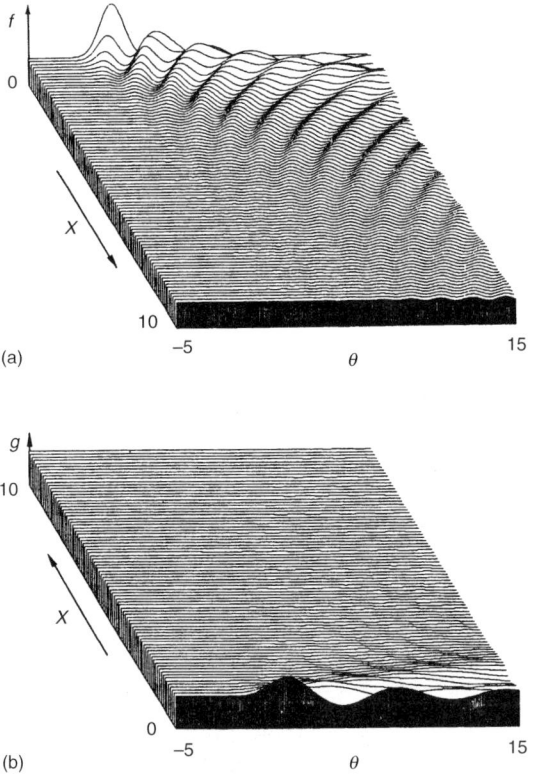

Fig. 8.13. Spatial evolution of a Gaussian-shaped pulse in the case with $\Omega = 1$ and $K = 10$; (a) and (b) show the evolution of f and g, respectively, and the height of the vertical arrow corresponds to unity

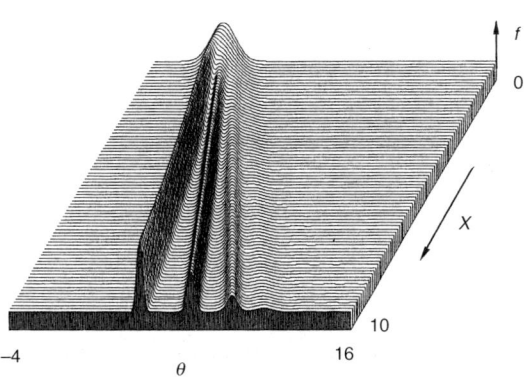

Fig. 8.14. Spatial evolution of a Gaussian-shaped pulse in the case with $\Omega = 200$ and $K = 1$; only f is displayed, but g is almost equal to f from equation (8.54), and the height of the vertical arrow corresponds to unity

disturbances no longer evolve into shock waves, but into a sequence of solitons, asymptotically as $X \to \infty$ [8.27]. Each soliton is expressed in the following form:

$$f = A \operatorname{sech}^2\left[\left(\frac{A}{12\Gamma}\right)^{1/2} (\theta - KX + \tfrac{1}{3} AX - \theta_0)\right] \tag{8.56}$$

where A and θ_0 are constants to be determined by solving an initial-value problem for equation (8.55). Figure 8.14 shows the evolution in the case where $\Omega = 200$ and $K = 1$. Although no shock wave emerges in this case, the rapid changes due to the solitons would give rise to problems on radiation from an exit. But it is worth noting the following point. The dispersion appears in equation (8.55) through the higher-order derivative (of the third order) rather than in the non-linear term. This 'higher-order dispersion' can compete with the non-linear steepening to inhibit shock waves.

On the other hand, the dispersion for $\Omega \leq 1$ yields only 'lower-order dispersion'. In fact, Ωg and ε in equation (8.52) may be dropped, and elimination of g in equation (8.51) leads to

$$\left(\frac{\partial}{\partial \theta} + \delta_r \frac{\partial^{1/2}}{\partial \theta^{1/2}}\right)\left(\frac{\partial f}{\partial X} - f \frac{\partial f}{\partial \theta} + \delta_R \frac{\partial^{1/2} f}{\partial \theta^{1/2}} - \beta \frac{\partial^2 f}{\partial \theta^2}\right) = -K\Omega f \tag{8.57}$$

Because the term due to the array appears on the right-hand side in a form without differentiation, its effect remains weak even when the non-linear steepening of the profile is occurring. Thus, it fails to counteract the non-linearity and allows the emergence of shock waves eventually [8.10].

By examining both limiting cases of Ω, it has been found that dispersion appears in two different ways. For a large but finite value of Ω, it is known that an acoustic solitary wave free from a shock can be propagated for $\Omega > 4\pi^2$ and that the acoustic soliton (equation (8.56)) is one limit of the acoustic solitary wave as $\Omega \to \infty$ [8.28]. Here we note that the typical frequency ω used in the definition of Ω is defined for the solitary wave, not for the initial value. In view of these results, shock-free propagation can be achieved only if the dispersion acts in a way similar to higher-order dispersion. Obviously, this is the case with $\Omega \gg 1$, and then no damping can be expected.

In order to bring both dispersion and damping into play, Ω may be chosen in an intermediate range, $1 < \Omega < 10$. In fact, the case of $\Omega = 3$–5 is found to be very effective. To make both mechanisms more effective, a double array of Helmholtz resonators may be devised. Using the subscript m ($m = 1, 2$) to distinguish quantities pertinent to the two arrays specified by m, two size parameters κ_m, defined by V_m/Ad_m, and two natural angular frequencies ω_m are introduced. Also, the coupling parameters K_m and the tuning parameters Ω_m are defined by $\kappa_m/2\varepsilon$ and $(\omega_m/\omega)^2$, respectively. For the double array, the non-linear wave equations (8.51) and (8.52) are modified as follows:

$$\frac{\partial f}{\partial X} - f \frac{\partial f}{\partial \theta} = -\delta_R \frac{\partial^{1/2} f}{\partial \theta^{1/2}} + \beta \frac{\partial^2 f}{\partial \theta^2} - K_1 \frac{\partial g_1}{\partial \theta} - K_2 \frac{\partial g_2}{\partial \theta} \tag{8.58}$$

and

$$\frac{\partial^2 g_m}{\partial \theta^2} + \delta_{rm}\frac{\partial^{3/2} g_m}{\partial \theta^{3/2}} + \Omega_m g_m = \Omega_m f + \varepsilon\left[\left(\frac{\gamma-1}{\gamma+1}\right)\frac{\partial^2 g_m^2}{\partial \theta^2} - \frac{2V_m}{(\gamma+1)B_m L_{em}}\left|\frac{\partial g_m}{\partial \theta}\right|\frac{\partial g_m}{\partial \theta}\right] \quad (8.59)$$

for $m = 1, 2$, where εg_m denotes the excess pressure $[(\gamma + 1)/2\gamma]p'_{cm}/p_0$ in the cavity of the array m. By examining the evolution for the double array, it is found that an array with $\Omega_1 = 10$ and $\Omega_2 = 1$ for $K_1 = K_2 = 1$ is very effective.

8.6. Experimental verification

The preceding sections have shown in theory that an array of Helmholtz resonators is very effective in the suppression of shock formation. To confirm the theoretical findings, experiments have been done in the laboratory. This section is devoted to describing the results of the experiments [8.29, 8.30].

8.6.1. Experimental set-up

The experimental set-up is shown in Fig. 8.15. It consists of a tube with an array of Helmholtz resonators connected along its axis, and a high-pressure chamber connected to one end of the tube through a valve, the other end being closed by a flat plate. The tube is 7.4 m long and the inner diameter D is 80 mm, while the high-pressure chamber is 0.3 m long with the same diameter. The tube, high-pressure chamber and resonators are made of stainless steel.

Each Helmholtz resonator has a cavity of volume V ($= 4.94 \times 10^{-5}$ m^3) and a throat of length L ($= 35.6$ mm) and diameter $2r$ ($= 7.11$ mm). The natural frequency of the resonator, with the end corrections, is 238 Hz. The cavity is connected to the tube through a throat, where the opening of the throat is flush with the tube wall and the throat's axis is normal to the wall. The resonators are connected in an array along the tube with an axial spacing d ($= 50$ mm), but set in a staggered manner on both sides of the tube as shown in Fig. 8.15. The total number of resonators is 148. For this array, the size parameter κ takes a value of 0.197.

The experiments were performed using air under atmospheric pressure, at a room temperature of 15°C. Pressure disturbances were generated in the tube by releasing pressurized air stored in the high-pressure chamber by opening the valve. The valve (Mac Valves, Inc., type 59B-32–726AA) was opened by operation of a solenoid, driven by nitrogen gas supplied externally. As long as the initial pressure in the chamber before opening the valve is fixed, the pressure disturbances generated in the tube are almost the same each time, so that the experiments can be repeated with high accuracy. The pressure in the tube was measured by high-pressure microphones (Brüel & Kjær, type 4136/WH-2967) spaced 0.4 m apart along the axis, beginning at the end of the tube with the valve. The temporal variation of the excess pressure was observed and stored with an oscilloscope.

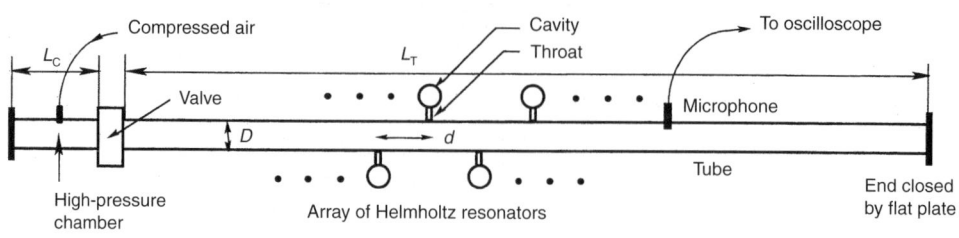

(a)

(b)

Fig. 8.15. (a) The experimental set-up: the length of the tube L_T is 7.4 m and that of the high-pressure chamber L_C is 0.3 m, while the inner diameter D is 80 mm in both of these; the Helmholtz resonators, with a natural frequency of 238 Hz, are connected to the tube with an axial spacing d (50 mm), and the parameter κ takes a value of 0.197. (b) The high-pressure chamber shown on the left in (a) is seen in the foreground, and via the valve, the tube with the array of Helmholtz resonators on both sides extends towards the background

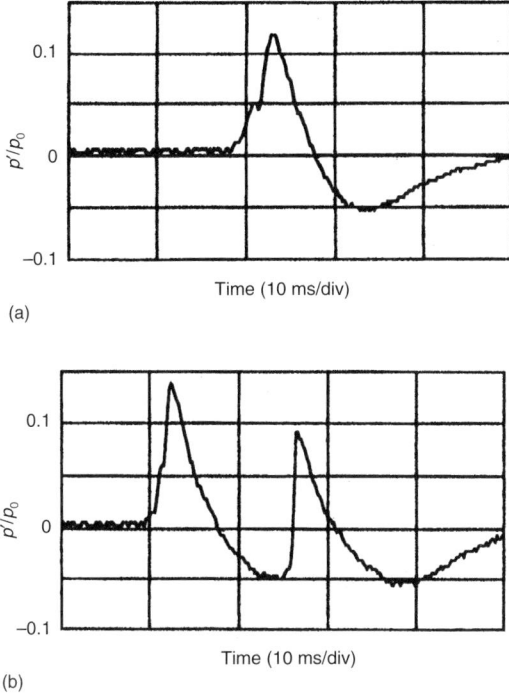

Fig. 8.16. Temporal profiles of the excess pressure p′ relative to p₀ in the tube without the array, measured at locations 0.4 m (a) and 4.8 m (b) distant from the valve

8.6.2. Experimental results

We first describe the results for the tube without the array of resonators. Figure 8.16 shows the temporal profiles of the excess pressure p' relative to the atmospheric pressure p_0 measured at locations 0.4 m and 4.8 m distant from the valve, when the initial pressure in the high-pressure chamber was $0.8p_0$ above p_0. As can be seen in Fig. 8.16(a), the pressure profile is smooth and free from any shock. The negative pressure behind the pulse is due to the fact that the diameter of the valve is small compared with the tube's diameter, so that the pressurized air will spread spherically on entering the tube.

The peak pressure is 1.2×10^4 Pa, so that ε is about 0.1. The coupling parameter K has a value of unity. On the other hand, the typical angular frequency ω is defined by the inverse of the half-value width of the pulse, which is the duration between the times when the pressure has a value equal to half the peak value. The half-value width was measured to be 2.9 ms. For this value of ω, the tuning parameter Ω has a value of 18.8. If the profile observed at the 0.4 m location is taken as the initial value, this value of Ω lies in the range where no shock formation takes place. As for the dissipative effects, the values of δ_R and δ_r are about 0.08 and 0.1, respectively.

Figure 8.16(b) shows two pulses, the left one propagating towards the closed end and the right one propagating towards the valve after being reflected at the closed

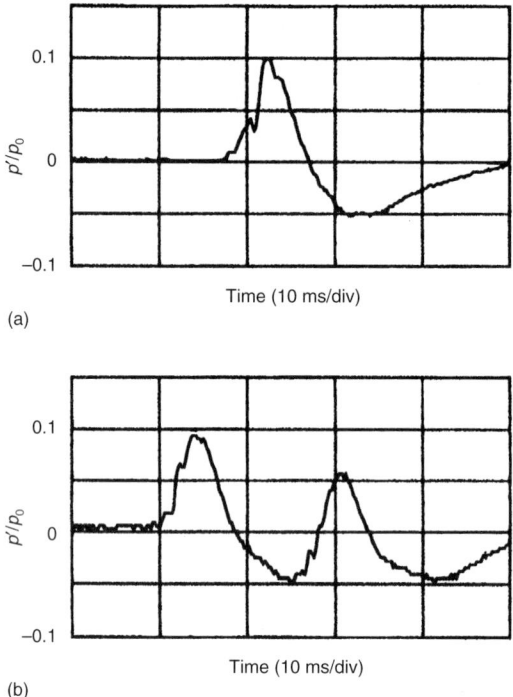

Fig. 8.17. Temporal profiles of the excess pressure p' relative to p_0 in the tube with the array, measured at locations 0.4 m (a) and 4.8 m (b) distant from the valve

end. The left pulse is steepened forward (leftward), and a shock appears clearly in the right pulse. The shock formation point is estimated to be at 8.8 m from the valve. Figure 8.17 shows the temporal profiles of the excess pressure in the tube with the array connected, measured at the same locations as used for the results shown in Fig. 8.16. The initial pressure in the high-pressure chamber was set equal to $0.8\,p_0$. The profile in Fig. 8.17(a) appears to be spread slightly compared with that in Fig. 8.16(a) because of the dispersive effects due to the several resonators located between the valve and this location. A remarkable difference can be seen in Fig. 8.17(b), where no non-linear steepening has taken place, so that the profile remains smooth. No shock emerges in further propagation. It is thus confirmed that an array of Helmholtz resonators can actually suppress shock formation. In addition, the reflected pulse in Fig. 8.17(b) can be identified as an acoustic solitary wave [8.30].

8.7. Conclusion

The pressure field generated in a tunnel by entry of a train has been considered in detail. The physical processes leading to shock formation have been elucidated, and a mathematical method to treat the evolution of the pressure wave into a shock has been given. As the magnitude of the pressure disturbance becomes larger owing to an increase of the train speed, the emergence of a shock becomes unavoidable in shorter tunnels. In order to realize shock-free propagation in tunnels, it has been proposed to connect an array of Helmholtz resonators to the tunnel. From the results of the theory and the experiments, it is concluded that shock formation can be suppressed by connecting the suitable array. Figure 8.2 shows simply an idealized model of the tunnel. In reality, there will be various ways to arrange the resonators, depending on their geometry. Besides the Helmholtz resonators, an array of acoustically compact and closed side branches of any shape, such as a quarter-wavelength tube or a lining of the tunnel wall containing small cavities, may serve to suppress shock formation [8.31].

In the Introduction, we mentioned briefly that no shock emerges in tunnels with ballasted track, although this is observed for train speeds limited to below 210 km/h. From the viewpoint of the present study, we speculate about the effects of ballasted track as follows. The damping will undoubtedly be increased because the surface becomes rough. In addition, the interstices in the tracks will constitute a kind of cavity. Their natural frequency will be very high, so that the tuning parameter Ω will take a large value. Provided that the ballasted track is regarded as a kind of lining with small cavities, then the size parameter κ will take a very small value. Unless the coupling parameter K were too small, it is suspected that no shock will emerge. Such a situation is the one that occurs in tunnels with ballasted tracks. But if the magnitude of the pressure disturbance becomes large and K becomes small ($K \ll 1$), it is an open question whether or not shock formation will be kept suppressed, even in tunnels with ballasted tracks. In the same context, Ozawa *et al.* [8.32] have examined the effects of dispersion on the basis of the K–dV–Burgers equation. In their results, the wavefront appears to be split and the dispersion seems to be too small, just as in the case shown in Fig. 8.14.

Finally, this chapter will be concluded by mentioning some problems related to implementation of an array of Helmholtz resonators in real tunnels. One problem is to clarify the effects of the airflow induced in the tunnel owing to the passage of a train or to ventilation. In the presence of the flow, how will the array of Helmholtz resonators behave? Another problem is that of what effects the array will have on the stability of a moving train. The study of these problems has just been started [8.33]. The shock-free tunnel still has many interesting topics to be examined in store.

8.8. References

8.1. OZAWA, S. *Studies of Micropressure Wave Radiated from a Tunnel Exit.* The Japan National Railways, Tokyo, 1979, Railway Technical Research Report, No.1121, pp. 1–92 [in Japanese].

8.2. TOLLMIEN, W. Luftwiederstand und Druckverlauf bei der Fahrt von Zügen in einem Tunnnel. *Zeitschrift für gesamte Technik/Verein Deutscher Ingenieure (VDI-Z)*, 1927, **71**, 199–203.

8.3. HARA, T. Aerodynamic force acting on a high speed train at tunnel entrance. *Bulletin of the Japan Society of Mechanical Engineers*, 1961, **4**, 547–553.

8.4. EDWARDS, L. K. High-speed tube transportation. *Scientific American*, 1965, **213**, 30–40.

8.5. HAMMITT, A. G. Aerodynamic analysis of tube vehicle systems. *AIAA Journal*, 1972, **10**, 282–290.

8.6. FOX, J. A. and HENSON, D. A. The prediction of the magnitudes of pressure transients generated by a train entering a single tunnel. *Proceedings of the Institution of Civil Engineers*, 1971, **49**, 53–69.

8.7. SOCKEL, H. Aerodynamik des Eisenbahntunnels. *Zeitschrift für angewante Mathematik und Mechanik*, 1989, **69**, 540–551.

8.8. OGAWA, T. and FUJII, K. Numerical simulation of compressible flows induced by a train moving into a tunnel. *Computational Fluid Dynamics Journal*, 1994, **3**, 63–82.

8.9. OGAWA, T. and FUJII, K. Numerical investigation of three-dimensional compressible flows induced by a train moving into a tunnel. *Computers and Fluids*, 1997, **26**, 565–585.

8.10. SUGIMOTO, N. Propagation of nonlinear acoustic waves in a tunnel with an array of Helmholtz resonators. *Journal of Fluid Mechanics*, 1992, **244**, 55–78.

8.11. PIERCE, A. D. *Acoustics. An Introduction to its Physical Principles and Applications.* Acoustical Society of America, New York, 1991, pp. 315–316.

8.12. HAMILTON, M. F. and MORFEY, C. L. Model equations. In: *Nonlinear Acoustics* (eds M. F. Hamilton and D. T. Blackstock), Academic Press, San Diego, 1998, p. 45.

8.13. NAYFEH, A. H. *Perturbation Methods.* Wiley, New York, 1973, pp. 110–158.

8.14. GLASS, I. I. *Shock Waves and Man.* University of Toronto Press, 1974, p. 54.

8.15. SUGIMOTO, N. Sound field in a tunnel generated by traveling of a high-speed train. In: *Theoretical and Computational Acoustics* (eds J. E. Ffowcs Williams, D. Lee and A. D. Pierce), World Scientific, Singapore, 1994, vol.1, pp. 45–56.

8.16. SUGIMOTO, N. and OGAWA, T. Acoustic analysis of the pressure field in a tunnel, generated by entry of a train. *Proceedings of the Royal Society of London A*, 1998, **454**, 2083–2112.

8.17. HOWE, M. *Acoustics of Fluid–Structure Interactions.* Cambridge University Press, Cambridge, 1998, pp. 246–252.

8.18. HOWE, M. The compression wave produced by a high-speed train entering a tunnel. *Proceedings of the Royal Society of London A*, 1998, **454**, 1523–1534.

8.19. RAYLEIGH, LORD. *The Theory of Sound.* Dover, New York, 1945, vol. 2, pp. 487–491.

8.20. CHESTER, W. Resonant oscillations in closed tubes. *Journal of Fluid Mechanics*, 1964, **18**, 44–64.

8.21. SUGIMOTO, N. 'Generalized' Burgers equations and fractional calculus. In: *Nonlinear Wave Motion* (ed. A. Jeffrey), Longman, Harlow, 1989, pp. 162–179.

8.22. SUGIMOTO, N. Evolution of nonlinear acoustic waves in a gas-filled pipe. In: *Frontiers of Nonlinear Acoustics* (ed. M. F. Hamilton and D. T. Blackstock), Elsevier, London, 1990, pp. 345–350.

8.23. SUGIMOTO, N. Burgers equation with a fractional derivative: hereditary effects on nonlinear acoustic waves. *Journal of Fluid Mechanics*, 1991, **225**, 631–653.

8.24. SUGIMOTO, N. and HORIOKA, T. Dispersion characteristics of sound waves in a tunnel with an array of Helmholtz resonators. *Journal of the Acoustical Society of America*, 1995, **97**, 1446–1459.

8.25. BRILLOUIN, L. *Wave Propagation in Periodic Structures.* Dover, New York, 1953.

8.26. KITTEL, C. *Introduction to Solid State Physics.* Wiley, New York, 1976.

8.27. DRAZIN, P. G. and JOHNSON, R. S. *Solitons: an Introduction.* Cambridge University Press, Cambridge, 1989.

8.28. SUGIMOTO, N. Acoustic solitary waves in a tunnel with an array of Helmholtz resonators. *Journal of the Acoustical Society of America*, 1996, **99**, 1971–1976.

8.29. SUGIMOTO, N., MASUDA, M., OHNO, J. and MOTOI, D. Suppression of shock formation and generation of acoustic solitary wave in an air-filled tube with an array of Helmholtz resonators. *Nonlinear Acoustics at the Turn of the Millenium* (eds W. Lauterborn and T. Kurtz). American Institute of Physics, New York, 2000, vol. S24, pp. 523–526.

8.30. SUGIMOTO, N., MASUDA, M., OHNO, J. and MOTOI, D. Experimental demonstration of generation and propagation of acoustic solitary waves in an air-filled tube. *Physical Review Letters*, in press.

8.31. SUGIMOTO, N. 'Shock-free tunnel' for future high-speed trains. *Proceedings of the International Conference on Speed-up Technology for Railway and Maglev Vehicles, Japan Society of Mechanical Engineers*, Yokohama, 1993, vol. 2, pp. 284–292.

8.32. OZAWA, S., MAEDA, T., MATSUMURA, T. and UCHIDA, K. Effect of ballast on pressure wave propagating through tunnel. *Proceedings of the International Conference on Speed-up Technology for Railway and Maglev Vehicles, Japan Society of Mechanical Engineers*, Yokohama, 1993, vol. 2, pp. 299–304.

8.33. SUGIMOTO, N. Aeroelastic stability of a beam traveling in a tunnel lined with resonators. *AIAA Journal*, 1996, **34**, 2005–2013.

Part 4
Generation of ground vibrations by surface trains

9. Generation of ground vibration boom by high-speed trains

V. V. Krylov
*Department of Civil and Structural Engineering, The Nottingham Trent University, Burton Street, Nottingham NG1 4BU, UK**

9.1. Introduction

Railway-generated ground vibrations cause significant disturbance to residents of nearby buildings even when generated by conventional passenger or heavy-freight trains [9.1, 9.2]. If train speeds increase, the intensity of railway-generated vibrations generally becomes larger. For modern high-speed trains the increase in ground vibration intensity is especially high when the train speeds approach certain critical velocities of waves propagating in the track–ground system. The two most important critical velocities are the velocity of Rayleigh surface wave in the ground and the minimal phase velocity of bending waves propagating in the track supported by ballast, the latter velocity being referred to as the track critical velocity. Both these velocities can easily be exceeded by modern high-speed trains, especially in the case of very soft soil, where both critical velocities become very low.

As has been theoretically predicted by the present author [9.3, 9.4], if the train speed v exceeds the Rayleigh wave velocity c_R in the supporting soil a ground vibration boom occurs. It is associated with a very large increase in the generated ground vibrations, as compared with the case of conventional trains. The phenomenon of the ground vibration boom is similar to the sonic boom for aircraft crossing the sound barrier, and its existence has been recently confirmed experimentally [9.5, 9.6] (see also Chapter 11). Measurements were carried out on behalf of the Swedish railway authorities when their West Coast Main Line from Gothenburg to Malmö was opened for the X2000 high-speed train. The speeds achievable by the X2000 train (up to 200 km/h) can be larger than lowest Rayleigh wave velocities in this part of Sweden, which is characterized by very soft ground. In particular, at a location near Ledsgård the Rayleigh wave velocity in the ground was

---

*From March 2001, with the Department of Aeronautical and Automotive Engineering, Loughborough University, Loughborough, Leicestershire LE11 3TU, UK.

around 45 m/s, so an increase in train speed from 140 to 180 km/h led to an increase of about ten times in the generated ground vibrations [9.5] (see Chapter 11). The above-mentioned first observations of the ground vibration boom indicate that now one can speak about 'supersonic' ('superseismic') or, more precisely, 'trans-Rayleigh' trains [9.7–9.9]. The increased attention of railway companies and local authorities to ground vibrations associated with high-speed trains stimulated a growing number of theoretical and experimental investigations in this area (see, e.g., [9.10–9.13]).

If the train speeds increase further and approach the track critical velocity, then the rail deflections due to the applied wheel loads may become much larger. The possible very large rail deflections around this speed may even result in train derailment, thus representing a serious problem for train and passenger safety [9.6, 9.14–9.16]. From the point of view of generation of ground vibrations outside the track, these large rail deflections can be responsible for an additional growth of ground vibration amplitudes, as compared with the above-mentioned case of the ground vibration boom [9.7, 9.9, 9.17].

In the present chapter, we review the current status of the theory of the ground vibration boom from high-speed trains. Among the problems to be discussed are the quasi-static pressure generation mechanism, the effects of the Rayleigh wave velocity and track wave resonances on the generated ground vibrations, the effects of a layered geological structure of the ground, and the waveguide effects of embankments. The results of theoretical calculations for TGV and Eurostar high-speed trains travelling along typical tracks are compared with the existing experimental observations.

9.2. Quasi-static pressure mechanism of generating ground vibrations

In what follows, an idealized model of a train comprising N carriages is considered. It is assumed that the train is travelling at a speed v along a track with a sleeper periodicity d (Fig. 9.1(a)). One can distinguish several mechanisms of railway-generated ground vibrations which may contribute to the total ground vibration level in different frequency bands. Among these mechanisms, one can mention the quasi-static wheel axle pressure on the track, the effects of joints in unwelded rails, the unevenness of wheels or rails and the dynamically induced forces of carriage and wheel axle vibrations excited mainly by the unevenness of wheels and rails.

Among the above-mentioned mechanisms, we consider only the most common one, which is present even for ideally flat rails and wheels – a quasi-static pressure of wheel axles onto the track. The quasi-static pressure generation mechanism causes downward deflections of the track beneath each wheel axle (Fig. 9.1(b)). These result in distribution of the axle loads over the deflection distance of the sleepers [9.3, 9.4, 9.18]. Thus, each sleeper acts as a dynamic vertical force applied to the ground during the time necessary for a deflection curve to pass over the sleeper. In the framework of the wheel–axle pressure mechanism, it is these forces that result in generation of ground vibrations by passing trains. As will be demonstrated below, this mechanism is also responsible for the railway-generated ground vibration boom. The role of other generation mechanisms is discussed elsewhere [9.9].

GENERATION OF GROUND VIBRATION BOOM

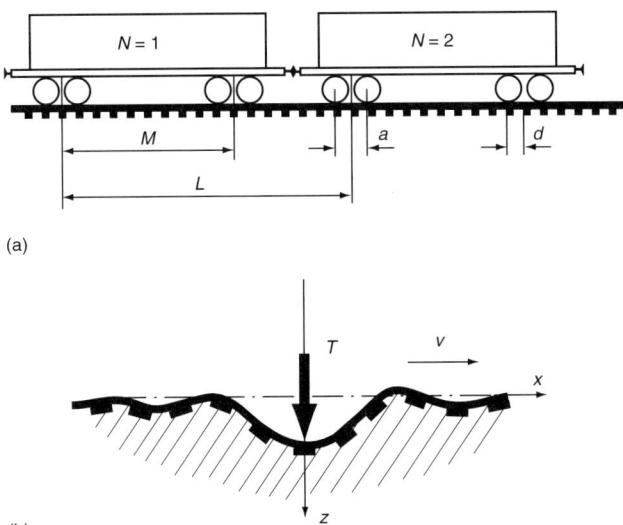

(a)

(b)

Fig. 9.1. (a) Geometry of track and train; (b) wheel axle pressure mechanism of generation of ground vibrations

9.2.1. Dynamic properties of the track

An essential aspect of analysing the above-mentioned quasi-static pressure generation mechanism is the calculation of the track deflection curve as a function of the applied axle load and train speed. One can treat each rail as an Euler–Bernoulli elastic beam of uniform mass m_0 (m_0 includes the contribution of the sleepers) lying on a viscoelastic half-space $z > 0$ and use the following dynamic equation to describe its vertical deflection (see, e.g., [9.19]):

$$EI\frac{\partial^4 w}{\partial x^4} + m_0 \frac{\partial^2 w}{\partial t^2} + 2m_0 \omega_b \frac{\partial w}{\partial t} + \alpha w = T\,\delta(x - vt) \tag{9.1}$$

Here w is the magnitude of the beam deflection, E and I are the Young's modulus and the cross-sectional moment of the beam, ω_b is the circular frequency of damping, α is the proportionality coefficient of the equivalent Winkler elastic foundation modelling the elastic ground, x is the distance along the beam, T is the wheel load applied to each rail and considered as a vertical point force, v is the train speed, and $\delta(x)$ is the Dirac delta function.

It is useful first to discuss free wave propagation in the supported beam without damping, i.e. to analyse equation (9.1) with $T = 0$ and $\omega_b = 0$. In this case, the substitution of the solution in the form of harmonic bending waves

$$w = A\exp(ikx - i\omega t) \tag{9.2}$$

into equation (9.1) gives the following dispersion equation for track waves propagating in the system:

$$\omega = \frac{(\alpha + EIk^4)^{1/2}}{m_0^{1/2}} \qquad (9.3)$$

Here k is the wavenumber of the track waves, and ω is the circular frequency. In the quasi-static (long-wave) approximation ($k = 0$) the dispersion equation (9.3) reduces to the well-known expression for the so-called track-on-ballast resonance frequency: $\omega_{tb} = \alpha^{1/2}/m_0^{1/2}$. For example, for the typical soil and track parameters $\alpha = 52.6$ MN/m^2 [9.20] and $m_0 = 300$ kg/m, this gives $F_{tb} = \omega_{tb}/2\pi = 67$ Hz. The frequency F_{tb} represents the minimal frequency at which track waves can propagate. It also follows from equation (9.3) that the frequency-dependent velocity of track wave propagation $c = \omega/k$ is determined by the expression

$$c = \frac{(\alpha/k^2 + EIk^2)^{1/2}}{m_0^{1/2}} \qquad (9.4)$$

which shows that at $k = (\alpha/EI)^{1/4}$ the velocity c has a minimum

$$c_{min} = \left(\frac{4\alpha EI}{m_0^2}\right)^{1/4} \qquad (9.5)$$

The value c_{min} is often referred to as the track critical velocity. For the above-mentioned typical track and ballast parameters and for a value of EI equal to 4.85 MNm^2, it follows from equation (9.5) that $c_{min} = 326$ m/s (1174 km/h), which is much larger than the speeds of the fastest modern trains. However, for very soft soils, e.g. alluvial soils, characterized by very low α, the value of c_{min} can be as low as 60–70 m/s and can be easily exceeded by even relatively moderate high-speed trains.

In practice, the value of c_{min} for a particular location can be estimated using equation (9.5), in which the stiffness of the equivalent Winkler foundation α is expressed in terms of the real elastic moduli of the ground. There are different theoretical models that give such expressions (see, e.g., [9.16]). Generally, it follows from these models that the track critical velocity is normally larger by 10–30% than the Rayleigh wave velocity for the same ground.

The solution of equation (9.1), with the right-hand side different from zero, has different forms for small and large values of time t. In the problem under consideration we are interested in an 'established' solution for large values of t which describes the track deflections as being at rest relative to a coordinate system moving at the train speed v – the so-called stationary solution. Obviously, this solution must depend only on the combination $x - vt$. Using the notation $p = \beta(x - vt)$, where $\beta = (\alpha/4EI)^{1/4}$, it is easy to obtain the stationary solution of equation (9.1) in the Fourier domain, $W(p)$ (see, e.g., [9.19]), where $W(p) = \int_{-\infty}^{\infty} w(z) e^{-ipz} dz$.

Taking the inverse Fourier transform of $W(p)$ gives analytical expressions for $w(x - vt)$ which have different forms depending on whether $v < c_{min}$, $v = c_{min}$ or $v > c_{min}$. In particular, if the train speed v approaches the minimal phase velocity c_{min} from below, the rail deflection amplitudes w experience a large resonance increase limited by track damping. Assuming for simplicity that there is no damping in the

GENERATION OF GROUND VIBRATION BOOM

system ($\omega_b = 0$), one can obtain the exact solution for $v < c_{min}$ in a very simple form [9.19]:

$$w(x-vt) = \frac{T}{8EI\beta^3\delta}\exp(-\beta\delta|x-vt|)\left(\cos(\beta\eta(x-vt))+\frac{\delta}{\eta}\sin(\beta\eta|x-vt|)\right) \quad (9.6)$$

where $\delta = (1-v^2/c_{min}^2)^{1/2}$ and $\eta = (1+v^2/c_{min}^2)^{1/2}$. One can see that, since the factor $\delta = (1-v^2/c_{min}^2)^{1/2}$ is present in the denominator of the expression in equation (9.6), the track deflection w increases as the train speed approaches the minimal track wave velocity. Note that in the example considered, without damping, it follows from equation (9.6) that $w \to \infty$ when $v \to c_{min}$. The transition of v through c_{min} can be considered only by taking damping into account. In this case approximate analytical expressions for $w(x-vt)$ in the three cases $v < c_{min}$, $v = c_{min}$ or $v > c_{min}$ can be found [9.19]. It follows from these expressions that the amplitudes of the rail deflections near track wave resonance, $v = c_{min}$, are determined by the influence of damping. For typical values of damping these amplitudes are two to three times larger than the corresponding static values. The corresponding large rail deflections at train speeds approaching the track critical velocity may affect the operation of the railway line and even result in train derailment. Different aspects of this problem are now being widely investigated (e.g. [9.6, 9.14–9.16]).

9.2.2. Forces applied from sleepers to the ground

To calculate the dynamic forces $P(t-x/v)$ applied from the sleepers to the ground, e.g. $P(t)$ for a sleeper located at $x = 0$, one should take into account the fact that $P(t)$ is proportional to the track deflection $w(vt)$ and to the sleeper periodicity d:

$$P(t) = 2\alpha w(vt)d \quad (9.7)$$

where α is the constant of the Winkler foundation, and the presence of two rails has been taken into account. It is convenient to eliminate α and d from equation (9.7). To do so one can use integration of the quasi-static equation (9.1), i.e. with $m_0 \partial^2 w/\partial t^2 = 0$, over x. The integration results in the formula

$$\alpha w_{max}^{st} dN_{eff}^{st} = T \quad (9.8)$$

which, combined with equation (9.7), gives the following expression for $P(t)$:*

$$P(t) = 2T\frac{w(vt)}{w_{max}^{st} N_{eff}^{st}} \quad (9.9)$$

Here index 'st' corresponds to the quasi-static solution of equation (9.1) (in particular, w_{max}^{st} is the maximum value of $w(vt)$ in the quasi-static approximation),

---

*In the author's earlier papers [9.7–9.9], instead of the sleeper periodicity d, the sleeper thickness Δd was used erroneously in equations (9.7) and (9.8). Fortunately, this did not affect equation (9.9), following from equations (9.7) and (9.8), because of the elimination of α and d. The author is grateful to G. Degrande for pointing this out.

and $N_{\text{eff}}^{\text{st}}$ is the effective number of sleepers equalizing the applied quasi-static wheel load T:

$$\sum_{m=-\infty}^{\infty} \frac{T}{N_{\text{eff}}^{\text{st}}} \frac{w(md)}{w_{\text{max}}^{\text{st}}} = T \qquad (9.10)$$

where m is a number labelling a current sleeper. Numerical solution of equation (9.10) shows that the value of $N_{\text{eff}}^{\text{st}}$ may be approximated with good accuracy by the simple analytical formula $N_{\text{eff}}^{\text{st}} = 0.625\pi/\beta d = 0.625 x_0^{\text{st}}/d$, where $x_0^{\text{st}} = \pi/\beta$ is the effective quasi-static track deflection distance. Using this formula in equation (9.9) results in the following expression:

$$P(t) = 3.2T \frac{w(vt)}{w_{\text{max}}^{\text{st}}} \frac{d}{x_0^{\text{st}}} \qquad (9.11)$$

As will be shown below, to describe the generation of ground vibrations by moving trains one needs to know the frequency spectrum of the force applied from each sleeper to the ground, $P(\omega)$, rather than its time dependence, $P(t)$. Note that, whereas the time-domain solution $P(t)$ has different forms for $v < c_{\text{min}}$, $v = c_{\text{min}}$ and $v > c_{\text{min}}$ [9.19], its Fourier representation $P(\omega)$ is described by the same formula for all these cases. Keeping in mind that for $x = 0$ the relationship $W(p) = -2\pi\beta v W(\omega)$ holds, where $\omega = p\beta v$, one can derive the following expression for $P(\omega)$:

$$P(\omega) = \frac{-12.8 T d/v\pi^2}{\omega^4/\beta^4 v^4 - 4\omega^2/c_{\text{min}}^2 \beta^4 - 8ig\omega/c_{\text{min}}\beta + 4} \qquad (9.12)$$

where $g = (m_0/\alpha)^{1/2}\omega_b$ is a non-dimensional damping parameter. Typical forms of the vertical-force spectra $P(\omega)$ calculated for a train travelling on very soft soil at speeds $v = 20$, 50 and 70 m/s (corresponding to the cases $v < c_R$, $c_R < v < c_{\text{min}}$, and $v > c_{\text{min}}$, respectively) are shown in Fig. 9.2.

Calculations were performed for the soft soil using the following parameters of train, track and soil: $T = 50$ kN, $d = 0.7$ m, $\beta = 1.28$ m^{-1}, $c_R = 45$ m/s, $c_{\text{min}} = 65$ m/s and $g = 0.1$. For relatively low train speeds, i.e. for $v < c_R$, the dynamic solution (equation (9.12)) for the force spectrum $P(\omega)$ goes over to the quasi-static solution [9.18]. As the train speed increases and approaches or exceeds the minimal track wave velocity, the spectrum $P(\omega)$ becomes broader and has larger amplitudes, and a second peak appears at higher frequencies.

The values of $P(\omega)$ in the model under consideration are limited by track wave damping, described by the non-dimensional damping parameter g. The effect of track damping on the spectra $P(\omega)$ is more pronounced for $v > c_{\text{min}}$. For low train speeds, $v < c_R$, the effect of track damping is negligibly small.

9.3. Green's function for the problem

As the next step, one has to derive the Green's function for the problem under consideration. This function describes ground vibrations generated by individual sleepers, which can be regarded as point sources in the low-frequency band.

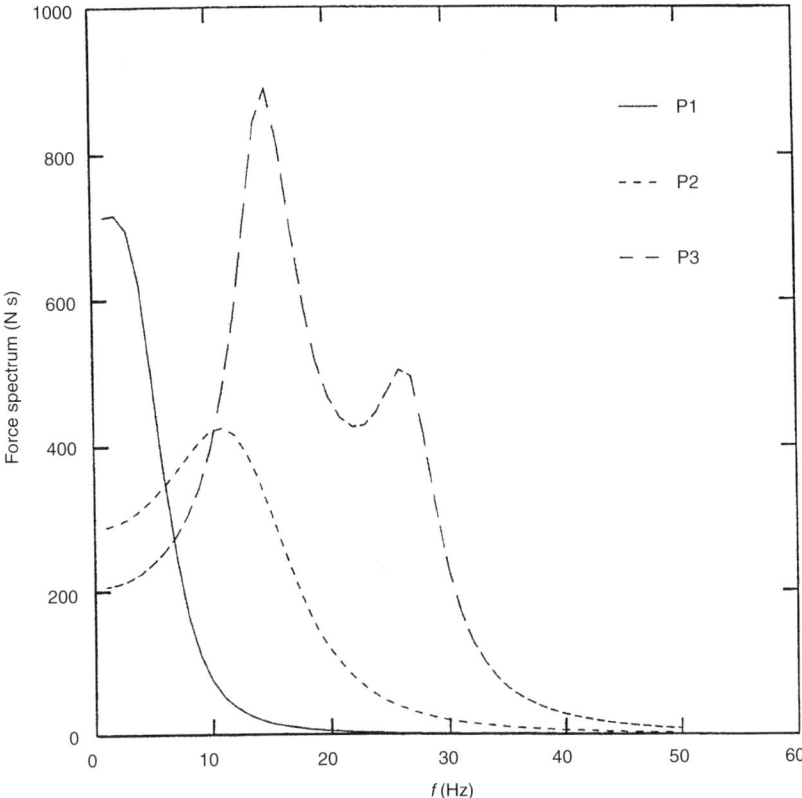

Fig. 9.2. Spectra of vertical forces applied from each sleeper to the ground (in N s) for three values of train speed v corresponding to the cases $v < c_R$, $c_R < v < c_{min}$, and $v > c_{min}$: $v = 20$ m/s (curve P1), 50 m/s (curve P2) and 70 m/s (curve P3)

9.3.1. Homogeneous elastic half-space

We recall that for a homogeneous elastic half-space, the corresponding Green's function can be derived using the results from the well-known axisymmetric problem of the excitation of an elastic half-space by a vertical point force applied to the surface (see, e.g., [9.21, 9.22]). The solution of this problem, which should satisfy the dynamic equations of elasticity for a homogeneous medium subject to stress-free boundary conditions on the surface outside the point of application of the force, gives the components of the dynamic Green's tensor (or, for simplicity, the Green's function) G_{zi} for an elastic half-space. For the problem under consideration, only the Rayleigh surface wave contribution (the Rayleigh part of the Green's function) is considered, since Rayleigh waves transfer most of the vibration energy to remote locations. For these waves, the spectral density of the vertical vibration velocity at the surface of the homogeneous half-space ($z = 0$) may be written in the form (see also [9.18])

$$v_z(\rho, \omega) = P(\omega)G_{zz}(\rho, \omega) = P(\omega)D(\omega)(1/\sqrt{\rho})\exp(ik_R\rho - \gamma k_R\rho) \tag{9.13}$$

where

$$D(\omega) = (1/2\pi)^{1/2}(-i\omega)qk_R^{1/2}k_t^2 \exp(-i3\pi/4)/\mu F'(k_R) \tag{9.14}$$

Here $\rho = [(x-x')^2 + (y-y')^2]^{1/2}$ is the distance between the source (with coordinates x', y') and the point of observation (with coordinates x, y), $\omega = 2\pi F$ is the circular frequency; $k_R = \omega/c_R$ is the wavenumber of the Rayleigh surface wave; c_R is the Rayleigh wave velocity; $k_l = \omega/c_l$ and $k_t = \omega/c_t$ are the wavenumbers of the longitudinal and shear bulk elastic waves, where $c_l = [(\lambda + 2\mu)/\rho_0]^{1/2}$ and $c_t = (\mu/\rho_0)^{1/2}$ are the longitudinal and shear wave velocities; λ and μ are the elastic Lamé constants; ρ_0 is the mass density of the ground; and $q = (k_R^2 - k_l^2)^{1/2}$. The factor $F'(k_R)$ is the derivative of the Rayleigh determinant

$$F(k) = (2k^2 - k_t^2)^2 - 4k^2(k^2 - k_t^2)^{1/2}(k^2 - k_l^2)^{1/2} \tag{9.15}$$

taken at $k = k_R$, and $P(\omega)$ is the Fourier transform of $P(t)$ (see equation (9.12)).

To describe the spectrum for the successive passage of two axle loads separated by a distance a (the case of a bogie), $P_b(\omega)$, one should use the following relationship between $P_b(\omega)$ and $P(\omega)$ [9.18]:

$$P_b(\omega) = 2P(\omega)\cos\frac{\omega a}{2v} \tag{9.16}$$

In writing equation (9.13), we have accounted for attenuation in the soil by replacing $1/c_R$ in the exponential of the Green's function by the complex value $1/c_R + i\gamma/c_R$, where $\gamma = 0.001 - 0.1$ is a constant describing the 'strength' of dissipation of Rayleigh waves in the soil [9.23].

9.3.2. Effect of layered ground structure

To consider the influence of a layered geological structure of the ground on generation of ground vibrations in a rigorous way, we would have to use the Green's function for a layered elastic half-space, instead of that for a homogeneous half-space. As a rule, such a function, which contains information about the total complex elastic field generated in the layered half-space considered (including the various modes of surface waves and modes radiating energy into the bulk (leaky waves)), cannot be obtained analytically (see also Chapter 10). However, for the description of railway-generated ground vibrations, the problem can be simplified by using an approximate engineering approach which takes into account the effects of the layered structure on the amplitude and phase velocity of only the lowest-order surface mode, which goes over to a Rayleigh wave in the limit of a homogeneous half-space. The propagating modes of higher orders and leaky modes are generated less efficiently by the surface forces associated with sleepers.

We recall that in layered media, surface waves become dispersive, i.e. their phase velocities c_R depend on frequency: $c_R = c_R(\omega)$. For a shear modulus of the ground μ that has larger values at larger depths, as is normal, there may be several surface modes, characterized by different phase velocities and cut-off frequencies. As a rule, these velocities increase at lower frequencies; this is associated with deeper penetration of the surface wave energy into the ground (see, e.g., [9.24]). For simplicity, we shall assume in the further considerations that the Poisson's ratio σ of the layered ground and the mass density ρ_0 are constant. Taking the above considerations into account and starting from the Green's function for a homogeneous half-space $G_{zz}(\rho, \omega)$ (see equations (9.13) and (9.14)), we shall construct its modification $G_{zz}^L(\rho, \omega)$ describing approximately the effects of a layered medium on the generation and propagation of the lowest-order surface Rayleigh-type mode. It can be shown that such an approximate Green's function can be written in the form [9.8]

$$v_z(\rho, \omega) = P(\omega)G_{zz}^L(\rho, \omega) = P(\omega)D^L(\omega)(1/\sqrt{\rho})\exp(ik_R^L\rho - \gamma k_R^L\rho) \quad (9.17)$$

$$D^L(\omega) = (1/2\pi)^{1/2}(-i\omega)q^L(k_R^L)^{1/2}(k_t^L)^2 \exp(-i3\pi/4)/\mu^L(\omega)F_L'(k_R^L) \quad (9.18)$$

Here $k_R^L = \omega/c_R(\omega)$ is the wavenumber of the lowest-order Rayleigh mode, propagating with the frequency-dependent velocity $c_R(\omega)$; the terms $k_l^L = \omega/c_l^L(\omega)$ and $k_t^L = \omega/c_t^L(\omega)$ are 'effective' wavenumbers of the longitudinal and shear bulk elastic waves at the given frequency ω (these wavenumbers are inversely proportional to the longitudinal ($c_l^L(\omega)$) and shear ($c_t^L(\omega)$) wave velocities averaged over the 'effective' depth of Rayleigh wave penetration into the ground, which is close to Rayleigh wavelength). In the model under consideration, these velocities and the corresponding 'effective' shear modulus $\mu^L(\omega)$ are expressed in terms of the frequency-dependent Rayleigh wave velocity $c_R(\omega)$ using the well-known relations:

$$\frac{c_R(\omega)}{c_t^L(\omega)} = \frac{0.87 + 1.12\sigma}{1+\sigma} \quad (9.19)$$

$$\frac{c_t^L(\omega)}{c_l^L(\omega)} = \left(\frac{1-2\sigma}{2(1-\sigma)}\right)^{1/2} \quad (9.20)$$

$$\mu^L(\omega) = \rho_0[c_t^L(\omega)]^2 \quad (9.21)$$

The term q^L is defined as $q^L = [(k_R^L)^2 - (k_l^L)^2]^{1/2}$, and the factor $F_L'(k_R^L)$ is determined according to the following relationship [9.24]:

$$F_L'(k_R^L) = N(\sigma)(k_R^L)^3 \quad (9.22)$$

where $N(\sigma)$ is a dimensionless function of the Poisson's ratio σ (e.g. for $\sigma = 0.25$, the function $N(\sigma)$ takes the value -2.3).

The dependence of the Rayleigh wave velocity on frequency, $c_R(\omega)$, is determined by the particular profile of the layered ground, characterized by the dependence of its elastic moduli λ and μ and its mass density ρ_0 on the vertical coordinate z. Figure 9.3(a) shows a widely used simplified model of layered ground, consisting of a single, top, elastic layer with parameters λ_1, μ_1 and ρ_{01} placed on an elastic half-space with parameters λ_2, μ_2 and ρ_{02}. Figure 9.3(b) demonstrates some more complicated layered structures on the assumption that they are characterized mainly by the shear stiffness μ as a function of z.

For all ground profiles, the determination of the velocity $c_R(\omega)$ is a complex boundary-value problem which, generally speaking, requires numerical calculation. In the engineering approach described above, we consider published values of the wave velocity functions $c_R(\omega)$, using where possible simple analytical approximations of these functions. In particular, the frequency-dependent Rayleigh wave velocity for a layered medium characterized by a monotonic change of the mechanical parameters with depth (e.g. for a simple two-layer systems) can be approximated by the function

$$c_R(\omega) = (c_1 - c_2)\exp\left(-\frac{\alpha'\omega}{2\pi}\right) + c_2 \tag{9.23}$$

where c_1 and c_2 are the values of $c_R(\omega)$ for $\omega = 0$ and $\omega = \infty$, respectively, and the parameter α' describes the 'strengths' of the dispersion (this depends on the thickness of characteristic layer and on the difference between the elastic moduli at depth and at the surface of the ground). Figure 9.4 shows typical functions $c_R(\omega)$ calculated according to equation (9.23) for $\alpha' = 0.1$. Curve cd1 corresponds to a soft

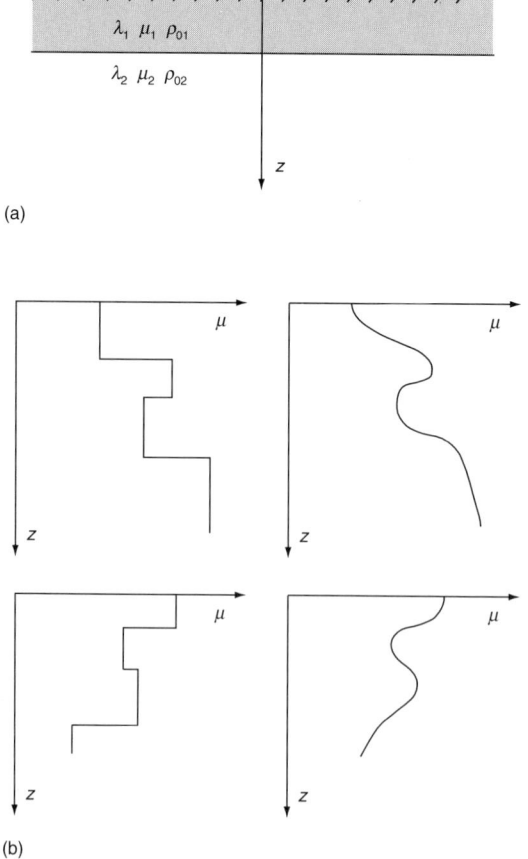

Fig. 9.3. *Some typical profiles of layered ground*

GENERATION OF GROUND VIBRATION BOOM

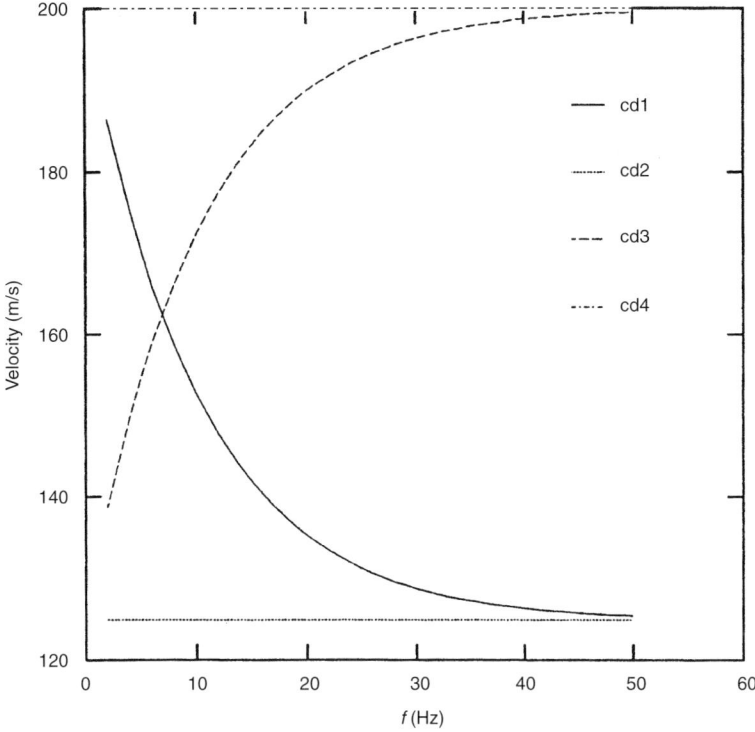

Fig. 9.4. Approximate analytical functions describing frequency-dependent Rayleigh wave velocities $c_R(\omega)$ for monotonic ground profiles: curve cd1, $(c_1 - c_2) \exp(-\alpha'_1 f) + c_2$; curve cd2, $(c_1 - c_2) \exp(-\alpha'_2 f) + c_2$; curve cd3, $(c_2 - c_1) \exp(-\alpha'_1 f) + c_1$; curve cd4, $(c_2 - c_1) \exp(-\alpha'_2 f) + c_1$; $\alpha'_1 = 0.1$ and $\alpha'_2 = 10$

layer ($c_2 = 125$ m/s) placed on a stiff ground ($c_1 = 200$ m/s). Curve cd3 describes a stiff layer ($c_2 = 200$ m/s) placed on a soft ground ($c_1 = 125$ m/s). The case $\alpha' = 10$ describes the situation when the effect of the ground substrate can be neglected and the functions $c_R(\omega)$ are determined entirely by the upper layer (curves cd2 and cd4 for soft and stiff layers, respectively).

For non-monotonic layered media (see Fig. 9.3(b)), which result in Rayleigh-type mode velocities $c_R(\omega)$ that are non-monotonic functions of frequency (with maxima or minima at certain frequencies), the simple formula of equation (9.23) is no longer applicable, and other approximate analytical expressions must be sought.

One can expect that the most significant effect of a layered structure on the generation of ground vibrations by high-speed trains is due to the phase variations of the waves caused by the frequency-dependent Rayleigh wave velocity, rather than to the changes in wave generation efficiency due to the stratification. In particular, for the case considered of a monotonic layered system with a soft upper layer, the increase in the Rayleigh wave velocity at low frequency might violate the trans-Rayleigh condition $v > c_R$ responsible for the generation of very intense ground vibrations associated with the ground vibration boom, thus causing a reduction in the low-frequency components of the generated ground vibration spectra.

9.4. Calculation of generated ground vibrations

To calculate the ground vibrations generated by a train, one needs to take into account the superposition of the waves generated by each elementary source (sleeper) activated by the wheel axles of all carriages, with the time and space differences between sources (sleepers) being taken into account. Using the Green's function, this may be written in the form [9.4–9.9, 9.18]

$$v_z(x, y, \omega) = \int_{-\infty}^{\infty}\int_{-\infty}^{\infty} P(x', y', \omega) G_{zz}^L(\rho, \omega) \, dx' \, dy' \tag{9.24}$$

where $P(x', y', \omega)$ describes the space distribution of all load forces acting along the track in the frequency domain. This distribution can be found by taking a Fourier transform of the time- and space-dependent load forces $P(t, x', y' = 0)$ applied from the track to the ground. Note that the function $P(t, x', y' = 0)$ does not depend on the layer structure of the ground and remains the same for both homogeneous and inhomogeneous half-spaces. In the model under consideration, all properties of the track and train which determine the generation of ground vibrations are described by the above-mentioned function of load forces $P(t, x', y' = 0)$.

9.4.1. Vibrations from a single axle load

It is useful first to consider ground vibrations generated by a single axle load. For a single axle load moving at speed v along the track, the load function has the form [9.4, 9.18]:

$$P(t, x', y' = 0) = \sum_{m=-\infty}^{\infty} P(t - x'/v)\delta(x' - md)\delta(y') \tag{9.25}$$

where the delta-function $\delta(x' - md)$ takes the periodic distribution of sleepers into account. Taking the Fourier transform of equation (9.25), substituting it into equation (9.24) and taking equations (9.17) and (9.18) into account results in the following formula for the vertical vibration velocity of Rayleigh waves generated at $z = 0, x = 0, y = y_0$ by a single axle load moving along the track at speed v:

$$v_z(x = 0, y = y_0, \omega) = P(\omega) D^L(\omega) \sum_{m=-\infty}^{\infty} \frac{\exp\{i(\omega/v)md + (i - \gamma)[\omega/c_R(\omega)]\rho_m\}}{\sqrt{\rho_m}} \tag{9.26}$$

where $\rho_m = [y_0^2 + (md)^2]^{1/2}$.

9.4.2. Vibrations from a complete train

To take account of all axles and carriages, one should use a more complicated load function [9.4, 9.18] (see also Chapter 10, which gives a modified form of this function):

$$P(t, x', y' = 0) = \sum_{m=-\infty}^{\infty}\sum_{n=0}^{N-1} A_n \left[P\left(t - \frac{(x' + nL)}{v}\right) + P\left(t - \frac{(x' + M + nL)}{v}\right) \right] \delta(x' - md)\delta(y') \tag{9.27}$$

where N is the number of carriages, M is the distance between the centres of the bogies in each carriage and L is the total carriage length. The dimensionless quantity A_n is an amplitude weight factor to account for different carriage masses. For simplicity, we assume all carriage masses to be equal ($A_n = 1$).

Taking the Fourier transform of equation (9.27), substituting it into equation (9.24) and making simple transformations, we obtain the following expression for the frequency spectrum of the vertical vibrations at $z = 0, x = 0$ and $y = y_0$ generated by a moving train [9.4, 9.7–9.9]:

$$v_z(x=0, y=y_0, \omega) = P(\omega)D^L(\omega) \sum_{m=-\infty}^{\infty} \sum_{n=0}^{N-1} \frac{\exp[-\gamma\omega\rho_m/c_R(\omega)]}{\sqrt{\rho_m}} \times$$

$$\left[1 + \exp\left(\frac{iM\omega}{v}\right)\right] \exp\left(i\frac{\omega}{v}(md + nL) + i\frac{\omega}{c_R(\omega)}\rho_m\right)$$

(9.28)

Note that equations (9.26) and (9.28) are applicable to trains moving at arbitrary speeds. In particular, for trains travelling at conventional speeds ($v < c_R$) they describe peaks in the ground vibration spectra corresponding to the well-known sleeper passage frequencies and other train-speed dependent combination frequencies [9.18]. Indeed, the peaks correspond to the frequencies determined by the condition $(\omega/v)(sd + qL) = 2\pi l$, where $s, q, l = 1, 2, 3, \ldots$ Obviously, $q = 0$ corresponds to the passage frequencies $f_p s = (v/d)l$. Other, more frequently observed maxima are determined either by the carriage length L ($s = 0$) or by a combination of both parameters (for $q \neq 0, s \neq 0$) (for experimental validation of these predictions for high-speed trains travelling at speeds below the Rayleigh wave velocity, see Chapter 10).

9.5. Trans-Rayleigh trains
9.5.1. General discussion
For 'trans-Rayleigh trains', i.e. trains travelling at speeds higher than the Rayleigh wave velocity in the ground, a separate analytical treatment can be performed to elucidate the special features of the problem [9.4, 9.9]. It is easy to show that maximum radiation of ground vibrations takes place if the train speed v and the frequency-dependent Rayleigh wave velocity $c_R(\omega)$ satisfy the following condition:

$$\cos\Theta = \frac{1}{K} = \frac{c_R(\omega)}{v}$$

(9.29)

where Θ is the observation angle. This relationship implies that the elastic surface waves radiated by all sleepers activated by the moving load are combined in phase at the point of observation and, as a result, a *ground vibration boom* takes place. Since the observation angle Θ must be real ($\cos\Theta \leq 1$), the value of $K = v/c_R(\omega)$ must be larger than 1, i.e. the train speed v should be larger than the Rayleigh wave velocity $c_R(\omega)$. In this case ground vibrations are generated as quasi-plane Rayleigh surface waves propagating symmetrically at angles Θ with respect to the track, and with amplitudes much larger than in the case of 'sub-Rayleigh trains'. The formation of a

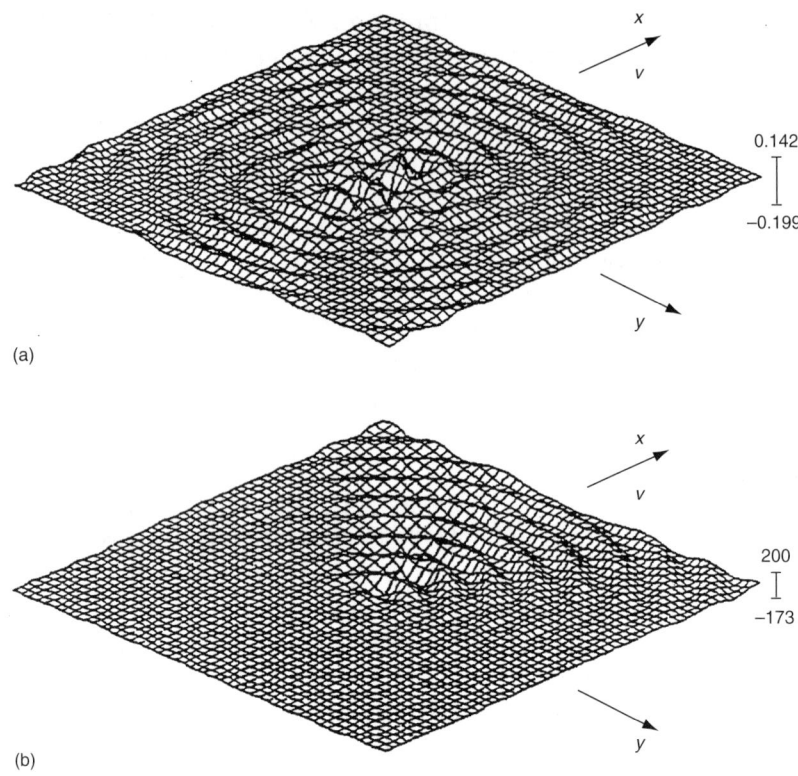

Fig. 9.5. Spatial distribution of the surface vertical displacements (in arbitrary units) generated at the frequency component f = 31.4 Hz by a single axle load moving over a small part of a track consisting of ten sleepers; the results are shown for (a) a sub-Rayleigh speed and (b) for trans-Rayleigh speed

ground vibration boom is illustrated in Fig. 9.5, which shows the spatial distribution of the ground surface vertical displacements generated at a chosen frequency component by a single axle load moving at different speeds over a small length of track consisting of ten sleepers.

The results are shown for a sub-Rayleigh speed (a) and a trans-Rayleigh speed (b). It is seen that in the first case the generated ground vibrations propagate almost in all directions, whereas in the second case they are concentrated around the angles Θ determined by equation (9.29).

The amplitudes of railway-generated ground vibrations for $v > c_R$ are determined by two factors. The first factor is that under this condition the surface waves radiated by different sleepers are combined in phase. Therefore, an increase by a factor equal to the number of effectively radiating sleepers of the track, i.e. about 200 for typical values of $\gamma = 0.05$, can be expected compared with the average vibration level for conventional trains. The second factor is the dependence of the function $P(\omega)$, determined by equation (9.12), on the train speed v. The analysis shows that the function $P(v, \omega)$ provides an average increase of about 10 times for an increase in the train speed v from 13.88 m/s (50 km/h) to 138.8 m/s (500 km/h).

Thus, a total increase of ground vibration amplitude by 1000–2000 times (60–66 dB) can be expected in the case of trans-Rayleigh trains for such an increase in train speeds.

It also follows from equations (9.26) and (9.28) that the amplitude of the generated ground vibration boom does not depend on the sleeper periodicity d. The amplitude is determined only by the distance along the track considered [9.4, 9.9]. Note that this conclusion remains valid also in the limiting case $d \to 0$. This means that radiation of a ground vibration boom by trans-Rayleigh trains may take place also on tracks without sleepers, i.e. on slab tracks. However, for conventional low-speed trains ($v \ll c_R$), the exponential functions inside the sums in equations (9.26) and (9.28) oscillate quickly as $d \to 0$, and the sums themselves become close to zero, indicating that ground vibrations in the form of waves are not generated. This agrees with the well-known result of elasticity theory [9.25] indicating that, for loads moving along the free surface of an elastic half-space at speed $v < c_R$, radiated waves do not appear (only localized quasi-static fields can accompany the moving load). Thus, the presence of sleepers is essential for generation of ground vibrations by conventional trains by the mechanism of quasi-static pressure considered here.

If the train speed increases even further, then the second critical velocity, c_{min}, makes its contribution to the amplitude of the generated ground vibrations, through the function $P(\omega)$. However, since this function can increase by only two to three times for realistic values of track wave dissipation (see Section 9.2.2), its effect on the generated vibrations may result only in an increase of the amplitude by the same amount. The second critical velocity is therefore less important for generating ground vibrations outside the track than the first critical velocity c_R.

9.5.2. Ground vibrations from TGV and Eurostar trains

Numerical calculations of ground vibrations generated by high-speed trains can be carried out according to equation (9.26) or (9.28) for different values of train speed, for different parameters characterizing the Rayleigh wave dispersion in layered ground, and for different geometrical and physical parameters of both track and train. For relatively short five-carriage trains it is normally sufficient to carry out the summation over m in equations (9.26) and (9.28) from $m = -150$ to $m = 150$, the corresponding length of track being greater than the total train length NL and the attenuation distance of Rayleigh waves in the frequency band considered. In many practical situations, the frequency-dependent Rayleigh wave velocity for layered media can be approximated by equation (9.23).

Figure 9.6 shows surface plots of the ground vibration spectra (in linear units, relative to the reference level of 10^{-9} m/s) generated by a single axle load $2T = 200$ kN moving at speeds ranging from 10 m/s to 320 m/s for homogeneous (a) and layered (b) ground. The results are given for the frequency band 0–50 Hz and for a value of ground attenuation $\gamma = 0.05$. The Poisson's ratio of the soil was set at 0.25, and the mass density ρ_0 was 2000 kg/m³. The units of calculation were $\Delta v = 10$ m/s and $\Delta F = 1$ Hz. The values of the parameter α' in equation (9.23) corresponding to the homogeneous and layered grounds were 10 and 0.1, respectively. Other parameters were $c_{min} = 326$ m/s, $g = 0.1$, $\beta = 1.28$ m^{-1} and $y_0 = 30$ m, where y_0 is the distance from the track to the observation point.

One can see that with an increase of train speed, the ground vibration level generally grows. For relatively low train speeds, the peaks corresponding to the train passage frequencies are almost invisible because of the huge increase of ground vibration level in the trans-Rayleigh range ($v \geq c_R(\omega)$). This increase proceeds even further for train speeds approaching the minimal track wave velocity ($c_{min} = 326$ m/s in this example). Comparison of Figs 9.6(a) and 9.6(b) shows that a two-layered structure results in a decrease of generated vibrations at low frequencies.

Figure 9.7 illustrates the ground vibration spectra (in dB, relative to a reference level of 10^{-9} m/s) generated by complete TGV or Eurostar trains travelling on homogeneous ground for both sub-Rayleigh and trans-Rayleigh train speeds ($v = 50$ km/h, curve Vz1, and $v = 500$ km/h, curve Vz2), and for layered ground at

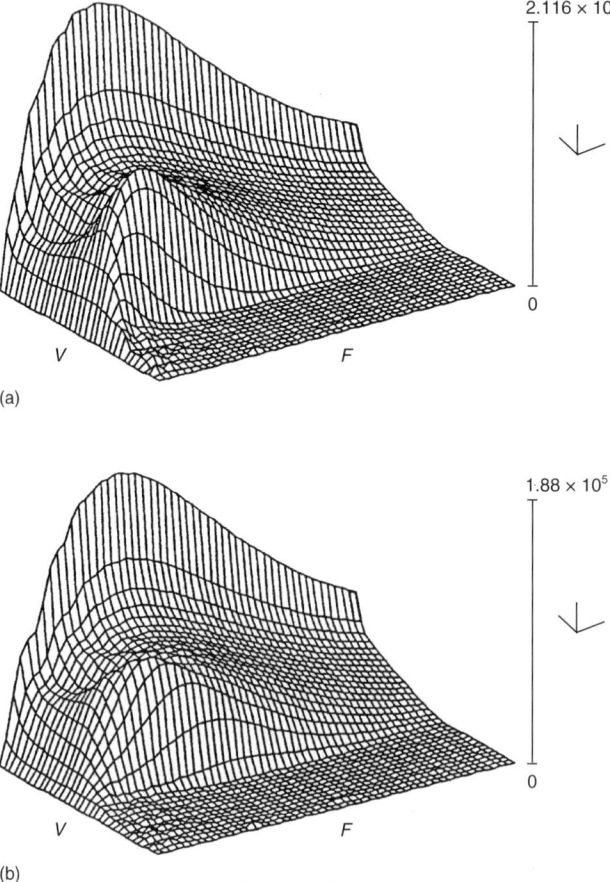

Fig. 9.6. Spectra of ground vibration velocity (in linear units, relative to a reference level of 10^{-9} m/s) for a single axle load moving along the track at speeds from 10 m/s to 320 m/s on the surface of (a) a homogeneous and (b) a layered half-space. The results are shown in the form of surface plots for the frequency band 0–50 Hz. Mesh: $\Delta v = 10$ m/s and $\Delta F = 1$ Hz

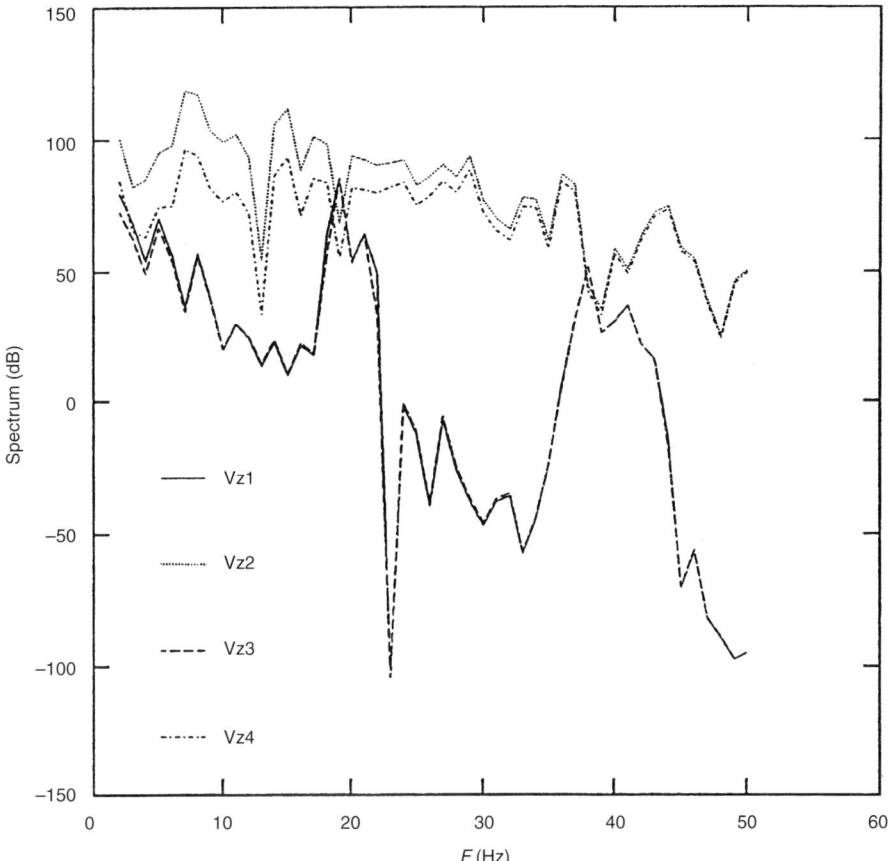

Fig. 9.7. Ground vibration spectra (in dB, relative to a reference level of 10^{-9} m/s) generated by complete TGV or Eurostar trains travelling on homogeneous ground at sub-Rayleigh and trans-Rayleigh speeds (v = 50 km/h, curve Vz1, and v = 500 km/h, curve Vz2) and on layered ground at the same speeds (curves Vz3 and Vz4)

the same train speeds (curves Vz3 and Vz4, respectively). The train consists of $N = 5$ equal carriages with the parameters $L = 18.9$ m and $M = 15.9$ m. Since the bogies of TGV and Eurostar trains have a wheel spacing of 3 m and are placed between carriage ends, i.e. they are shared between two neighbouring carriages, to use equation (9.28) one should consider each carriage as having one-axle bogies ($a = 0$) separated by the distance $M = 15.9$ m. The other parameters of the track and ground used in the calculations were the same as in Fig. 9.6. One can see that for homogeneous ground (curves Vz1 and Vz2), the average ground vibration level for a train moving at the trans-Rayleigh speed 500 km/h (138.8 m/s) is approximately 70 dB higher than for a train travelling at a speed of 50 km/h (13.8 m/s). Including the effect of layered ground, however, results in a decrease of the ground vibration level for the trans-Rayleigh train at low frequencies (curve Vz4). Note that for trains travelling at low speeds the effect of the layered ground structure is small (curves Vz1 and Vz3 are almost indistinguishable).

Calculated ground vibration frequency spectra $v_z(F)$ (in dB, versus a reference level of 10^{-9} m/s) generated by complete TGV or Eurostar trains travelling on a very soft, homogeneous ground at different speeds are shown in Fig. 9.8 for three values of train speed: $v = 20$, 50 and 70 m/s (curves V1, V2 and V3, respectively). The parameters used in the calculations were: $T = 100$ kN, $\gamma = 0.05$, $\beta = 1.28$ m^{-1}, $d = 0.7$ m, $c_R = 45$ m/s, $c_{min} = 65$ m/s, $g = 0.1$ and $y_0 = 30$ m. One can see that for the trans-Rayleigh train speed of 50 m/s, corresponding to the case $c_R < v < c_{min}$, the overall level of generated ground vibrations is much higher than for the sub-Rayleigh train speed of 20 m/s. For a train speed of 70 m/s, exceeding the value of the track critical velocity c_{min}, a significant increase of the spectra takes place at higher frequencies. However, since the amplitudes of the high-frequency components are generally low, the overall increase due to the track wave resonance is not very large in comparison with the increase associated with the ground vibration boom.

It is interesting to compare the theory described above with recent observations made near Ledsgård on the high-speed railway line from Gothenburg to Malmö (see

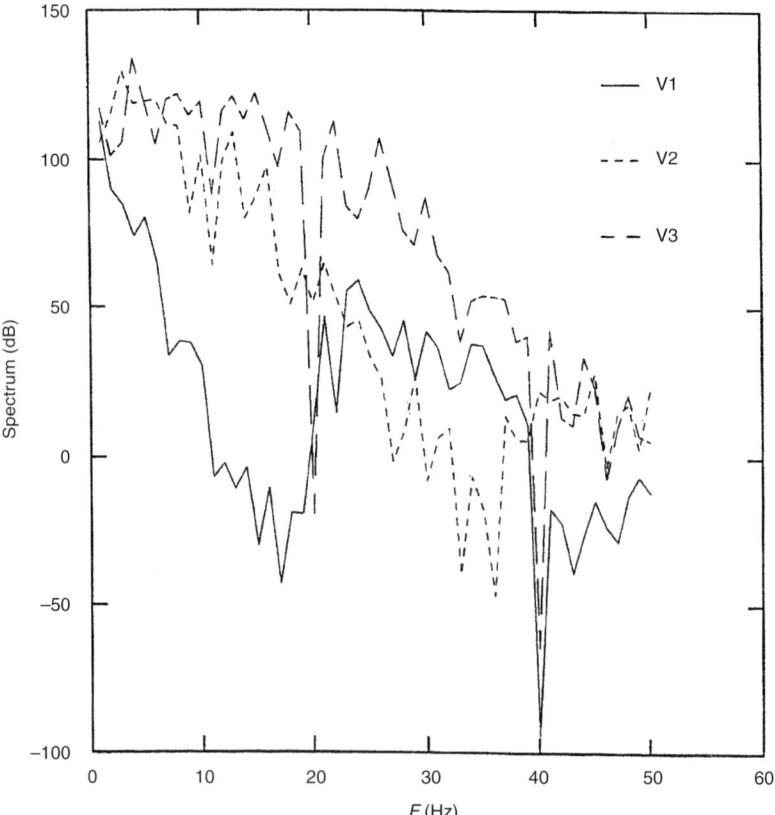

Fig. 9.8. Ground vibration spectra (in dB, relative to a reference level of 10^{-9} m/s) generated by TGV trains comprising $N = 5$ equal carriages for three values of train speed v corresponding to the cases $v < c_R$, $c_R < v < c_{min}$ and $v > c_{min}$: $v = 20$ m/s (curve v1), 50 m/s (curve v2) and 70 m/s (curve v3)

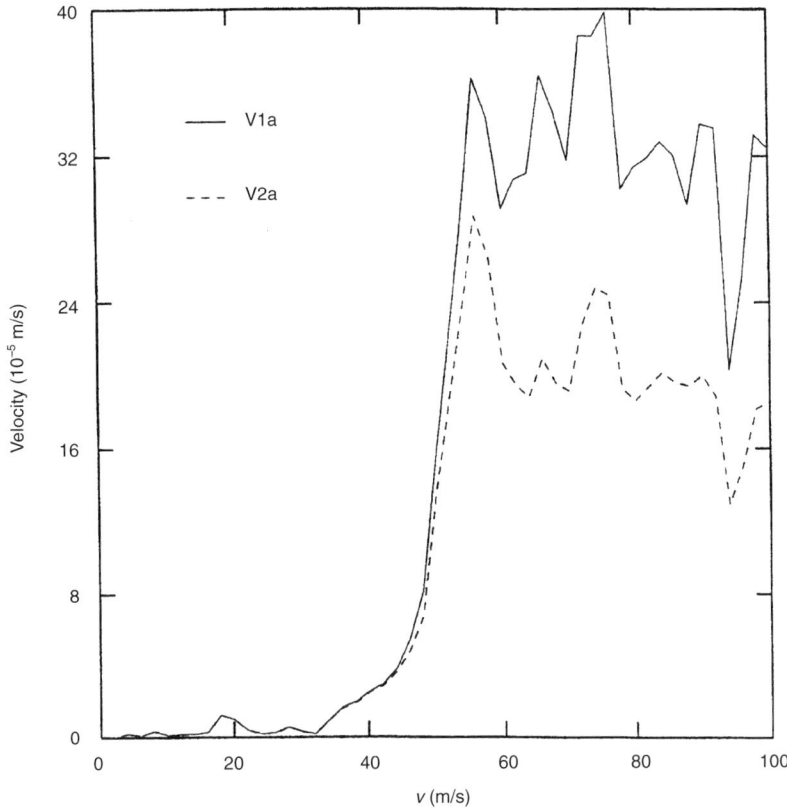

Fig. 9.9. Ground vibration velocities averaged over the frequency range 0–50 Hz (in 10^{-5} m/s) for TGV trains comprising N = 5 equal carriages; the resulting amplitudes are shown as functions of the train speed v for two values of track critical velocity: c_{min} = 65 m/s (curve V1a) and c_{min} = 10 000 m/s (curve V2a) (the latter, very large value of c_{min} describes the hypothetical case when the effects of track dynamics can be ignored)

the introduction to this chapter). For a very rough estimate, the layered structure of the ground can be ignored, and the parameters of the TGV trains can be used (instead of those of the X2000 train). Calculations of the vertical ground vibration velocity averaged over the frequency range 0–50 Hz were carried out (this corresponds roughly to the peak level of vibration velocity in the observations). We also used the reported low value of the Rayleigh wave velocity in the ground (c_R = 45 m/s), assuming that the Poisson's ratio of the ground σ was 0.25. To facilitate the comparison of the predicted increase in ground vibration level with the observed increase, the amplitudes of the generated ground vibrations were calculated in linear units (m/s).

The resulting amplitudes, as functions of train speed, are shown in Fig. 9.9 for two values of track critical velocity: c_{min} = 65 m/s (curve V1) and c_{min} = 10000 m/s (curve V2) (the latter very large value of c_{min} describes the hypothetical case when the effects of the track dynamics can be completely ignored). One can see that in both cases the predicted amplitude of the peak vertical velocity of the generated ground vibrations changes from 2×10^{-5} m/s at $v = 140$ km/h (38.8 m/s) to 16×10^{-5} m/s at $v = 180$ km/h (50 m/s). Thus, the estimated eightfold increase in ground vibration level following from the above theory for the given train speeds and Rayleigh wave velocity is in reasonable agreement with the tenfold increase recently observed experimentally for the Swedish high-speed railway line built on soft ground mentioned earlier [9.5].

According to the conclusions of Section 9.3.2, if one had taken into account the actual ground stratification at Ledsgård [9.6] (which can be approximated by a non-monotonic four-layered system, with the 'slowest' layer being positioned beneath the top layer), the predicted values of peak vibration velocity would have been roughly the same. The only difference would be expected in the form of the spectrum of the generated ground vibration, which would have lower values at low and high frequencies, in agreement with the fact that the Rayleigh wave velocity on the site has a minimum at a frequency around 4 Hz.

If we assume that the train speed increases further and approaches or exceeds the track critical velocity ($c_{min} = 65$ m/s), then a comparison of curves V1 and V2 shows that the level of generated ground vibrations also becomes larger (by approximately 1.5–2 times, as compared with the case in the absence of effects of the track dynamics). This increase is not as large as in the case of the ground vibration boom. However, since it occurs in combination with the latter, it gives a noticeable amplification of the resulting ground vibration impact.

9.5.3. High-speed trains travelling underground

For high-speed underground trains, the contribution of bulk shear and longitudinal elastic waves (S and P waves, respectively) is often more important than that of Rayleigh waves considered in the previous sections [9.26]. Obviously, the radiated S and P waves can also be significantly amplified if the train speed is high enough and the conditions $v > c_t$ or even $v > c_l$ hold, in addition to the trans-Rayleigh condition $v > c_R$ considered so far (we recall that $c_R < c_t < c_l$). In such cases one can expect that these waves will be radiated into the ground as conical Mach waves propagating at the angles $\Theta_t = \arccos(c_t/v)$ and $\Theta_l = \arccos(c_l/v)$ relative to the track, in addition to the relatively low-amplitude Rayleigh waves radiated as quasi-plane waves along the surface at the angle $\Theta = \arccos(c_R/v)$. The most likely contribution to the ground vibration boom from underground trains might be that of radiated S waves since their velocity c_t, being only about 10% higher than the velocity of Rayleigh waves, can be more easily exceeded by moving trains than the velocity of longitudinal (compressive) waves c_l can be. In the presence of a layered structure in the ground around the tunnel, the S waves initially radiated at the angle $\Theta_t = \arccos(c_t/v)$ relative to the track may experience total internal reflection from lower layers characterized by higher shear wave velocities and return to the surface. Repeated reflections from the ground surface and from lower layers may cause a waveguide propagation of S and P waves, which will affect the total vibration field in relatively remote locations.

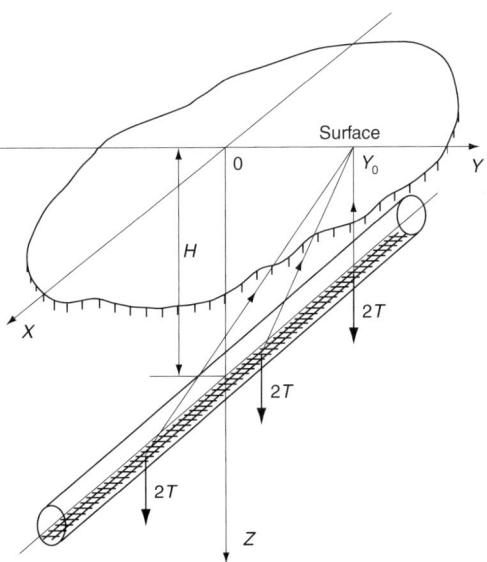

Fig. 9.10. A train travelling in underground tunnel

In comparison with surface trains, a theoretical description of the case of underground trains is more difficult, partly because the influence of the tunnel geometry makes the problem of constructing the corresponding Green's function extremely complex. We recall that the physical meaning of the Green's function for the problem under consideration is that it describes the ground vibrations generated by an individual sleeper, which may be regarded as a point source in the low-frequency band. In the case of underground trains, i.e. for sleepers placed on the bottom of a tunnel, bulk acoustic waves usually make a major contribution to the ground vibration field near the surface (especially if the tunnel is deep enough), in contrast to the case of above-ground trains, where Rayleigh surface acoustic waves prevail.

The approximate analytical approach described here considers the problem in the low-frequency approximation, i.e. the characteristic wavelengths of the bulk acoustic waves generated in the ground are assumed to be much larger than the diameter of the tunnel (Fig. 9.10). For simplicity, we consider only the case of homogeneous ground. In this low-frequency approximation, the formal expression for the vertical component of the particle velocity of ground vibrations generated on the ground surface by a train travelling underground may be written as follows [9.26]:

$$v_z(x, y, \omega) = \int_{-\infty}^{\infty} \int_{-\infty}^{\infty} \int_{-\infty}^{\infty} P(x', y', \omega) G_{zz}(r, \omega) dx' dy' dz' \qquad (9.30)$$

Here $G_{zz}(r, \omega)$ is the component of the elastic Green's tensor (Green's function) satisfying the boundary conditions on the ground surface and describing the vertical component of the particle vibration velocity due to a vertical point force located on the bottom of the tunnel, $r = [(x - x')^2 + (y - y')^2 + (z - z')^2]^{1/2}$ is the distance from the elementary source to the observation point, and $P(x', y', z', \omega)$ describes the Fourier transform of the total distribution of vertical load forces along the underground track.

For low frequencies of radiated waves (around 10–15 Hz), an approximate expression for the far-field Green's function $G_{zz}(r, \omega)$ at an observation point on the surface ($z = 0$) may be written, in the approximation in which the tunnel diameter a is taken as zero (i.e. independent of a), with only bulk elastic waves being taken into account [9.26], as

$$G_{zz}(r, \omega)|_{z=0} = \frac{i\omega}{4\pi\rho_0 r} \times$$
$$\{e^{i(\omega/c_l)r}[1 + R_l(\omega, \varphi)]\cos^2\varphi - e^{i(\omega/c_t)r}[1 + R_t(\omega, \varphi)]\sin^2\varphi\} \quad (9.31)$$

Here the terms $e^{i(\omega/c_l)r}$ and $e^{i(\omega/c_t)r}$ describe the contributions of radiated longitudinal and shear bulk waves, $R_l(\omega, \varphi)$ and $R_t(\omega, \varphi)$ are the corresponding reflection coefficients from the surface for incident longitudinal and shear waves, respectively (note that each of these coefficients takes account of both waves reflected from the surface – longitudinal and shear), and φ is the observation angle relative to the vertical direction ($\cos\varphi = (z - z')/r$). The dependence on frequency in $R_l(\omega, \varphi)$ and $R_t(\omega, \varphi)$ takes account of the impedance load resulting from the influence of buildings or other engineering structures on the surface. In what follows we assume, without loss of generality, that $R_l(\omega, \varphi) = 0$ and $R_t(\omega, \varphi) = 0$, i.e. we consider that all the energy of the radiated waves is absorbed by the structure.

The load distribution along an underground track may be written in a form similar to that for a surface track (see equation (9.27)). For a complete train moving at speed v along a track lying underground at depth H, this distribution has the form [9.26]

$$P(t, x', y', z') = \sum_{m=-\infty}^{\infty} \sum_{n=0}^{N-1} A_n \left[P\left(t - \frac{(x' + nL)}{v}\right) + P\left(t - \frac{(x' + M + nL)}{v}\right) \right] \delta(x' - md)\, \delta(y')\, \delta(z' - H) \quad (9.32)$$

All other notations in equation (9.32) are the same as in the previous sections.

Taking the Fourier transform of equation (9.32) and substituting it into equation (9.30), taking account of equation (9.31), one can obtain the following expression for the frequency spectrum of the vertical vibrations at $z = 0, x = 0$ and $y = y_0$ generated by an underground train [9.26]:

$$v_z(0, y_0, \omega) = \frac{i\omega P(\omega)}{4\pi\rho_0} \sum_{m=-\infty}^{\infty} \sum_{n=0}^{N-1} \left[1 + \exp\left(\frac{iM\omega}{v}\right) \right] \exp\left(i\frac{\omega}{v}(md + nL)\right) \times$$
$$\frac{1}{r_m} \left[\exp\left(-\frac{\gamma_l \omega r_m}{c_l} + i\frac{\omega}{c_l} r_m\right) \cos^2\varphi_m - \exp\left(-\frac{\gamma_t \omega r_m}{c_t} + i\frac{\omega}{c_t} r_m\right) \sin^2\varphi_m \right] \quad (9.33)$$

Here $r_m = [y_0^2 + (md)^2 + H^2]^{1/2}$, $\cos(\varphi_m) = H/r_m$ and $P(\omega)$ is the Fourier transform of $P(t)$ described by equation (9.12). In writing equation (9.33) we have taken account of attenuation in the soil by replacing $1/c_l$ and $1/c_t$ in the exponentials by the complex values $1/c_l + i\gamma_l/c_l$ and $1/c_t + i\gamma_t/c_t$, where $\gamma_{l,t} \ll 1$ are constants describing the 'strength' of the dissipation of longitudinal and shear waves in the soil. To generalize equation (9.33) to describe the action of two axle loads separated by a distance a (the case of a bogie), one should replace $P(\omega)$ by $P_b(\omega)$, according to equation (9.16).

It follows from equation (9.33) that, similarly to the case of a surface train (see Section 9.4.2), the spectrum of ground vibrations from an underground train is quasi-discrete, with maxima at frequencies determined by the condition $(\omega/v)(sd + qL) = 2\pi l$, where $s, q, l = 1, 2, 3, \ldots$.

Calculations according to equation (33) show that for underground trains travelling at conventional speeds, the main contribution to the vertical component v_z of the total ground vibration field at the ground surface is usually due to the radiated shear bulk waves rather than to the longitudinal bulk waves.

In the case of high-speed underground trains, an analysis of equation (9.33) shows that it has two maxima, for values of the observation angle satisfying the conditions $\cos\Theta_t = c_t/v$ for radiated shear waves, and $\cos\Theta_l = c_l/v$ for radiated longitudinal waves. Since $\cos\Theta_t$ and $\cos\Theta_l$ must be less than 1, it follows from these conditions that the maxima can be achieved if the train speed is high enough and the condition $v > c_t$ or even $v > c_l$ holds. In such cases the corresponding waves are radiated into the ground as conical waves propagating at angles $\Theta_t = \arccos(c_t/v)$ and $\Theta_l = \arccos(c_l/v)$ relative to the track.

The results of calculations of the amplitudes of generated ground vibrations at a frequency of 15 Hz (in dB re 10^{-9} m/s) as a function of the tunnel depth H for $Y_0 = 30$ m are shown in Fig. 9.11 for two values of train speed: $v = 13.8$ m/s (conventional speed) and $v = 80$ m/s (trans-shear speed for the ground considered). The vertical components of the ground vibration velocity on the ground surface are indicated by Vz1 and Vz2 for $v = 13.8$ m/s and $v = 80$ m/s, respectively. The corresponding separate contributions of longitudinal and shear elastic waves are labelled as VzL1, VzS1, VzL2 and VzS2. The parameters used here are those of a TGV high-speed train with $N = 5$ carriages, as in the previous examples. The other parameters are the following: the elastic parameters of the ground are $c_t = 76$ m/s, $c_l = 129$ m/s and $c_R = 70$ m/s (corresponding to a Poisson's ratio $\sigma = 0.25$); the mass density of the soil ρ_0 is 2000 kg/m^3; and the wheel load T is 100 kN. The soil attenuation parameters $\gamma_{l,t}$ were chosen as 0.05, and the effect of track wave velocity was neglected.

It can be seen from Fig. 9.11 that the ground vibrations generated by an underground train travelling at a speed v larger than the shear wave velocity in the ground c_t are much larger than those generated by the same train moving at a conventional speed (the speed-related amplification of the total ground vibration field varies from about 50 dB to 20 dB when the tunnel depth H is changed from 2 m to 100 m). For practical values of H (less than 60–70 m), the contribution of shear waves for the given Poisson's ratio is much higher than the contribution of longitudinal waves. For larger depths, the contributions of shear and longitudinal waves first become comparable to each other, causing an oscillatory behaviour of the resulting field versus H, and then longitudinal waves prevail. Note, in this connection, that the difference between the ground vibration levels corresponding to the contribution of longitudinal waves at $v = 13.8$ m/s and $v = 80$ m/s (about 20 dB, or ten times) is attributed entirely to the effect of the function $P(v, \omega)$ determined by equation (12) (see also Section 9.5.1) since no ground vibration boom is experienced with regard to longitudinal waves in the case considered. Therefore, the observed additional increase of up to 30 dB in the resulting ground vibration field is associated with a ground vibration boom for radiated shear waves.

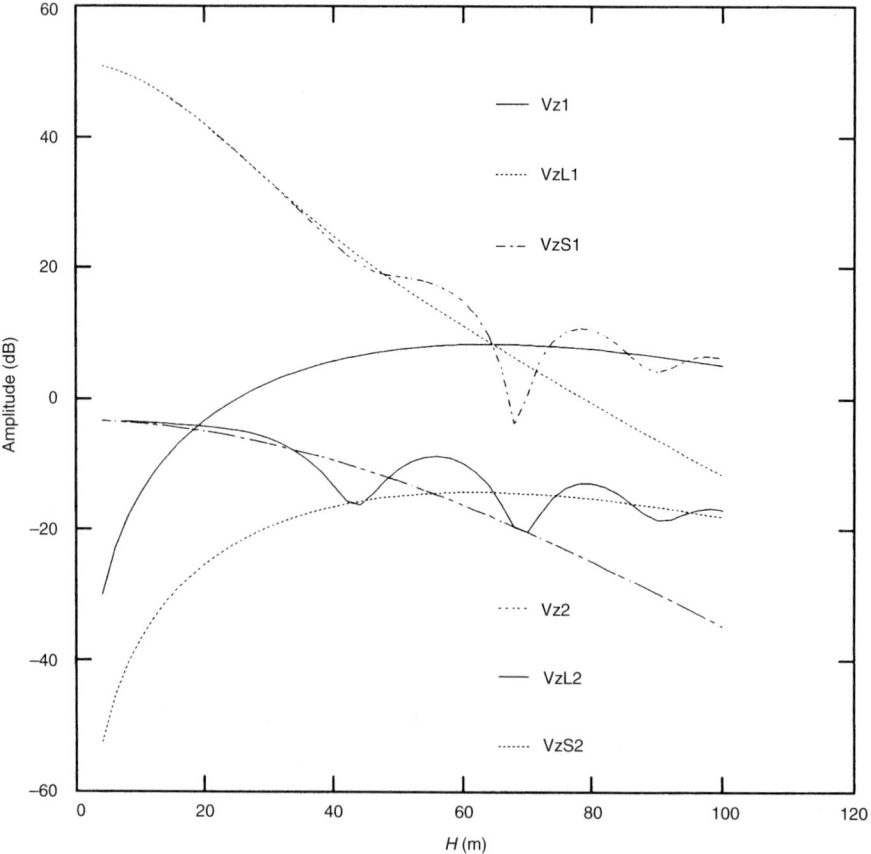

Fig. 9.11. Ground vibration amplitudes (in dB, relative to a reference level of 10^{-9} m/s) generated by underground TGV trains comprising $N = 5$ equal carriages at frequency $f = 15$ Hz as functions of the tunnel depth H; the results are shown for two values of train speed, $v = 13.8$ m/s (conventional speed) and $v = 80$ m/s (trans-shear speed for the ground considered); the vertical components of the ground vibration velocity on the ground surface are indicated by Vz1 and Vz2 for $v = 13.8$ m/s and $v = 80$ m/s, respectively. The corresponding separate contributions of longitudinal and shear elastic waves are indicated by VzL1, VzS1 and VzL2, VzS2

GENERATION OF GROUND VIBRATION BOOM

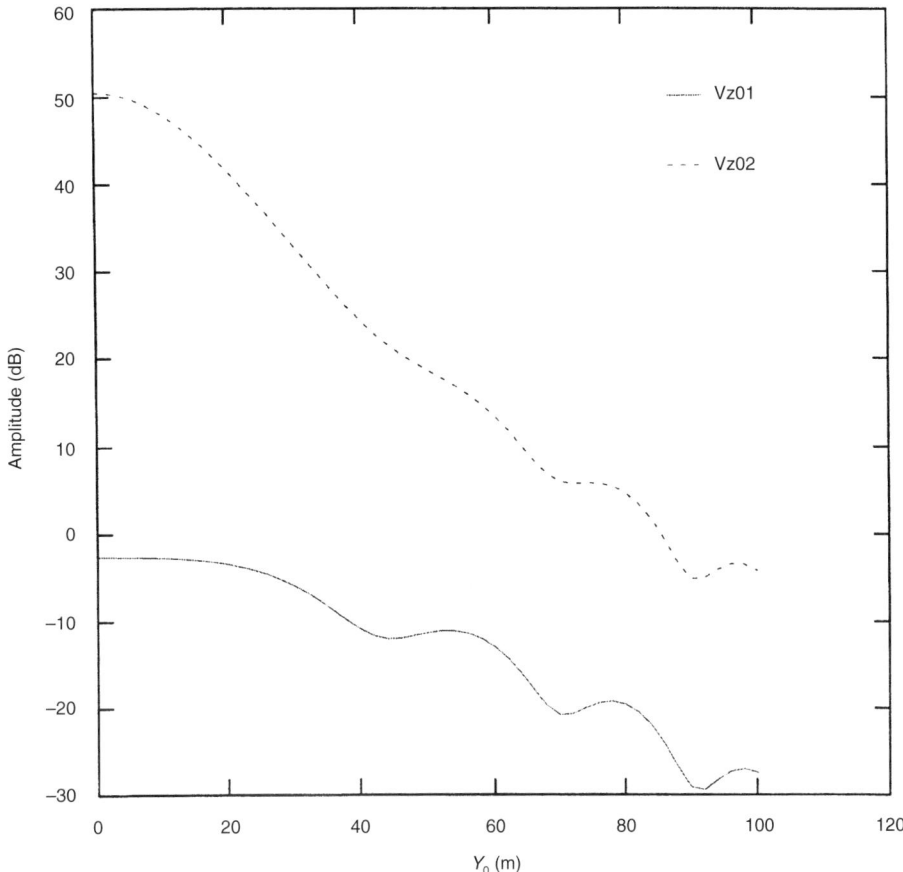

Fig. 9.12. Ground vibration amplitudes (in dB, relative to a reference level of 10^{-9} m/s) generated by underground TGV trains comprising N = 5 equal carriages at frequency f = 15 Hz as functions of the observation distance from the track Y_0; the results are shown for two values of train speed, v = 13.8 m/s (conventional speed) and v = 80 m/s (trans-shear speed for the ground considered); the vertical components of the ground vibration velocity on the ground surface are indicated by Vz01 and Vz02 for v = 13.8 m/s and v = 80 m/s, respectively

The generated ground vibrations (in dB re 10^{-9} m/s), as functions of the observation distance from the track Y_0 for H = 30 m at a frequency of 15 Hz, are shown in Fig. 9.12 for the same values of train speed (v = 13.8 m/s and 80 m/s) and the same train, track and ground parameters – curves Vz01 and Vz02, respectively. One can see that in both cases the fields decrease with the distance Y_0, especially in the case of the ground vibration boom associated with shear waves.

The results of calculations of the total ground vibration spectrum generated by the same TGV train, comprising five carriages (N = 5) and travelling underground at speeds v = 13.8 m/s and 80 m/s, are shown in Fig. 9.13 (curves Vz1 and Vz2, respectively). The train, track and ground parameters are the same as in Figs 9.11 and 9.12; H = 30 m and Y_0 = 30 m. For comparison, in the same figure the spectra

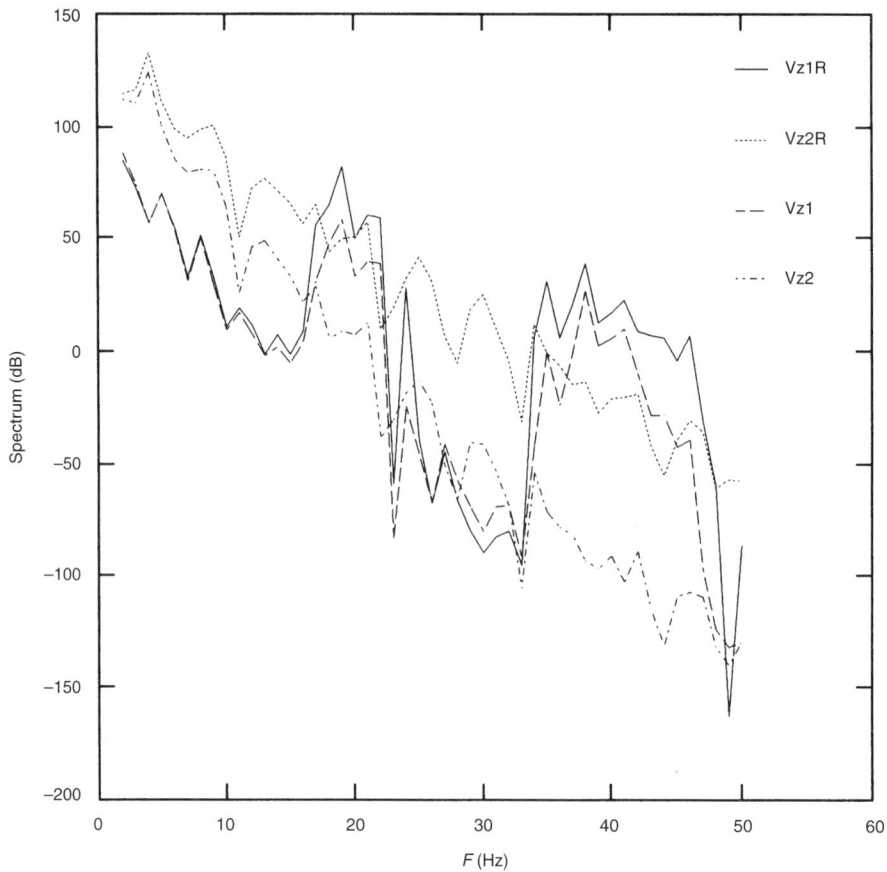

Fig. 9.13. Ground vibration spectra (in dB, relative to a reference level of 10^{-9} m/s) generated by underground TGV trains comprising N = 5 equal carriages for two values of train speed, v = 13.8 m/s (conventional speed) and v = 80 m/s (trans-shear speed for the ground considered) – curves Vz1 and Vz2, respectively. For comparison, the ground vibration spectra generated by the same train travelling above the ground (curves Vz1R and Vz2R, respectively) are also shown

generated by the same train travelling above the ground at Y_0 = 30 m (curves Vz1R and Vz2R) are also shown. In the latter case, which was extensively discussed in the previous sections, the Rayleigh wave contribution prevails.

One can see that in the case of a ground vibration boom, which is present at v = 80 m/s for both underground and surface trains, a very large increase in generated ground vibrations is observed, although for an underground train this increase is less pronounced, especially at higher frequencies. One can see that the shapes of the ground vibration spectra for underground and above-ground trains travelling at the same speed are very similar. This implies that the shapes of ground vibration spectra are determined mainly by geometrical parameters of the track and train rather than by the tunnel depth and, consequently, by the types of elastic waves predominantly generated.

We remind the reader that the low-frequency approximation used in this section to describe ground vibrations from underground trains is inaccurate for frequencies higher than 10–15 Hz. To improve the situation, one can take into account the next term in the series expansion of the Green's function for the problem under consideration [9.27]. This term is proportional to the product of the tunnel diameter and the characteristic wavenumber of the radiated ground vibrations.

9.5.4. Waveguide effects of embankments on generated ground vibration fields

Many interesting phenomena can be expected when high-speed railway lines are built on embankments. In particular, one can mention possible waveguide effects of embankments on the ground vibration boom, which were first discussed in [9.4]. The possibility of an embankment acting as a waveguide for generated ground vibrations is closely related to the fact that the Rayleigh surface waves associated with the ground vibration boom are usually radiated at small angles relative to the track. For this reason, a dominant part of the radiated energy can be expected to be trapped and dissipated within the embankment itself, without significant leakage to the area outside. In what follows we briefly discuss the process of generation of a ground vibration boom by a high-speed train travelling along the top of an embankment.

The effect of railway embankments on generated ground vibrations can be considered by constructing a specific Green's function for an elastic half-space with an embankment. Such a Green's function must take into account the internal reflections of generated surface Rayleigh waves from the geometric boundaries between the top and the side surfaces of the embankment, and between the side surfaces and the ground below. For simplicity, we assume that the elastic properties of the embankment are the same as those of the supporting ground. Also, we consider only reflections from the boundaries between the top surface of the embankment, of width $2W$, and the side surfaces, assuming that the side surfaces join smoothly to the ground so that no additional reflections take place.

The theory of Rayleigh wave reflection from the boundary between two surfaces intersecting at an obtuse angle Θ has been developed earlier by the present author (see, e.g., [9.24], p. 277). The corresponding expression for the Rayleigh wave reflection coefficient has the form

$$R(\alpha) = \frac{ik_t^4(k_R^2 - k_t^2)^{1/2}}{2k_R^2 F'(k_R)} \beta \frac{1}{\cos\alpha}\left(1 - 4\frac{k_R^2}{k_t^2}\sin^2\alpha\right) \quad (9.34)$$

where $\beta = \pi - \Theta$, and α is the incidence angle of the Rayleigh wave relative to the normal to the boundary between the two surfaces; other notations are the same as in the previous sections. It follows from equation (9.34) that R is independent of frequency and is defined only by the angle β (measured in radians), the Poisson's ratio of the ground σ and the angle of incidence α. Note that equation (9.34) has been derived for relatively small β and α. Therefore, it becomes invalid when α approaches $\pi/2$, where $|R(\alpha)|$ becomes larger than 1. To extrapolate $|R(\alpha)|$ for

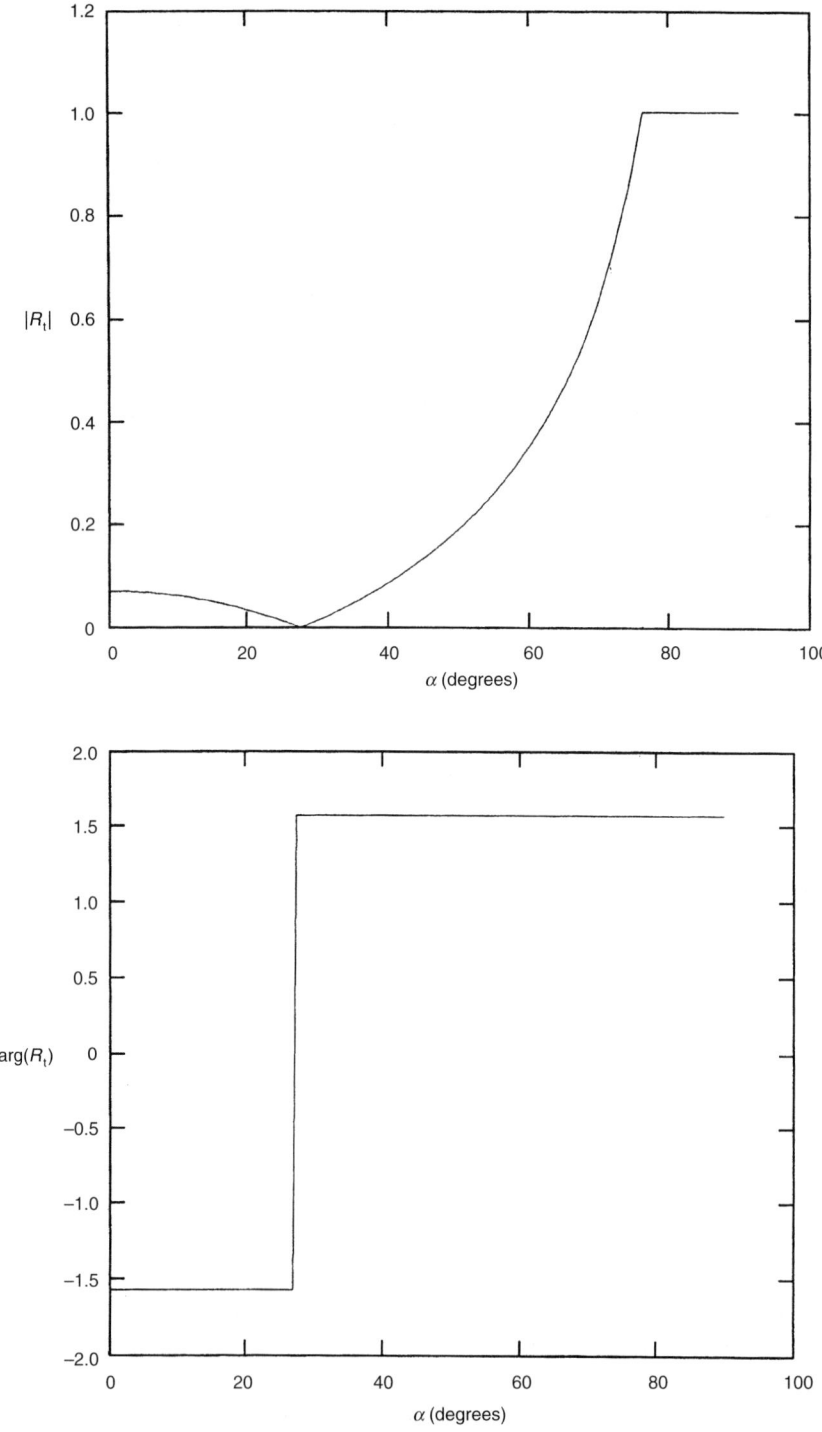

Fig. 9.14. Amplitude and phase of the truncated Rayleigh wave reflection coefficient $R_t(\alpha)$ for the boundary between two surfaces crossing at an angle $\Theta = 150°$ ($\beta = 30°$) as functions of the incidence angle α

arbitrary α, we introduce a truncated reflection coefficient $R_t(\alpha)$ in the following way:

$$R_t(\alpha) = R(\alpha)\Phi[1-|R(\alpha)|] + i\Phi[|R(\alpha)|-1] \tag{9.35}$$

where $\Phi(z)$ is the Heaviside step function. The behaviour of the amplitude and phase of $R_t(\alpha)$ is shown in Fig. 9.14 for $\beta = 30°$ and Poisson's ratio $\sigma = 0.25$.

For the problem under consideration, one can use the above-mentioned reflection coefficient to represent the Green's function for the top area of the embankment. This can be done in a way similar to the ray approach to the representation of the field in an ideal waveguide, i.e. by considering the waveguide contribution to the Green's function in question as an infinite sum of Rayleigh waves radiated by imaginary sources with amplitudes defined by multiples of the reflection coefficient $R_t(\alpha)$. Then, for a point source located at $y' = 0$ and for observation points located on the flat top of the embankment, i.e. for $y < |W|$, the resulting Green's function has the form

$$G_{zz}(\rho, \omega) = D(\omega)\Bigg[(1/\sqrt{\rho^{(0)}})\exp(ik_R\rho^{(0)} - \gamma k_R \rho^{(0)}) + $$
$$\sum_{n=1}^{\infty}(R_{tn}^{(1)})^n(1/\sqrt{\rho_n^{(1)}})\exp(ik_R\rho_n^{(1)} - \gamma k_R\rho_n^{(1)}) + \tag{9.36}$$
$$\sum_{n=1}^{\infty}(R_{tn}^{(2)})^n(1/\sqrt{\rho_n^{(2)}})\exp(ik_R\rho_n^{(2)} - \gamma k_R\rho_n^{(2)})\Bigg]$$

Here $D(\omega) = (1/2\pi)^{1/2}(-i\omega)qk_R^{1/2}k_t^2\exp(-i3\pi/4)/\mu F'(k_R)$ (see also equations (9.13) and (9.14)), and $\rho^{(0)} = [(x-x')^2 + (y)^2]^{1/2}$, $\rho_n^{(1)} = [(x-x')^2 + (2Wn-y)^2]^{1/2}$, $\rho_n^{(2)} = [(x-x')^2 + (-2Wn-y)^2]^{1/2}$, and $R_{tn}^{(1)}$ and $R_{tn}^{(2)}$ are the values of the reflection coefficient (equation (9.35)) taken for the angles $\alpha = \alpha_{n1}$ and $\alpha = \alpha_{n2}$ defined by the expressions $\alpha_{n1} = \arccos[(2H-y)/\rho_n^{(1)}]$ and $\alpha_{n2} = \arccos[(2H+y)/\rho_n^{(2)}]$, respectively; other notations are the same as in the previous sections. For points of observation outside the embankment, i.e. for $y > |W|$, the expressions for the Green's function are constructed in a similar way and include also the transmission coefficient $T_t(\alpha)$ for Rayleigh waves passing through the boundary between the top and side areas of the embankment. For brevity, these expressions are not displayed here. The spatial distribution of surface vibration velocity (in arbitrary units) inside and outside the embankment corresponding to the above-mentioned Green's function is shown in Fig. 9.15. The calculations have been performed for a frequency of 30 Hz within an area of 48 × 48 m², the width of the embankment 2H being equal to 8 m. One can see that the embankment results in a redistribution of the Rayleigh waves radiated at small angles α relative to the embankment's direction. The corresponding Rayleigh waves become trapped and propagate as waveguide modes, rather than as free cylindrically divergent waves.

Substitution of this Green's function into equation (9.30) in the way described in the previous sections allows calculation of ground vibrations generated by high-

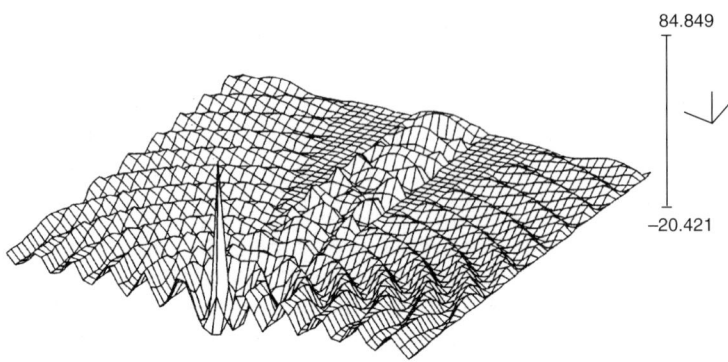

Fig. 9.15. Spatial distribution of the surface vertical displacements (in arbitrary units) generated at the frequency component f = *30 Hz by a single axle load moving over a single sleeper placed on an embankment (the Green's function for the system considered); the calculations have been performed for a surface area of 48 × 48 m^2 and for an embankment of width 2W = 8 m*

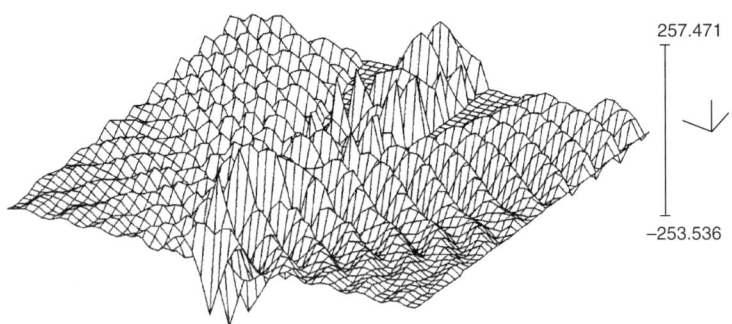

Fig. 9.16. Spatial distribution of the surface vertical displacements (in arbitrary units) generated at the frequency component f = *30 Hz by a single axle load moving at the trans-Rayleigh speed v = 127 m/s over a length of track consisting of 20 sleepers placed on an embankment; the calculations have been performed for a surface area of 48 × 48 m^2 and for an embankment of width 2W = 8 m*

speed trains travelling along embankments. Since the resulting expressions are too bulky and contain a triple summation – over sleepers, carriage axles and imaginary sources – we consider only the simplest example, of a single axle load travelling along the embankment at a speed which is slightly higher than the Rayleigh wave velocity in the ground.

Figure 9.16 shows the spatial distribution of the ground vibration field (in arbitrary units) generated by a single axle load travelling at a trans-Rayleigh speed $v = 47$ m/s along an embankment of 8 m wide and with a slope angle $\beta = 30°$, over a length of track consisting of 20 sleepers. The area considered is 48 × 48 m^2. The other parameters used in the calculation were: $c_R = 45$ m/s, $\sigma = 0.25$ and $\rho_0 = 2000$ kg/m^3. The track parameters are the same as in the previous figures, and the effect of track bending waves has been neglected.

One can see from Fig. 9.16 that the generated ground vibrations propagate predominantly along the embankment, where their amplitudes are much larger than in the outside area. This demonstrates that embankments can act as waveguides for a generated ground vibration boom, reducing the hazardous impact of the very intense vibrations on the built environment. At the same time, the waveguide effects may result in a large increase of ground vibration inside the embankment. These waveguide-induced additional vibrations of the embankment, which propagate at the speed of the train, are expected to amplify the quasi-static bending deflections in the track–ground system accompanying the moving axle loads. As a result of this, very large vibrations of the embankment might be observed at train speeds around the velocity of Rayleigh waves, i.e. before achieving the value of the track critical velocity.

9.6. Conclusions

The theory of generation of ground vibrations by high-speed trains described in this chapter shows that if the train speed exceeds the velocity of Rayleigh surface waves in the supporting soil, a ground vibration boom occurs, associated with a very large increase in the amplitude of the generated vibrations. Crossing the critical velocity for track waves results in a further increase of generated ground vibrations, albeit not as dramatic as in the case of the ground vibration boom.

Recent experimental observations of a ground vibration boom generated by Swedish X2000 trains confirm the main predictions of the theory. This implies that the railway-generated ground vibration boom is no longer an exotic theoretical effect. It is today's reality for high-speed railways crossing soft soil, and so are 'supersonic', or 'trans-Rayleigh', trains. Builders and operators of high-speed railways must be aware of the possible consequences of the ground vibration boom, as well as of dynamic rail deflections. One can expect that, with the general trend of increasing operating speeds of trains, this phenomenon will occur in many countries, especially in those with railways built on soft soil.

Problems associated with track–soil critical velocities may occur also for underground trains. In this case the first critical velocity is the velocity of shear bulk elastic waves, and the ground vibration boom for train speeds exceeding this velocity corresponds to a Mach cone of shear waves radiated from the tunnel.

Waveguide effects of embankments may cause concentration of the ground vibration energy radiated by trans-Rayleigh trains. This can reduce ground vibration outside the embankment. However, the vibrations of the embankment itself may increase significantly. Since these guided vibrations propagate at the speed of the train, they may amplify the quasi-static bending deflections in the track–ground system moving with the train.

It is too early, at this stage, to foresee how the phenomenon of the railway-generated ground vibration boom and its amplification due to effects of track dynamics will be reflected in future standards on noise and vibration from high-speed trains. However, one can expect that such an important parameter as the Rayleigh wave velocity in the ground for the sites considered will be present in such standards, indicating a maximum train speed beyond which excessive ground vibrations can be expected.

9.7. Acknowledgements

The author is grateful to Prof. G. Degrande for useful discussions about part of the material described in this chapter and for drawing the author's attention to some inaccuracies in some previous publications.

9.8. References

9.1. REMINGTON, P. J., KURZWEIL, L. G. and TOWERS, D. A. Low-frequency noise and vibrations from trains. In: *Transportation Noise. Reference Book* (ed. P. Nelson). Butterworths, London, 1987, chapter 16.

9.2. NEWLAND, D. E. and HUNT, H. E. M. Isolation of buildings from ground vibration: a review of recent progress. *Proceedings of the Institution of Mechanical Engineers*, 1991, **205**, 39–52.

9.3. KRYLOV, V. V. On the theory of railway-induced ground vibrations. *Journal de Physique IV*, 1994, **4**(C5), 769–772.

9.4. KRYLOV, V. V. Generation of ground vibrations by superfast trains. *Applied Acoustics*, 1995, **44**, 149–164.

9.5. MADSHUS, C. Public announcement. *Ground Dynamics and Man-made Processes: Prediction, Design, Measurement* [conference]. London, 1997.

9.6. MADSHUS, C. and KAYNIA, A. M. High speed railway lines on soft ground: dynamic behaviour at critical train speed. *Proceedings of the 6th International Workshop on Railway and Tracked Transit System Noise*. Ile des Embiez, France, 1998, pp. 108–119.

9.7. KRYLOV, V. V. Vibrational impact of high-speed trains. I. Effect of track dynamics. *Journal of the Acoustical Society of America*, 1996, **100**(5), 3121–3134; erratum, 1997, **101**(6), 3810.

9.8. KRYLOV, V. V. Spectra of low-frequency ground vibrations generated by high-speed trains on layered ground. *Journal of Low Frequency Noise and Vibration*, 1997, **16**(4), 257–270.

9.9. KRYLOV, V. V. Effect of track properties on ground vibrations generated by high-speed trains. *Acustica – Acta Acustica*, 1998, **84**(1), 78–90.

9.10. TAKEMIYA, H. Lineside ground vibrations induced by high-speed train passage. *Proceedings of the Workshop on Effect of High-Speed Vibration on Structures and Equipment*. Tainan, Taiwan, 1998, pp. 43–49.

9.11. DEGRANDE, G. and SCHILLEMANS, L. Free field vibrations during the passage of HST. *Proceedings of ISMA 23, Noise and Vibration Engineering*. Leuven, Belgium, 1998, vol. 3, pp. 1563–1570.

9.12. PETYT, M. and JONES, C. J. C. Modelling of ground-borne vibrations from railways. *Proceedings of the 4th European Conference on Structural Dynamics (EURODYN '99)*. Prague, 1999, pp. 79–87.

9.13. SHENG, X., JONES, C. J. C. and PETYT, M. Ground vibration generated by a load moving along a railway track. *Journal of Sound and Vibration*, 1999, **228**(1), 129–156.

9.14. WOLDRINGH, R. F. and NEW, B. M. Embankment design for high speed trains on soft soils. *Proceedings of the 12th European Conference on Soil Mechanics and Geotechnical Engineering, 'Geotechnical Engineering for Transportation Infrastructure'*. Amsterdam (eds F. B. J. Barends, J. Lindenberg, H. J. Luger, L. de Queleriy and A. Verruiyt). Balkema, Rotterdam, 1999, vol. 3, pp. 1703–1712.

9.15. ADOLFSSON, K., ANDREASSON, B., BENGTSSON, P.-E. and ZACKRISSON, P. High speed train X2000 on soft organic clay – measurements in Sweden. *Proceedings of the 12th European Conference on Soil Mechanics and Geotechnical Engineering, 'Geotechnical Engineering for Transportation Infrastructure'*. Amsterdam (eds F. B. J. Barends, J. Lindenberg, H. J. Luger, L. de Queleriy and A. Verruiyt). Balkema, Rotterdam, 1999, vol. 3, pp. 1713–1718.

9.16. HEELIS, M. E., COLLOP, A. C., DAWSON, A. R., CHAPMAN, D. N. and KRYLOV, V. V. Transient effects of high speed trains crossing soft soil. *Proceedings of the 12th European Conference on Soil Mechanics and Geotechnical Engineering, 'Geotechnical Engineering for Transportation Infrastructure'*. Amsterdam (eds F. B. J. Barends, J. Lindenberg, H. J. Luger, L. de Queleriy and A. Verruiyt). Balkema, Rotterdam, 1999, vol. 3, pp. 1809–1814.

9.17. KRYLOV, V. V., DAWSON, A. R., HEELIS, M. E. and COLLOP, A. C. Rail movement and ground waves caused by high-speed trains approaching track–soil critical velocities. *Proceedings of the Institution of Mechanical Engineers, Part F: Journal of Rail and Rapid Transit*, 2000, **214**, 107–116.

9.18. KRYLOV, V. V. and FERGUSON, C. C. Calculation of low-frequency ground vibrations from railway trains. *Applied Acoustics*, 1994, **42**, 199–213.

9.19. FRYBA, L. *Vibration of Solids and Structures under Moving Loads*. Noordhoff, Groningen, 1973.

9.20. BROCKLEY, C. A. The influence of track support structure and locomotive traction characteristics on short wavelength corrugations. *Wear*, 1992, **153**, 315–322.

9.21. EWING, W. M., JARDETZKY, W. S. and PRESS, F. *Elastic Waves in Layered Media*. McGraw-Hill, New York, 1957.

9.22. GRAFF, K. F. *Wave Motion in Elastic Solids*. Clarendon Press, Oxford, 1975.

9.23. GUTOVSKI, T. G. and DYM, C. L. Propagation of ground vibration: a review. *Journal of Sound and Vibration*, 1976, **49**, 179–193.

9.24. BIRYUKOV, S. V., GULYAEV, Y. V., KRYLOV, V. V. and PLESSKY, V. P. *Surface Acoustic Waves in Inhomogeneous Media*. Springer, Berlin, 1995.

9.25. COLE, J. and HUTH, J. Stresses produced in a half plane by moving loads. *Journal of Applied Mechanics*, 1958, **25**, 433–436.

9.26. KRYLOV, V. V. Low-frequency ground vibrations from underground trains. *Journal of Low Frequency Noise and Vibration*, 1995, **14**(1), 55–60.

9.27. LIN, Q. and KRYLOV, V. V. Effect of tunnel diameter on ground vibrations generated by underground trains. *Journal of Low Frequency Noise, Vibration and Active Control*, 2000, **19**(1), 17–25

10. Free-field vibrations during the passage of a high-speed train: experimental results and numerical predictions

G. Degrande
Department of Civil Engineering, Katholieke Universiteit Leuven, Kasteelpark Arenberg 40, B-3001 Heverlee, Belgium

10.1. Introduction

Vibrations caused by the passage of a high-speed train (HST) are an environmental concern, as waves propagate through the soil and interact with nearby buildings, where they may cause malfunctioning of sensitive equipment and discomfort to people.

Six weeks before the inauguration of the HST track between Brussels and Paris in December 1997, the Belgian railway company organized homologation tests during the passage of a Thalys HST at a speed varying between 160 and 330 km/h. As the available experimental data are scarce, especially regarding the influence of the train speed on the vibration amplitude, the opportunity was taken to perform free-field vibration measurements on the track and in the free field at distances from the track varying from 4 m to 72 m. The *in situ* measurements were performed near Ath, 55 km south of Brussels, where the train can reach maximum speed. The results obtained are complementary to *in situ* vibration measurements performed during the passage of a Thalys HST on the track between Amsterdam and Utrecht in the Netherlands, at speeds between 40 and 160 km/h [10.2], as well as to data reported earlier by Auersch [10.1] for the German ICE train at speeds varying between 100 and 300 km/h.

Apart from clarifying the physical phenomena involved, free-field vibration measurements can be used to validate the analytical and numerical models that are presently under development to predict the propagation of waves due to the passage of a train. The final objective of these modelling efforts is to forecast vibration levels

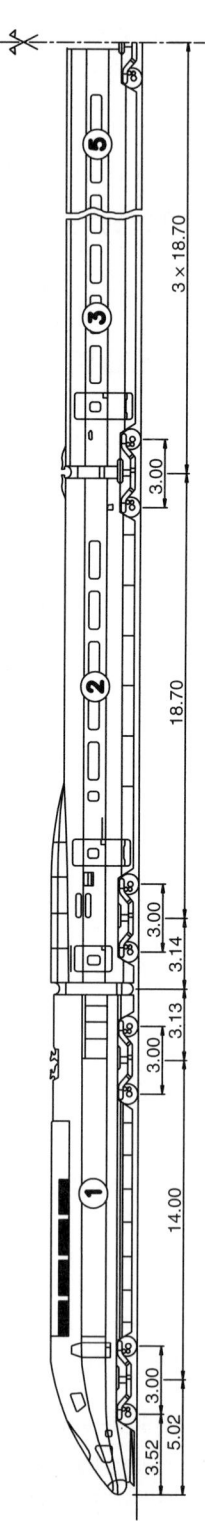

Fig. 10.1. Configuration of the Thalys HST (dimensions in m)

in structures and to assess the efficiency of vibration isolation measures for both existing and newly built structures. Krylov and Ferguson [10.4] and Krylov [10.11, 10.12] have proposed an analytical model where the quasi-static force transmitted by a sleeper is derived from the deflection curve of the track, modelled as a beam on an elastic foundation; wave propagation through the soil is represented by the surface wave contribution to the Green's function of a half-space. This model can easily be extended to incorporate the Green's functions of a layered half-space [10.3]. A more advanced track model has been proposed by Van den Broeck and De Roeck [10.16, 10.17] that, besides quasi-static loading due to the moving axles, also incorporates parametric excitation, transient excitation due to rail joints and wheel flats and excitation due to wheel and rail roughness. Through-soil coupling of the sleepers, resting on a layered half-space, is accounted for as well. Similar developments have recently been presented by Knothe and Wu [10.10].

The objectives of this chapter are the following. First, the characteristics of the train, the track and the subsoil, as well as the experimental set-up, are briefly described. Second, the time history and the frequency content of the vertical response of the rail, the sleeper and the free field at various distances from the track are discussed. Special emphasis is placed on the influence of the train speed on the peak particle velocity (PPV) and the frequency content of the response in the free field. Third, the essential elements of a relatively simple analytical prediction model, proposed by Krylov, are briefly recapitulated. We illustrate how the estimated forces, transmitted by the individual sleepers, depend on the characteristics of the track and the train and how the free-field vibrations are subsequently predicted, using the force distribution along the track and the Green's functions of a horizontally stratified subsoil. This finally results in the fourth objective of the chapter, the comparison of the experimental results and the numerical predictions. Although a large degree of uncertainty remains and some of the required input parameters and a quantitative validation of the model is not really possible, the comparison of experimental data and numerical results allows for a qualitative validation of the analytical prediction model.

10.2. The *in situ* measurements

10.2.1. The train

Figure 10.1 shows the configuration of the Thalys HST, consisting of two locomotives and eight carriages; the total length of the train is 200.18 m. The locomotives are supported by two bogies and have four axles. The carriages next to the locomotives share one bogie with the neighbouring carriage, while the six other carriages share both bogies with neighbouring carriages. The total number of bogies equals 13 and, consequently, the number of axles on the train is 26. The carriage length L_t, the distance L_b between bogies, the axle spacing L_a, the total axle mass M_t, the sprung axle mass M_s and the unsprung axle mass M_u of all carriages are summarized in Table 10.1.

10.2.2. The track

Continuously welded UIC60 rails with a mass per unit length of 60 kg/m and a moment of inertia $I = 0.3038 \times 10^{-4}$ m^4 are fixed with a Pandroll E2039 rail-fixing system on precast, prestressed concrete monoblock sleepers of length $l = 2.5$ m, width $b = 0.285$ m, height $h = 0.205$ m (under the rail) and mass 300 kg. Flexible rail pads with a thickness $t = 0.01$ m and a static stiffness of about 100 MN/m, for a load varying between 15 kN and 90 kN, are placed under the rail. The track is supported by a porphyry ballast layer (calibre 25/50, layer thickness $d = 0.3$ m), a limestone or porphyry layer (0/32, $d = 0.2$ m) and a limestone supporting layer (0/80 to 0/120, $d = 0.5$–0.7 m).

During the homologation tests, access to the track was limited to the time needed for the installation of the accelerometers on the rails and the sleepers. No forced-vibration test on the track could be performed to measure the dynamic impedance of the track, which turned out to be a major shortcoming in this study.

10.2.3. The soil

Cone penetration tests and triaxial tests on undisturbed samples taken from borehole experiments revealed the presence of a Quaternary loam layer (0–1.5 m) on a transition layer (Quaternary loam and/or Ypresian clay, 1.5–4.0 m) on a Tertiary Ypresian clay layer (4.0–12.0 m).

During the homologation tests, additional geophysical prospection tests were performed. Owing to time and budget limitations, non-intrusive geophysical prospecting methods were used, allowing an estimate of the variation of stiffness in the top layers and the determination of a material damping ratio for an 'equivalent' homogeneous half-space that exhibits similar attenuation at the surface to that at the test site.

A spectral analysis of surface waves (SASW) was performed to determine the dynamic soil characteristics of the site [10.7]. A transient excitation was generated by dropping a mass of 110 kg from a height of 0.9 m on a square (0.7 m × 0.7 m) steel foundation with a mass of 600 kg. A dashpot was used to control the frequency content of the loading and to prevent rebound of the mass. The vertical response was measured at nine points on the surface at distances from 2 m to 48 m from the source. The inversion procedure minimizes the difference between the experimental and theoretical dispersion curves of the first surface wave, as illustrated in Fig. 10.2.

Table 10.1. Geometrical and mass characteristics of the Thalys HST

	No. of carriages	No. of axles	L_t (m)	L_b (m)	L_a (m)	M_t (kg)	M_s (kg)	M_u (kg)
Locomotives	2	4	22.15	14.00	3.00	17 000	15 267	1 733
End carriages	2	3	21.84	18.70	3.00	14 500	12 674	1 830
Central carriages	6	2	18.70	18.70	3.00	17 000	15 170	1 826

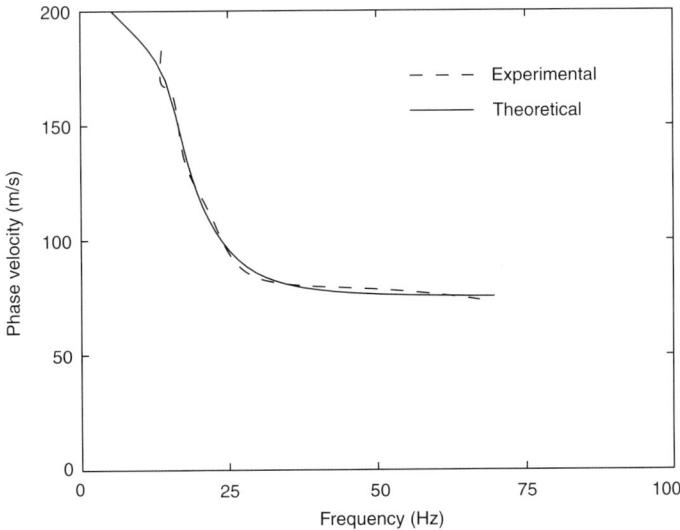

Fig. 10.2. Comparison of the experimental and theoretical dispersion curves from the SASW test

The theoretical curve corresponds to a horizontally stratified half-space with a top layer of thickness $d = 1.4$ m and a shear wave velocity $C_s = 80.0$ m/s, and another layer ($d = 1.9$ m, $C_s = 133.0$ m/s) on top of a half-space ($C_s = 226.0$ m/s), in good agreement with the layering revealed by the borehole experiments.

It must be noted that the track was constructed by excavation to a depth of a few metres, and the soil under the ballast was stabilized before construction. On the other hand, the SASW test was performed on top of the unexcavated soil away from the track. It is therefore realistic to assume that the soil under the track is stiffer than the soft superficial layer revealed by the SASW test. In the numerical calculations presented below, a shear wave velocity $C_s = 100.0$ m/s was therefore used for the top layer.

Apart from the variation of the stiffness with depth, a good estimate of the hysteretic material damping is needed. As the cone penetration tests and the SASW test revealed that the site was not homogeneous, the material damping ratio β^s is expected to vary with depth (it usually decreases with depth). In the following, however, the classical Barkan expression for a homogeneous half-space is used. This expression relates the vertical velocity $\hat{v}_z(r, \omega)$ at a distance r from the source to the vertical velocity $\hat{v}_z(r_1, \omega)$ at a reference distance r_1:

$$\hat{v}_z(r, \omega) = \hat{v}_z(r_1, \omega) \left(\frac{r}{r_1}\right)^{-n} \exp\left(-\frac{2\pi\beta^s}{\lambda_R}(r - r_1)\right) \tag{10.1}$$

where the exponent n represents the radiation damping in the soil and the coefficient β^s is the hysteretic material damping ratio; λ_R is the wavelength of the Rayleigh wave in the half-space. A simple inversion problem is obtained, from which a material damping ratio β^s can be derived for an equivalent homogeneous half-space that exhibits similar attenuation at the surface to that at the test site.

Figure 10.3 shows the dimensionless vertical velocity $\hat{v}_z(r, \omega)/\hat{v}_z(r_1, \omega)$ as a function of the dimensionless distance r/r_1, derived from the frequency content of the transient signals during the SASW test. The average Rayleigh wave velocity in the equivalent homogeneous half-space was taken as 80 m/s, the high-frequency limit of the first surface mode. A least-squares fit of these data reveals a radiation damping coefficient $n = 2.10$ and a material damping ratio $\beta^s = 0.03$. The estimated contribution of body waves is relatively high, as the radiation damping coefficient n is close to the theoretical value $n = 2$ for body waves and much higher than the theoretical value $n = 0.5$ that prevails when surface waves dominate the response. In practice, the material damping ratio is expected to decrease with depth and for deeper layers may be lower than the proposed value.

10.2.4. The experimental set-up

Vertical accelerations were measured at 14 locations (Fig. 10.4). On both tracks, a Dytran piezoelectric accelerometer was glued to the rail and the sleeper. In the free field, ten seismic piezoelectric PCB accelerometers were placed at distances 4, 6, 8, 12, 16, 24, 32, 40, 56 and 72 m from the centre of track 2. They were mounted on steel or aluminium stakes with a cruciform cross-section to minimize dynamic soil–structure interaction effects [10.6]. A Kemo VBF 35 system was used as a power supply, amplifier and anti-aliasing filter, with a low-pass frequency fixed at 500 Hz for the measurements on the track and 250 Hz in the free field. The signals were recorded with an analogue 14-channel TEAC tape recorder. The analogue to digital conversion was performed using a 16-bit Daqbook 216 data acquisition system at a sampling rate of 1000 Hz.

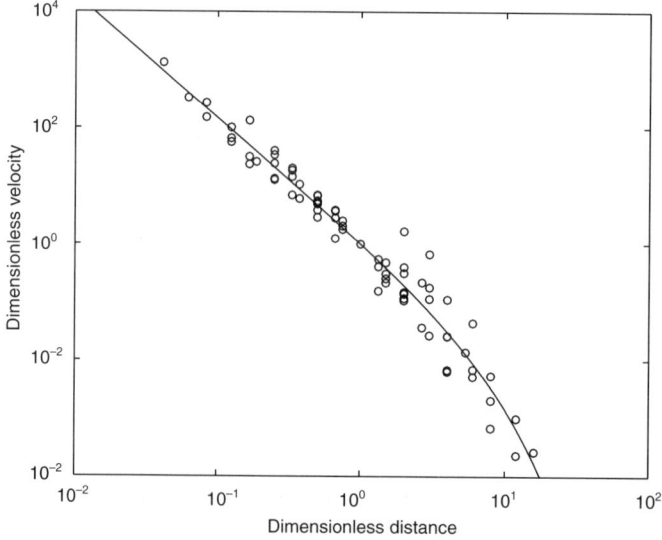

Fig. 10.3. Dimensionless representation of the amplitude decay in the free field

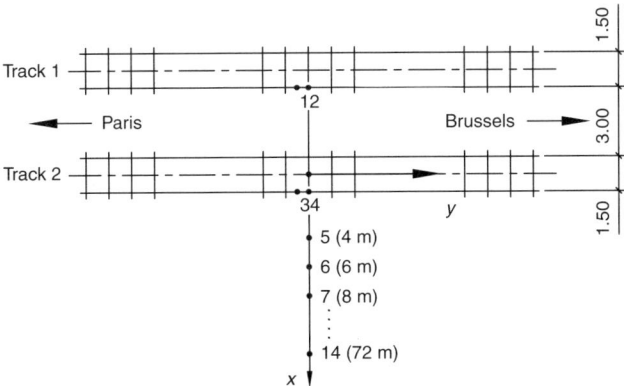

Fig. 10.4. Location of the measurement points (dimensions in m)

10.3. Experimental results

Nine train passages at speeds varying between 223 km/h and 314 km/h were recorded, as summarized in Table 10.2. As an example, the track and free-field response will be discussed in detail for the passage of the Thalys HST on track 2 with a speed $v = 314$ km/h. Results for other train speeds are summarized afterwards, so that conclusions can be drawn regarding the influence of the train speed on the peak particle velocity and frequency content of the response.

10.3.1. The passage of a Thalys HST at a speed v = 314 km/h

10.3.1.1. Track response

Figure 10.5 shows the time history and frequency content of the vertical acceleration of the rail and the sleeper during the passage of the Thalys HST at a speed $v = 314$ km/h.

Very high accelerations (Fig. 10.5(a)), up to 250 m/s², are observed on the rail. The passage of individual bogies and axles can clearly be noticed in the acceleration time history. The time separation between the passage of bogies 2 and 3, and of 11 and 12 is small, however. The frequency content of the response is broad (Fig. 10.5(b)).

The acceleration time history of the sleeper (Fig. 10.5(c)) show much lower amplitudes than for the rail. It clearly allows one to identify the passage of all individual axles. The acceleration has a quasi-discrete spectrum (Fig. 10.5(d)) with peaks at the fundamental bogie passage frequency $f_b = v/L_b = 4.66$ Hz and its higher-order harmonics, modulated at the axle passage frequency $f_a = v/L_a = 29.07$ Hz.

10.3.1.2. Free-field response

Figures 10.6 and 10.7 summarize the time history and frequency content of the vertical velocity of the rail, sleeper and soil at all distances from the centre of track 2. The velocities were obtained by integration of the accelerations, after removal of electrical noise; an Ormsby digital filter [10.15] with $N = 500$ weighting functions, a

Table 10.2. Overview of Thalys HST passages recorded

Event	Track	Date	Passage	Direction*	Speed (km/h)
t11	1	Nov. 3	4	P–B	223
t12	1	Nov. 4	3	B–P	265
t13	1	Nov. 3	6	P–B	272
t14	1	Nov. 4	4	P–B	302
t21	2	Nov. 3	5	B–P	256
t22	2	Nov. 4	2	P–B	271
t23	2	Nov. 4	1	B–P	289
t24	2	Nov. 4	6	P–B	300
t25	2	Nov. 4	5	B–P	314

*P, Paris; B, Brussels.

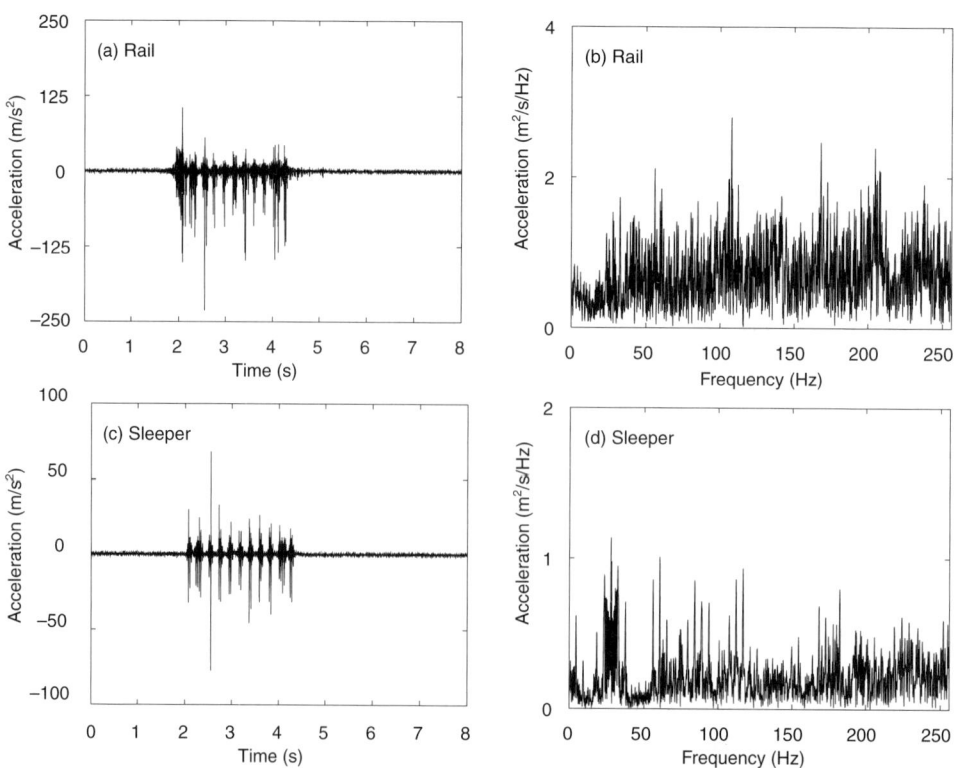

Fig. 10.5. Measured time history (left) and frequency content (right) of the vertical acceleration of the rail and the sleeper during the passage of a Thalys HST on track 2 with a speed v = 314 km/h

roll-off termination frequency $f_t = 2.5$ Hz and a cut-off frequency $f_c = 3.0$ Hz was used to eliminate the DC component and avoid drift. In Figs 10.6 and 10.7, the vertical scale is kept constant for those channels where the same type of accelerometer has been used, so that the effect of radiation and material damping in the soil on the amplitudes with increasing distance and excitation frequency can be better appreciated.

The time history of the vertical response $v_z(x^R = 6, t)$ at a receiver position x^R 6 m from the track (Fig. 10.6(d)), for example, still allows one to detect the passage of the bogies, whereas the passage of the individual axles can no longer be distinguished. The PPV is about 2.5 mm/s. Owing to the specific train composition, the observed velocity spectrum $\hat{v}_z(x^R = 6, \omega)$ (Fig. 10.7(d)) is quasi-discrete, with a maximum at the fundamental bogie passage frequency $f_b = v/L_b = 4.66$ Hz. The sleeper passage frequency $f_s = v/d = 145.37$ Hz is still noticeable in the spectrum.

Considering the response at a larger distance from the track, at 40 m for example, allows one to observe the effect of radiation and material damping in the soil. The time history $v_z(x^R = 40, t)$ (Fig. 10.6(j)) shows a PPV of about 0.2 mm/s. The velocity spectrum $\hat{v}_z(x^R = 40, \omega)$ (Fig. 10.7(j)) is dominated by the bogie passage frequency and its second harmonic. Higher frequencies are attenuated by radiation and material damping in the soil. The sleeper passage frequency, for example, can no longer be observed.

10.3.2. The influence of the train speed

An important question is how the train speed affects the time history and frequency content of the free-field response at various distances from the track.

As an example, Fig. 10.8 shows the time history and frequency content of the vertical velocity at 8 m from track 2 during the passage of a Thalys HST on track 2 at five different train speeds, varying from 256 km/h to 314 km/h. Although the duration of the signal is evidently affected by the train speed, the time histories do not show a strong dependence of the PPV on the train speed.[*] It will be demonstrated in a later section how the dynamic force transmitted by a single sleeper during the passage of an HST depends on the train speed. The frequency content of the signals shows a shift towards higher frequencies with increasing train speed; this is especially clear for the sleeper passage frequency, which varies from 118.52 Hz at a speed $v = 256$ km/h to 145.37 Hz at a speed of $v = 314$ km/h.

The previous observations, based on the response at a single receiver at 8 m from track 2, are confirmed in Fig. 10.9, where the PPV is shown as a function of the receiver distance to track 2 for Thalys HST passages on track 2 at different speeds. The decrease of PPV with distance due to radiation and material damping in the soil can clearly be observed. Figure 10.9 shows only a very moderate tendency towards increasing vibration levels with increasing train speed.

---

[*] Note that the Rayleigh wave velocity in the top layer of the site (around 90 m/s, i.e. 326 km/h; see Section 10.2.3) has not been exceeded even at a train speed of 314 km/h – Ed.

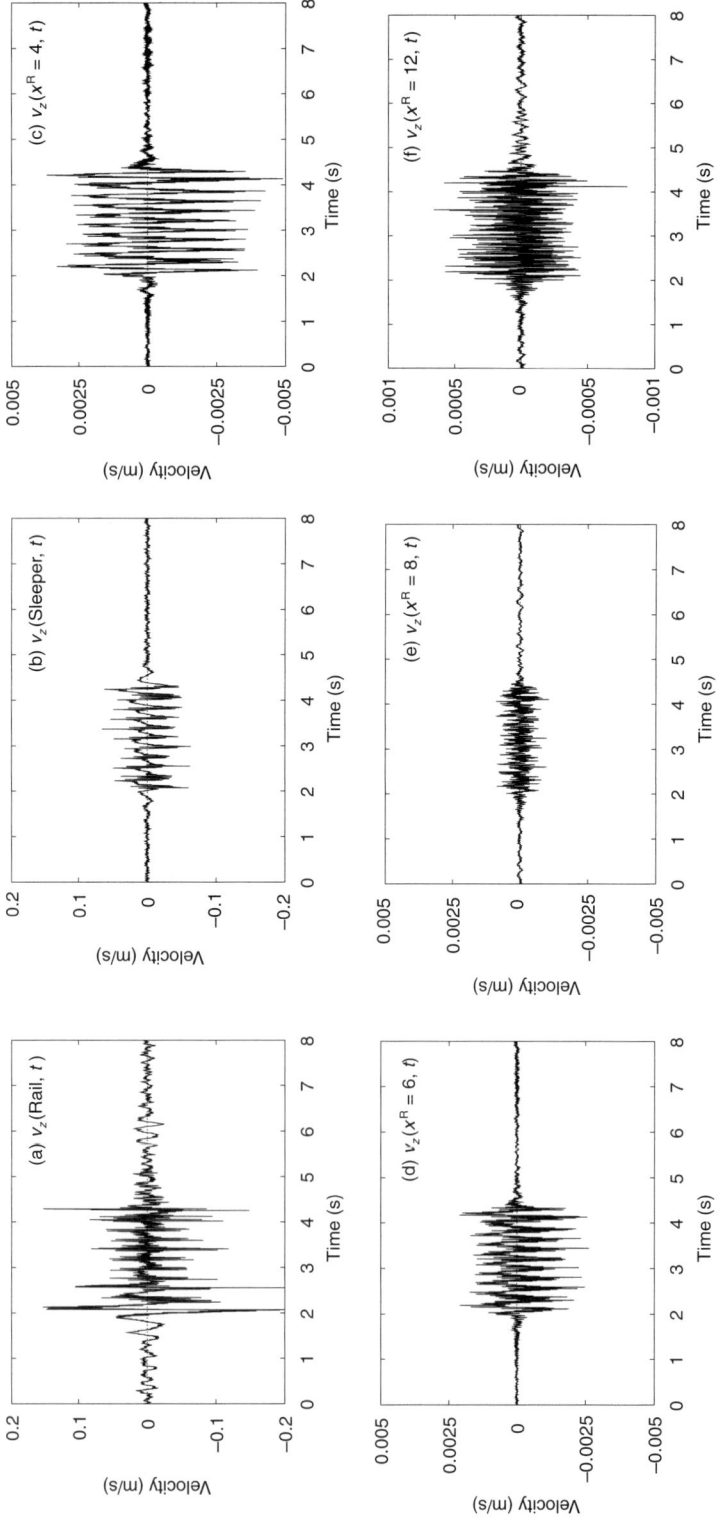

FREE-FIELD VIBRATIONS DURING PASSAGE OF A TRAIN 295

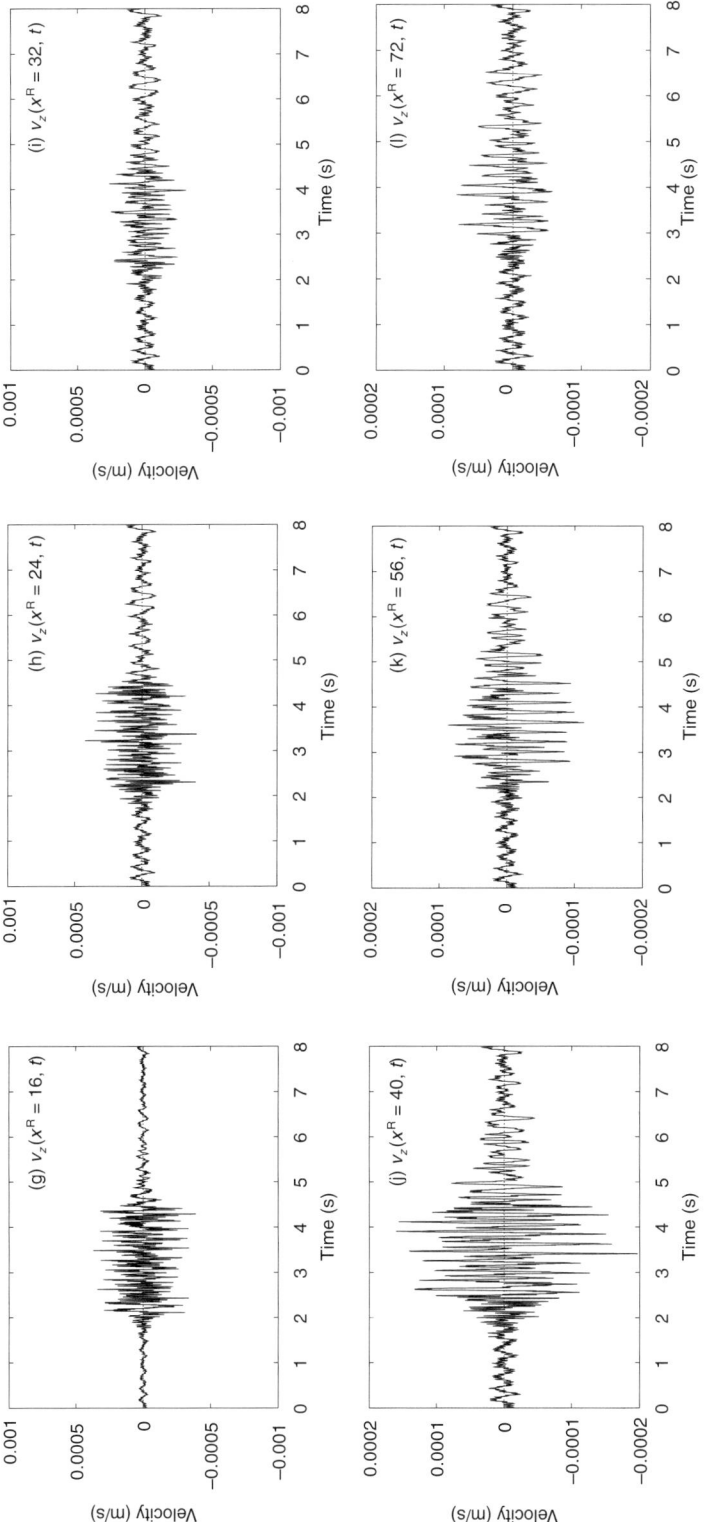

Fig. 10.6. Measured time history of the vertical velocity on the track and in the free field for the passage of a Thalys HST on track 2 with v = 314 km/h

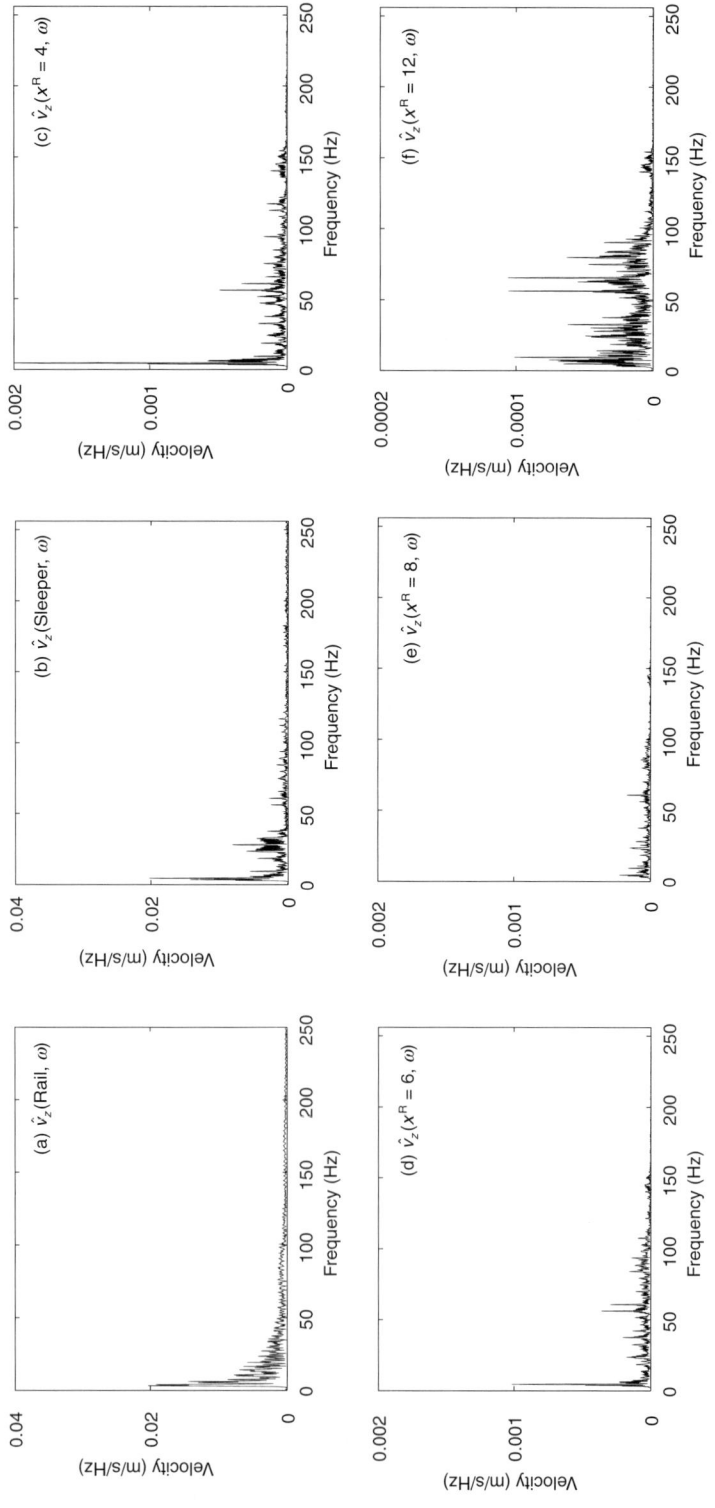

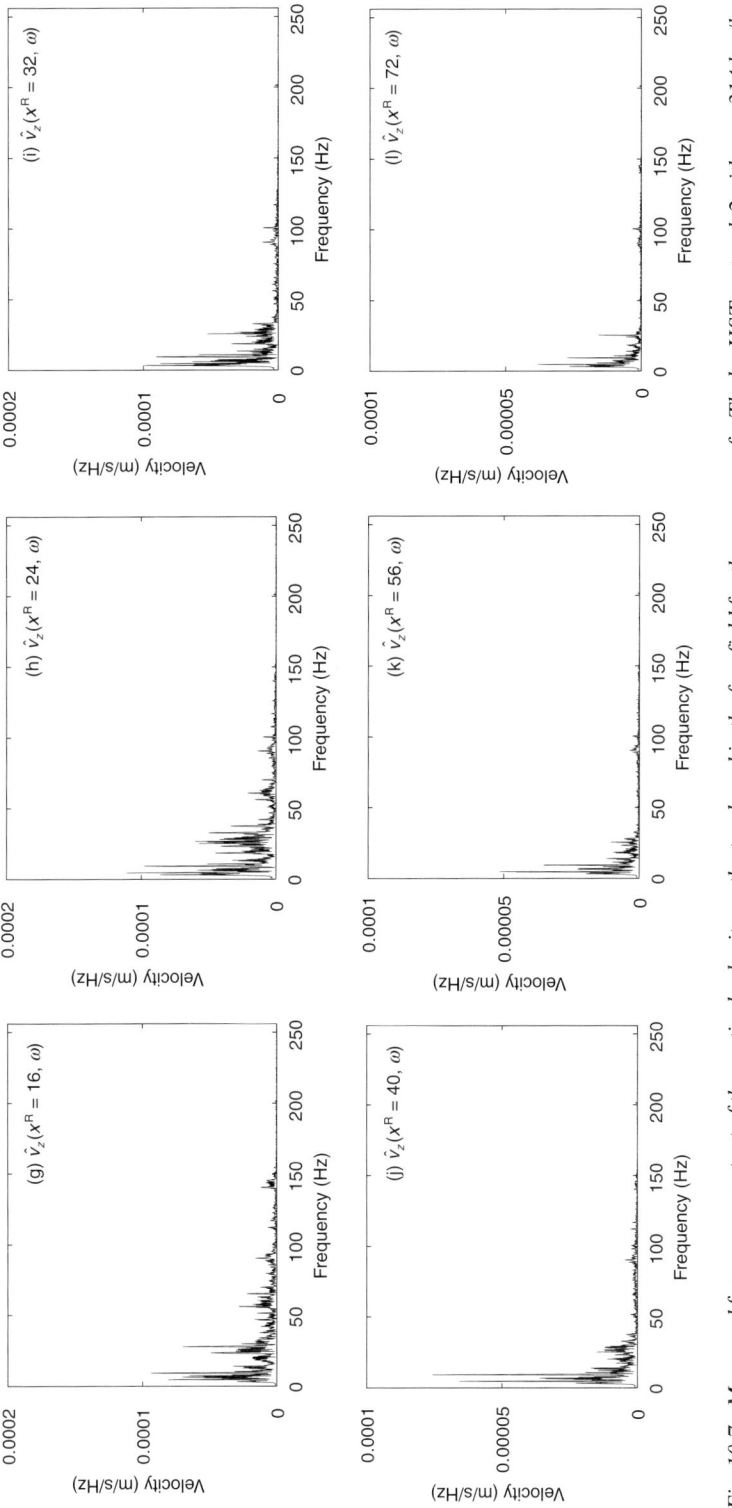

Fig. 10.7. Measured frequency content of the vertical velocity on the track and in the free field for the passage of a Thalys HST on track 2 with $v = 314$ km/h

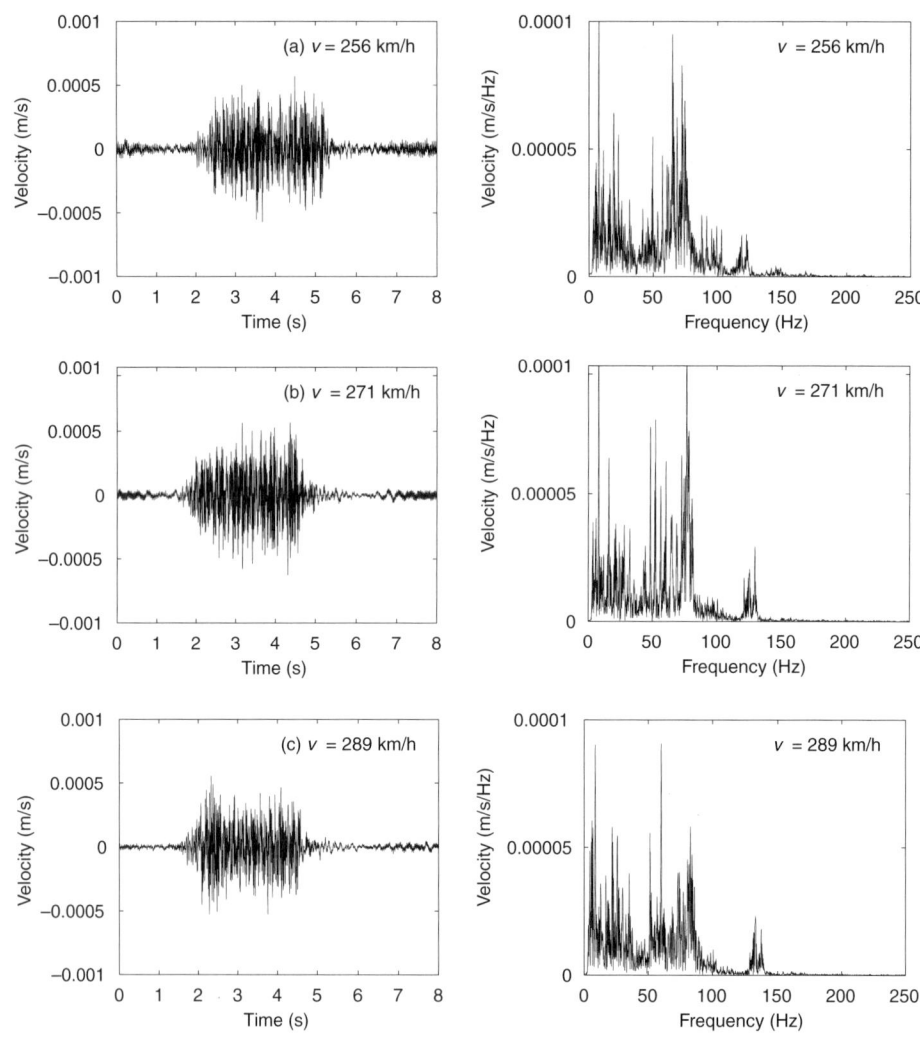

Fig. 10.8. Measured time history (left-hand plots) and frequency content (right-hand plots) of the vertical velocity at 8 m from track 2 during the passage of a Thalys HST on track 2 at varying speed

10.4. Krylov's analytical prediction model

A brief review is given here of an analytical prediction model proposed by Krylov [10.13] (see also Chapter 9), that accounts for the effect of quasi-static loading due to moving axle loads. The model disregards other important sources of vibration, such as parametric excitation and track or wheel irregularities. The model will subsequently be validated by means of the aforementioned experimental results. It is assumed that the track is directed along the y direction, with the vertical z axis pointing downwards and the horizontal x axis perpendicular to the (y, z) plane (Fig. 10.10).

FREE-FIELD VIBRATIONS DURING PASSAGE OF A TRAIN

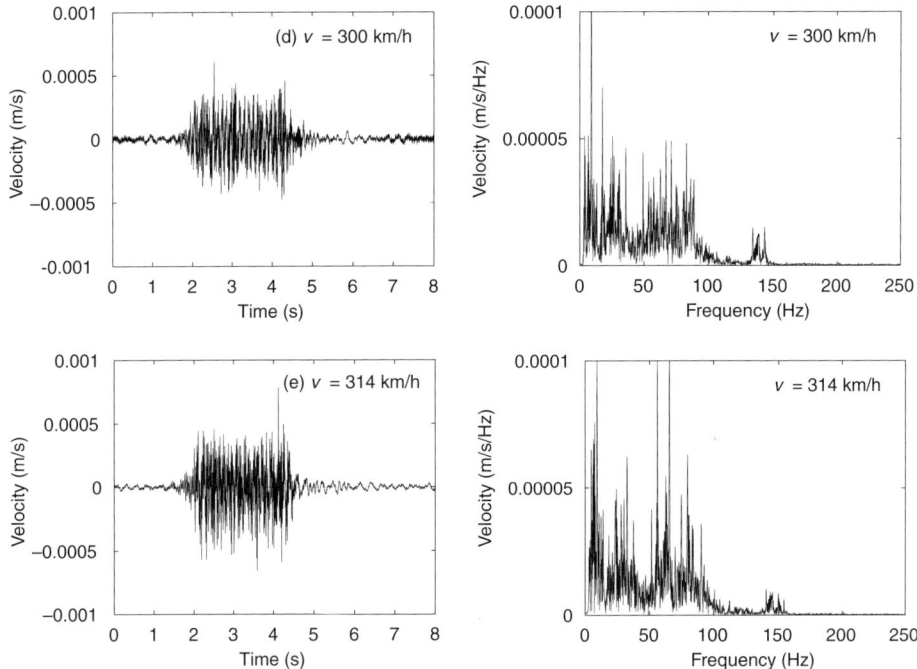

Fig. 10.8. Contd

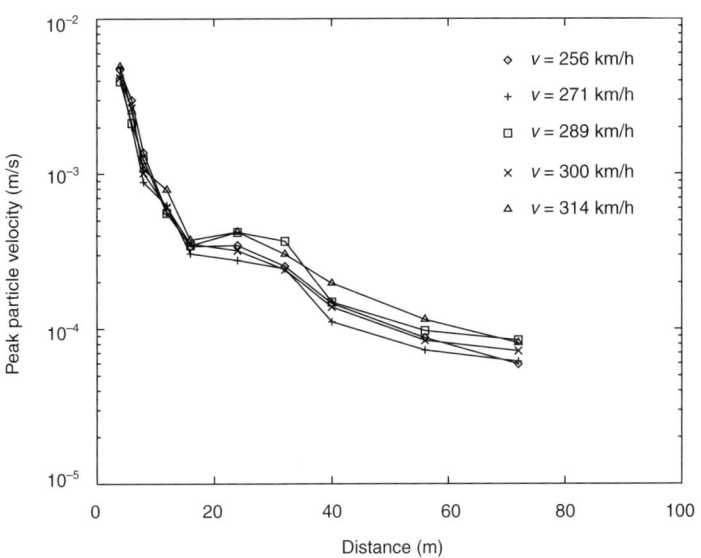

Fig. 10.9. Measured PPV as a function of the distance to track 2 for Thalys HST passages on track 2 at different speeds

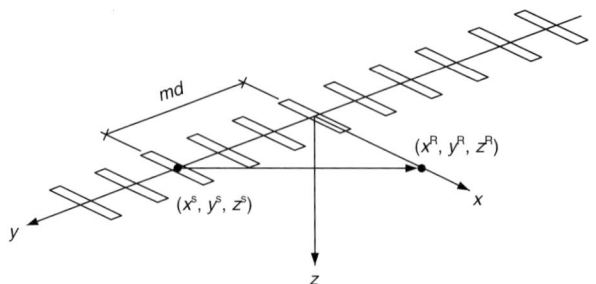

Fig. 10.10. *Definition of Cartesian coordinate system on the track*

10.4.1. The force transmitted by a sleeper due to a single axle load

The track is modelled as a beam with bending stiffness EI and mass m per unit length on an elastic foundation with subgrade stiffness k_s:

$$EI\frac{\partial^4 w}{\partial y^4} + m\frac{\partial^2 w}{\partial t^2} + k_s w = T\delta(y-vt) \qquad (10.2)$$

where $w(y)$ is the vertical deflection of the track, T is the load from a single axle and v is the speed of propagation of the load. The solution of this equation is [10.13]:

$$w(y-vt) = \frac{T}{8EI\beta^3\delta}\exp(-\beta\delta|y-vt|)\left(\cos\beta\eta(y-vt) + \frac{\delta}{\eta}\sin\beta\eta|y-vt|\right) \qquad (10.3)$$

Here, $\beta = (k_s/4EI)^{0.25}$, $\delta = [1-(v/c_{min})^2]^{0.5}$ and $\eta = [1+(v/c_{min})^2]^{0.5}$; $c_{min} = (4k_s EI/m^2)^{0.25}$ is the velocity of free track waves.

The force $P(t)$ transmitted by a sleeper located at $y = 0$, when a single axle load T initially located at $y = 0$ moves with speed v along the track, is estimated as

$$P(t) = TF(t) = \frac{T}{N_{eff}^{st}}\frac{w(vt)}{w_{max}^{st}} \qquad (10.4)$$

Here $F(t)$ is the force transmitted by a sleeper for a single moving unit load, and N_{eff}^{st} is the effective number of sleepers that would be needed to support the quasi-static axle load T if all sleepers were to take the maximum load corresponding to the maximum quasi-static deflection w_{max}^{st} of the track. N_{eff}^{st} is equal to $2y_0^{st}/\pi d$, where $y_0^{st} = \pi/\beta$ is the effective quasi-static track deflection distance and d is the sleeper spacing. Equation (10.4) expresses the fact that every sleeper transmits a fraction of the axle load T that is proportional to the instantaneous deflection $w(vt)$ of the track at time t (Fig. 10.11). The Fourier transform $\hat{F}(\omega)$ of the force $F(t)$ is equal to [10.13]

$$\hat{F}(\omega) = \frac{1}{N_{eff}^{st}}\frac{1}{\beta v}\left(\frac{\delta+\eta+\omega/\beta v}{\delta^2+(\eta+\omega/\beta v)^2} + \frac{\delta+\eta-\omega/\beta v}{\delta^2+(\eta-\omega/\beta v)^2}\right) \qquad (10.5)$$

and depends on the characteristics of the track, the subgrade stiffness and the speed of the train. When v is much lower than c_{min}, the beam inertial forces can be

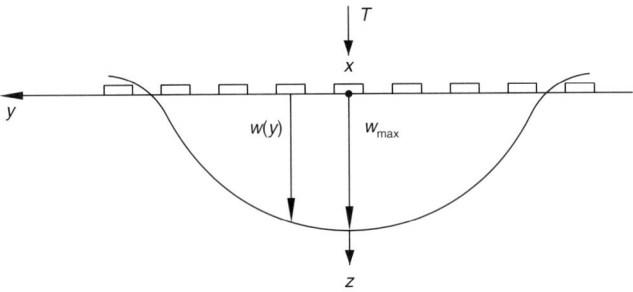

Fig. 10.11. Quasi-static deformation of the track

neglected and δ and η tend to 1, resulting in the original expression of Krylov and Ferguson [10.14] for low train speeds.

10.4.2. The forces transmitted by all sleepers due to a train passage

The force per unit length along the track ($x^S = 0$ and $z^S = 0$) transmitted by a single sleeper located at $y = y^S$ when a train with multiple axles moves with speed v along the track is equal to

$$p(y,t) = \delta(y - y^S) \sum_{n=1}^{N} \sum_{k=1}^{K_n} T_{nk} F\left(t - \frac{y + L_{nk}}{v}\right) \tag{10.6}$$

where N is the number of carriages, K_n is the number of axles on carriage n, T_{nk} is the load on axle k of carriage n and L_{nk} is the distance from axle k on carriage n to the front end of the train. In the frequency domain, this force distribution becomes

$$\hat{p}(y, \omega) = \delta(y - y^S) \exp\left(-i\omega \frac{y}{v}\right) \hat{F}(\omega) \hat{C}(\omega) \tag{10.7}$$

The function $\hat{C}(\omega)$ represents the composition of the train in the frequency domain:

$$\hat{C}(\omega) = \sum_{n=1}^{N} \sum_{k=1}^{K_n} T_{nk} \exp\left(-i\omega \frac{L_{nk}}{v}\right) \tag{10.8}$$

If all carriages have the same length L_t, axle spacing L_a, distance L_b between bogies and distance L_1 between the centre of the first bogie and the front end of the carriage, and all axles on carriage n carry the same axle load T_n, equation (10.8) simplifies to

$$\hat{C}(\omega) = 2\cos\left(\omega \frac{L_a}{2v}\right)\left[1 + \exp\left(-i\omega \frac{L_b}{v}\right)\right] \exp\left(-i\omega \frac{L_1}{v}\right) \times$$
$$\sum_{n=1}^{N} T_n \exp\left(-i\omega \frac{(n-1)L_t}{v}\right) \tag{10.9}$$

The distribution of forces transmitted by all sleepers when a train with multiple axles moves with speed v along the track can now be written as

$$p(y,t) = \sum_{m=-\infty}^{\infty} \delta(y-md) \sum_{n=1}^{N} \sum_{k=1}^{K_n} T_{nk} F\left(t - \frac{y + L_{nk}}{v}\right) \qquad (10.10)$$

In the frequency domain, this force distribution becomes

$$\hat{p}(y,\omega) = \hat{F}(\omega)\hat{C}(\omega) \sum_{m=-\infty}^{\infty} \delta(y-md) \exp\left(-i\omega \frac{y}{v}\right) \qquad (10.11)$$

10.4.3. Response of the soil

In the frequency domain, the vertical displacement $\hat{u}_z(x^R, \omega)$ at a receiver with coordinates x^R is equal to

$$\hat{u}_z(x^R, \omega) = \int_{-\infty}^{\infty} \hat{p}(y,\omega) \hat{u}_z^G(\rho, \omega) dy \qquad (10.12)$$

where $\hat{u}_z^G(\rho, \omega)$ is the vertical Green's function of the layered soil domain, which can be calculated with a direct stiffness formulation [10.4, 10.8], and ρ is the distance between a point x^S on the source line and the receiver point x^R. Using equation (10.11) for the force distribution $\hat{p}(y, \omega)$ along the track, the response becomes

$$\hat{u}_z(x^R, \omega) = \hat{F}(\omega)\hat{C}(\omega) \int_{-\infty}^{\infty} \sum_{m=-\infty}^{\infty} \delta(y-md) \exp\left(-i\omega \frac{y}{v}\right) \hat{u}_z^G(\rho, \omega) dy \qquad (10.13)$$

After the discrete nature of the support of the sleepers is taken into account, the integral along the source line disappears. Furthermore, the resulting infinite summation over the sleepers can be truncated so that only the contribution of the N_s sleepers behind and ahead of the receiver is accounted for:

$$\hat{u}_z(x^R, \omega) = \hat{F}(\omega)\hat{C}(\omega) \left[\hat{u}_z^G(\rho_0, \omega) + \sum_{m=1}^{N_s} 2\cos\left(\omega \frac{md}{v}\right) \hat{u}_z^G(\rho_m, \omega) \right] \qquad (10.14)$$

Equation (10.14) illustrates that the response at the receiver x^R is governed by three factors, as follows.

(1) The frequency content $\hat{F}(\omega)$ of the force transmitted by a single sleeper when a unit axle load moves with speed v along the track.
(2) The function $\hat{C}(\omega)$, determined by the composition and the speed of the train. The product $\hat{F}(\omega)\hat{C}(\omega)$ is the frequency content of the force transmitted by a single sleeper when a train moves with speed v along the track.
(3) The third factor represents the frequency content of the response at the receiver when each sleeper is consecutively loaded by a Dirac delta function impulse, shifted in time by d/v. This factor depends on the sleeper spacing, the speed of the train and the Green's functions of the soil. If y^R and z^R are zero, ρ_m is equal to $\sqrt{[(x^R)^2 + (md)^2]}$.

10.5. Analytical predictions

10.5.1. Track response

Calculations were performed for a track with a bending stiffness $EI = 12.76 \times 10^6$ N m² (both rails) and a mass per unit length $m = 620.0$ kg/m (including both of the rails and the sleepers).

It has been mentioned before that no experimental data were available for the subgrade stiffness k_s. According to a review by Knothe and Grassie [10.9], the values for the ballast stiffness may vary within a wide range. Figure 10.12(a) illustrates the influence of the subgrade stiffness k_s (for values ranging between 100 MN/m² and 500 MN/m²) on the modulus of the function $\hat{F}(\omega)$ for a train speed $v = 315$ km/h. The quasi-static value $\hat{F}(\omega = 0)$ depends only weakly on k_s. This can be understood by considering equation (10.6), where β is proportional to $k_s^{0.25}$. As $N_{\text{eff}}^{\text{st}}$ is inversely proportional to β, the product $N_{\text{eff}}^{\text{st}} \beta$ in the denominator of equation (10.6) is not affected by the subgrade stiffness k_s, and the effect of k_s on $\hat{F}(\omega = 0)$ is small. For increasing k_s, the frequency content of $\hat{F}(\omega)$ increases, resulting in larger forces transmitted by an individual sleeper. Owing to the lack of experimental data,

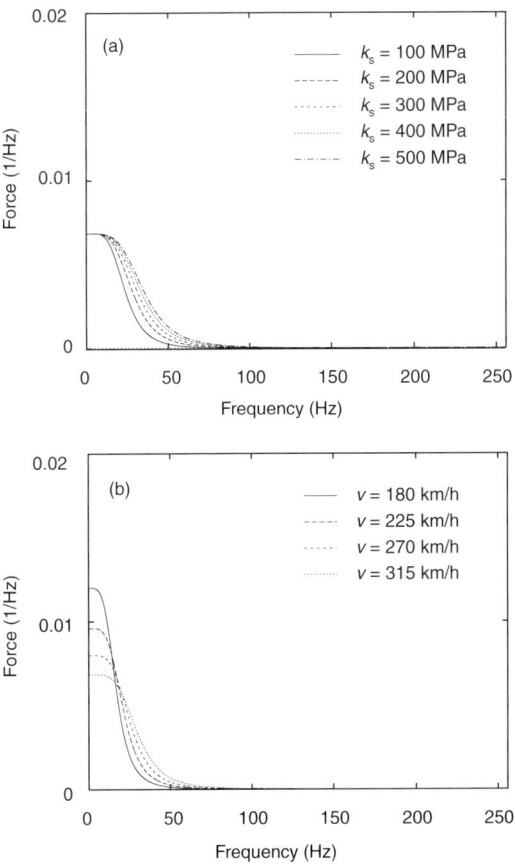

Fig. 10.12. Modulus of $\hat{F}(\omega)$ for (a) varying subgrade stiffness k_s and (b) varying train speed v

subsequent calculations were performed for a track with a subgrade stiffness $k_s = 250$ MPa.

Figure 10.12(b) shows the modulus of $\hat{F}(\omega)$ for a track with subgrade stiffness $k_s = 250$ MPa and a train speed v varying between 180 km/h and 315 km/h. For increasing v, the quasi-static value $\hat{F}(\omega = 0)$ decreases, while the frequency content of $\hat{F}(\omega)$ increases; the former follows immediately from the appearance of the train speed v in the denominator of equation (10.6).

Figure 10.13(b) shows the modulus of $\hat{C}(\omega)$ for a Thalys HST moving at a speed $v = 315$ km/h. The quasi-static value $\hat{C}(\omega = 0)$ is equal to the sum of all axle loads. If all carriages were identical, equation (10.9) would apply; the first term in this equation is responsible for discrete maxima at $f_a = v/L_a = 29.07$ Hz and higher harmonics, while the second term is responsible for discrete maxima at $f_b = v/L_b = 4.66$ Hz and higher harmonics. The locomotives and the end carriages of the Thalys HST have a different axle composition from the six central carriages, however, and the more general equation (10.8) applies. A quasi-discrete spectrum results, where the aforementioned harmonics remain dominant (Fig. 10.13(b)). Note that the measured vertical acceleration of a sleeper during the passage of a train (Fig. 10.5(d)) has a similar quasi-discrete spectrum.

The spectrum in Fig. 10.13(c) is obtained by multiplication of the functions $\hat{F}(\omega)$ (Fig. 10.13(a)) and $\hat{C}(\omega)$ (Fig. 10.13(b)) and thus represents the frequency content of

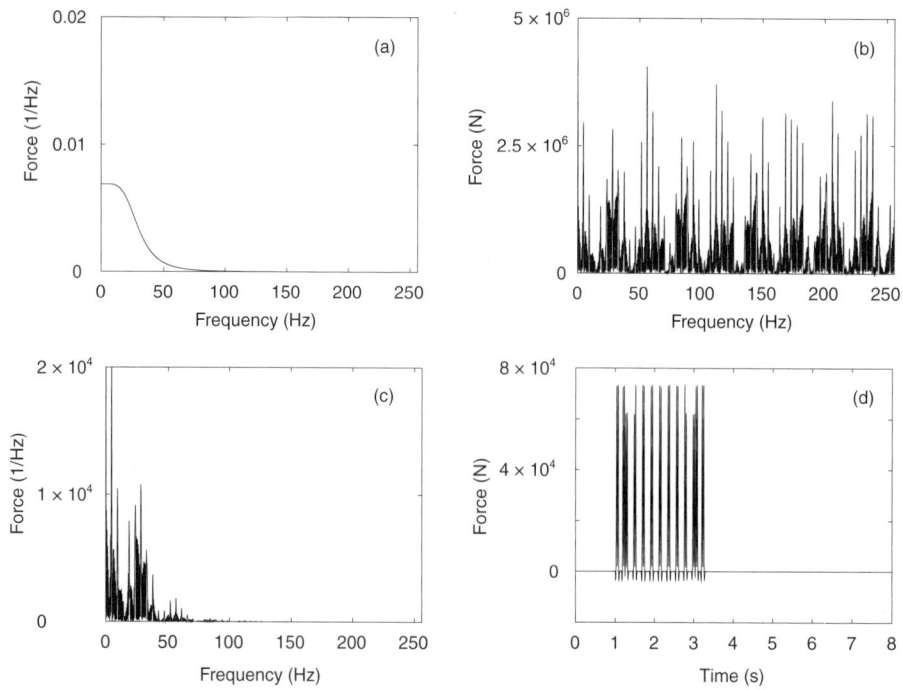

Fig. 10.13. Modulus of the functions (a) $\hat{F}(\omega)$, (b) $\hat{C}(\omega)$ and (c) frequency content $\hat{F}(\omega)\hat{C}(\omega)$, and (d) history of the force transmitted by a single sleeper during the passage of a Thalys HST at a speed $v = 315$ km/h

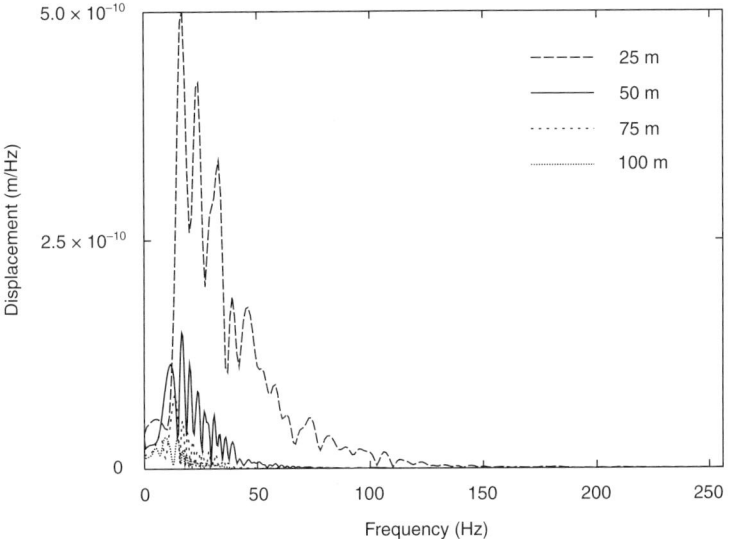

Fig. 10.14. Modulus of the Green's function $\hat{u}_z^G(\rho_m, \omega)$ of the layered half-space for source–receiver distances of 25, 50, 75 and 100 m

the force transmitted by a single sleeper during the passage of a Thalys HST at a speed $v = 315$ km/h. The decay of the function $\hat{F}(\omega)$ with frequency governs the frequency content of the transmitted load.

Figure 10.13(d), finally, shows the time history of the force transmitted by a single sleeper during the passage of a Thalys HST at a speed $v = 315$ km/h, as obtained with an inverse Fourier transformation.

10.5.2. Green's functions

Results are presented for a layered half-space, as revealed by the SASW test, with a top layer ($d = 1.4$ m and $C_s = 100$ m/s) and a second layer ($d = 1.9$ m and $C_s = 133$ m/s) on top of a half-space ($C_s = 226$ m/s). The Poisson's ratio was taken as $\nu^s = 0.3$ and the density as $\rho^s = 1850.0$ kg/m³ for all layers. The hysteretic material damping ratio was estimated to be uniform with depth and equal to $\beta^s = 0.03$.

The Green's functions of the layered half-space were calculated using a direct stiffness formulation for wave propagation in multilayered poroelastic media [10.4]. Figure 10.14 shows the modulus of the vertical Green's function $\hat{u}_z^G(\rho_m, \omega)$ for source–receiver distances of 25, 50, 75 and 100 m. The effect of radiation and material damping on the frequency content of the response, as well as of the layered structure of the subsoil, is apparent. The response is low for frequencies lower than 15 Hz, which is due to the assumed stratification of the soil.

Figure 10.15 illustrates the third factor of equation (10.14) for a receiver at 16 m from the track. The displacement $\hat{u}_z(x^R, \omega)$ was obtained by a summation of $N_s = 999$ Green's functions $\hat{u}_z^G(\rho_m, \omega)$, accounting for 999 sleepers behind and ahead of the receiver line $y^R = 0$, multiplied by a cosine function $\cos(\omega m d/v)$,

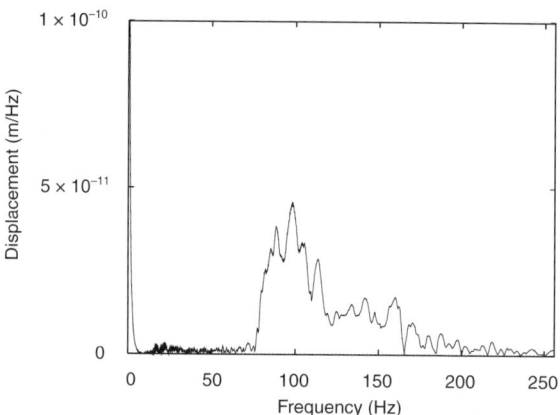

Fig. 10.15. Modulus of the vertical displacement at 16 m from the track when each sleeper is consecutively loaded by a Dirac impulse

sampled at discrete locations md. Figure 10.15 shows that, after an important quasi-static contribution in the low-frequency range, the evaluation of the term $\cos(\omega md/v)$ for discrete values of m causes small vibration amplitudes between 20 and 80 Hz. Around the sleeper passage frequency $f_d = v/d = 145.37$ Hz, larger values prevail (all cosine terms are equal to 1.0 at the sleeper passage frequency). The fact that considerable response is found for frequencies below and above the sleeper passage frequency is an illustration of the Doppler effect.

10.5.3. Free-field response

The vertical displacement $\hat{u}_z(x^R, \omega)$ at a receiver due to the passage of a Thalys HST was obtained by evaluation of equation (10.14). The three factors in this equation have been illustrated already in Figs 10.12, 10.13(b) and 10.15. The vertical soil displacements $u_z(x^R, t)$ in the time domain were finally obtained by evaluating the inverse Fourier transform with an FFT algorithm.

Figures 10.16 and 10.17 show the computed time history and frequency content of the vertical velocity at the ten receivers during the passage of a Thalys HST at 314 km/h. These results should be compared with the experimental data, presented in Figs 10.6 and 10.7, respectively. In the following, we shall discuss the response at a few of the receiver positions.

At 4 m from the track, the passage of the bogies can be observed, while the passage of the individual axles is no longer observable; this is true for both the observed (Fig. 10.6(c)) and the predicted (Fig. 10.16(a)) velocity time histories. While the predicted PPV has the same order of magnitude as the measured PPV, the time history already clearly shows that the frequency content of the predicted response (Fig. 10.17(a)) differs from the measured spectrum (Fig. 10.7(c)). Owing to the specific train composition, the observed velocity spectrum (Fig. 10.7(c)) is quasi-discrete, with a maximum at the fundamental bogie passage frequency $f_b =$

$v/L_b = 4.66$ Hz. A similar behaviour can be observed at low frequencies in the predicted spectrum (Fig. 10.17(a)), although the contribution at the fundamental bogie passage frequency is underestimated. This is probably due to the rather sharp cut-off or high-pass frequency introduced by the assumed soil stratification. The predicted spectrum has low amplitudes between 30 and 75 Hz, while the frequency components around the sleeper passage frequency $f_d = 145.37$ Hz are more pronounced. This is due to the fact that Krylov's analytical prediction model includes only the effect of quasi-static loading and sleeper passage, while other excitation mechanisms, such as rail or wheel irregularities, are not accounted for. The model also overestimates the sleeper passage effect at high frequencies; this is probably due to the fact that the sleeper forces are assumed to be transmitted as point forces, an assumption that is open to challenge when the frequency increases and the wavelengths in the soil have the same order of magnitude as the sleeper dimensions.

Figure 10.16(e) shows the vertical velocity at 16 m from the track. The PPV is about 0.25 mm/s and corresponds well with the measured PPV (Fig. 10.6(g)). The predicted time history reveals, however, that its frequency content is too high. The measured velocity spectrum (Fig. 10.7(g)) is dominated by the bogie passage frequency and its second harmonic. Higher frequencies are attenuated by radiation and material damping in the soil. The predicted velocity spectrum (Fig. 10.17(d)) is more pronounced around 25 Hz, while the low-frequency components are much lower. This is due to the assumed stratification of the soil, which introduces filtering of frequencies below 15 Hz. Furthermore, the predicted frequency content is much higher than that observed. Apart from the overestimation of the response at frequencies related to sleeper passage, this may also indicate that the material damping ratio $\beta^s = 0.03$ underestimates the damping in the top layers. Comparing the measured and predicted responses at larger distances (e.g. at 72 m) reveals that the reverse is true for the deeper layers, i.e. damping seems to be overestimated.

Apart from some shortcomings of the analytical prediction model, the previous discussion has also illustrated the need for a more elaborate *in situ* geophysical prospecting survey, to determine the layering and the variation of dynamic soil characteristics (shear and longitudinal wave velocities, damping ratios) down to sufficient depth (20 m, for example). It is therefore judged not to be very useful at the moment to start adjusting parameters such as subgrade stiffness and dynamic soil characteristics in order to match the numerical predictions to the measurements.

Figure 10.18, finally, summarizes the PPV at all points in the free-field for the five train passages on track 2. The decrease of PPV with distance due to radiation and material damping in the soil is apparent. The experimental results of Fig. 10.9 demonstrate a rather weak dependence of PPV on train speed, whereas this dependence is more pronounced in the numerical results (Fig. 10.18); although the PPV is predicted with rather good accuracy, it has been demonstrated above that the same is not true regarding the frequency content of the response.

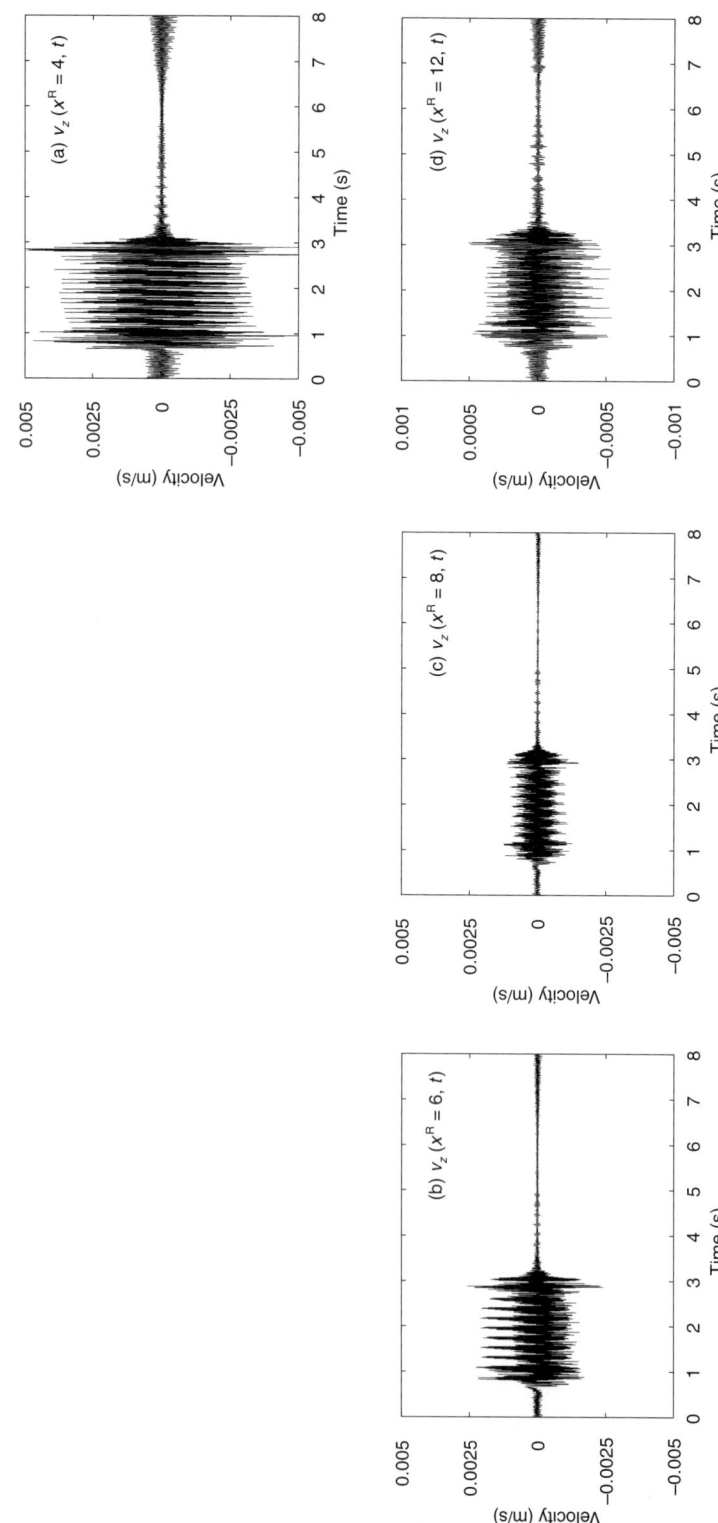

FREE-FIELD VIBRATIONS DURING PASSAGE OF A TRAIN

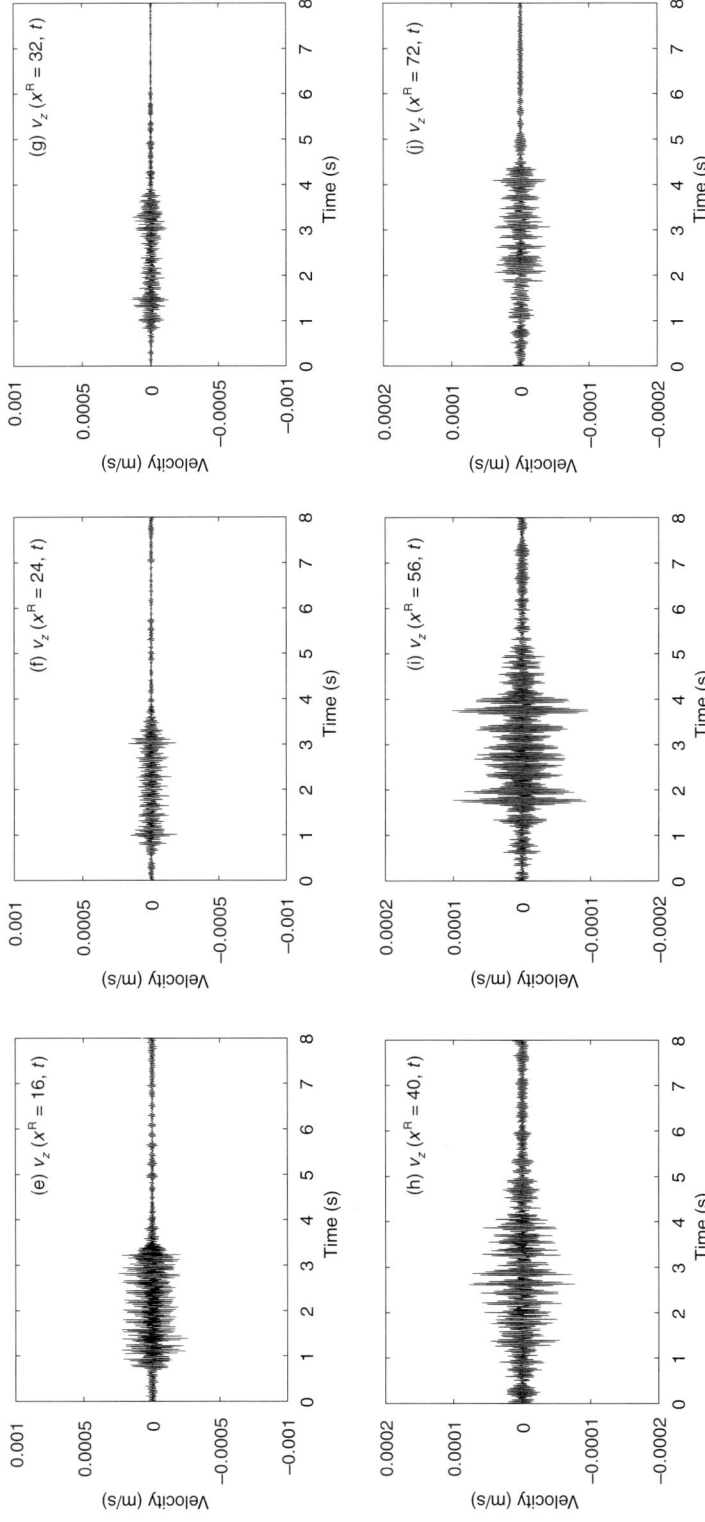

Fig. 10.16. Computed time history of the vertical velocity at varying distances from the track during the passage of a Thalys HST at $v = 314\ km/h$

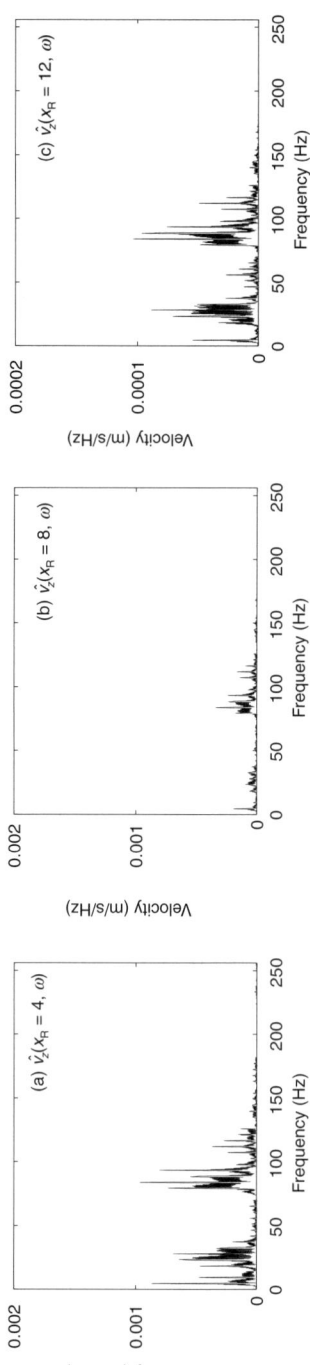

FREE-FIELD VIBRATIONS DURING PASSAGE OF A TRAIN

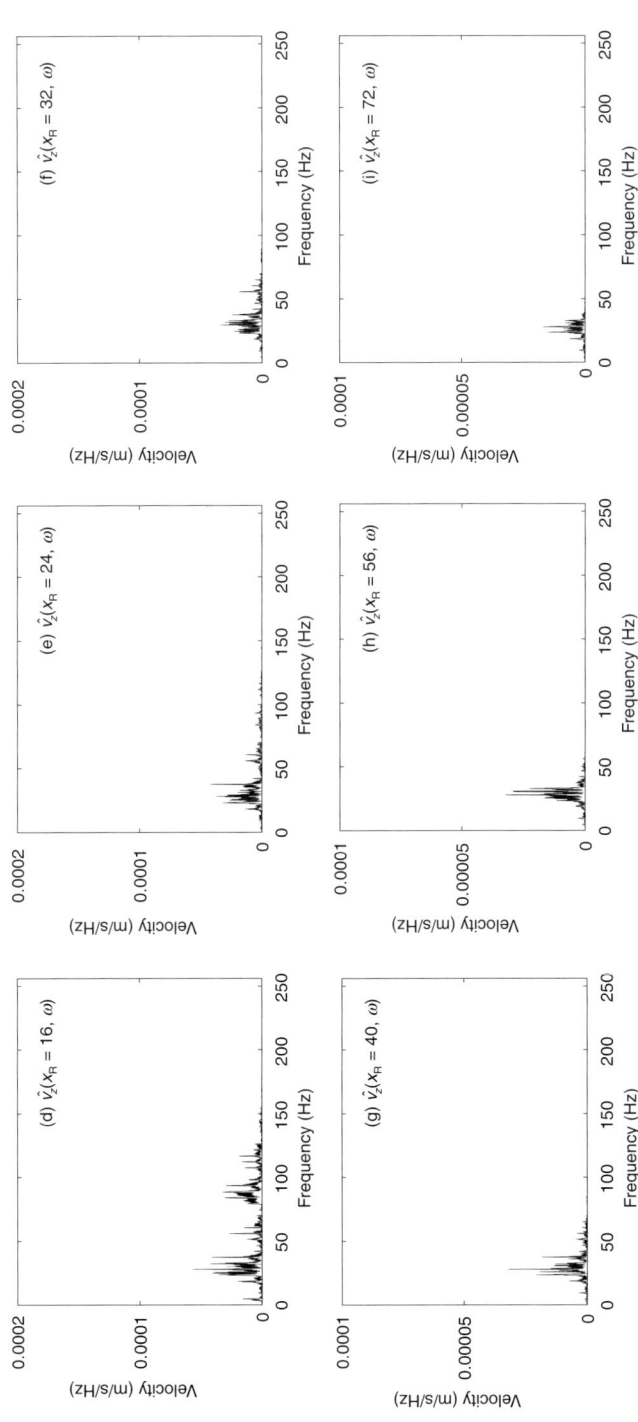

Fig. 10.17. Computed frequency content of the vertical velocity at varying distances from the track during the passage of a Thalys HST at $v = 314$ km/h

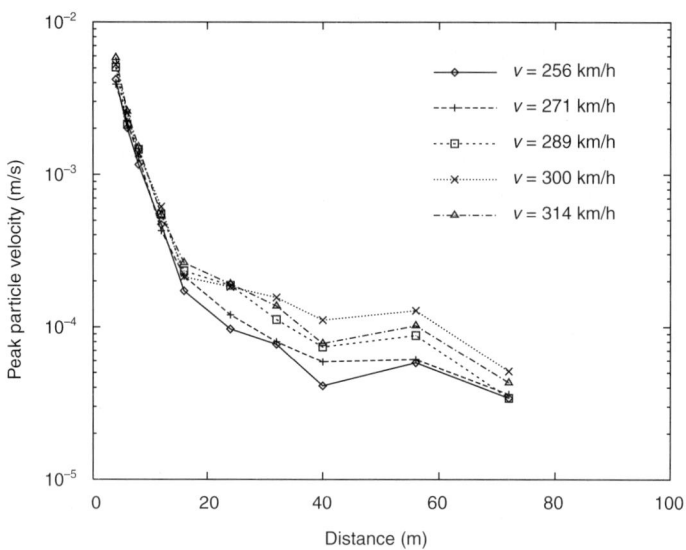

Fig. 10.18. Computed PPV as a function of distance and train speed

10.6. Conclusion

The results of free-field vibration measurements during the passage of a Thalys HST at varying speed have been compared with numerical results obtained with Krylov's analytical model, which has been extended to incorporate the Green's functions of layered media.

The experimental data presented in this chapter are complementary to the few other data sets published in the literature. In particular, the fact that measurements have been made at nine different train speeds between 223 km/h and 314 km/h makes this data set unique. Although the frequency content of the free-field vibrations shifts towards higher frequencies with increasing train speed, only a very weak dependence of the PPV on the train speed has been revealed.

A major shortcoming of the present data set is that, owing to time and budget limitations, no *in situ* experiments were performed to determine the subgrade stiffness of the track. Furthermore, only limited data were available on the stratification of the soil and the variation of dynamic soil characteristics with depth. This is especially the case for the material damping. These shortcomings compromise the quantitative validation of numerical prediction models, as the quality of the predictions is strongly related to the quality of the data.

Instead of trying to match the experimental results by modifying the input parameters in a 'trial and error' procedure, a qualitative assessment of the predictions based on Krylov's analytical prediction model has been made. The model has proved to have good predictive capabilities at low (quasi-static) and high (sleeper passage frequency and its higher harmonics) frequencies, but seems to underestimate the response in the mid-frequency band. Apart from the incomplete input data, this is due to the fact that the model incorporates only quasi-static loading, while dynamic loading due to rail and wheel irregularities, for example, is disregarded. Our present research is concentrated on the development of a prediction model that accounts for different excitation mechanisms and through-soil coupling of the sleepers.

If similar vibration measurements are planned in the near future for the validation of numerical models, railway companies should be persuaded to invest in more elaborate geophysical prospecting of the site, using cross-hole experiments for example, and an *in situ* determination of the track impedance.

10.7. Acknowledgements

The *in situ* experiments were performed in cooperation with L. Schillemans of Technum and with the assistance of K. Peeraer. The collaboration of P. Godart and W. Bontinck of the NMBS is kindly acknowledged. W. Dewulf inverted the SASW data. Professor V. V. Krylov of The Nottingham Trent University is gratefully acknowledged for feedback on the theory and the numerical results.

10.8. References

10.1. AUERSCH, L. *Zur Entstehung und Ausbreitung von Schienenverkehrserschutterungen – theoretische Untersuchungen und Messungen an Hochgeschwindigkeitszug Intercity Experimental.* Bundesanstalt für Materialforschung und -prüfung, Berlin, 1989, Forschungsbericht 155.

10.2. BRANDERHORST, J. Modellen voor het boeggolfprobleem bij hogesnelheidstreinen. Ontwerp en validatie met behulp van de resultaten van de proef Amsterdam-Utrecht. Master's thesis, University of Twente, The Netherlands, 1997.

10.3. DEGRANDE, G. Free field vibrations during the passage of a high speed train: validation of a numerical model by means of experimental results. *Proceedings of the 13th ASCE Engineering Mechanics Division Speciality Conference.* Baltimore, MD, 1999 [CD-ROM].

10.4. DEGRANDE, G., DE ROECK, G., VAN DEN BROECK, P. and SMEULDERS, D. Wave propagation in layered dry, saturated and unsaturated poroelastic media. *International Journal of Solids and Structures*, 1998, **35**(34–35), 4753–4778.

10.5. DEGRANDE, G. and SCHILLEMANS, L. Free field vibrations during the passage of a HST. *Proceedings of ISMA 23, Noise and Vibration Engineering.* Leuven, Belgium, 1998, vol. 3, pp. 1563–1570.

10.6. DEGRANDE, G., VAN DEN BROECK, P. and CLOUTEAU, D. A critical appraisal of in situ vibration measurements. *Proceedings of the 3rd European Conference on Structural Dynamics: Eurodyn '96* (eds G. Augusti, C. Borri and P. Spinelli). Balkema, Rotterdam, 1996, pp. 1107–1114.

10.7. DEWULF, W., DEGRANDE, G. and DE ROECK, G. Spectral analysis of surface waves: an automated inversion technique based on a Gauss–Newton inversion algorithm. *Conference on Inverse Problems of Wave Propagation and Diffraction.* Aix-les-Bains, France, 1996.

10.8. KAUSEL, E. and ROËSSET, J. M. Stiffness matrices for layered soils. *Bulletin of the Seismological Society of America*, 1981, **71**(6), 1743–1761.

10.9. KNOTHE, K. and GRASSIE, S. L. Modelling of railway track and vehicle/track interaction at high frequencies. *Vehicle Systems Dynamics*, 1993, **22**, 209–262.

10.10. KNOTHE, K. and WU, Y. Receptance behaviour of railway track and subgrade. *Archive of Applied Mechanics*, 1998, **68**, 457–470.

10.11. KRYLOV, V. V. On the theory of railway-induced ground vibrations. *Journal de Physique IV*, 1994, **4**(C5), 769–772.

10.12. KRYLOV, V. V. Generation of ground vibrations by superfast trains. *Applied Acoustics*, 1995, **44**, 149–164.
10.13. KRYLOV, V. V. Effects of track properties on ground vibrations generated by high-speed trains. *Acustica – Acta Acustica*, 1998, **84**(1), 78–90.
10.14. KRYLOV, V. V. and FERGUSON, C. C. Recent progress in the theory of railway-generated ground vibrations. *Proceedings of the Institute of Acoustics*, 1995, **17**(4), 55–68.
10.15. ORMSBY, J. F. A. Design of numerical filters with application to missile data processing. *Journal of the Association for Computing Machinery*, 1961, **8**(3), 440–466.
10.16. VAN DEN BROECK, P. and DE ROECK, G. Dynamic behaviour of railway track on layered media. *Proceedings of the 3rd European Conference on Structural Dynamics: Eurodyn '96* (eds G. Augusti, C. Borri and P. Spinelli). Balkema, Rotterdam, 1996, pp. 1083–1089.
10.17. DVAN DEN BROECK, P. and DE ROECK, G. The vertical receptance of track including soil–structure interaction. *Proceedings of the 4th European Conference on Structural Dynamics: Eurodyn '99* (eds L. Frýba and J. Náprstek). Balkema, Rotterdam, 1999, pp. 837–842.

11. High-speed trains on soft ground: track–embankment–soil response and vibration generation

C. Madshus and A. M. Kaynia
Norwegian Geotechnical Institute, PO Box 3930, Ullevaal Stadion, N-0806 Oslo, Norway

11.1. Introduction

This chapter deals with the phenomena and problems met with when the train speed approaches an apparent critical value determined by the wave propagation velocities in the ground and in an embankment. On sites with soft soils like clay and peat, these speeds may be as low as 40 to 50 m/s.

The presentation in the chapter is based on an actual case in Sweden, where these problems were encountered and thoroughly investigated and where countermeasures are being planned. After an introduction to this case, topics related to vibration measurements, dynamic properties of soil and embankment materials, laboratory and *in situ* methods for the determination of these properties, the physical explanation of the phenomenon, numerical methods for simulation and prediction, diagnostic methods, and countermeasures will be discussed and illustrated with examples from the actual case. Finally, the chapter will give a brief discussion of ground-borne trackside vibration, which is a severe environmental problem for railways on soft soil through urban areas, both for lower train speeds and for high speeds; for high speeds, countermeasures are found to overcome the critical speed limitation.

11.2. Case study

In 1997 a high-speed passenger train service was opened on an existing track along the west coast of southern Sweden. The line passes through several sites containing soft clay. On some of these sites large rolling vibrations of the track, embankment and trackside ground were observed when trains passed at speeds near 200 km/h. It was feared that the vibrations might pose an immediate threat of train derailment,

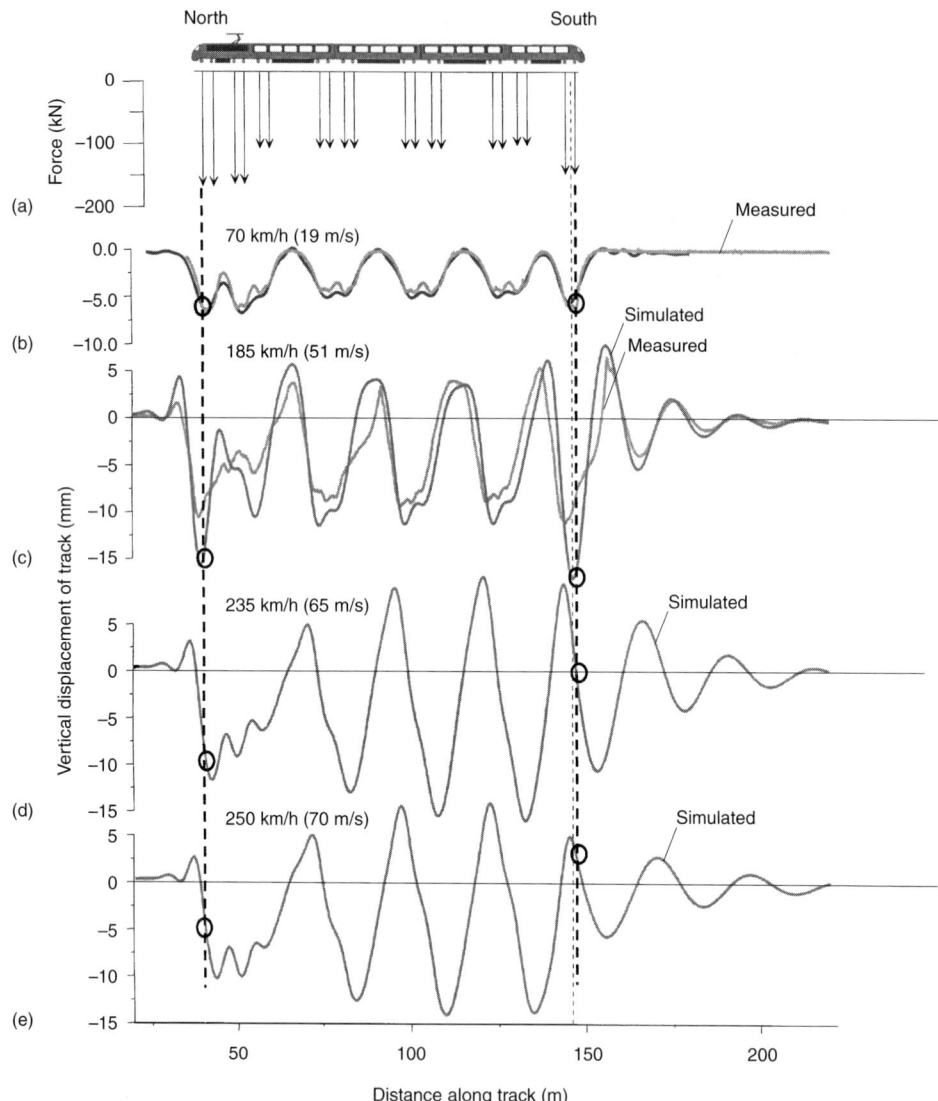

Fig. 11.1. Measured and simulated vertical displacement of track in the speed range from 70 to 250 km/h

ground failure or excessive distortion of the embankment and the track. Additionally, vibration of the contact wire masts caused instability of the power supply to the locomotive. As a temporary solution speed limitations were imposed on the soft-soil sections of the line, until permanent countermeasures could be implemented.

An extensive investigation programme was initiated by the Swedish Railway Administration (Banverket) with the purpose of explaining the phenomenon, determining the severity of the problems, developing simulation and prediction methods, and recommending countermeasures. This actual case will be used for guidance throughout the rest of this chapter when dealing with measurement techniques, soil properties, investigation methods, numerical tools and countermeasures.

11.2.1. Test site and test programme

The most important task in the investigation programme was to perform instrumented test runs to record and interpret the actual response of the track, the embankment and the surrounding soil at one of the sites where the most severe vibrations had been observed. The chosen site was at Ledsgård, located about 24 km south of the city of Gothenburg.

The ground at the site is characterized by a 1.5 m thick crust layer of weathered clay over 3.0 m of soft organic clay. Under these layers is a thick deposit of marine clay, displaying increased stiffness with depth down to bedrock at about 70 m below the ground surface. The groundwater table is close to the ground surface.

The railway embankment at the site is about 1.4 m thick and is constructed from sand with a 0.8 m top layer of crushed rock ballast. The rails are of type UIC60, resting on concrete sleepers at 0.67 m spacing. The rails have an elevation of about 1 m above the surrounding ground surface. There are three parallel tracks at the site. The one furthest west was used for the tests.

An X-2000 Swedish high-speed tilting train was used for the tests. The train used had a locomotive and four passenger cars. The last passenger car was equipped with a driver's cabin. The axle loads varied between 120 and 180 kN. Figure 11.1(a) illustrates the geometry of the train and the loads. The fundamental vertical natural frequency of the car-body/primary suspension is about 1 Hz, and the fundamental rocking-mode natural frequency is about 0.5 Hz.

A total of 20 test runs were made, covering speeds from 10 km/h to 202 km/h. The running direction of the train alternated between northbound and southbound. When northbound the train had its locomotive at the front end. When the train was southbound, the locomotive was at the rear. In addition to these test runs, observations were made when the train was at a standstill. Some preliminary observations were also made at another site closer to Gothenburg, where the soil profile contains a more homogeneous marine clay.

11.2.2. Observations

Careful examination of the recorded data from the test runs at Ledsgård revealed that the vertical displacement of the track and embankment was the most informative single parameter for the purpose of describing and further analysing the phenomena arising as the train speed increases. Figures 11.1(b) and 11.1(c) show typical examples of recorded traces of the vertical displacement from two train passages, at 70 km/h and 185 km/h. To make the traces directly comparable, their original time axes have been transformed to a common length axis by multiplication by the train speed. This length axis has been made the same as the one used for the train geometry and loads in Fig. 11.1(a). The loads and the resulting displacement patterns can therefore be directly compared. Evaluation and interpretation of all recorded displacement traces from the test runs reveals important findings about the nature of the high-speed phenomena that appear at the Ledsgård test site, as described.

When presented on a common length scale as in Fig. 11.2(a), the recorded displacement traces for all train speeds up to 70 km/h turn out to be virtually identical, and also agree with the recording for the stationary train. The displacement pattern synchronously follows the train motion, and its shape and amplitude do not change with train speed. The displacements are consistently downwards; no upward motions are seen. The displacement pattern appears as a 'footprint' of the train loads, and is symmetrical between the front of the train and the rear of the train to the extent the load pattern is symmetrical. The dominant wavelengths of the displacement pattern correspond to the bogie spacings in the train. When viewed in the time domain, the corresponding frequencies are therefore proportional to the train speed. This displacement pattern can be interpreted as belonging to a quasi-static stress/deformation field which is in static equilibrium with the dead load of the train and moves with the train. The peak displacements under

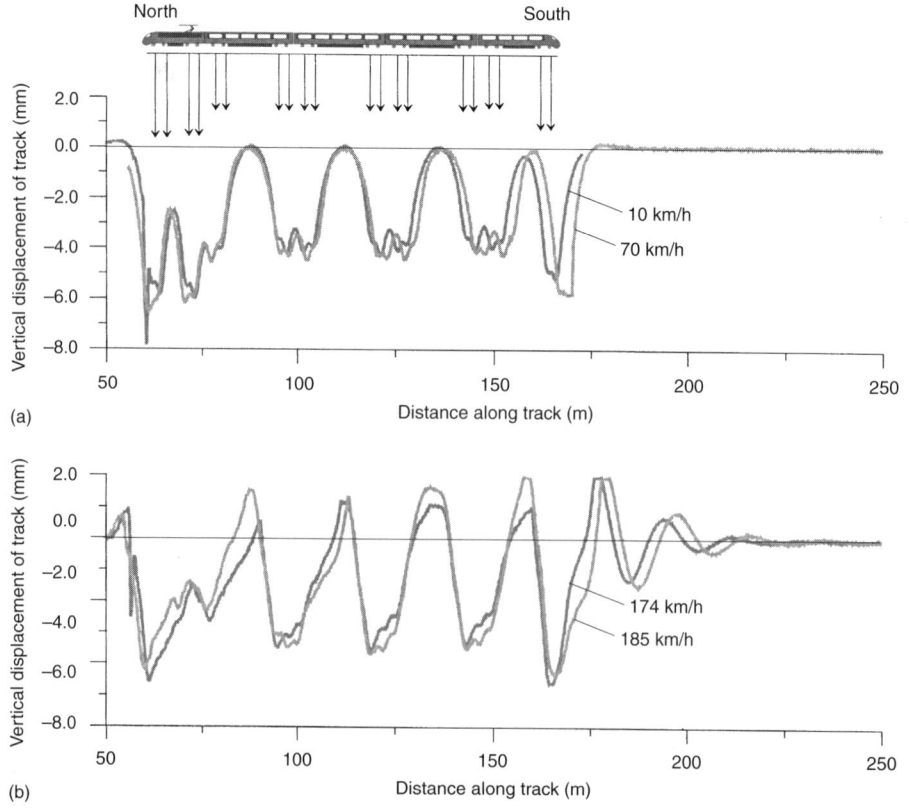

Fig. 11.2. Measured vertical displacement of track: quasi-static and dynamic response

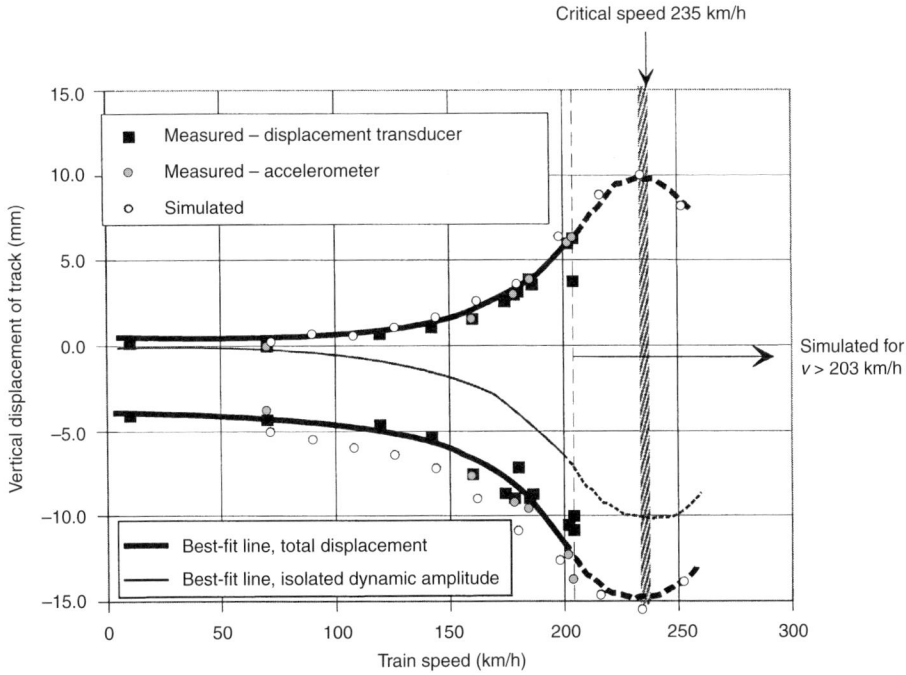

Fig. 11.3. Peak vertical displacement of track versus train speed: measurement and simulation

the central part of the train are about 4.5 mm. This high value indicates the softness of the ground and embankment at the test site.

As train speed exceeds 70 km/h, a change gradually appears in the displacement pattern, which becomes more pronounced as the speed increases. The trace in Fig. 11.1(c) is typical of those recorded at these higher speeds. The pattern now involves both downward and upward motions and the displacement amplitude increases drastically with increasing train speed. Figure 11.3 shows a plot of the peak upward and downward amplitudes versus train speed. At 200 km/h the recorded peak-to-peak amplitude is nearly 20 mm. This is four times the deformation for low train speeds. The displacement pattern is no longer symmetrical with respect to the direction of motion of the train. The motions are more abrupt under the front of the train, they increase towards the end of the train and there is a tail of oscillations following behind the train, much like the water waves behind a boat. A careful interpretation reveals that the frequency of the oscillations in the tail is independent of the train speed, and is about 2.7 Hz for the Ledsgård site. How this frequency is related to the properties of the site will be discussed later in this chapter. The phase velocity of the tail oscillation is identical to the train speed. The whole displacement pattern, including the tail, therefore moves with the train and is stationary when viewed from the train. The decay of the oscillations behind the train has a logarithmic decrement corresponding to a damping between 20% and 30%.

It is apparent from the interpretation of the recorded data that the displacement pattern for the higher train speeds can be decomposed into two parts: a quasi-static component and a dynamic component. The quasi-static component is identical to the one described for speeds below 70 km/h and corresponds to the stress/deformation field which balances the static load of the train. It does not change with train speed when viewed on the length scale. The dynamic component, on the other hand, can be interpreted as corresponding to a stress/deformation field composed of waves, mainly of Rayleigh type, propagating along the embankment and the ground. Since it is a propagating wave phenomenon, this field is internally in dynamic equilibrium without a direct contribution from the train load. Its energy is gained and kept in balance with the loss mechanisms through interaction with the moving train. The nature of the dynamic component can be studied, after careful subtraction of the speed-transformed quasi-static displacements from the total displacements recorded at higher speeds. It is apparent that the dynamic displacement component is, overall, symmetrical with respect to upward and downward motion, as it should be if it represents propagating waves. Except for some transient motions near the front of the train, its oscillations gradually build-up from zero to a maximum at the end of the train, and then die out in the tail behind the train. Figure 11.3 also shows the maximum amplitude of the dynamic displacement component plotted versus train speed.

Numerical simulations of the Ledsgård tests, which will be presented later in this chapter, show that the amplitude of the dynamic displacement component is expected to continue to rise if the train speed is increased beyond 200 km/h up to about 235 km/h, where, for this site, a maximum is reached. For speeds higher than 235 km/h, the amplitude will decrease. This is shown by the dotted curves for train speeds above 200 km/h in Fig. 11.3. The speed at which the maximum dynamic response appears will be denoted here as the 'critical speed'.* How the critical speed is related to the properties of the site and of the train will be discussed later in this chapter. The amplification of the dynamic response component has similarities to the response of a system with a single degree of freedom with respect to both amplitude and phase. The simulated displacement traces plotted in Figs 11.1(b)–(e) show that at low speeds the train loads ride close to the bottom of the troughs in the displacement pattern. As the train speed approaches the critical speed, the loads start to 'climb', until at the critical speed they ride about halfway between trough and crest, at the steepest inclination. At this point the loads and displacements are about 90° out of phase. For even higher speeds the loads 'climb' further towards the crests, and are 180° out of phase with the displacement. This behaviour explains the energy transfer from the moving train to the dynamic displacement component. When the wheels are forced to run on the permanent 'uphill slope' created by the moving displacement pattern of the track, a horizontal force component adds to the rolling resistance of the train, and thus drains energy from the train to the propagating waves.

---

*For definitions of critical velocities, see also Chapter 9 – *Ed.*

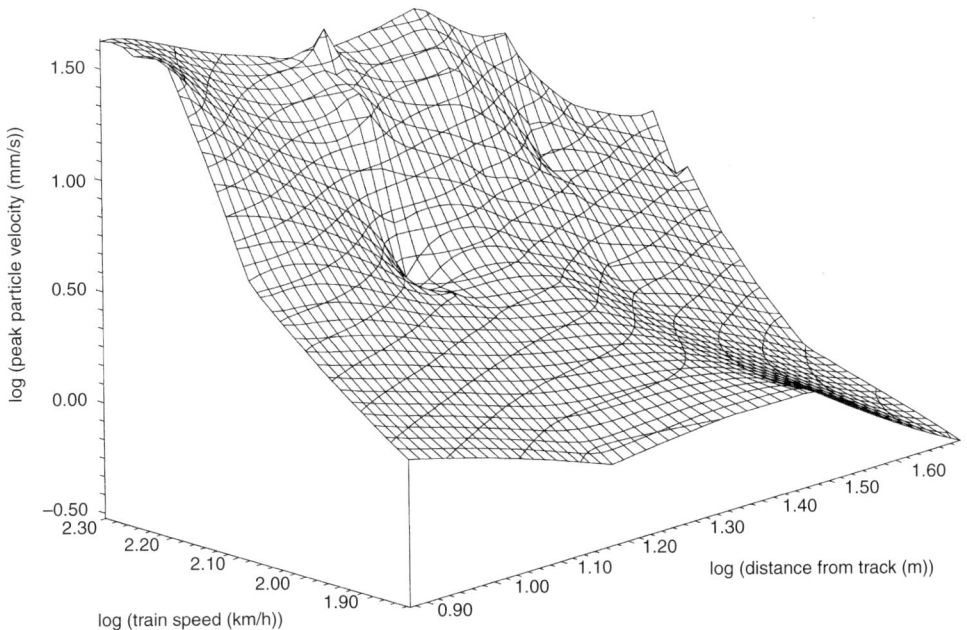

Fig. 11.4. Trackside vibration: peak particle velocity versus distance from track and train speed (numbers on axes are logarithmic values)

The part of the dynamic displacement component under the train has wavelengths corresponding to the bogie spacings in the train and therefore does not change with train speed, and is thus largely load controlled. In contrast, the part representing the displacement tail behind the train has a wavelength which increases proportionally with the train speed since its frequency is speed independent. Between these parts is a transition zone. There is a trend in the recorded data that with increasing train speed, the constant-frequency tail increasingly takes control under the train also, starting from the rear and progressing forward with increasing speed, and thus violating the load control which dominates here at lower speeds. In interaction with the quasi-static displacement pattern, this change in the dynamic pattern is the mechanism that causes the total displacements to gradually get out of phase with the axle loads as the speed approaches and passes the critical speed.

If one looks at the vibrations recorded on the ground surface at some distance from the track during the Ledsgård test runs, it appears that they are not well correlated with the displacements recorded on the embankment. These trackside vibrations, expressed in terms of particle velocity rather than displacement, increase with increasing train speed even in the speed range up to 70 km/h, but their dominant frequency in this case is fairly constant, independent of the train speed. The vibrations gradually increase as the train approaches the measurement site and die gradually out after the train has passed. At high train speeds, closer to the critical speed, the character of the trackside vibration is different and is better correlated

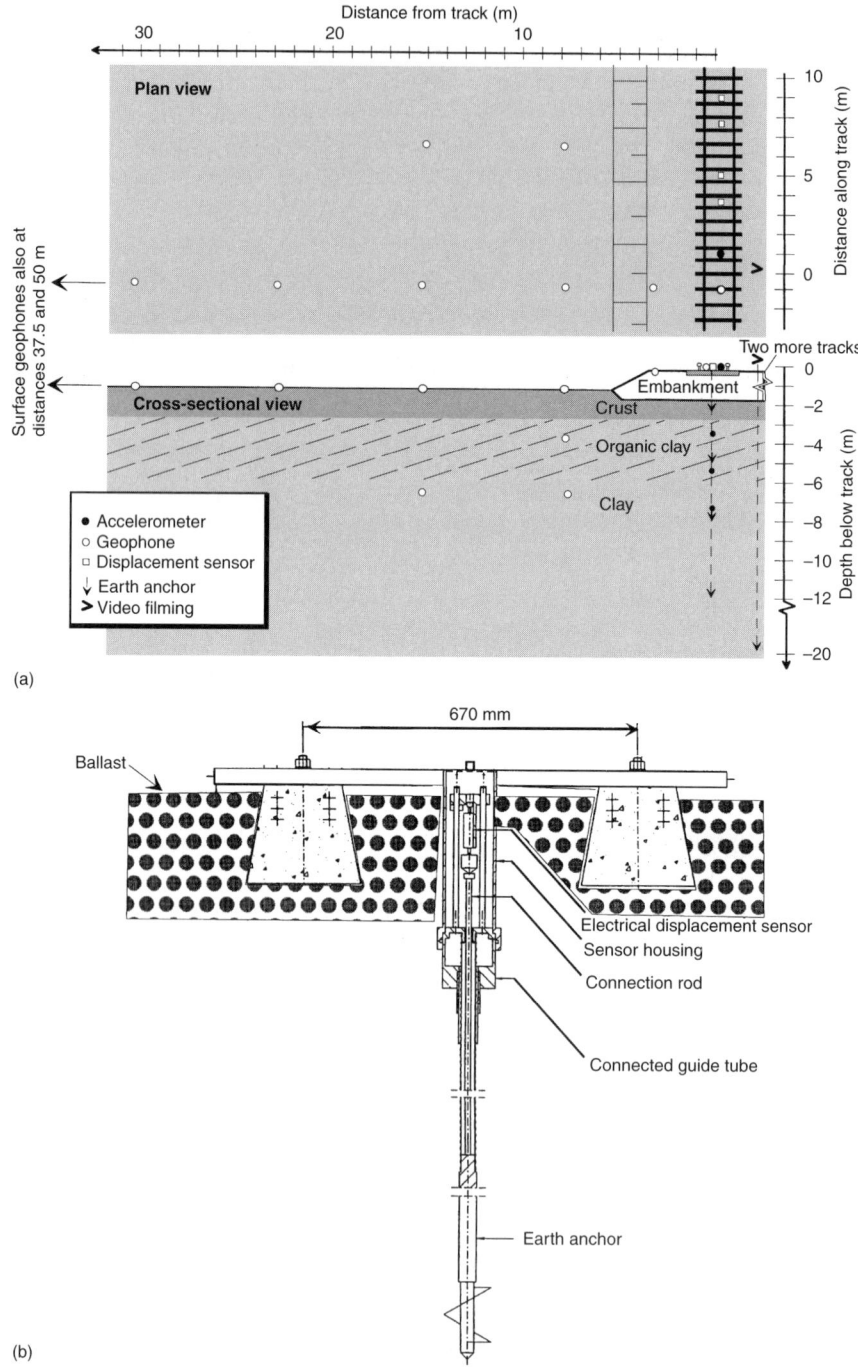

Fig. 11.5. Instrumentation at the Ledsgård test site: (a) overview in plan and cross-section; (b) details of displacement sensor

with the vibration of the track and embankment. For these speeds, precursor vibrations ahead of the train are also observed, but the major tremor defines a wavefront following the train. These wavefronts are not well defined in the recorded data, but the existence of Mach fronts moving with the train can be traced. The vibration amplitudes at high train speeds are much higher and decay at a lower rate with increased distance from the track than for the lower train speeds. Figure 11.4 illustrates this in a 3-D presentation, where the peak particle velocity measured on the trackside ground is plotted versus the train speed and the distance from the track. Note that all three axes in the figure are logarithmic.

11.3. Measurements

For studying the phenomenon of an excessive response of the track, embankment and ground to high-speed trains, field tests involving high-quality measurements are essential. Data from such measurements are needed for the purposes of gaining fundamental physical understanding of the mechanisms involved, diagnosing the problem at a specific site and validating simulation models. This section presents the instrumentation used in the Ledsgård tests, discusses its performance and gives general advice on how to perform such measurements to obtain the necessary data of sufficient quality.

During the Ledsgård test runs the following measurements were made:

- vertical displacement of track and deflection of soil layers: measured by sensors attached to the sleepers at four positions along the track
- vertical displacement of rail: measured through video-filming of the rail against a reference rod in one position
- acceleration and velocity of embankment: measured by one accelerometer and one geophone placed in the ballast and one geophone on a sleeper
- acceleration within soil: measured by one accelerometer pressed into the ground under the centre of track
- pore pressure: measured by sensors in the soil at three elevations under the centre of the track
- velocity of trackside ground: measured by ten geophones placed on the ground surface and in the ground out to 50 m from the track
- train speed: read by the driver from the train's speedometer.

Figure 11.5(a) presents a plan view and a cross-sectional view of the instrumentation.

The interpretation of the Ledsgård tests clearly showed that the vertical displacement of the track–embankment system was the most important single parameter for understanding and diagnosing the phenomenon. The highest priority in any set of instrumentation should therefore be to obtain high-quality records of this parameter. When choosing sensors for this purpose, it is important to keep in mind the low frequencies involved. The interpretation showed the importance of being able to separate quasi-static from dynamic displacements; to obtain this, it is necessary not only to capture the right peak-to-peak displacements, but also to capture the right absolute values of both the upward and the downward motion. The lowest frequency to be covered is therefore not the bogie repetition frequency of the trains at their lowest speed in the test sequence, but the frequency that represents the passage time of a large portion of the entire train. This may well be as low as 0.1 to 0.01 Hz.

Four resistance-based electrical sensors [11.1] were used for the vertical-displacement measurements at Ledsgård. The sensors were mounted between sleepers and measured against connection rods attached to earth anchors installed in the ground at depths of 2.5, 5, 8 and 12 m under the top of the embankment. The rods were fed through corrugated plastic guide tubes to avoid distortion of the measurements owing to friction between the rod and the surrounding soil and embankment material. Figure 11.5(b) illustrates the arrangement. The sensor positions were separated by distances of 1.34 or 2.68 m along the track. As an auxiliary system, one additional connection rod with an earth anchor at 20 m depth was installed close to one of the rails. Its top was video-filmed against a ruler attached to the rail, and the displacements were automatically determined by digital image processing of the video frames.

The sensor system had excellent low-frequency performance all the way down to static displacements, and the signals had high resolution and good quality for recording. The sensors also monitored settlements of the track. The deepest earth anchors represented a reliable basis for absolute displacement measurements. The installation of the anchors to various depths enabled determination of how much the various soil layers contributed to the total displacement and gave a rough estimate of the dynamic strains in the soil. The offset between the sensors enabled determination of the propagation speed of the displacement patterns along the embankment. The image processing of the video recordings was not easy, owing to bad picture quality, but sufficient agreement between the results from the two systems was obtained.

Double integration of the accelerometer signals revealed the peak-to-peak displacements with satisfactory accuracy. However, even when sophisticated signal processing was applied, these sensors did not have sufficient low-frequency performance to capture the correct absolute upward and downward displacements during train passage. The geophones, which had natural frequencies at 2.0 and 4.5 Hz, did not even capture the peak-to-peak displacements correctly. For rating trackside vibrations with respect to their disturbing effect on people, however, the geophones gave satisfactory results.

For fundamental studies of the critical-train-speed problem, instrumentation for absolute displacement measurements should be included in the system used. Systems like the one at Ledsgård, based on an anchored reference, are the most reliable. The main disadvantage is the installation cost. Drilling and the installation of guide tubes and anchors, particularly through an embankment, are tedious and also interfere with the traffic on the line. The alternative of using even the most sophisticated low-frequency/high-accuracy accelerometer system should be adopted with extreme care, and will probably be unreliable. Optical systems may be an alternative, but even these have problems in obtaining a reliable absolute reference on soft-soil sites, where high levels of vibration extend to long distances from the track.

The optimum instrumentation may be based on an anchored reference system at one point, supplemented by accelerometers at other points. Conventional, sensitive, low-frequency accelerometers on the ballast, placed at some positions along the track, will be satisfactory for capturing propagating displacement patterns. The spacing should be greater than that used at Ledsgård, ideally one-half to one-third of the longest wavelength to be captured, which requires a spacing of about 5 to 10 m. Accelerometer placement in the ballast is preferred over placement on sleepers or rails, to reduce high frequency disturbance. The same type of accelerometer, inserted into the ground at different elevations or in an array under the embankment, will also be satisfactory for determining the dynamic strains in the ground.

To study trackside vibrations, accelerometers in an array on the ground surface are the best solution. For fundamental studies, an anchored reference system could be added at one point as well. If the objective is strictly limited to the study of vibration annoyance to people, geophones may be satisfactory.

For pore pressure measurement at Ledsgård, filter elements were installed under the centre of the track, in the clay at about 3.5, 6 and 8 m below the top of the embankment. The filters were connected to pressure sensors on the ground surface through thin, water-filled hydraulic hoses. Pore pressure measurements are relevant at sites susceptible to soil degradation due to dynamic strains, which are reflected as pore pressure build-up. Only minor, hardly detectable excess pore pressures appeared during the Ledsgård tests. Despite some disagreement among the different sensors, the pore pressure system seemed to capture those excess pressures which appeared during the test runs and later dissipated. An attempt to record the dynamic pore pressure fluctuations was, however, not successful. It appeared that the pore pressure sensor system was sensitive to accelerations, leading to false signals with magnitudes higher than what the real pressure fluctuations could have been. The measurement of short-term pore pressure in low-permeability soils, such as clay, is extremely demanding. To be successful, the pressure sensors must be integrated in filter units installed in the ground. The sensors must have an extremely low compliance and the water volume within the system must be at a minimum. Additionally, extreme care must be taken to ensure that the whole system is completely saturated. If these requirements are met, the build-up and dissipation of excess pore pressure should be reliably measured. It is, however, unrealistic to expect reliable measurement of dynamic pore pressure fluctuations in such soils. Soil degradation may also be detected through short-term settlement observations.

The transformation of recorded displacement traces to a spatial coordinate for direct comparison between various trains speeds, as done in Fig. 11.1, requires accurate train speed determination. As seen for example in Fig. 11.2(b), the speed reading provided by the train driver, as used in the Ledsgård tests, is not as accurate as it should be for this purpose. It is recommended that one should include in the instrumentation set-up for this type of measurement a pair of photocell stations or other detectors, placed in two positions along the track. In addition to precise train speed readings, such a detector pair will give confirmation of the train direction and, importantly, provide a key to synchronize the parameters recorded during the train passage with the instantaneous train position.

During the Ledsgård tests, different teams were responsible for various parts of the instrumentation system, and the signals were recorded on four different systems that were not synchronized. For this type of measurement, where wave propagation velocities, train speeds and instantaneous positions are essential, all signals should be registered on one recording system, facilitating simultaneous sampling from the various channels. All signal processing should be performed so as to maintain the common time frame of the recording.

When designing instrumentation systems for this type of measurement, it is important to be aware of the fact that railways, particularly those that are electrically powered, represent a demanding environment for precision measurements. There are high electrical voltages and high electromagnetic fields and there are strong stray currents in the ground, all appearing at frequencies of interest for the measurements. To reduce disturbance, measurement systems based on differential signals should be preferred over single-ended systems. Cables need good and continuous shielding. All sensors must be electrically isolated from the ground and the track, and ground loops must be avoided. To reduce electrical disturbances, battery-operated recording systems are preferred over systems powered from the mains supply.

11.4. Dynamic properties of soil and embankment materials

From the interpretation of the measurements made during the Ledsgård tests, it is obvious that the track and ground responses to passing high-speed trains are mainly controlled by the interaction between the train loads and the Rayleigh-type surface waves in the soil and embankment. The Rayleigh wave velocity of the site is therefore the most important parameter to determine if one wishes to be able to understand the phenomenon, simulate or predict the response, and design countermeasures.

For a homogeneous half-space, the Rayleigh and shear wave velocities are closely related, in that the Rayleigh wave propagates at a velocity which is practically somewhere between 0.91 and 0.96 times that of the shear wave, depending on the Poisson's ratio of the soil. The shear wave velocity is determined by the properties of the soil, and is thus the key to the Rayleigh wave velocity. However, most soil deposits are layered, and confining stress and density tend to increase with depth. This leads to a shear wave velocity which varies and usually increases with depth. For such deposits, the Rayleigh waves will be dispersive, and more than one Rayleigh mode may appear. Often, existing top layers of a harder weathered crust and a frozen crust in winter will break the monotonic trend of increased shear wave velocity with depth and make the dispersion more complicated. The same effect appears when Rayleigh waves propagate along a railway embankment of stiffer material interacting with the natural surrounding ground. In these more complicated layered systems, the Rayleigh wave velocities are still uniquely related to the shear wave velocities, but the relations are more complicated. For more information, the reader is referred to specific literature in the field, e.g. [11.2, 11.3].

The most common soft soil types are clay, silt and peat. Embankments are usually made of sand and crushed rock. All these materials are porous assemblies of grains which are in contact with each other and form a grain skeleton. The voids between the grains may be filled with air or partly or fully saturated with water. The strength and stiffness of soils are related to the contact forces between the particles in the grain skeleton and thus to the effective confining stress, which is the total stress minus the pressure in the pore fluid. For all the above soils, except peat, where the particles consist of soft organic material, the grains are made of hard minerals. Clay and silt may contain a substantial amount of organic particles. In the following discussion all the above materials will be termed soils.

At low dynamic strains, below about 10^{-5} to 10^{-4}, soils behave nearly elastically and the shear wave velocity is determined by the elastic shear modulus of the grain skeleton, G_{max}, and the overall mass density including grains and pore fluid, ρ, according to the equation

$$C_s = \sqrt{\frac{G_{max}}{\rho}} \tag{11.1}$$

The mass density of most soils is in the range 1500 to 2000 kg/m³, and is therefore not a sensitive parameter for the determination of C_s. The elastic shear modulus G_{max}, on the other hand, depends on the soil type, its past geological history, its porosity and the effective confining stress, and may vary over a large range. At these low strains most soils have a small amount of viscous damping, about 2% to 4%.

Rough estimates of the elastic shear modulus and shear wave velocity of soils may be obtained from empirical expressions. An expression suitable for most soils is [11.4]:

$$G_{max} = Af(e)p_a \left(\frac{\sigma'_{oct}}{p_a}\right)^n \tag{11.2}$$

where σ'_{oct} is the mean effective confining stress of the soil at the depth under consideration; p_a is a reference pressure, defined as 100 kPa; n is an exponent, about 0.5 for most soils; $f(e)$ is a monotonically decreasing function of the soil void ratio e (a measure of porosity), where $f(e) = 1/(0.3 + 0.7e^2)$; and A is a dimensionless constant depending on the soil type and its geological history.

For sands and crushed rock, A is typically in the range of 500 to 700; $f(e)$ varies from about 2.2 for the most dense constitutions to about 1.4 for the most loose.

For most non-organic soft clays A is usually in the range of 500 to 2500, with the lowest values for the most plastic clays and for clays with the highest water content.

The void ratio may be between 2.0 and 3.0, leading to $f(e)$ values from 0.30 to 0.15. For such clays, experience shows that G_{max} correlates reasonably well with the undrained shear strength of the clay, S_u, which may often be known from the results of ordinary geotechnical tests. It may therefore be more convenient to estimate G_{max} from

$$G_{max} = cS_u \qquad (11.3)$$

where c can be approximated as $c \approx (208/I_p) + 250$, and I_p is the plasticity index of the clay, as is usually available from standard geotechnical laboratory testing. With a range of I_p typically being 0.80–0.15, the scalar constant c will have values in the range 500–1500. To obtain more precise estimates of G_{max}, data from dynamic *in situ* or laboratory tests on the actual clay are needed [11.5].

For peat and soils with a high content of organic material, G_{max} and C_s may take unexpected values, and the use of empirical relations like those above is not recommended. Usually, moduli and velocities will be substantially overpredicted by these equations. For the organic clay layer at Ledsgård, the c factor was found to be as low as 100.

For fully saturated soft soils, the bulk modulus of the pore water will completely dominate the compressibility, leading to compressional wave velocities close to 1500 m/s, fairly independent of other soil parameters. However, even a minute amount (less than 1% by volume) of gas bubbles in the pore water will reduce its volumetric stiffness so as to make the grain skeleton control the compressional-wave velocity, which will then usually be 2 to 2.5 times the shear wave velocity.

Owing to the many uncertain factors involved, empirical equations like those above can only give rough and often uncertain estimates of the shear wave velocity of the ground. To obtain more reliable velocities at a particular site, actual measurements are needed, preferably performed *in situ* at the site or in the laboratory on soil samples from the site.

There are three commonly used methods for *in situ* determination of the shear wave velocity versus depth:

- the cross-hole method
- the down-hole or seismic CPT method
- the SASW (spectral analysis of surface waves) method.

The cross-hole method needs at least three vertical holes to be drilled into the soil deposit and lined with plastic tubes. An impact source is clamped in one of the outermost holes at the depth where measurements are to start. Sensors are clamped at the same depth in the other two holes. The source generates horizontally propagating, vertically polarized shear waves. The wave velocity is determined from the arrival time at the sensor holes and the interhole spacing. The measurements are repeated at regular depth intervals through the whole deposit.

The down-hole method uses a source on the surface to generate vertically propagating, horizontally polarized shear waves. The wave arrival is detected by a sensor built into an instrumented conical tip attached to a steel tube pushed into the ground. The cone also measures the penetration resistance and pore pressure as it is pushed down. This is a standardized *in situ* soil-testing method, termed CPT (cone penetration testing). By detecting the difference in wave arrival times as the cone is stopped at regular depth intervals, the shear wave velocity versus depth can be determined for the soil profile. Details of these methods can be found in [11.4] and [11.6].

The SASW method is a non-intrusive method based on measuring the velocity and dispersion of Rayleigh waves in the soil deposit. The Rayleigh waves are excited by a vertical-impact (or vibratory) surface source and detected by two or several sensors along the soil surface. By analysing the signals in the frequency domain, i.e. propagation velocity versus frequency, the dispersion curve of the site can be determined. From the dispersion curve, the shear wave velocity versus depth can be determined through an iterative inversion process based on forward modelling of wave propagation in a layered system. Details of the method can be found in [11.7] and [11.8]. The computer program WinSASW [11.9] may be used for data processing and inversion.

Before determining the wave velocities of a site, it may be useful to obtain an overall picture of the soil types and the layering. This is best obtained from general geological knowledge of the area, supplemented by sounding at some points. CPT sounding, without the above wave detection option, is an efficient tool for this purpose. Soil layering can be detected easily, and through established empirical correlations, soil types together with other characteristics can be determined from the recordings [11.6].

In recent years, down-hole/seismic CPT has been developed into a cost-effective and reliable method for shear wave velocity determination of soft-soil sites. The cross-hole method may be slightly more accurate, but is far more expensive since it requires the drilling of lined holes. None of the methods will give useful data for the top 1–2 m. The SASW method is efficient since it does not require any drilling. To obtain deep measurements a fairly heavy wave source is needed. The practical depth limit for soft-soil sites is between 10 and 20 m. The SASW method has the advantage over the other methods that it measures average properties over a larger soil volume and gives reliable velocities all the way up to the surface. On the other hand, it may be less accurate. The inversion process may be cumbersome for complicated sites with stiff layers over softer soil. The SASW method is most efficiently used to obtain a rough overview of the shear wave velocities of a site or used in combination with a few measurements by one of the other methods to map a large site. When used with care, the SASW method can also be used to determine the shear wave velocity in railway embankments.

All the above methods were used to determine the wave velocities and soil profile at the Ledsgård test site. In addition, soil samples were taken and tested in the laboratory. Figure 11.6 shows the soil profile and the results of all shear wave velocity measurements. The 3 m thick organic clay layer under the weathered crust should be noticed in particular. It has a shear wave velocity of 40 m/s, and thus has a dominant effect on the response to high-speed trains. The lower clay layer extends down to bedrock at about 70 m depth. The weathered crust and the organic clay were unsaturated. The rest of the clay deposit was saturated. The scatter in the results reflects the spatial variability over the site and the uncertainty in the measurements.

For dynamic strains higher than 10^{-5} to 10^{-4}, soils display a non-linear hysteretic behaviour that becomes more pronounced as the strain increases. This sort of behaviour can only be determined in the laboratory. The cyclic triaxial device is the apparatus most commonly used for such soil testing. Here, a cylindrical soil sample is enclosed in a thin rubber membrane and placed on a pedestal in a pressurized chamber to bring it back to the *in situ* stress conditions. A dynamic loading is applied in the axial direction. To better reproduce the dynamic load situation in the soil volume under a railway, tests are also needed in a direct simple shear (DSS) device, where a dynamic shear stress is applied to the end faces of a cylindrical sample. Peat, clay, silt and sand can be tested in standardized devices with a typical sample diameter of around 50 mm. Ballast materials need much larger devices, which are not standard. Such materials are therefore more cumbersome to test. Undisturbed samples can usually be obtained from peat, clay and silt materials. Tests on sand and ballast materials have to be made on samples reconstituted from remoulded material.

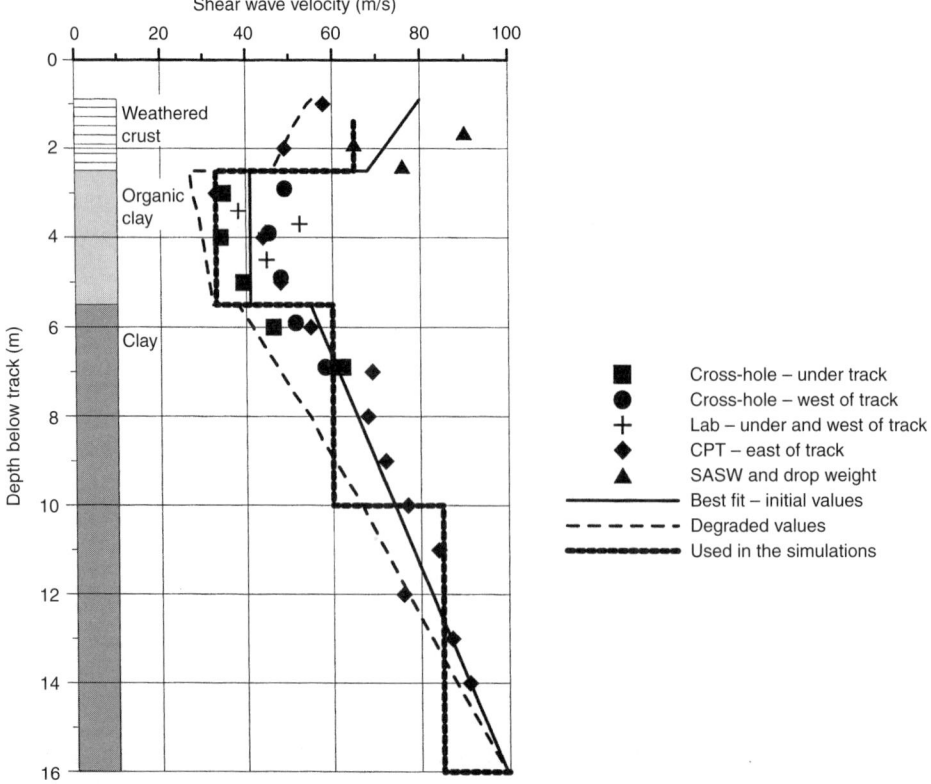

Fig. 11.6. *Soil profile of the Ledsgård test site: shear wave velocity versus depth*

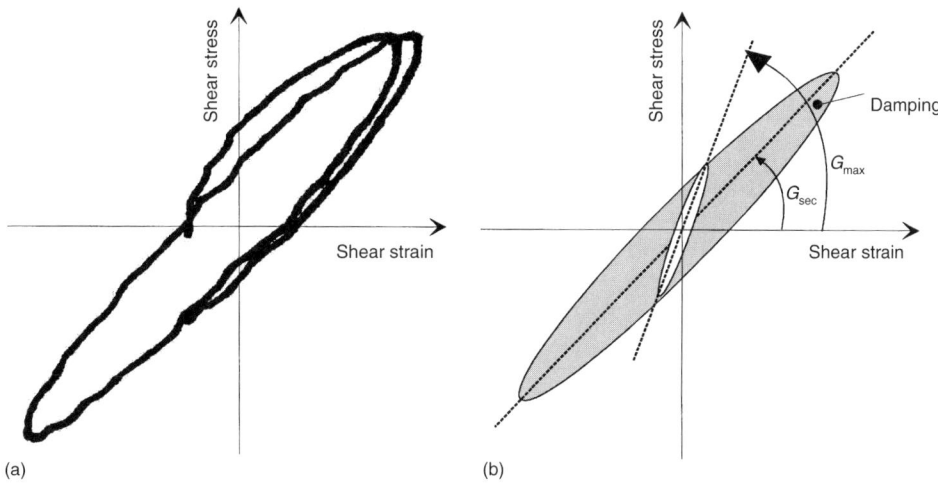

Fig. 11.7. Hysteric, non-linear response of soil

The triaxial and DSS devices may be equipped with piezoceramic elements enabling the measurement of shear wave velocity and G_{max} in laboratory specimens [11.10]. To measure damping, a resonant-column device may be used. Here a cylindrical soil sample is tuned into resonance, and the secant modulus and damping can then be back-calculated from the recordings.

Figure 11.7(a) shows a plot of typical measured hysteretic loops from the cyclic triaxial testing of the organic clay from the Ledsgård test site. Figure 11.7(b) shows an idealized loop defined by two parameters: a secant modulus, G, and a hysteretic damping ratio, D. It has been demonstrated [11.11] that the hysteretic non-linear soil behaviour may be unified in two simple plots, where the secant modulus G, normalized through division by the elastic shear modulus G_{max}, i.e. G/G_{max}, and the damping ratio D are plotted versus the cyclic strain amplitude. These plots turn out to be unique for any given soil type, independent of confining stress and frequency of cycling. Figure 11.8 presents such plots for the organic clay, the marine clay and the embankment material at the Ledsgård site.

It has been further demonstrated [11.12, 11.13] that the dynamic non-linear response of soil systems can be modelled with sufficient accuracy by applying these idealized loops, instead of having to reproduce the real loops. This approach is termed the equivalent linear approach, and can be applied in the frequency domain through an iterative process, where a match is obtained between the actual dynamic strains at any point throughout the soil volume and the secant modulus and damping ratio, as shown in plots like Fig. 11.8. This approach was applied in the numerical simulations presented in this chapter.

The measurements and numerical simulations of the highest-speed train passages during the Ledsgård tests reveal that the maximum cyclic strains were about 1% in the embankment, 0.8% in the organic clay and 0.2% in the upper part of the marine clay. These values are indicated on the curves in Fig. 11.8. As can be seen from the

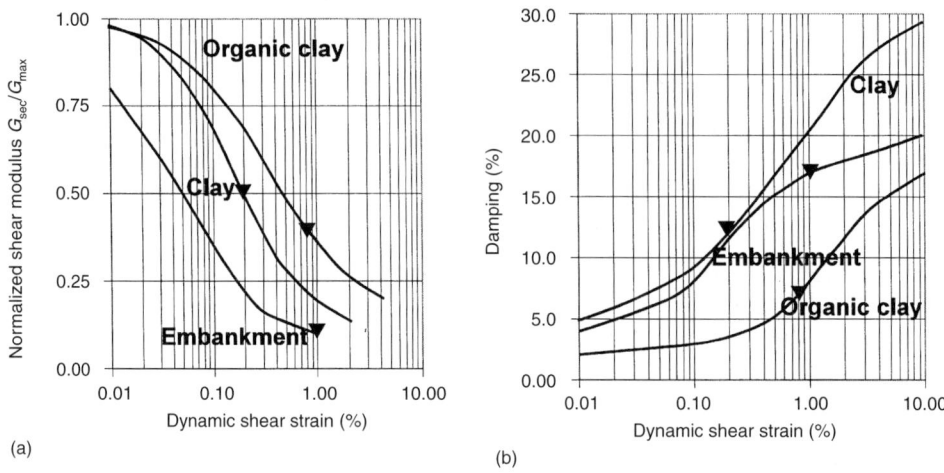

Fig. 11.8. Normalized secant shear modulus and damping for embankment material, organic clay and clay, from the Ledsgård test site

plots, the non-linearity of the soil materials leads to equivalent secant moduli which are reduced to 10%, 40% and 50% of the elastic values for the embankment material, organic clay and marine clay, respectively. Correspondingly, the damping ratios matching the strains in these material layers are 16%, 7% and 12%, respectively. Owing to these non-linearly reduced moduli, the actual wave velocities in the ground and embankment are lower during train passage than those measured by the seismic methods. This is indicated in Fig. 11.6. The numerical simulations have shown that it is vital to take account of this modulus reduction and increased damping if one is to be able to reproduce the responses measured during the high-speed train passages.

During dynamic (cyclic) loading above shear strains of about 10^{-4}, soils tend to densify. However, this tendency varies highly among soil types. For saturated soils, densification requires pore water to be squeezed out. For soils such as clay with a low permeability, there will not be enough time for the water to escape during a train passage. This will lead to a build-up of excess pore pressure in the soil. Since the total stress is determined by the overburden pressure, this leads to a reduction of the effective stress. The effective stress reflects the interparticle forces in the grain skeleton and thus determines both its strength and its stiffness. Therefore, such a pore pressure build-up will reduce the bearing capacity of the foundation soil and may, in the worst case, lead to a soil failure. This is what happens in the liquefaction often observed during earthquakes. Although the cyclic strains which appear in the ground when a high-speed train passes at near its critical speed are lower than those occurring during destructive earthquakes, they may be high enough to substantially reduce the bearing capacity of the ground. In particular, clays may be so impermeable that the excess pressure does not have time to dissipate between train passages, leading to an accumulation of pore pressure from train to train. As the excess pore pressure is released, the densification of the grain skeleton leads to accelerated settlement. The ability of a particular soil to densify or to build up pore pressure and reduce its strength can be determined in cyclic triaxial tests. Generally, the rate of pore pressure build-up and densification increases with increasing cyclic stress level, but it also depends on the mode of loading. More details can be found in [11.5].

The tendency of pore pressure to build up was investigated for the organic and marine clays at the Ledsgård test site. It was found that the organic clay had an extraordinary resistance against pore pressure build-up. Even with the high strains appearing in this soil layer, substantial pore pressure should not be expected in this layer. The pore pressures were monitored during the tests, but no excess pressures were observed. The marine clay is more prone to pore pressure build-up but was exposed to lower strains. The laboratory tests indicate that if this clay had been exposed to the same high strains as in the organic clay layer, a critical reduction in bearing capacity could have been expected.

11.5. Numerical simulation

Moving loads, such as high-speed trains and propagating air-pressure waves, have long been recognized as potential sources of ground vibration. The response of ground to a moving load is dictated largely by the relation between the speed of the load and the characteristic wave velocities of the ground medium. At low speeds, the response of the ground to a moving load is essentially quasi-static. That is, the displacements and stress fields are essentially the static fields under the load and simply move with it. However, as the speed of the load increases, dynamic phenomena gradually take over and dominate the response. In the realm of elastodynamic theory it has become common to categorize moving-load problems as *subseismic*, *superseismic* and *transeismic*, depending on whether the speed of the load is less than the Rayleigh wave velocity of the ground, greater than the pressure wave velocity or intermediate between these velocities, respectively.

Analytical studies of the response of a half-space to moving loads have revealed that, whereas the subseismic regime represents a quasi-static condition, both the transeismic and superseismic cases are characterized by large dynamic effects associated with the development of Mach lines and Mach surfaces in the ground response [11.14–11.17]. Various formulations have been proposed for the prediction of train-induced ground vibration taking account of an embankment [11.18–11.21]. The most recent of these solutions modelled the embankment as a beam on a uniform half-space and solved the problem by the application of Fourier transformation in space and time [11.20, 11.21]. In addition, attempts have been made to make numerical simulations of the problem using general-purpose numerical tools, including the finite-difference code FLAC-3D [11.22, 11.23], the finite-element code ABAQUS [11.24] and the spectral-element method [11.25]. In addition to shedding light on the mechanism of wave generation for high-speed loads, these studies have further confirmed the existence of a critical speed beyond which the dynamic phenomena may become dramatic. This section aims at

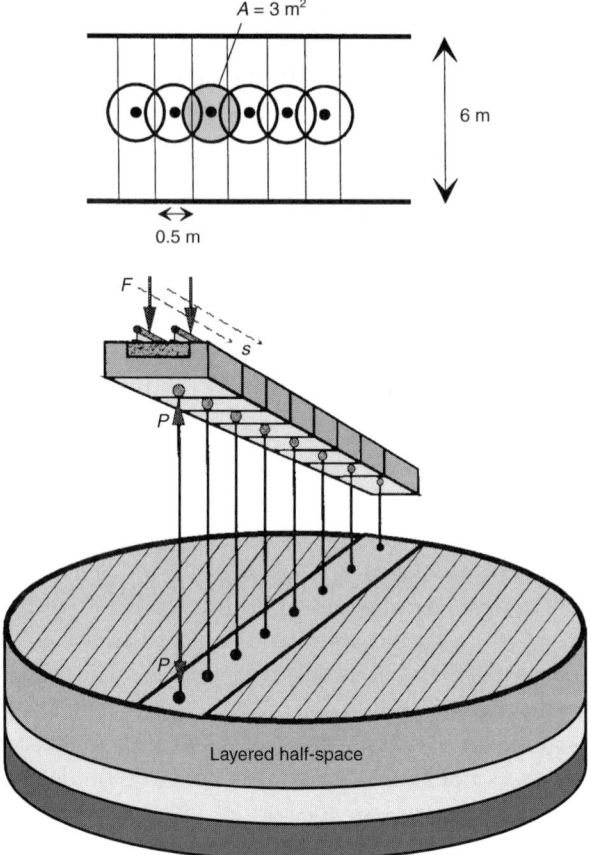

Fig. 11.9. The VibTrain composite discrete Green's function/finite-element numerical model

presenting results of a robust calculation model called VibTrain [11.26, 1127] aimed at prediction of the response of the ground to high-speed trains. The work essentially extends that in [11.20] by accounting for soil layering. To achieve this, a different methodology based on the discrete impedance matrix of the ground and a substructuring principle has been utilized. An outline of the proposed formulation is given in the following.

Figure 11.9 shows schematically the problem under investigation. The track/embankment structure is modelled as a beam resting on a layered half-space. The track/embankment is represented by its bending rigidity EI, mass per unit length m and hysteretic damping ratio β. Each soil layer is characterized by its shear wave velocity V_s, pressure wave velocity V_P (or, alternatively, the Poisson's ratio ν), mass density ρ and hysteretic damping ratio D. It is assumed that the embankment is bonded to the supporting half-space at discrete points along the embankment, referred to as nodes. These points coincide with the location of the sleepers.

The excitation is a series of concentrated loads representing the wheel loads of a train moving with constant speed V. Figure 11.1(a) shows the static load of the X-2000 train used in the numerical simulations. The rail is not explicitly included in the calculation model; however, to preserve its role in distributing the train load, each concentrated axle load is distributed over the embankment according to the displacement variation under the rail. Using the theory of a beam on an elastic foundation, one can show that for a beam with bending rigidity EI and modulus of subgrade reaction k, the displacement of the beam can be calculated from the following expression:

$$\phi(x) = \frac{1}{\sqrt{2}L} \exp\left(-\left|\frac{x}{L}\right|\right) \sin\left(\left|\frac{x}{L}\right| + \frac{\pi}{4}\right) \tag{11.4}$$

where $L = \sqrt[4]{4EI/K}$ and $K = kb$; b represents the equivalent width of the beam. This way of defining the load helps to avoid the extra computational effort required in representing unnecessarily high frequencies in the load variation. The loads are applied at the nodes with time shifts corresponding to the train speed.

As the loads travel along the track, interaction forces develop between the embankment (beam) and the supporting ground. The ground and the embankment can be considered as separate substructures subjected to the nodal interaction forces, as illustrated in Fig. 11.9. If, at a given frequency ω, $\boldsymbol{P}$ denotes the vector of the interaction forces and $\boldsymbol{W}$ represents the associated vector of vertical ground–embankment displacements, then one can relate these vectors through the notion of Green's functions by

$$\boldsymbol{W} = \boldsymbol{GP} \tag{11.5}$$

where $\boldsymbol{G}$ is a symmetric matrix with frequency-dependent complex entries G_{ij} defining the ground response at node i due to a unit load at node j. Inverting this relation, one can write

$$\boldsymbol{P} = \boldsymbol{G}^{-1}\boldsymbol{W} = \boldsymbol{K}_s \boldsymbol{W} \tag{11.6}$$

where $\boldsymbol{K}_s$ is the stiffness matrix of the layered ground corresponding to the interaction nodes.

A similar relation can be established by considering the equilibrium of the embankment substructure. The stiffness matrix of the embankment can be assembled from the stiffness matrices of the individual beam elements. However, this matrix involves rotational degrees of freedom (θ) in addition to translational degrees. If the vector of these nodal motions is denoted by $\boldsymbol{U} = [\boldsymbol{W}, \boldsymbol{\theta}]^T$, then one can write

$$\boldsymbol{F} - \boldsymbol{P} = \boldsymbol{K}_B \boldsymbol{U} \tag{11.7}$$

where $\boldsymbol{F}$ is the vector of applied forces and $\boldsymbol{K}_B$ is the dynamic stiffness matrix of the beam. This matrix is assembled from the classical stiffness matrix $\boldsymbol{K}^i$ and the consistent mass matrix $\boldsymbol{M}^i$ of a generic element i as follows (see, e.g., [11.28]):

$$\mathbf{K}_B^i = \mathbf{K}^i - \omega^2 \mathbf{M}^i \qquad (11.8)$$

Finally, eliminating the interaction force vector from equations (11.6) and (11.7), one obtains

$$F = (\mathbf{K}_s + \mathbf{K}_B)U \qquad (11.9)$$

where it is assumed that the matrix $\mathbf{K}_s$ has been augmented with the necessary number of rows and columns of zeros to match the size of the matrix $\mathbf{K}_B$. Alternatively, one could condense out the rotational degrees of freedom in $\mathbf{K}_B$ and directly assemble it with $\mathbf{K}_s$.

Essential to the above formulation is the implementation of a routine for the derivation of the Green's functions. In the present study, the Kausel–Roësset Green's functions for disc loads in layered media [11.29] have been used. Because the response of the layered ground and of the embankment is frequency-dependent, the problem is formulated in the frequency domain. To this end, the loads are resolved into their frequency components by Fourier transformation and the responses are calculated for the individual frequencies. The final time-domain responses are then calculated by inverse Fourier transformation.

11.5.1. Simulations and comparisons

The numerical model VibTrain [11.26] was used to simulate the response of the track to the passages of X-2000 trains at different speeds. The estimated soil parameters, which are consistent with the dynamic strains for low and high train speeds, are given in Figs 11.6 and 11.8. The embankment's bending rigidity EI for such speeds was estimated at 200 MN m^2 and 80 MN m^2, respectively, and its mass per unit length was taken as 10 800 kg/m.

Figure 11.1(b) shows the time history of the simulated vertical track displacements (together with the recorded values) for a train passage at 70 km/h in the southbound direction. This speed is smaller than the lowest Rayleigh wave velocity in the ground, and represents a subseismic condition. Therefore, the displacement field is quasi-static. The figure shows that the numerical simulation reproduces the measurements not only with the correct amplitude but also in the details of the response.

Figure 11.1(c) shows the same comparison for a train speed of about 185 km/h. The speed is now greater than the shear wave velocity in the organic clay layer, and the transeismic regime is expected to develop. This is indeed verified by the time histories in the figure, which exhibit considerably larger motions in the embankment. The figure shows again that the proposed numerical model captures the salient features of the embankment response. In particular, it reproduces well the vibration tail that is generated behind the train, a phenomenon also observed in other theoretical solutions [11.20, 11.21]. Figure 11.1(d) displays the predicted (simulated) track deformations for a train speed of 235 km/h. This is the maximum response that can be expected at this site. Beyond this speed the response of the track starts to drop. Figure 11.1(e) shows the predicted track deformations for a train speed of 250 km/h.

11.6. Countermeasures

A central question in this problem is the design of a countervibration scheme. Two potential solutions can be envisaged:

- strengthening the ground under the embankment
- stiffening the embankment.

These solutions have been tried in connection with train-induced ground vibration resulting from other excitation sources, such as rail roughness. Strengthening of the ground can be achieved by a variety of techniques, such as the use of lime–cement piles [11.30]. Stiffening of the embankment can be realized by either installing a concrete slab under the embankment or replacing the embankment by a concrete box girder, if a larger stiffness is required. The effectiveness of the latter approach for the test site was investigated with the help of the VibTrain simulation model. To this end, two alternative embankments with bending rigidities $EI = 800$ MN m^2 (medium stiffness) and $EI = 4000$ MN m^2 (high stiffness) were selected, and the results were compared with those for the existing embankment with $EI = 80$ MN m^2 (a representative stiffness accounting for the non-linearity corresponding to the displacements at 200 km/h train speed). These two alternatives roughly represent a 0.4 m thick concrete slab under the embankment and a 1.2 m high concrete box girder forming the main body of embankment.

Figure 11.10(a) displays the time histories of the track deformations for the three embankment cases (i.e. existing, medium stiffness and high stiffness). The simulations are for a northbound passage at $V = 30$ m/s (108 km/h). For consistency of comparison, the degraded soil parameters have been used. The figure reveals that the medium-stiffness and high-stiffness embankments may reduce the cyclic displacement of the track (defined as the average peak-to-peak displacement) by about 25% and 50%, respectively. The effectiveness of stiffening, however, improves as the train speed increases and enters the transeismic range. This can be verified by examining the responses of the same embankments for $V = 60$ m/s (216 km/h) in Fig. 11.10(b). Reduction of about 40% for the medium-stiffness embankment and 75% for the high-stiff embankment are achieved. It is also worth noting the shift from one-way to two-way oscillation in the track response as the stiffness increases. This may have an important implication for the degradation and liquefaction potential of the foundation material. It should also be noted that these calculations did not take account of the larger soil moduli that are expected to prevail at smaller deformations. Therefore, the actual reduction in the track/embankment response may be even larger than the above estimates.

The preceding results point to the potential of embankment stiffening as a counter vibration scheme for trains running near or in the transeismic regime. Plans for test runs to examine the performance of stiffened embankments are being made by several railway organizations. Further calibration of simulation codes such as VibTrain against these tests would provide a valuable calculational tool for high-speed trains. At the Ledsgård site, the test installation of lime–cement piles under

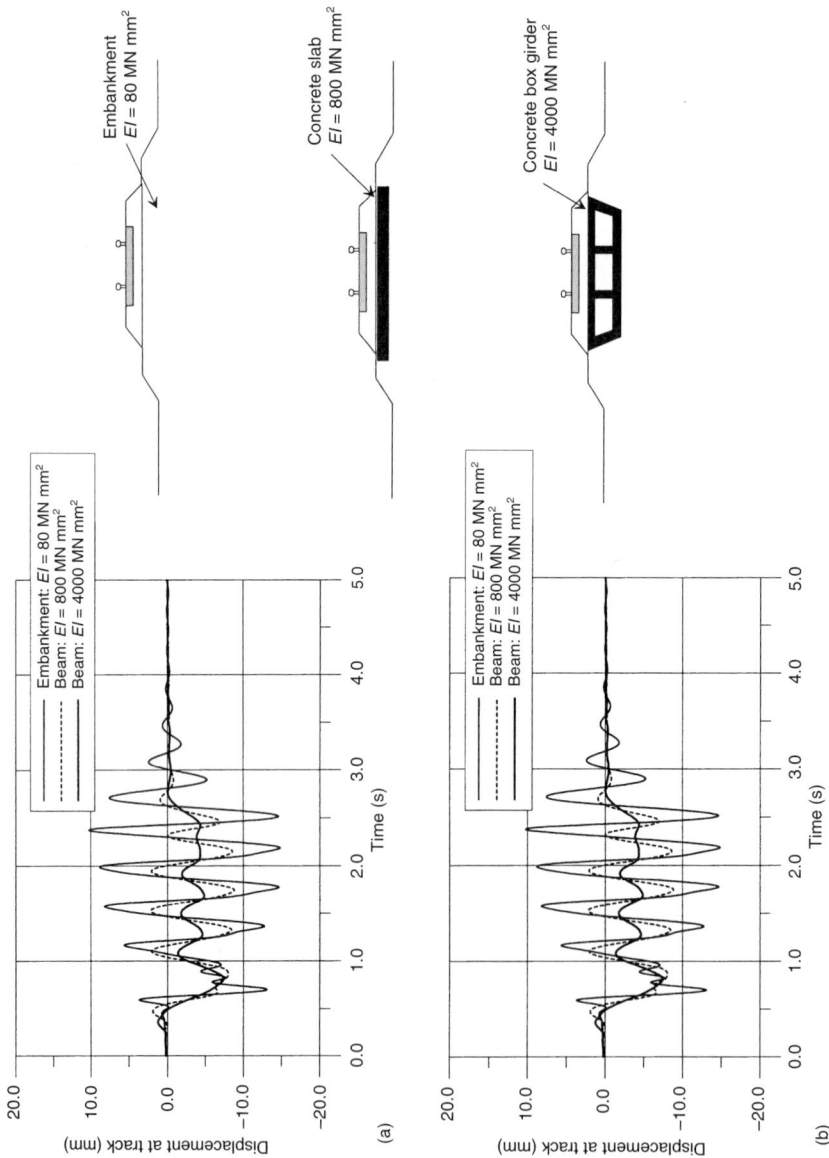

Fig. 11.10. Predicted effect of increased bending stiffness of the railway embankment as a countermeasure

the track is planned as a countermeasure. Extensive measurements to document the effect of these countermeasures will be performed.

11.7. Physical model

The rail/embankment/ground response to high-speed trains is easier to understand physically, when viewed in the frequency–wavenumber (ω–k) domain than in the time–space domain. Here the response $R(\omega, k)$ can be expressed as the product

$$R(\omega, k) = H(\omega, k)P(\omega, k) \tag{11.10}$$

where $H(\omega, k)$ is a site-specific dynamic transfer function and $P(\omega, k)$ is the train load excitation function.

Figure 11.11(a) illustrates the magnitude of $H(\omega, k)$ for the Ledsgård test site in a three-dimensional contour plot. The ridge in ω–k space (dotted curve) which forms the locus where $H(\omega, k)$ and thus the dynamic amplification have their highest values follows the dispersion curve for the first Rayleigh mode of the soil/embankment profile of the test site, as determined by SASW measurements. The points corresponding to freely propagating waves from simulated drop weight tests and to the natural frequency of the rail/embankment/ground system for stationary excitation, determined numerically with the VibTrain program, also lie on this locus. Also, the freely oscillating displacement tails observed behind the trains during the tests are close to this locus. The steepest tangent to the dispersion curve through the origin defines the slowest Rayleigh wave possible at this site. Its phase velocity is 51 m/s. The hatched region in the upper left part of the ω–k plane is a quasi-static region. Moving loads here will not set up waves; they will only give quasi-static displacements.* The border between this region and the 'dynamic region' is the locus of what may be termed the 'cut-off speeds'. Observe that although construction of $H(\omega, k)$ is a straightforward operation using a suitable numerical model, such as VibTrain as explained in this chapter, the plotted $H(\omega, k)$ function is only a schematic representation and is meant as an illustration of the features of the response.

The train load excitation function $P(\omega, k)$ is formed from the Fourier transform $P(k)$ of the load sequence $p(x)$, as shown in Fig. 11.1(a). When the train moves at speed S, $P(\omega, k)$ becomes

$$P(\omega, k) = P(k)\delta\left(k - \frac{1}{S}\omega\right) \tag{11.11}$$

where δ is the Dirac delta function. Figure 11.11(b) illustrates $P(\omega, k)$ for the X-2000 train when stationary, at a low speed and at a high speed. The marked peaks in the function correspond to wavelengths equal to the distances between the main loads in the train, i.e. the longer bogie spacings, the shorter bogie spacings etc. Other trains will have their peaks at other wavelengths and, when moving, at other frequencies.

---

*This is true if the periodicity of the sleepers is not taken into account – Ed.

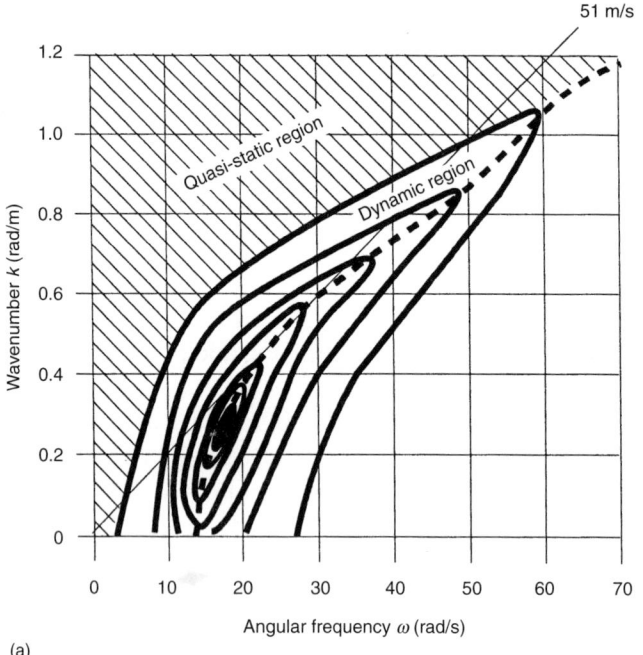

(a)

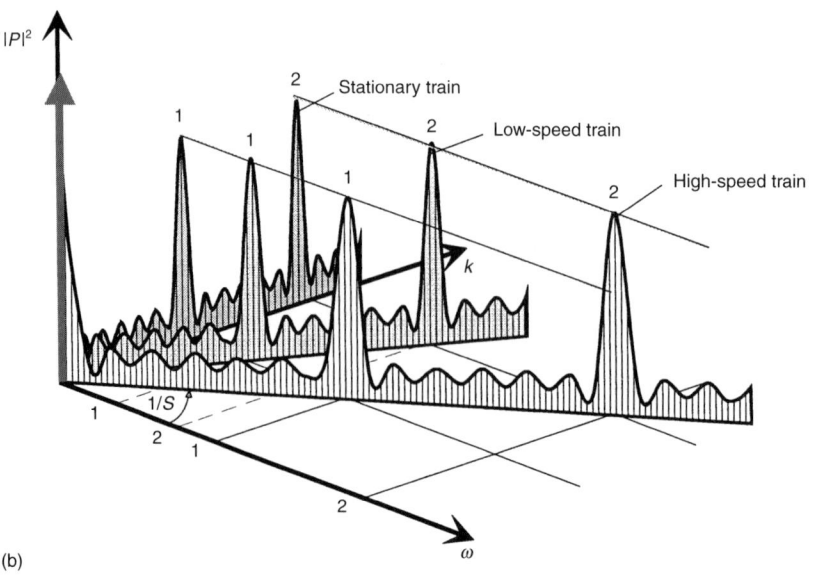

(b)

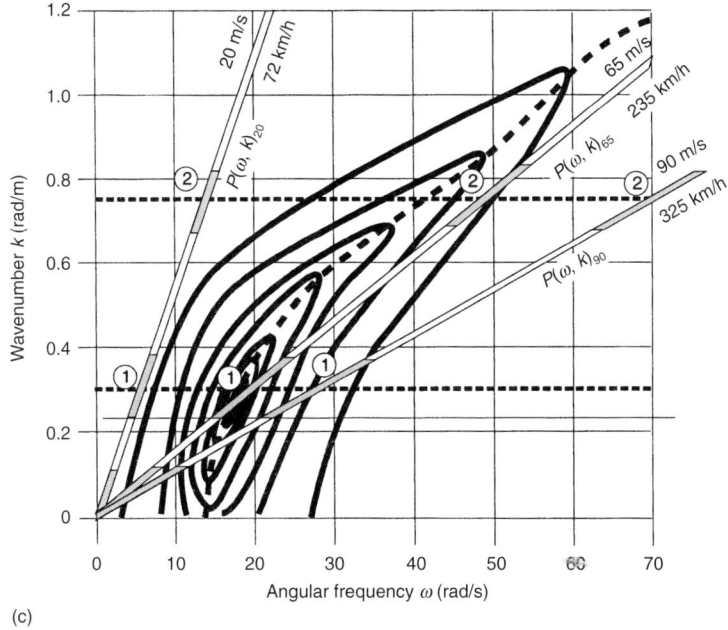

(c)

Fig. 11.11. Frequency–wavenumber domain representation of response to trains at subcritical, critical and supercritical speeds

In Fig. 11.11(c), the load functions $P(\omega, k)$ which were shown in a three-dimensional view in Fig. 11.11(b), are seen projected onto the ω–k plane for three train speeds 72, 235 and 325 km/h. They thus appear as straight lines. The ω–k regions where the $P(\omega, k)$ functions have their main peaks are visualized in Fig. 11.11(c) by shading of the lines in the appropriate regions. Figure 11.11(c) illustrates how $P(\omega, k)$ and the $H(\omega, k)$ shown in Fig. 11.11(a) are multiplied to form the response $R(\omega, k)$. As can be seen, $R(\omega, k)$ receives contributions only along the line $k = (1/\omega)S$, since $P(\omega, k)$ only has non-zero values here and is zero elsewhere. From the figure it can be observed that for low train speeds, i.e. around 70 km/h and below, no dynamic amplification takes place, since $P(\omega, k)$ falls entirely in the quasi-static region of $H(\omega, k)$. For increasing speed, the first and then the second peak in $P(\omega, k)$ start to enter into the dynamic region of $H(\omega, k)$, and the response $R(\omega, k)$ starts to gain dynamic amplification. The response has its maximum at about 235 km/h (65 m/s), where the first and largest peak in $H(\omega, k)$ nearly coincides with the peak in $H(\omega, k)$. A further increase in train speed brings the peaks in $P(\omega, k)$ out of the regions where $H(\omega, k)$ gives high dynamic amplification, and the response will thus decrease for a further increase in train speed.

Figure 11.11 illustrates the important feature that a given site does not have one unique critical speed. Which speed is the most critical depends on the characteristic bogie and axle spacings of the train relative to the dynamic transfer properties of the site, and may therefore vary between different trains. For example, observe that for the X-2000 train and the test site as shown in Fig. 11.11(c), the lowest Rayleigh wave phase velocity, 51 m/s, is not the critical speed, since it does not lead to a coincidence between peaks in $H(\omega, k)$ and $P(\omega, k)$. Instead, at about 235 km/h (65 m/s), the peaks are at their closest and the response reaches its maximum. Also, observe that the amount of dynamic amplification at the most unfavourable speed may be different for different trains.

11.8. Environmental vibration

For railway lines on soft soil, the issue of the critical train speed is not the only vibration-related problem that appears. Trains running below the critical speed may still produce low-frequency vibration that is annoying to people living and working in buildings along the track, out to distances of more than 100–200 m. Heavy freight trains and trains at high speed pose the major problem.

The mechanisms of how low-frequency vibrations in soft soils are generated by trains below the critical speed are complex and are far from fully understood. Factors such as rail corrugation, resonance in the railcar and locomotive suspension systems, the discontinuous support of the rail on sleepers, irregularities in the track and inhomogeneities in the embankment and supporting soil are probably among the most important ones. The fact that the vibration excitation is not understood and that the dynamic systems involved are generally complicated makes the use of sophisticated numerical models unreliable for vibration prediction. The prediction tools mostly used for planning of new lines, for upgrading of lines and for planning building development along railway lines are therefore mainly empirically based [11.31–11.33]. These models are founded on large amounts of vibration data recorded under various conditions.

Models of this type usually account for the following factors: ground condition, train type, train speed, condition of track, distance from track to building and building type. For example, the model in [11.33] is formulated as

$$V = V_\text{T} \left(\frac{S}{S_0}\right)^A \left(\frac{D}{D_0}\right)^{-B} F_\text{R} F_\text{B} \tag{11.12}$$

where V is the predicted vibration in the specified building, and V_T is a reference vibration, defined as the vibration on the ground surface for the specified ground conditions at a reference distance $D_0 = 15$ m from the track, caused by a train of the specified type running at a reference speed $S_0 = 70$ km/h on a track of 'standard' quality. S is the train speed, D is the distance from the track to the building, F_R is a factor accounting for the quality and type of the track and embankment, and F_B accounts for the building type and its ability to amplify vibration. The factors $V_\text{T}, A, B, F_\text{R}$ and F_B have been determined through regression analyses of a large amount of recorded vibration data, and are tabulated for various train types, ground condition categories, track–embankment types and building types. The above model takes into consideration the statistical variation in the background data shown by the regression analyses. On this basis, not only does the model give the best-estimate vibration predictions, but it can also give estimates at any desired confidence level. The model can predict either single vibration values or $\frac{1}{3}$ octave-band values.

To form the basis for empirical prediction models, vibration data and all associated data about the train, track, ground and building must be stored in a structured way in a database. Several databases of this kind exist in various countries. The one that forms the basis for the above prediction model is described in [11.33].

Humans do not perceive vibrations in the same way at all frequencies. For this reason, recorded vibrations must be filtered to account for the characteristics of human perception. The vibration has also to be averaged over one second. The filter and the averaging are defined in the international standard ISO 2631, Part 2 [11.34]. The ISO standard does not, however, define how non-stationary and repeated vibrations (such as from railways) are to be measured. Neither does it specify the acceptance levels for such vibrations. Recently, a national Norwegian standard, NS 8176 [11.35] has been issued. This standard recommends a method for measuring and quantifying vibrations from railways. The standard also defines vibration classes. Class C is the one that should be aimed at when designing new lines or new buildings. The vibration threshold for this class is 0.3 mm/s for the ISO-weighted average vibration. Class C is designed to correspond to the vibration level where 10% of a normal population feel the vibration as annoying or highly annoying. The basis for determining the vibration classes was an extensive sociological study on how people experience vibration in their homes. Germany has a corresponding standard [11.36], DIN 4150.

The vibration predicted by the empirical model described above can be frequency-weighted and averaged according to the ISO, DIN and NS standards. The database [11.33] also contains entries that register human response to the measured vibrations. The intention is that over the years, the database will provide an even better background for improved prediction models and definitions of vibration classes.

11.9. Conclusions

This chapter has presented results from an instrumented test where trains have been running near the critical speed over a soft-soil site. On the basis of the findings from the test, the physical mechanisms controlling the critical speed and the response of the track, embankment and soil have been studied. A numerical prediction tool is presented, and its ability to simulate the test results has been demonstrated. The dynamic properties of the soil and embankment materials have been discussed, and *in situ* and laboratory techniques have been described. Methods for performing measurements of the response to critical-speed trains and of the general vibrations from trains on soft ground, as well as the performance of track stiffening as a countervibration scheme, have been presented. Finally, prediction of vibrations in soft ground from trains running at under the critical speed, and the way in which such vibrations affect people have been briefly discussed.

11.10. Acknowledgements

The authors wish to thank the Swedish Rail Administration (Banverket) for their willingness to release results from the project 'High-speed lines on soft ground' for use in this chapter. The authors would also like to thank the teams from Banverket, the Royal Institute of Technology (KTH), the Swedish Geotechnical Institute (SGI), Jacobson and Widmark and Geo-Engineering AB, for their cooperation throughout the project. The financial support from the Norwegian Rail Administration (Jernbaneverket) and the Norwegian Geotechnical Institute (NGI) that made the writing of this chapter possible is gratefully acknowledged. We wish to thank our colleagues at NGI, H. Heyerdahl and J. K. Holme, for their support in producing the figures.

11.11. References

11.1. ADOLFSSON, K., ANDRÉASSON, B., BENGTSSON, P. E. and ZACKRISSON, P. High speed train X2000 on soft organic clay – measurements in Sweden. *Proceedings of the 12th European Conference on Soil Mechanics and Geotechnical Engineering*. Amsterdam, 1999, pp. 1713–1718.

11.2. AKI, K. and RICHARDS, P. G. *Quantitative Seismology*, Freeman, San Francisco, 1980.

11.3. KRAMER, S. L. *Geotechnical Earthquake Engineering*. Prentice Hall, Englewood Cliffs, 1996.

11.4. LARSSON, R. and MULABDIC, M. *Shear Moduli in Scandinavian Clays. Measurements of Initial Shear Modulus with Seismic Cone. Empirical Correlations for Initial Shear Modulus in Clay*. Swedish Geotechnical Institute, Linköping, 1991, Report 40.

11.5. ANDERSEN, K. H., KLEVEN, A., and HEIEN, D. Cyclic soil data for design of gravity structures. *Journal of Geotechnical Engineering Division of the ASCE*, 1988, **114**(5), 517–539.

11.6. LUNNE, T., ROBERTSON, P. K. and POWEL, J. J. M. *Cone Penetration Testing in Geotechnical Practice*. Spon, London, 1997.

11.7. NAZARIAN, S. and STOKOE, K. H. In situ shear wave velocities from spectral analysis of surface waves. *Proceedings of the 8th World Conference on Earthquake Engineering*. San Francisco, 1984, vol. 3, pp. 31–38.

11.8. MADSHUS, C. and WESTERDAHL, H. Surface wave measurements for construction control and maintenance planning of roads and airfields. *Proceedings of the International Conference on Bearing Capacity of Roads and Airfields*. Trondheim, 1990, vol. 1, pp. 233–243.

11.9. OFFSHORE TECHNOLOGY RESEARCH CENTER. *WinSASW. Data reduction program for spectral-analysis-of-surface-waves (SASW) tests*. Offshore Technology Research Center, University of Texas at Austin, Austin, 1993.

11.10. DYVIK, R. and MADSHUS, C. Lab measurements of G_{max} using bender elements. *Proceedings of the ASCE Convention, Advances in the Art of Testing Soils under Cyclic Conditions*. Detroit, 1985, pp. 186–196.

11.11. SEED, H. B. and IDRISS, I. M. Influence of soil conditions on ground motions during earthquakes. *Journal of the Soil Mechanics and Foundations Division of the* ASCE, 1965, **1**, 99–137.

11.12. SCHNABELL, P. B., LYSMER, J. and SEED, H. B. *SHAKE – A Computer Program for Earthquake Response Analysis of Horizontal Layered Sites*. University of California, Berkeley, 1972, Report EERC 72-12.

11.13. MADSHUS, C. Soil nonlinearity and its effect on the dynamic behaviour of offshore platform foundations. PhD thesis, University of Oslo, 1997.

11.14. COLE, J. and HUTH, J. Stress produced in a half-space by moving loads. *Journal of Applied Mechanics*, 1958, **25**, 433–436.

11.15. PAYTON, R. G. An application of the dynamic Betti–Rayleigh reciprocal theorem to moving point load in elastic media. *Quarterly Journal of Applied Mathematics*, 1964, **21**, 299–313.

11.16. DE BARROS, F. C. P. and LUCO, J. E. Response of a layered viscoelastic half-space to a moving point load. *Wave Motion*, 1994, **19**, 189–210.

11.17. VERRUIJT, A. Dynamics of soils with hysteretic damping. *Proceedings of the 12th European Conference on Soil Mechanics and Geotechnical Engineering*. Amsterdam, 1999, vol. 1, pp. 3–14.

11.18. AUBRY, D., CLOUTEAU, D. and BONNET, G. Modeling of wave propagation due to fixed or mobile dynamic sources. In: *Proceedings of Wave '94* (eds N. Chouw and G. Schmid). Berg, Bochum, 1994, pp. 79–93.

11.19. KRYLOV, V. V. Generation of ground vibrations from superfast trains. *Applied Acoustics*, 1995, **44**, 149–164.

11.20. DIETERMAN, H. A. and METRIKINE, A. V. Steady-state displacements of a beam on an elastic half-space due to a uniformly moving constant load. *European Journal of Mechanics, A/Solids*, 1997, **16**(2), 295–306.

11.21. LIEB, M. and SUDRET, B. A fast algorithm for soil dynamics calculations by wavelet decomposition. *Archive of Applied Mechanics*, 1998, **68**, 147–157.

11.22. ANDREASSON, B. *High Speed Lines on Soft Ground: Numerical Analysis by the FLAC and FLAC3D Programmes*. Jacobson and Widmark AB, Gothenburg, Sweden, 1998, internal report.

11.23. KALINSKI, M. *Finite Difference Modelling of Ground Vibration from Trains*. Norwegian Geotechnical Institute, Oslo, 1999, Research Report 514063-1.

11.24. HALL, L. Frequency characteristics in railway traffic induced ground vibrations. *Proceedings of the 14th International Conference on Soil Mechanics and Foundation Engineering*. Hamburg, 1997, vol. 1, pp. 677–682.

11.25. FACCIOLI, F., PAOLUCCI, R. and VANINI, M. (eds). *TRISEE: 3D Site Effects and Soil–Structure Interaction in Earthquake and Vibration Risk Evaluation*. European Commission, Directorate General XII for Science, Research and Development, Brussels, 1999.

11.26. KAYNIA, A. M. *VibTrain: A Computer Code for Numerical Simulation of Train-induced Ground Vibration*. Norwegian Geotechnical Institute, Oslo, 1999.Report 514063-2.

11.27. KAYNIA, A. M., MADSHUS, C. and ZACKRISSON, P. Ground vibration from high speed train: prediction and countermeasure. *Journal of the Geotechnical and Geoenvironmental Engineering Division of the ASCE*, 2000, **126**(6), 531–537.

11.28. CHOPRA, A. K. *Dynamics of Structures*. Prentice Hall, Englewood Cliffs, 1995.

11.29. KAUSEL, E. and ROESSET, J. M. Stiffness matrices for layered soils. *Bulletin of the Seismological Society of Am*erica, 1981, **71**(6), 1743–1761.

11.30. KARLSRUD, K. General aspects of transportation infrastructure. *Proceedings of the 12th European Conference on Soil Mechanics and Geotechnical Engineering*. Amsterdam, 1999, vol. 1, pp. 17–30.

11.31. KURTZEIL, L. G. Ground-borne noise and vibration from underground rail systems. *Journal of Sound and Vibration*, 1979, **66**, 363–370.

11.32. ACKVA, J. and NIEDERMEYER, S. Ganzheitliches Ausbreitungsgesetz für erschütterungen aus dem schienenverkehr-ausgangslage und möglichkeiten der prognose. In: *Proceedings of Wave '94* (eds N. Chouw and G. Schmid). Berg, Bochum, 1994.

11.33. MADSHUS, C., BESSASON, B. and HÅRVIK, L. Prediction model for low frequency vibration from high speed railways on soft ground. *Journal of Sound and Vibration*, 1996, **193**(1), 195–203.

11.34. INTERNATIONAL ORGANIZATION FOR STANDARDIZATION. *Evaluation of Human*

Exposure to Whole-body Vibration. Part 2: Continuous and Shock-induced Vibration in Buildings (1 to 80 Hz). ISO, Geneva, 1989, ISO 2631-2.

11.35. NORWEGIAN STANDARDS ASSOCIATION. *Vibration and Shock. Measurement of Vibration in Buildings from Landbased Transport and Guidance to Evaluation of its Effects on Human Beings*. Norwegian Standards Association, Oslo, 1999 [English version], NS 8176.

11.36. Deutsches Institut für Normung. *Vibration in Buildings: Effects on Persons in Buildings*. DIN, Berlin, 1992, DIN 4150, Part 2 [in German].

12. Ground vibrations alongside tracks induced by high-speed trains: prediction and mitigation

H. Takemiya
Department of Environmental and Civil Engineering, Okayama University, Okayama, Japan

12.1. Introduction

The passage of high-speed trains induces very intense motions of short duration on the track and vibrations in the nearby ground. Excessive track vibrations should be avoided for reasons of safety of train operations. Further, the induced ground vibrations, even if not damaging to nearby structures, are annoying to residents alongside the track and may have unfavourable effects on high-technology facilities near the track.

Assessment and prediction of the vibration should be performed properly and effective countermeasures, if necessary, should be taken. This issue has so far been treated mostly using a statistical regression formula based on field measurements. However, in view of recent developments of high-speed operations, the extrapolation of old data may not be applicable, and a scientific approach should be taken by applying elastodynamic theory.

The problems of track dynamics have been dealt with using a model of a beam on Winkler springs. In the full theory, the ground has been idealized by either a full-space, a half-space or a layered system. In the case of a half-space, the wave field is predominantly governed by Rayleigh waves, so that when the speed of the moving load approaches the velocity of these waves, a large response results. In the case of a layer, on the other hand, dispersion appears and the wave field is governed by modal waves. The modal waves can be characterized by different wave speeds for different frequencies. The main contribution to the energy transmission comes from the modes that are most important in the particular situation. There exists a situation in which higher modes give a major contribution and the response changes drastically from that in a homogeneous half-space.

The speed of motion of the load is a crucial factor to the understanding of wave generation and propagation in the medium. Eason [12.1] investigated wave motions in a full space for a constant load and obtained the transient response, which indicates divergence at the shear wave speed of the medium. Gakenheimer and Miklowitz [12.2], noting the crucial role of the free surface of the ground, formulated a half-space ground response directly in the time domain for a point load by classifying the subsonic, transonic and supersonic states. The solution method was based on the application of the Laplace transform to the time coordinate and the Fourier transform to the horizontal space coordinates. The inverse transform was evaluated analytically by the Cagniard–de Hoop technique [12.3].

In the field of track dynamics, Dieterman and Mertrikine [12.4, 12.5] analysed the track–ground system with a beam–half-space ground model, and clarified the features of the dynamic interaction affected by the speed of motion of the load. Jones and Block [12.6] modelled the track by a beam, supported by springs and masses to represent the ballast. Kaynia *et al.* [12.7] used the discrete Green's function, using the thin-layer method proposed by Kausel and Roesset [12.8].

Another important aspect is the generation mechanism of the vibration at the source. The sleeper spacing and the train geometry produce pseudo-harmonic vibrations. From measurements on a Shinkansen line, Yoshioka [12.9] noted that the frequencies of vibration were primarily generated by the moving axle loads. He also validated the source modelling, using the time delays for successive actions of the force [12.10]. Taking the train geometry into account, Krylov [12.11–12.14] analysed the ensuing ground response. In the frequency domain, the loads thus obtained have been used to incorporate Lamb's solution [12.15] for a half-space. On the basis of the results, he pointed out the importance of the moving train load, leading to a large response when the train speed approaches the Rayleigh wave velocity.

An alternative idealization of the ground is to take a layered model. Takemiya and Goda [12.16] developed the direct time domain solution by applying the Laplace transform with respect to time. They performed parametric studies and presented information to understand the nearby ground response [12.17]. In the work of Takemiya and others [12.18–12.21], in order to account for the source mechanism, a quasi-static track load distribution with simultaneously oscillating harmonics was assumed. Recently, Sheng *et al.* [12.22] investigated a layered beam on a layered half-space subjected to a point moving load oscillating simultaneously. The vibration characteristics were interpreted for the situations below, near to and above the lowest ground wave speed. In view of the topography of a railway line, Takemiya and Goda [12.18–12.20] and Takemiya and Shiotsu [12.21] developed the FEM–BEM solution method for the analysis of an embankment-type track. A recent publication by Takemiya [12.23] confirmed the effective use of quasi-static loads for a wide range of train speeds.

When the Fourier transform technique is used for the solution, the procedure for the inverse Fourier transform from the wavenumber domain to the space domain may affect the accuracy of the solution. The discrete-wavenumber technique proposed by Bouchon and Aki [12.24] is straightforward, but care must be taken over the wavelengths and wavenumbers to be used. Grundmann *et al.* [12.25] applied the wavelet decomposition scheme, which was proposed by Lieb and Sudret [12.26], effectively.

Comparison with measurements shows the usefulness of computer simulation. Jones and Block [12.6] investigated the track–ground system for heavy freight trains and noted that the quasi-static track deflection dominates the response in the lower frequency range, while the dynamic vibration is important at higher frequencies. Kaynia *et al.*[12.7], taking account of the action of forces due to the train geometry, compared the track displacement and found that it matched well the measured value containing very low frequencies only. Active incorporation of the measurement data has been performed by Takemiya *et al.* [12.20], who attempted vibration prediction for a Shinkansen train. The measurement data in the range up to relatively high frequencies corresponding to the sleeper spacing have been used for solving the inverse problem.

Vibration mitigation procedures have long been investigated by many people in order to reduce the excessive vibrations at source or at receiver locations. Open or filled-in trenches or concrete walls are typical measures traditionally used. In the case of a wave-impeding block or barrier (WIB), Takemiya and others [12.26–12.29] have given a detailed interpretation of the mechanism.

In what follows, the solution method is formulated (Section 12.2). The three-dimensional (3-D) transmission of vibration in layered soils is first described for a moving load at a particular driving frequency. The method is based on the Fourier transform with respect to the horizontal coordinates and time, with the layer stiffness being taken into account. The rigorous solution in the transformed wavenumber–frequency domain is then obtained. The inverse Fourier transformation is performed for the wavenumber along the axis of motion and time. Next, a 2.5-dimensional (2.5-D) formulation of the finite-element method (FEM) that can be employed for irregular topographies of the track is presented. The coupling of the 3-D Green's function for layered soils and the 2.5-D FEM is developed to provide a tractable procedure for the analysis of the total system.

In Section 12.3, the fundamental features of the responses are interpreted from an analysis of a homogeneous layer on a rigid foundation. Attention is addressed to the effects of the speed of the load and the driving frequency. The interpretation is obtained by referring to the dispersion characteristics of the wave field concerned.

In Section 12.4, measurement data from a site at a Shinkansen line for wave propagation in layered soils are presented. A vibration prediction for the site in the form of an inverse problem based on the measurement data is presented. The effect of a WIB as a vibration mitigation measure is investigated.

Finally, Section 12.5 makes some concluding remarks.

12.2. Basic theory
12.2.1. Solution method for a moving load
The Fourier transform method is applied for the solution of the problem. The Fourier transform pairs are defined as follows:

$$\bar{g}(\xi_x, \xi_y, z, \omega) = \int_{-\infty}^{\infty}\int_{-\infty}^{\infty}\int_{-\infty}^{\infty} g(x, y, z, t)\exp(i\xi_x x)\exp(i\xi_y y)\exp(-i\omega t)\,dx\,dy\,dt \qquad (12.1)$$

$$g(x, y, z, t) = \frac{1}{8\pi^3}\int_{-\infty}^{\infty}\int_{-\infty}^{\infty}\int_{-\infty}^{\infty} \tilde{\bar{g}}(\xi_x, \xi_y, z, \omega)\exp(-i\xi_x x)\exp(-i\xi_y y)\exp(i\omega t)\,d\xi_x\,d\xi_y\,d\omega \qquad (12.2)$$

where the functions $\bar{g}$ and $\tilde{g}$ represent the transformed quantities with respect to time (indicated by the overbar) and with respect to space (indicated by the tilde). The symbol ω represents the frequency, ξ_x the x component of the wavenumber, and ξ_y is the y component of the wavenumber, so that $\xi = \sqrt{(\xi_x^2 + \xi_y^2)}$ defines the wavenumber in the plane of wave propagation, and i is the imaginary unit.

In the case of a load moving with a constant speed c and simultaneously oscillating harmonically with a frequency ω_0, the wave field is associated with specific wavenumbers along the axis of motion. Suppose a moving load that is described by

$$P(x, y, z, t) = F(x - ct, y, z)\exp(i\omega_0 t) \qquad (12.3)$$

where the argument $(x - ct)$ indicates the movement in the x direction at the velocity c. The Fourier transform of equation (12.3) is

$$\tilde{\bar{P}}(\xi_x, \xi_y, z, \omega) = \tilde{F}(\xi_x, \xi_y, z)\frac{2\pi}{c}\delta\left(\xi_x - \frac{\omega - \omega_0}{c}\right) \qquad (12.4)$$

where $\delta(x)$ is the Dirac delta function. Note that in equation (12.4) the transformed function $\tilde{\bar{P}}(\xi_x, \xi_y, z)$ is defined by the stationary load counterpart $\tilde{F}(\xi_x, \xi_y, z)$ and by the moving effect of $(2\pi/c)\delta[x - (\omega - \omega_0)/c]$. Using the expression for the moving load of equation (12.4), the linear response of an elastic track and ground is sought. Once the solution for the stationary forcing function, which is denoted by $u_{st}(\xi_x, \xi_y, z, \omega)$, has been found, then the solution for the moving force is obtained straightforwardly by taking into account the effect of the motion. The back transformation into the space and time domain is expressed by equation (12.2), with due consideration of the effect of motion, which restricts the wavenumbers along the x axis to the specific value $\xi_x = (\omega - \omega_0)/c$, as

$$u(x, y, z, t) = \frac{1}{4\pi^2 c}\int_{-\infty}^{\infty}\int_{-\infty}^{\infty} \tilde{u}_{st}\left(\frac{\omega - \omega_0}{c}, \xi_y, z, \omega\right)\exp\left(i\frac{\omega_0 x}{c}\right)\exp(-i\xi_y y) \times \exp\left[i\omega\left(t - \frac{x}{c}\right)\right]d\xi_y\,d\omega \qquad (12.5)$$

Note that the solution is only valid for the source moving towards the x-positive direction. The inverse operation back into the space and time domain is in general not expected to be performed analytically. Rather, we resort to an effective and efficient numerical approach. When the continuous inverse Fourier transform is replaced by the discrete inverse Fourier transform both for the wavenumber in the y direction (the wavenumber method) and for the frequency, equation (12.5) results

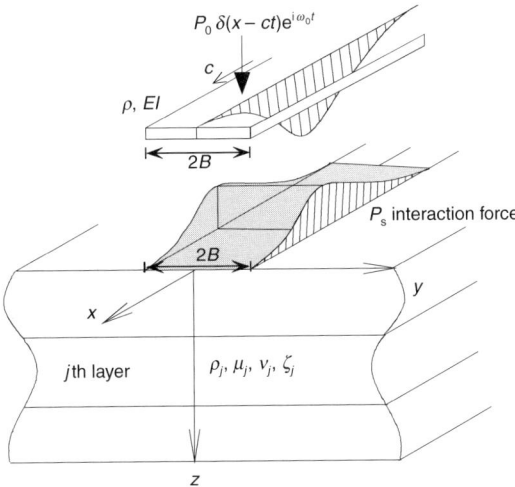

Fig. 12.1. Track–ground interaction

in the following:

$$u(x, y, z, t_m) = \frac{1}{cLT} \sum_{n=1}^{N} \sum_{k=0}^{K} \tilde{\bar{u}}_{st}\left(\frac{\omega_n - \omega_0}{c}, \xi_{y_k}, z\right) \times \exp\left(i\frac{\omega_0 x}{c}\right) \exp(-ik\Delta\xi_y\, y) \exp\left[i\omega_n\left(t_m - \frac{x}{c}\right)\right] \quad (12.6)$$

where the discrete wavenumbers are prescribed by $\xi_{yk} = 2\pi k/L$, with an increment of $\Delta\xi_y = 2\pi/L$, and with a fundamental wavelength L that should be chosen appropriately for the given frequency. The periodicity T in time gives a frequency increment of $\Delta\omega = 2\pi/T$, which leads to frequencies $\omega n = n\,\Delta\omega$ and $t_m = m\,\Delta t$. The inverse Fourier transformation with respect to frequency is carried out numerically by an efficient fast Fourier transform (FFT) algorithm.

12.2.2. Track–ground dynamic interaction [12.23]

Suppose that a railway track can be modelled by a Euler–Bernoulli beam resting on ballast on the site ground, for simplicity, as illustrated in Fig. 12.1. Consider a moving load P_0 simultaneously oscillating with frequency ω_0. The governing equation is then expressed by

$$EI\frac{\partial^4 u_b}{\partial x^4} + m\frac{\partial^2 u_b}{\partial t^2} + P_b(x,t) = P_0\,\delta(x - ct)e^{i\omega_0 t} \quad (12.7)$$

in which EI is the bending moment, m is the mass per unit length, and $P_b(x, t)$ is the soil reaction. The function $\delta(x)$ stands for the Dirac delta function. The Fourier integral transform (equation (12.1)) for equation (12.7) with respect to time and the space coordinate x results in

$$(EI\xi_x^4 - \omega^2 m)\tilde{\bar{u}}_b(\xi_x, \omega) + \tilde{\bar{P}}_b(\xi_x, \omega) = \frac{2\pi P_0}{c}\delta\left(\xi_x - \frac{\omega - \omega_0}{c}\right) \tag{12.8}$$

The soil reaction should be evaluated through an analysis of the dynamic interaction between the track and the subsoil (Fig. 12.1). For this purpose, we first take a vertical force $P_s(x, t)$ acting directly on the ground surface. The dynamic three-dimensional soil analysis (see Section 12.2.5) is referred to the associated displacement response. In the wavenumber–frequency domain we obtain

$$\tilde{\bar{u}}_s(\xi_x, \xi_y, z, \omega) = \tilde{\bar{G}}_{zz}(\xi_x, \xi_y, z, \omega)\tilde{\bar{P}}_s(\xi_x, \xi_y, z = 0, \omega) \tag{12.9}$$

where $G_{zz}(\xi_x, \xi_y, z, \omega)$ denotes the z-directional response of the transformed Green's function. In view of the one-dimensional motion (equation (12.7)) and the corresponding wavenumber ξ_x, we transform equation (12.9) for the surface in the two-dimensional ξ_x–ξ_y wavenumber domain to suit the counterpart quantity of the one-dimensional motion. The load distribution in the y direction is represented by $\Psi(y)$, which may be specified to be compatible with the one-dimensional equation (12.7) if it satisfies

$$\int_{-B}^{B} \Psi(y)\,dy = 1 \tag{12.10}$$

Then,

$$\tilde{\bar{u}}_s(\xi_x, y = 0, z = 0, \omega) = \tilde{\bar{G}}_s(\xi_x, y = 0, z = 0, \omega)\tilde{\bar{P}}_s(\xi_x, z = 0, \omega) \tag{12.11}$$

with

$$\tilde{\bar{G}}_s(\xi_x, y = 0, z = 0, \omega) = \int_{-\infty}^{\infty} \tilde{\bar{G}}_{zz}(\xi_x, \xi_y, z = 0, \omega)\tilde{\Psi}(\xi_y)\,d\xi_y \tag{12.12}$$

One approximation to $\Psi(y)$ is a uniform distribution over the beam width. Then,

$$\Psi(y) = \begin{cases} 1/2B; & B \geq |y| \\ 0; & B \leq |y| \end{cases} \tag{12.13}$$

$$\tilde{\Psi}(\xi_y) = \frac{\sin(\xi_y B)}{\xi_y B} \tag{12.14}$$

The displacement compatibility between the beam and the ground surface is specified by

$$\tilde{\bar{u}}_b(\xi_x, \omega) = \tilde{\bar{u}}_s(\xi_x, y = 0, z = 0, \omega) \tag{12.15}$$

and the interaction force equilibrium by

$$\tilde{\bar{P}}_b(\xi_x, \omega) = \tilde{\bar{P}}_s(\xi_x, y = 0, z = 0, \omega) \tag{12.16}$$

Therefore, solving equations (12.8) and (12.9) for the beam response, we obtain

$$\tilde{P}_s(\xi_x, z=0, \omega) = \frac{2\pi P_0}{c} \frac{\delta[\xi_z - (\omega - \omega_0)/c]}{(L_c \xi_z)^4 - (\omega/\Omega)^2 + 1} \tag{12.17}$$

The inverse Fourier transform into the space solution is

$$u_b(x,t) = \frac{1}{2\pi} \int_{-\infty}^{\infty} \frac{P_0}{c} \frac{\exp[-i(\omega - \omega_0)x/c]\exp(i\omega t)}{EI[(\omega - \omega_0)/c]^4 - \omega^2 m + \tilde{K}[(\omega - \omega_0)/c, \omega]} d\omega \tag{12.18}$$

in which a new parameter to define the ground stiffness,

$$\tilde{K}(\xi_x, \omega) = \frac{2B}{\tilde{G}_s(\xi_x, y=0, z=0, \omega)} \tag{12.19}$$

has been introduced for the inverse of the Green's function $\tilde{G}_s(\xi_x, y = 0, z = 0, \omega)$ for the soil.

Once the interaction force $P_s(\xi_x, y = 0, z = 0, \omega)$ is determined, then the ground motion at locations other than the interface at $y = 0, z = 0$ is obtained from

$$\tilde{u}_s(\xi_x, \xi_y, z, \omega) = \tilde{G}_z(\xi_x, \xi_y, z, \omega) \tilde{P}_s(\xi_x, z=0, \omega) \tilde{\Psi}(\xi_y) \tag{12.20}$$

where

$$\tilde{P}_s(\xi_x, y=0, z=0, \omega) = \frac{2\pi P_0}{c} \tilde{\Phi}(\xi_x, \omega) \delta\left(\xi_x - \frac{\omega - \omega_0}{c}\right) \tag{12.21}$$

$$\tilde{\Phi}(\xi_x, \omega) = \frac{\tilde{K}(\xi_x, \omega)}{EI\xi_x^4 - m\omega^2 + \tilde{K}(\xi_x, \omega)} \tag{12.22}$$

where the ground Green's function $\tilde{G}_z(\xi_x, \xi_y, z, \omega)$ is defined for the x-, y- and z-directional response components. The wave field is calculated for the in-plane and out-of-plane motions and the respective contributions are converted to Cartesian coordinates using equation (12.67). Therefore, the final results are functions of the wavenumbers ξ_x and ξ_y.

If the soil impedance is assumed to be constant, i.e. $K(\xi_x, \omega) = K$, then

$$\tilde{P}_s(\xi_x, z=0, \omega) = \frac{2\pi P_0}{c} \frac{\delta[\xi_x - (\omega - \omega_0)/c]}{(L_c \xi_x)^4 - (\omega/\Omega)^2 + 1} \tag{12.23}$$

where

$$L_c = \sqrt[4]{\frac{4EI}{K}} \tag{12.24}$$

$$\Omega = \sqrt{\frac{K}{m}} \tag{12.25}$$

For the frequency range $\omega/\Omega \ll 1$, the soil reaction is approximated by

$$\tilde{P}_s(\xi_x, y=0, z=0, \omega) = \frac{2\pi P_0}{c} \tilde{\Phi}(\xi_x) \delta\left(\xi_x - \frac{\omega - \omega_0}{c}\right) \tag{12.26}$$

$$\tilde{\Phi}(\xi_z) = \frac{4}{(L_c \xi_x)^4 + 4} \tag{12.27}$$

which describe the action of the moving force given by the reaction against the static beam deformation due to the load applied to the beam.

The ground response in the space–time domain is further transformed from equation (12.20), using equation (12.14), to

$$u_s(x, y, z, t) = \frac{1}{(2\pi)^2} \int_{-\infty}^{\infty} \int_{-\infty}^{\infty} G_s\left(\frac{\omega - \omega_0}{c}, \xi_y, z, \omega\right) \Psi(\xi_y) \exp(i\xi_y y) \times \\ \tilde{P}_s\left(\frac{\omega - \omega_0}{c}, z = 0, \omega\right) \exp\left(-i\frac{\omega - \omega_0}{c}\right) d\xi_y \exp(i\omega t) d\omega \tag{12.28}$$

The discrete Fourier transform is carried out following equation (12.6).

12.2.3. Modelling of a loading by train [12.20]

Consider a moving train comprising N cars. The loading due to the passage of cars at a particular point on the rail is described by

$$F_N(x - ct, y, z, t) = \sum_{n=0}^{N-1} F_1(x - ct - nL_t, y, z, t) \tag{12.29}$$

in which the function $F_1(x - ct - nL_t, y, z, t)$ indicates the contribution of a single car, evaluated from

$$F_1(x - ct, y, z, t) = [\Phi(x - ct) + \Phi(x - ct - a) + \Phi(x - ct - a - b) + \\ \Phi(x - ct - 2a - b) + \Phi(x - ct)] \times \\ \Psi(y)\delta(z) T \sum_j A_k(\omega_j) \exp(i\omega_j t) \tag{12.30}$$

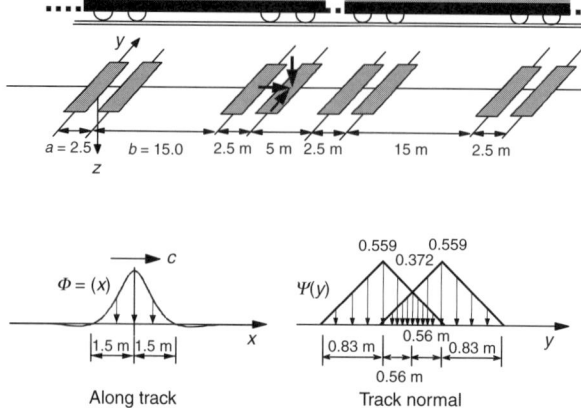

Fig. 12.2. *Assumptions about Shinkansen train loading*

Table 12.1. Shinkansen trains

Type	Weight per vehicle (kN)	Velocity (km/h)	Weight per axle (kN)
0	630	220–222	160
100	620	203–214	160
300	440	252–254	110

where the terms in the braces denote the successive axle loads distributed via the track onto the ground, prescribed by $\Phi(x_i)$ along the direction of motion. The quantities a and b are axle spacings and refer to Fig. 12.2. The term $\Psi(y)$ is related to the load distribution in the plane normal to the track. The notation $\delta(z)$ means the Dirac delta function for the track surface. The periodic terms with frequencies ω_j arise from the transfer of successive axle loads T (assumed equal; see Table 12.1) onto the ground via the sleepers. The spacing of the equidistant sleepers d and the speed of motion of the load c give rise to quasi-harmonic frequencies of

$$\omega_j = \pi C/jd \quad (j = 1, 2, 3, \ldots) \tag{12.31}$$

where the wavelengths are assumed to be even multiples of the sleeper spacing.

12.2.4. Ground vibration due to a quasi-static moving load

In the range of speeds of motion of the load below the critical wave velocity of the track, $C_{cr} = \sqrt[4]{4KEI/m^2}$ (see Kenny [12.30]), as we have seen in the solution above, the loading on the ground through the track may be approximated by equation (12.26). The assumption that a rail can be represented by a beam of bending rigidity EI on the Winkler springs of stiffness K representing the ballast mat leads to a quasi-static loading (equation (12.27)) in the frequency–wavenumber domain. The inverse Fourier transform gives

$$P_s(x, t) = P_0 \Phi(x - ct) \exp(i\omega_0 t) \tag{12.32}$$

$$\Phi(x) = \frac{1}{\sqrt{2}L_c} \exp\left(-\left|\frac{x}{L_c}\right|\right) \sin\left(\left|\frac{x}{L_c}\right| + \frac{\pi}{4}\right) \tag{12.33}$$

where L_c is given by the value of the major contributing characteristic width of $\Phi(x)$

$$q = \frac{3\pi L_c}{4} \tag{12.34}$$

at which equation (12.33) has its first zero. The distribution normal to the track, on the other hand, is approximated either by the uniform distribution of equation (12.13) or by two overlapping triangular distributions:

$$\Psi(y \pm y_n) = \begin{cases} \dfrac{1}{B}\left(1 - \left|\dfrac{y \pm y_0}{B}\right|\right) & |y \pm y_0| \le B \\ 0 & B \le |y \pm y_0| \end{cases} \tag{12.35}$$

in which y_0 gives the coordinate of the triangle apex, and B is its base width.

The Fourier transform of equation (12.29) can be expressed for a specific frequency ω_j, using equation (12.30), as

$$\tilde{\bar{F}}_N(\xi_x,\xi_y,z,\omega) = \frac{2\pi}{c}\tilde{\Phi}(\xi_x)\tilde{\Psi}(\xi_y)\chi(\xi_x)\sum_j A_k(\omega_j)\delta\left(\xi_x - \frac{\omega-|\omega_j|}{c}\right)\delta(z) \qquad (12.36)$$

where

$$\chi_N(\xi) = (1 + e^{ia\xi_x} + e^{i(a+b)\xi_x} + e^{i(2a+b)\xi_x})\frac{1-e^{iNL_T\xi_x}}{1-e^{iL_T\xi_x}} \qquad (12.37)$$

where $A_k(\omega_j)$ is the load intensity in the k direction at frequency ω_j. The Fourier transform of equation (12.35) is obtained as

$$\tilde{\Psi}(\xi_y) = \frac{4[1-\cos(B\xi_y)]}{(B\xi_y)^2}\cos(\xi_y y_0) \qquad (12.38)$$

The characteristic value is taken as $q = 1.5$ m in the later analysis in view of the conventional track structure. The coefficients $A_k(\omega_j)$ should be determined on the basis of matching the simulation results to measurement data (see, e.g., [12.20]). The response in the transformed domain can be solved for, as

$$\tilde{\bar{u}}(\xi_x,\xi_y,z,\omega) = G_z(\xi_x,\xi_y,z,\omega)\sum_j A(\omega_j)\delta\left(\xi_x - \frac{\omega-\omega_j}{c}\right) \qquad (12.39)$$

where $G_z(\xi_x,\xi_y,z,\omega)$ denotes the transformed-domain solution for the stationary load corresponding to equation (12.36). The computation of $G_z(\xi_x,\xi_y,z,\omega)$ is formulated in the next section. The response in the space and time domain is therefore obtained from the inverse transform of

$$u_s(x,y,z,t) = \frac{1}{8\pi^3}\sum_j A(\omega_j)\int_{-\infty}^{\infty}\int_{-\infty}^{\infty}\tilde{\bar{u}}\left(\frac{\omega-\omega_j}{c},\xi_{yk},z,\omega\right)\exp\left(i\frac{\omega_j x}{c}\right)\times$$
$$\exp(-i\xi_y y)\exp\left[i\omega\left(t-\frac{x}{c}\right)\right]d\xi_y\,d\omega \qquad (12.40)$$

The discretized solution then follows according to equation (12.6). Hence,

$$u_s(x,y,z,\omega) = \frac{1}{4\pi^2}\sum_{k=0}^{K}\sum_j \tilde{\bar{u}}_s\left(\frac{\omega-\omega_j}{c},\xi_{yk},z,\omega\right)A(\omega_j)\times$$
$$G_z\left(\frac{\omega-\omega_j}{c},\xi_{yk},z,\omega\right)\exp\left(-i\frac{\omega-\omega_j}{c}x\right)\Delta\xi_y \qquad (12.41)$$

where

$$\xi_{yk} = \frac{2\pi k}{L} \quad (k = 0, 1, 2, 3, \ldots, K)$$

where $\Delta\xi_y = 2\pi/L$, L is a specific length.

12.2.5. Elastodynamic analysis

12.2.5.1. Three-dimensional wave motions [12.17, 12.19]

An inhomogeneous layered medium for which the properties are constant within individual layers of depths h is defined by the density ρ and the complex Lamé constants $\lambda^c = \lambda(1 + 2\zeta i)$ and $\mu^c = \mu(1 + 2\zeta i)$, where ζ is the internal damping ratio of the focused layer. The governing equation of an elastic body under the force action f is described by

$$\mu^c \tilde{u}_{i,jj} + (\lambda^c + \mu^c)\tilde{u}_{j,ji} + \omega^2 \rho \tilde{u}_i + \tilde{f}_i = 0 \tag{12.42}$$

where the subscripts i and j denote the space coordinates and the comma convention is used for space derivatives.

The resolution of the three-dimensional wave equation into the SV–P and the SH wave fields is performed via the relationship

$$\begin{Bmatrix} \tilde{u}_x \\ \tilde{u}_y \\ \tilde{u}_z \end{Bmatrix} = \begin{bmatrix} i\xi_x/\xi & 0 & -i\xi_y/\xi \\ i\xi_y/\xi & 0 & i\xi_x/\xi \\ 0 & 1 & 0 \end{bmatrix} \begin{Bmatrix} \tilde{u}_1 \\ \tilde{u}_2 \\ \tilde{u}_3 \end{Bmatrix} \quad \text{or} \quad \tilde{u}_{x,y,z} = \mathbf{C}\tilde{u}_{1,2,3} \tag{12.43}$$

where the subscripts 1 and 2 correspond to the coordinates in the transformed domain. Similar expressions hold for the force vectors as well:

$$\tilde{f}_{x,y,z} = \mathbf{C}\tilde{f}_{1,2,3} \tag{12.44}$$

Hence, the decoupled in-plane motions comprising the SV and P waves are governed by

$$\mu^c \frac{d^2\tilde{u}_1}{dz^2} - (\lambda^c + 2\mu^c)k_\alpha^2 \tilde{u}_1 - (\lambda^c + \mu^c)\xi \frac{d\tilde{u}_2}{dz} = -\tilde{f}_1 \tag{12.45}$$

$$(\lambda^c + \mu^c)\xi \frac{d\tilde{u}_1}{dz} + (\lambda^c + 2\mu^c)\frac{d^2\tilde{u}_2}{dz^2} - \mu^c k_\beta^2 \tilde{u}_2 = -\tilde{f}_2 \tag{12.46}$$

The out-of-plane motion comprising the SH wave is governed by

$$\mu^c \frac{d^2\tilde{u}_3}{dz^2} - \mu^c k_\beta^2 \tilde{u}_3 = -\tilde{f}_3 \tag{12.47}$$

We define $V_P = \sqrt{[(\lambda^c + 2\mu^c)/\rho]}$ and $V_S = \sqrt{(\mu^c/\rho)}$ to denote the P-wave and the S-wave velocity, respectively. The notations

$$k_\alpha = \sqrt{\xi^2 - (\omega/V_P)^2}, \quad k_\beta = \sqrt{\xi^2 - (\omega/V_S)^2} \tag{12.48}$$

with $\xi^2 = \xi_x^2 + \xi_y^2$ have been introduced for defining the wavenumbers for the P-wave and the S-wave field, respectively.

The displacements obtained from the solution of equations (12.45) and (12.46) can be expressed as

$$\left\{ \begin{matrix} \tilde{u}_1 \\ \tilde{u}_2 \end{matrix} \right\} = \begin{bmatrix} -\xi & k_\beta & -\xi & k_\beta \\ k_\alpha & \xi & k_\alpha & -\xi \end{bmatrix} \begin{Bmatrix} A_P e^{-k_\alpha z} \\ A_{SV} e^{-k_\beta z} \\ B_P e^{-k_\alpha z} \\ B_{SV} e^{-k_\beta z} \end{Bmatrix} + \left\{ \begin{matrix} \tilde{f}_1/k_\alpha^2 (\lambda^c + 2\mu^c) \\ \tilde{f}_2/k_\beta^2 \mu^c \end{matrix} \right\} \tag{12.49}$$

and the associated stresses are

$$\left\{ \begin{matrix} \tilde{\sigma}_{12} \\ \tilde{\sigma}_{22} \end{matrix} \right\} = \mu^c \begin{bmatrix} 2\xi k_\alpha & -(\xi^2 + k_\beta^2) & -2\xi k_\alpha & \xi^2 + k_\beta^2 \\ \xi^2 + k_\beta^2 & -2\xi k_\beta^2 & \xi^2 + k_\beta^2 & -2\xi k_\beta \end{bmatrix} \begin{Bmatrix} A_P e^{-k_\alpha z} \\ A_{SV} e^{-k_\beta z} \\ B_P e^{-k_\alpha z} \\ B_{SV} e^{-k_\beta z} \end{Bmatrix} +$$

$$\left\{ \begin{matrix} -\xi \tilde{f}_2/k_\beta^2 \\ (\lambda^c/k_\alpha)^2 \tilde{f}_1/(\lambda^c + 2\mu^c) \end{matrix} \right\} \tag{12.50}$$

Similarly, from the solution of equation (12.47) for the SH wave field, the associated displacement and stress are

$$\tilde{u}_3 = A_{SH} e^{-k_\beta z} + B_{SH} e^{k_\beta z} + \tilde{f}_3/\mu^c k_\beta^2 \tag{12.51}$$

$$\tilde{\sigma}_{32} = \mu^c k_\beta (A_{SH} e^{-k_\beta z} + B_{SH} e^{k_\beta z}) \tag{12.52}$$

The unknown amplitudes A are introduced for the up-going waves and B for the down-going waves; both of these amplitudes are determined so as to fulfil the boundary conditions. The subscripts of the amplitudes A and B identify the waves concerned.

12.2.5.2. Multi-layered system

For the analysis of a multi-layered system, we apply the layer stiffness method. Eliminating the unknown amplitudes from the homogenous solutions given by equations (12.49) and (12.50) for the in-plane motion, as well as from equations (12.51) and (12.52) for the out-of-plane motion, we obtain the relationship between stresses and displacements for the individual layers. For instance, for the jth layer, of thickness h_j, we have for the in-plane motion

$$\tilde{\sigma}_j^T = \mathbf{K}_j^{SV-PV} \tilde{u}_j^T \tag{12.53}$$

where

$$\tilde{\sigma}_j^T = \{ \tilde{\sigma}_{12}^U \quad \tilde{\sigma}_{22}^U \quad \tilde{\sigma}_{12}^L \quad \tilde{\sigma}_{22}^L \}_j^T \tag{12.54}$$

$$\tilde{u}_j^T = \{ \tilde{u}_1^U \quad \tilde{u}_2^U \quad \tilde{u}_1^L \quad \tilde{u}_2^L \}_j^T \tag{12.55}$$

and for the out-of-plane motion,

$$\tilde{\sigma}_j^T = \mathbf{K}_j^{SV-PV} \tilde{u}_j^T \tag{12.56}$$

where

$$\tilde{\boldsymbol{\sigma}}_j^T = \{\tilde{\sigma}_{32}^U \quad \tilde{\sigma}_{32}^L\}_j^T \tag{12.57}$$

$$\tilde{\boldsymbol{u}}_j^T = \{\tilde{u}_3^U \quad \tilde{u}_3^D\}_j^T \tag{12.58}$$

where $\mathbf{K}_j^{SV\text{-}P}$ and $\mathbf{K}_j^{SH}$ define the layer stiffnesses for in-plane and out-of-plane motions, respectively.

$$\mathbf{K}_j^{SV\text{-}P} = \mu^c \begin{bmatrix} -2k_\alpha \xi & \xi^2 + k_\beta^2 & 2k_\alpha \xi & -(\xi^2 + k_\beta^2) \\ -(\xi^2 + k_\beta^2) & 2k_\beta \xi & -(\xi^2 + k_\beta^2) & 2k_\beta \xi \\ 2k_\alpha \xi e^{-k_\alpha h_j} & -(\xi^2 + k_\beta^2)e^{-k_\beta h_j} & -2k_\alpha \xi e^{k_\alpha h_j} & (\xi^2 + k_\beta^2)e^{k_\beta h_j} \\ (\xi^2 + k_\beta^2)e^{-k_\alpha h_j} & -2k_\beta e^{-k_\beta h_j} & (\xi^2 + k_\beta^2)e^{k_\alpha h_j} & 2k_\beta \xi e^{k_\beta h_j} \end{bmatrix} \times$$

$$\begin{bmatrix} -\xi & k_\beta & -\xi & k_\beta \\ -k_\alpha & \xi & k_\alpha & -\xi \\ -\xi e^{-k_\alpha h_j} & k_\beta e^{-k_\beta h_j} & -\xi e^{k_\alpha h_j} & k_\beta e^{k_\beta h_j} \\ -k_\alpha e^{-k_\alpha h_j} & \xi e^{-k_\beta h_j} & k_\alpha e^{k_\alpha h_j} & -\xi e^{k_\beta h_j} \end{bmatrix} \tag{12.59}$$

$$\mathbf{K}_j^{SV\text{-}P} = \mu^c \begin{bmatrix} k_\beta & -k_\beta \\ -k_\beta e^{-k_\beta h_j} & k_\beta e^{k_\beta h_j} \end{bmatrix}_j \begin{bmatrix} 1 & 1 \\ e^{-k_\beta h_j} & e^{k_\beta h_j} \end{bmatrix}_j^{-1} \tag{12.60}$$

In the half-space case, only the down-going waves exist for the radiation condition, so that only the elements associated with the down-going wave components with superscript U remain in equations (12.53) and (12.56). The details of the stiffness matrix are described in the Appendix (Section 12.6). Putting together equations (12.53) and (12.56) for the SV–P and SH wave fields and taking account of the continuity of displacements across layers,

$$\tilde{\boldsymbol{u}}_j^L = \tilde{\boldsymbol{u}}_{j+1}^U \tag{12.61}$$

and the corresponding stress equilibrium,

$$\tilde{\boldsymbol{\sigma}}_j^L - \tilde{\boldsymbol{\sigma}}_{j+1}^U = \tilde{\boldsymbol{f}}_j \tag{12.62}$$

we obtain the governing equation for all layers:

$$\mathbf{K}(\xi, \omega)\tilde{\boldsymbol{U}}(\xi, \omega) = \tilde{\boldsymbol{F}}(\xi, \omega) \tag{12.63}$$

where the elements are taken appropriately from the individual layer stiffness matrices for the SV–P and SH wave fields:

$$\mathbf{K}^{SV\text{-}P}(\xi, \omega) = \sum_j^{\text{Layers}} \mathbf{K}_j^{SV\text{-}P}(\xi, \omega) \tag{12.64}$$

$$\mathbf{K}^{SH}(\xi,\omega) = \sum_{j}^{\text{Layers}} \mathbf{K}_{j}^{SH}(\xi,\omega) \tag{12.65}$$

The solution of equation (12.63) is obtained as

$$\begin{Bmatrix} \tilde{\tilde{U}}_{1}(\xi,\omega) \\ \tilde{\tilde{U}}_{2}(\xi,\omega) \\ \tilde{\tilde{U}}_{3}(\xi,\omega) \end{Bmatrix} = \mathbf{K}(\xi,\omega)^{-1} \mathbf{F}(\xi,\omega)$$

$$= \begin{bmatrix} S_{11}(\xi,\omega) & S_{12}(\xi,\omega) & 0 \\ S_{21}(\xi,\omega) & S_{22}(\xi,\omega) & 0 \\ 0 & 0 & S_{33}(\xi,\omega) \end{bmatrix} \begin{Bmatrix} \tilde{\tilde{F}}_{1}(\xi,\omega) \\ \tilde{\tilde{F}}_{2}(\xi,\omega) \\ \tilde{\tilde{F}}_{3}(\xi,\omega) \end{Bmatrix} \tag{12.66}$$

which can be back transformed into the original *xyz* coordinates as

$$\begin{Bmatrix} \tilde{\tilde{U}}_{x}(\xi_{x},\xi_{y},z,\omega) \\ \tilde{\tilde{U}}_{y}(\xi_{x},\xi_{y},z,\omega) \\ \tilde{\tilde{U}}_{z}(\xi_{x},\xi_{y},z,\omega) \end{Bmatrix} = \mathbf{C}(\xi_{x},\xi_{y}) \begin{bmatrix} S_{11}(\xi,\omega) & S_{12}(\xi,\omega) & 0 \\ S_{21}(\xi,\omega) & S_{22}(\xi,\omega) & 0 \\ 0 & 0 & S_{33}(\xi,\omega) \end{bmatrix} \times$$

$$\mathbf{C}(\xi_{x},\xi_{y})^{-1} \begin{Bmatrix} \tilde{\tilde{F}}_{x}(\xi_{x},\xi_{y},z,\omega) \\ \tilde{\tilde{F}}_{y}(\xi_{x},\xi_{y},z,\omega) \\ \tilde{\tilde{F}}_{z}(\xi_{x},\xi_{y},z,\omega) \end{Bmatrix} \tag{12.67}$$

or

$$\mathbf{G}(\xi_{x},\xi_{y},\omega) = \mathbf{C}(\xi_{x},\xi_{y}) \mathbf{S}(\xi,\omega) \mathbf{C}(\xi_{x},\xi_{y})^{-1} \tag{12.68}$$

Hence the numerical Green's function is obtained for the prescribed load patterns.

12.2.5.3. 2.5-D finite-element method [12.19]

Suppose that an FEM discretization is carried out in a *yz* cross-section whose nodal elements have the three degrees of freedom, while the wavenumber representation is used along the *x* axis. The associated formulation starts with the transformation of the displacement as follows:

$$\tilde{\tilde{u}}(\xi_{x},z,\omega) = \int_{-\infty}^{\infty}\int_{-\infty}^{\infty} u(x,y,z,t) \exp(i\xi_{x}x) \exp(-i\omega t) \, dx \, dt \tag{12.69}$$

$$u(x,y,z,t) = \int_{-\infty}^{\infty}\int_{-\infty}^{\infty} \tilde{\tilde{u}}(\xi_{x},z,\omega) \exp(-i\xi_{x}x) \exp(i\omega t) \, d\xi_{x} \, d\omega \tag{12.70}$$

Then, following the conventional formulation of the FEM, we obtain the governing equation in which the arguments are indicated only for the wavenumber, dropping the frequency ω,

$$[\bar{\mathbf{K}}^{\text{FEM}}(\xi_x)]\bar{U}^{\text{FEM}}(\xi_x) = \bar{F}^{\text{FEM}}(\xi_x) \tag{12.71}$$

where

$$\bar{\mathbf{K}}^{\text{FEM}}(\xi_x) = \mathbf{K}^{\text{FEM}}(\xi_x) - \omega^2 \mathbf{M}^{\text{FEM}} \tag{12.72}$$

where $\mathbf{M}^{\text{FEM}}$, $\mathbf{K}^{\text{FEM}}(\xi_x)$ and $\bar{F}^{\text{FEM}}(\xi_x)$ denote the mass and stiffness matrices and the force vector, respectively, for individual elements based on an assumed interpolation function $\mathbf{N}(y, z)$, as follows:

$$\mathbf{M}^{\text{FEM}} = \sum_{\text{element}} \rho \iint \mathbf{N}^{\text{T}} \mathbf{N} \det |\mathbf{J}| \, dy_e \, dz_e \tag{12.73}$$

$$\mathbf{K}^{\text{FEM}}(\xi_x) = \sum_{\text{element}} \iint [\mathbf{B}^*(\xi_x)\mathbf{N}]^{\text{T}} \mathbf{D}[\mathbf{B}(\xi_x)\mathbf{N}] \det |\mathbf{J}| \, dy_e \, dz_e \tag{12.74}$$

$$\bar{\mathbf{K}}^{\text{FEM}}(\xi_x) = \sum_{\text{element}} \iint \mathbf{N}^{\text{T}} \bar{f}(\xi_x)\mathbf{N}] \det |\mathbf{J}| \, dy_e \, dz_e \tag{12.75}$$

with ρ being the density of elements and $\mathbf{B}$ being the strain–displacement relation matrix

$$\mathbf{B} = \begin{bmatrix} -i\xi_x & 0 & 0 & \partial/\partial y & 0 & \partial/\partial z \\ 0 & \partial/\partial y & 0 & -i\xi_x & \partial/\partial z & 0 \\ 0 & 0 & \partial/\partial z & 0 & \partial/\partial y & -i\xi_x \end{bmatrix}^{\text{T}} \tag{12.76}$$

$\mathbf{D}$ is the generalized Hooke matrix,

$$\mathbf{D} = \begin{bmatrix} \lambda^c + 2\mu^c & \lambda^c & \lambda^c & 0 & 0 & 0 \\ & \lambda^c + 2\mu^c & \lambda^c & 0 & 0 & 0 \\ & & \lambda^c + 2\mu^c & 0 & 0 & 0 \\ & & & \mu^c & 0 & 0 \\ & \text{sym} & & & \mu^c & 0 \\ & & & & & \mu^c \end{bmatrix} \tag{12.77}$$

and $\det|\mathbf{J}|$ indicates the Jacobian arising from the coordinate transformation from the elements to total system. The vector F is the external force. In equation (12.71), we differentiate between the displacements for free nodes (denoted by the subscript S) from those for interface boundary nodes (denoted by the subscript I). The matrix partitioning is performed accordingly. Hence,

$$\begin{bmatrix} \mathbf{K}_{\text{SS}}(\xi_x) & \mathbf{K}_{\text{SI}}(\xi_x) \\ \mathbf{K}_{\text{IS}}(\xi_x) & \mathbf{K}_{\text{II}}^{\text{FEM}}(\xi_x) \end{bmatrix} \begin{Bmatrix} \bar{U}_S(\xi_x) \\ \bar{U}_I^{\text{FEM}}(\xi_x) \end{Bmatrix} = \begin{Bmatrix} 0 \\ \bar{F}_I^{\text{FEM}}(\xi_x) \end{Bmatrix} + \begin{Bmatrix} \bar{F}_S^0(\xi_x) \\ 0 \end{Bmatrix} \tag{12.78}$$

The first term of the force vector on the right-hand side corresponds to the interface nodal forces and the second term to the external forces.

12.2.5.4. Finite element–boundary element method

A hybrid method can be used for the analysis by combining the finite-element (FE) method for the irregular track section and the boundary element (BE) method for the underlying layered soil. Suppose a linear displacement variation for both displacement and traction between adjacent FE and BE nodes. The conversion of the traction distribution into the FE nodal forces are performed, based on the assumption of a linear interoperation function $N_j(\eta)$ compatible with that for the FE, such that

$$\bar{F}^{\text{BEM}}(\xi_x) = [\mathbf{W}\bar{\mathbf{G}}(\xi_x, \omega)^{-1}]\bar{U}^{\text{BEM}}(\xi_x) = \mathbf{K}^{\text{BEM}}(\xi_x)\bar{U}^{\text{BEM}}(\xi_x) \tag{12.79}$$

where $\mathbf{K}^{\text{BEM}}(\xi_x)$ denotes the stiffness matrix, and the associated transformation matrix is given by

$$G(\xi_x, \omega) = \frac{1}{2\pi}\int_{-\infty}^{\infty} G(f\grave{\mathbf{l}}_x, \xi_y, \omega)N(\xi_x)\exp(-i\xi_y y)\,\mathrm{d}\xi_y \tag{12.80}$$

$$\mathbf{W} = \sum_{\lambda=1}^{\text{Elements}} \int_{f_{i\lambda}} \mathbf{N}_\lambda^T(\eta)\mathbf{N}_\lambda(\eta)\det\mathbf{J}\,\mathrm{d}\eta \tag{12.81}$$

where $f_{i\lambda}$ is the length of the λth interface element.

For the sake of convenience in the formulation of the coupling, we differentiate between nodes for the interface with the FEM domain, indicated by the subscript I and the others indicated by the subscript B. The matrix partitioning for the dynamic stiffness matrix follows accordingly:

$$\begin{bmatrix} \bar{\mathbf{K}}_{\text{II}}^{\text{BEM}}(\xi_x) & \bar{\mathbf{K}}_{\text{IB}}(\xi_x) \\ \bar{\mathbf{K}}_{\text{BI}}(\xi_x) & \bar{\mathbf{K}}_{\text{BB}}(\xi_x) \end{bmatrix} \begin{Bmatrix} \bar{U}_{\text{I}}^{\text{BEM}}(\xi_x) \\ \bar{U}_{\text{B}}(\xi_x) \end{Bmatrix} = \begin{Bmatrix} \bar{F}_{\text{I}}^{\text{BEM}}(\xi_x) \\ 0 \end{Bmatrix} + \begin{Bmatrix} \bar{F}_{\text{I}}^0(\xi_x) \\ \bar{F}_{\text{B}}^0(\xi_x) \end{Bmatrix} \tag{12.82}$$

where $\bar{F}_S^0$, $\bar{F}_I^0$ and $\bar{F}_B^0$ are the forces acting directly, $\bar{F}_I^{\text{FEM}}$ and $\bar{F}_I^{\text{BEM}}$ and are interaction forces. The coupled motion is established by applying the displacement compatibility

$$\bar{U}_{\text{I}}^{\text{FEM}}(\xi_x) = \bar{U}_{\text{I}}^{\text{BEM}}(\xi_x) \tag{12.83}$$

and the force equilibrium

$$\bar{F}_{\text{I}}^{\text{FEM}}(\xi_x) + \bar{F}_{\text{I}}^{\text{BEM}}(\xi_x) = \mathbf{0} \tag{12.84}$$

The governing equations for the FEM–BEM system are then expressed by

$$\begin{bmatrix} \mathbf{K}_{\text{SS}}(\xi_x) & \mathbf{K}_{\text{SI}}(\xi_x) & 0 \\ \mathbf{K}_{\text{IS}}(\xi_x) & \mathbf{K}_{\text{II}}^{\text{FEM}}(\xi_x) + \mathbf{K}_{\text{II}}^{\text{BEM}}(\xi_x) & \mathbf{K}_{\text{IB}}(\xi_x) \\ 0 & \mathbf{K}_{\text{BI}}(\xi_x) & \mathbf{K}_{\text{BB}}(\xi_x) \end{bmatrix} \begin{Bmatrix} \tilde{\bar{U}}_{\text{S}}(\xi_x) \\ \tilde{\bar{U}}_{\text{I}}(\xi_x) \\ \tilde{\bar{U}}_{\text{B}}(\xi_x) \end{Bmatrix} = \begin{Bmatrix} \tilde{\bar{F}}_{\text{S}}^0(\xi_x) \\ \tilde{\bar{F}}_{\text{I}}^0(\xi_x) \\ \tilde{\bar{F}}_{\text{B}}^0(\xi_x) \end{Bmatrix} \tag{12.85}$$

where the element matrices for the BEM (subscript B) are chosen to suit the FEM

GROUND VIBRATIONS: PREDICTION AND MITIGATION

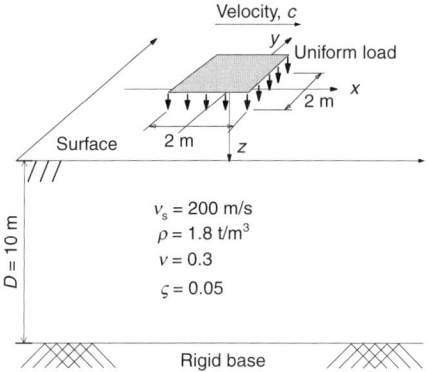

Fig. 12.3. *A uniform soil layer on a rigid base*

(subscript F) discretization. Solving equation (12.85) and taking the inverse transform according to equation (12.6), we obtain the displacement in the original coordinates.

12.3. Features of the response for a moving load

12.3.1. Dispersion characteristics of layers [12.16, 12.17]

For simplicity, the soil layer is assumed here to be uniform. A square moving load, oscillating at a frequency f_0 is applied to the surface (see Fig. 12.3). The wavenumber representation of the load is described by the form of equation (12.14) in both the x and the y directions. The soil properties are described by the mass density ρ, the shear velocity V_s, the Poisson's ratio ν and the internal damping δ. Although the layer depth D is fixed in the present computation, the results are presented in dimensionless form.

Since a layered model implies a dispersive nature of wave propagation, in contrast to a half-space model, the relationship between frequency f and wavenumber ξ will be investigated first. Figure 12.4 shows the dispersion characteristics of the modal frequencies versus the wavenumber in dimensionless form. The gradient of these curves corresponds to the velocity. Therefore, a constant moving load of velocity c is defined by a straight line passing through the origin. Depending on the speed of the load, a subdivision can be made into subsonic, transonic and supersonic ranges. These are indicated in Fig. 12.4. A moving load with a driving frequency f_0 is characterized by the lines $f = c\xi/2\pi \pm |f_0|$ in the figure. Since the driving frequency $\omega_0 = 2\pi f$ is related to the wavenumber by $\delta[\xi_x - (\omega - \omega_0)/c]$ in equation (12.4), the wave field may be strongly controlled by the wavenumber $\xi_x = (\omega - \omega_0)/c$. The crossing of the characteristic line with the modal curves means that those modal responses are predominantly induced by that moving speed. From Fig. 12.4, we may predict which modal waves will dominate the wave field. A constant, moving load or a moving load with a very low dimensionless frequency generates a Rayleigh wave. However, with a high dimensionless driving frequency, the higher modes may make a dominant contribution to the response. An increase in the speed of the moving load also increases the contribution from higher modes. This is the unique feature of the response of a layered ground, which is different from that of a half-space in which the Rayleigh wave dominates the wave field.

12.3.2. Transient responses [12.16, 12.17]

The time history of the transient response was computed to show the features of the response along the line of motion and at locations away from it. An interpretation is given with reference to the dispersion characteristic of the soil layer.

Figure 12.5 depicts the transient responses for different velocities of motion as normalized with respect to the shear velocity of the layer. The Mach number denoted by $M_\beta = c/V_s$ extends from the subsonic range of $M_\beta < 1$ to the transonic and supersonic range of $M_\beta > 1$. In the subsonic situation, for instance at $M_\beta = 0.5$, a symmetrically shaped large response appears, the peak of which is located at the time when the load passes by. Before it, the response has a similar nature to that of an impulse load at a fixed location, indicating the arrivals of the P wave, the Rayleigh wave and the S wave. In the transonic loading situation, a sharply rising impulse response appears at the time when the load passes by the location under consideration. Furthermore, in this range, after passage of the load, appreciable periodic motions appear. Figure 12.6 shows the Fourier transform of the time histories of Fig. 12.5. The subsonic case of $M_\beta = 0.5$ indicates very smooth frequency contents up to the dimensionless frequency $f(4D/V_s) = 1.0$, in contrast to the responses in the transonic range, which have their peak at the dimensionless frequency $f(4D/V_s) = 1.8$. It is interesting to note the result that at the offset

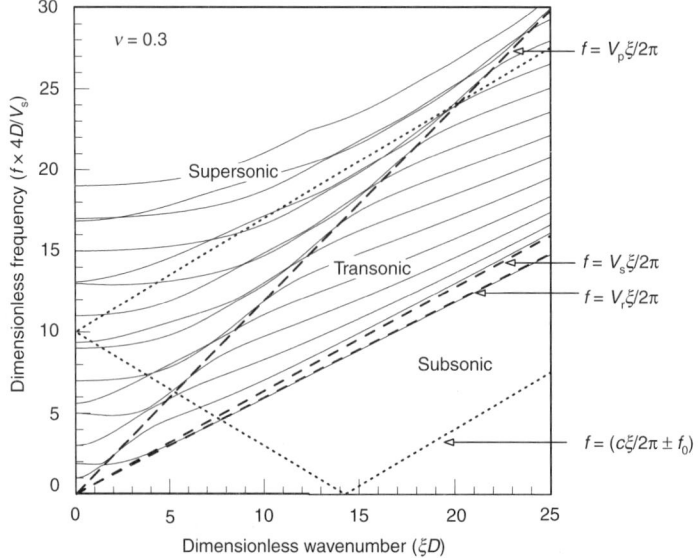

Fig. 12.4. Dispersion characteristics of a layer

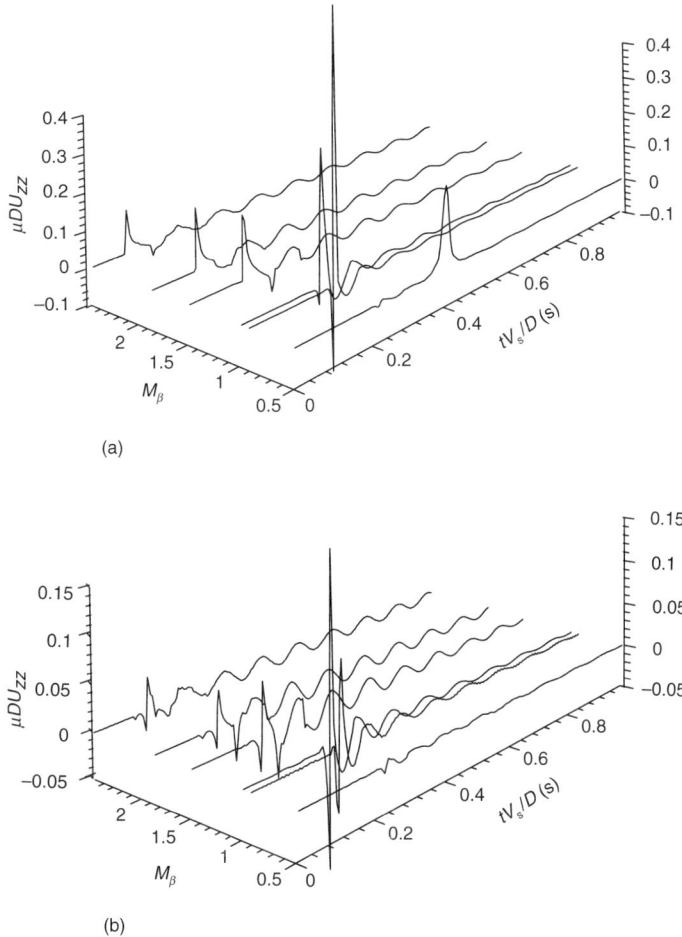

Fig. 12.5. Vertical surface response due to a moving vertical load (D = 10 m, ν = 0.3): (a) Ly = 0 m; (b) Ly = 10 m

location, $Ly = 10$ m, almost no frequency content exists below this frequency. In order to interpret this phenomenon, we must consider the characteristics of the layer with respect to the group velocity, shown in Fig. 12.7. The minimum group velocity yields the so-called Airy phase, at which the maximum wave energy is transferred. Two values can be read off as $f(4D/V_s) = 1.8$ and 5.0 in the dimensionless frequency. The first Airy phase is checked in the frequency–wavenumber relation. The lines corresponding to the speed of motion that pass through this Airy phase lie in the range of $M_\beta = 1.57$–1.87. The frequency at this phase is noted in Fig. 12.6.

The responses for a moving load accompanied by oscillation are drawn in Fig. 12.8 for the velocity ratio $M_\beta = 0.5$, with frequencies $f_0 = 5$, 10 and 12.5 Hz. The Doppler effect appears clearly before and after the load passes. Figure 12.9 shows the Fourier

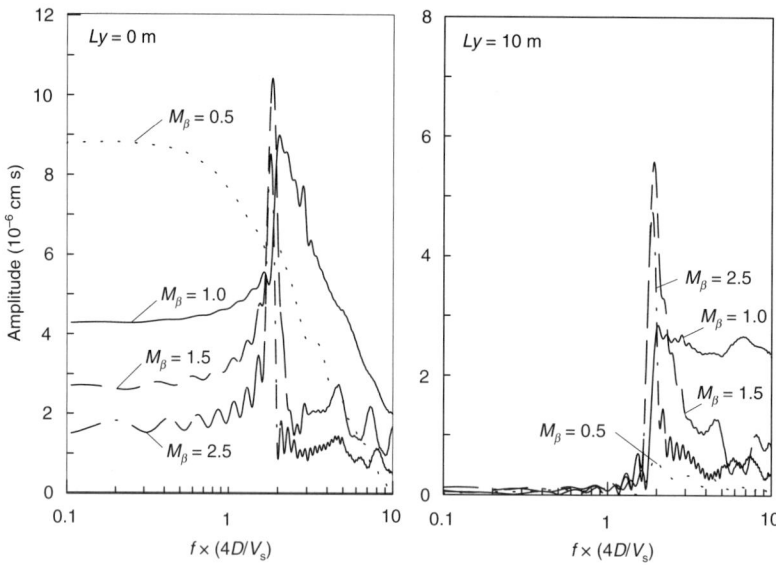

Fig. 12.6. Fourier amplitudes of the response time histories shown in Fig. 12.5

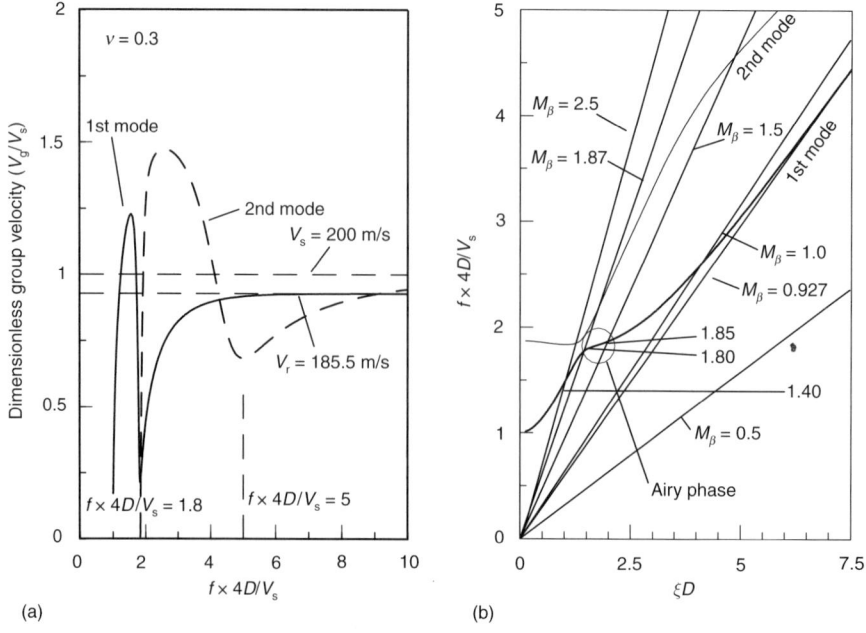

Fig. 12.7. Dispersion characteristics of a layered soil ($D = 10$ m, $\nu = 0.3$): (a) group velocity; (b) frequency

GROUND VIBRATIONS: PREDICTION AND MITIGATION

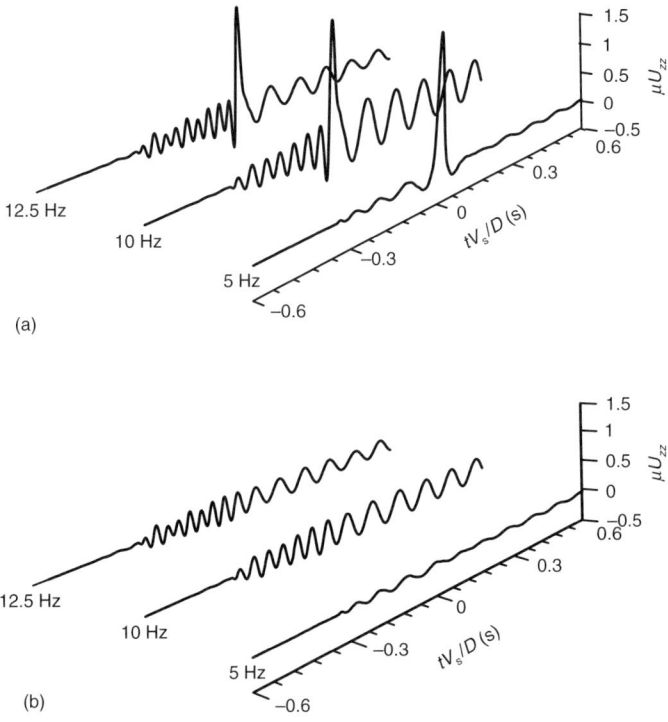

Fig. 12.8. Ground surface response due to an oscillating moving load (D = 10 m, ν = 0.3, M_β = 0.5): (a) Ly = 0 m; (b) Ly = 10 m

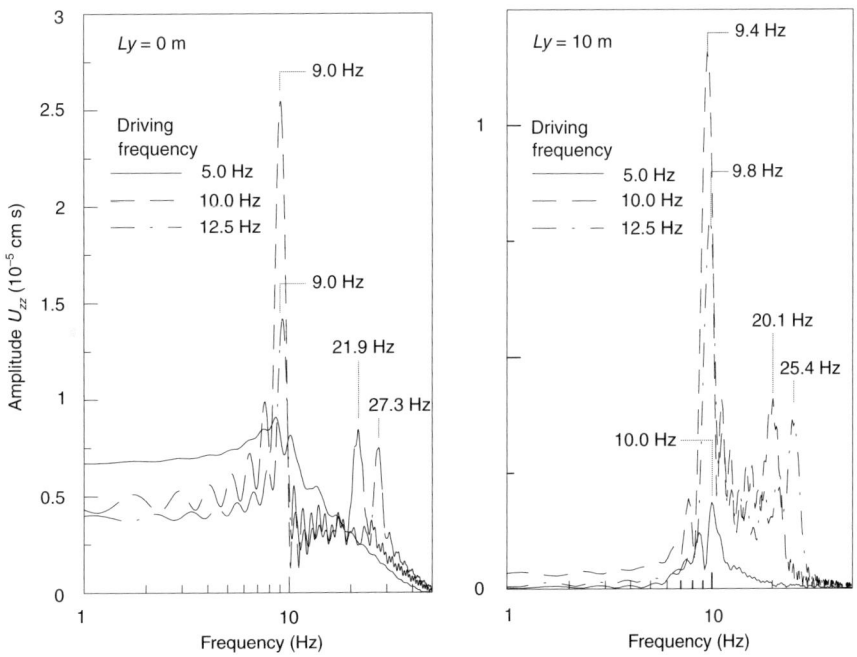

Fig. 12.9. Fourier amplitudes of the response time histories shown in Fig. 12.8

amplitudes of those responses. Note here that multiple peaks appear in the higher frequency range above the Airy phase owing to the driving frequency of the load. In order to interpret the response features, the characteristic lines of the moving load are depicted on the dispersion curves in Fig. 12.10. The intersections of these lines and curves result in significant response contributions. The peak responses around $f_0 = 21$ Hz and 27 Hz correspond to the first modal wave, which is close to the Rayleigh wave mode after taking account of the Doppler effect. The peak around 9 to 10 Hz is the first mode induced by the Airy phase.

Figure 12.11 shows the effect of the speed of the load when the driving frequency f_0 is fixed to 12.5 Hz. Note that as the speed increases, a sharper impulse response appears at the moment when the load passes by location under consideration. Figure 12.12 gives the Fourier amplitudes. The peak frequency, before the arrival of the load, i.e. 15.8 Hz for $c = 40$ m/s, 22.1 Hz for $c = 80$ m/s and 36.6 Hz for $c = 120$ m/s, corresponds to the intersection of the characteristic lines with the first-mode dispersion curve. When the driving frequency is high, for instance $f_0 = 12.5$ Hz, the lines for the moving load intersect the dispersion curve of the same mode more than once (see Fig. 12.10), so that this mode gives a response at different frequencies. The peak response at 10.5 Hz for $c = 40$ m/s, 9.6 Hz for $c = 120$ m/s, and 9.2 Hz for $c = 120$ m/s corresponds to the intersection of the characteristic lines with the first mode of the dispersion curves.

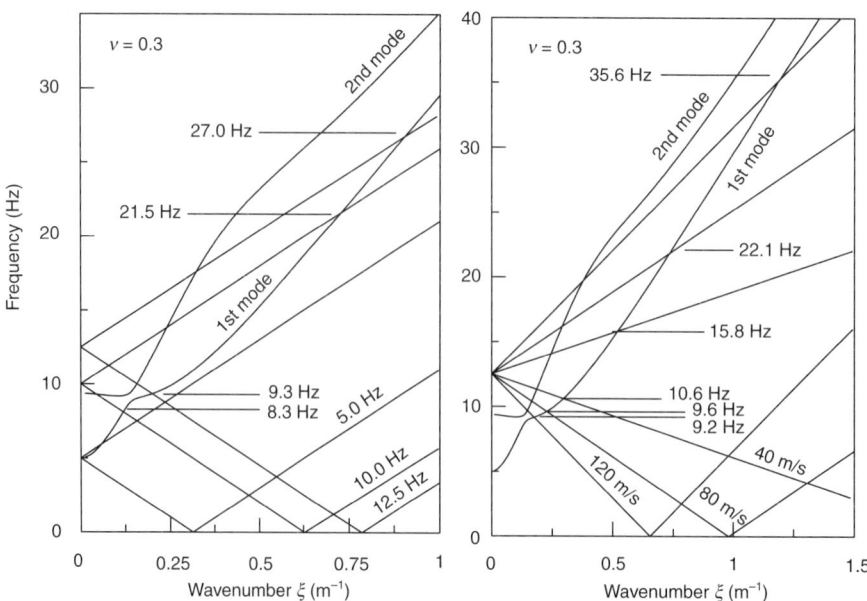

Fig. 12.10. Dispersion characteristics and characteristics of the motion: (a) driving frequency; (b) speed of motion

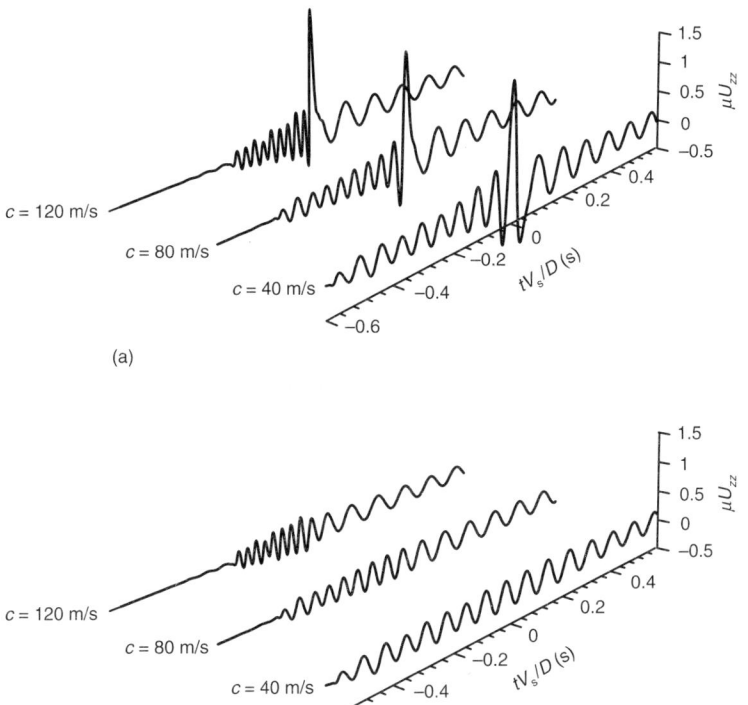

Fig. 12.11. Ground surface response to different load speeds (D = 10 m, ν = 0.3, f₀ = 12.5 Hz): (a) Ly = 0 m; (b) Ly = 10 m

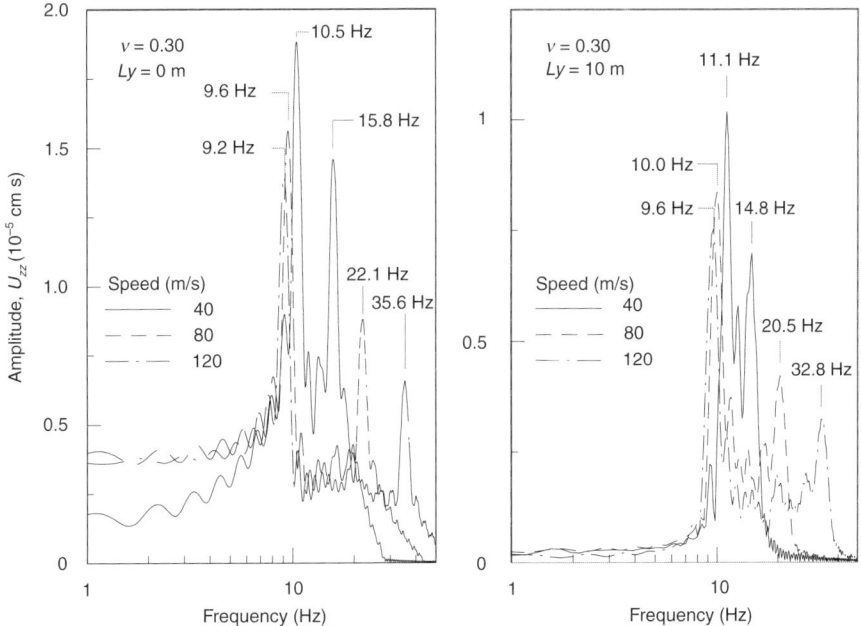

Fig. 12.12. Fourier amplitudes of the response time histories shown in Fig. 12.11

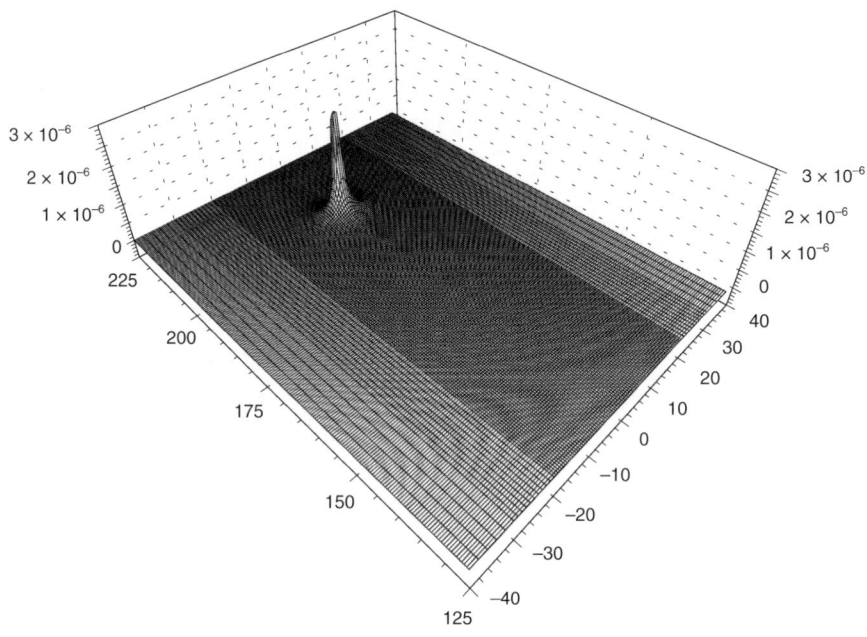

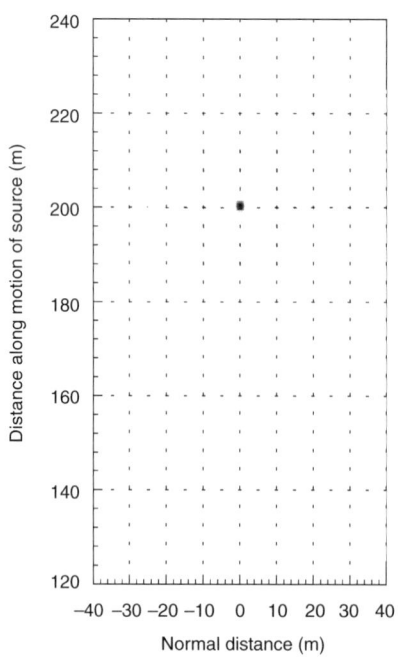

(a)

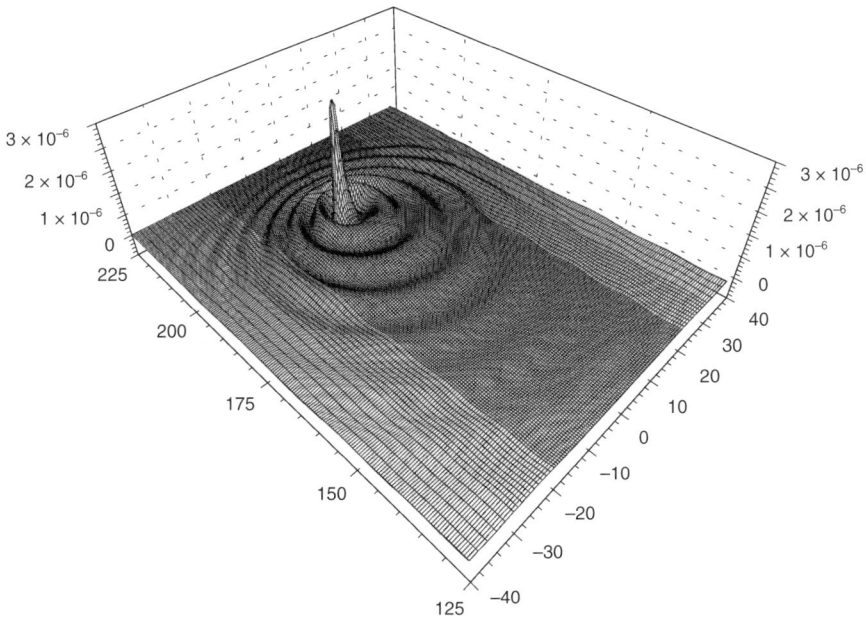

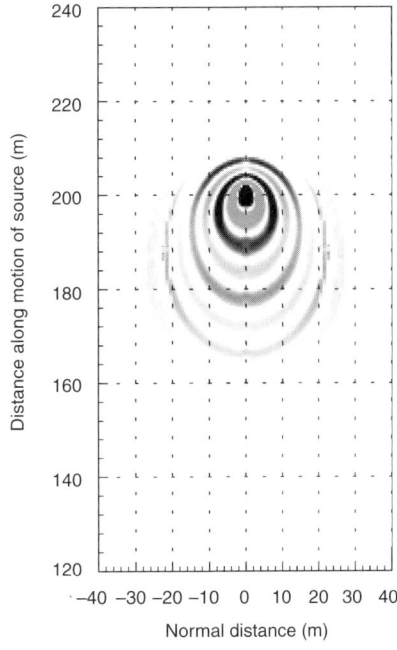

(b)

Fig. 12.13. Ground surface response to (a) a moving load in a subsonic range ($M_\beta = 0.5$, $f_0 = 0$ Hz) and (b) a moving oscillating load in a subsonic range ($M_\beta = 0.5$, $f_0 = 25$ Hz)

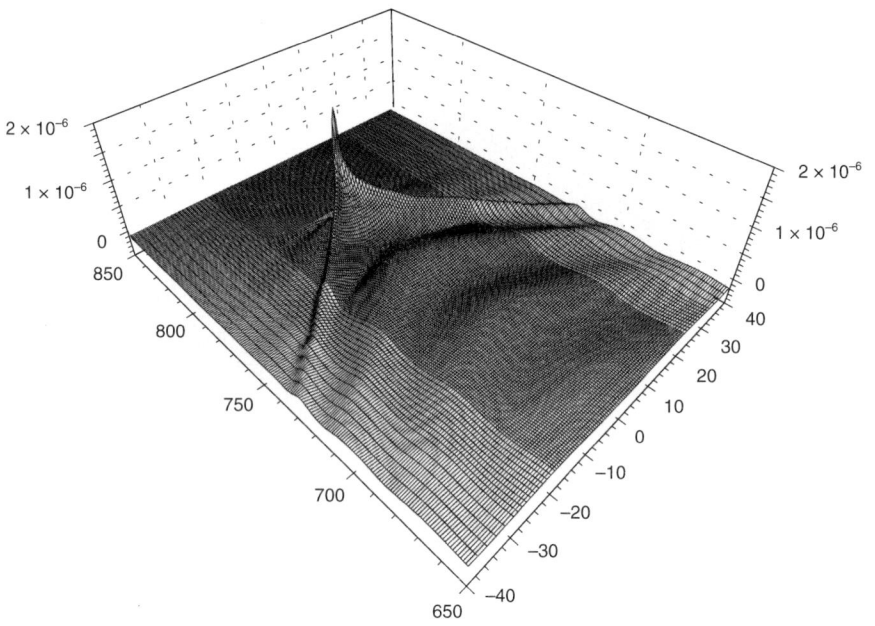

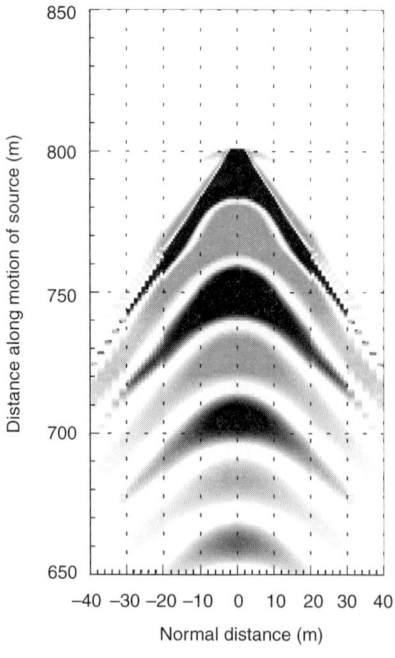

(a)

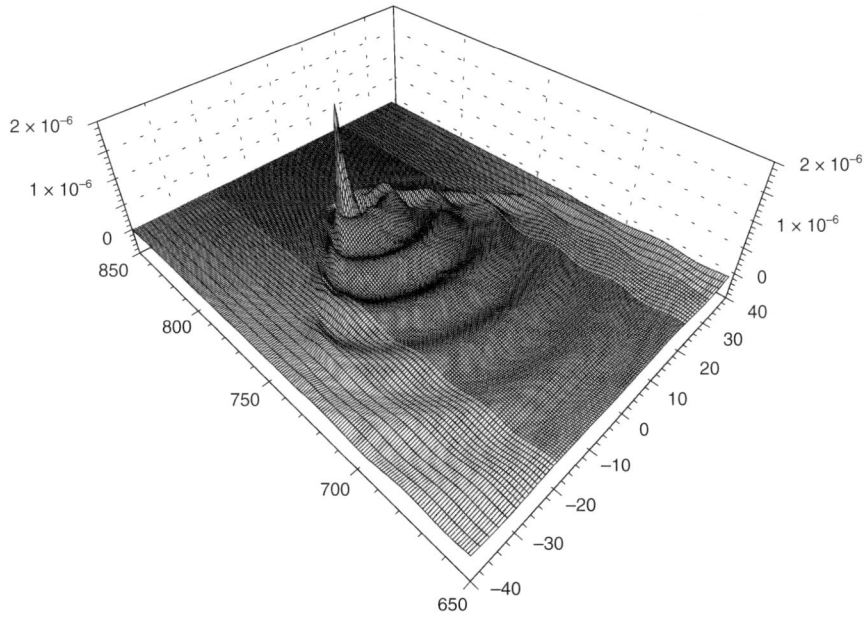

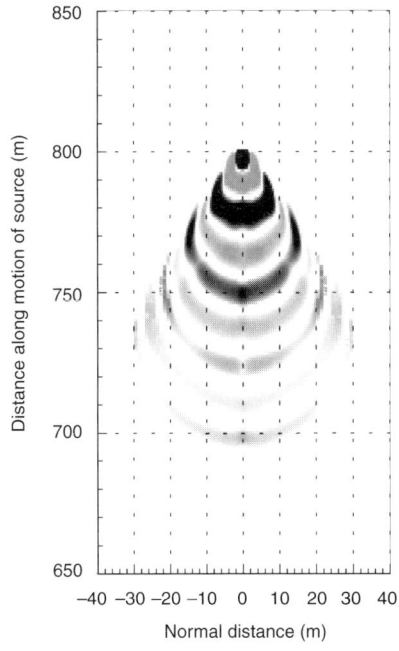

(b)

Fig. 12.14. Ground surface response to (a) a moving load in a supersonic range ($M_3 = 2.0$, $f_0 = 0$ Hz) and (b) a moving oscillating load ($M_3 = 2.0$, $f_0 = 25$ Hz)

12.3.3. Ground surface motions [12.23]

The speed of motion c of the load is normalized here by the soil shear velocity V_S of the ground and is indicated by a Mach value $M_\beta = c/V_S$. Of importance are the features of the wave on the ground surface ahead of the current load position and behind it.

In order to investigate the effect of the speed of the load, the driving frequency is first set to zero. The soil layer, when get a point moving load on the surface, generates waves propagation primarily at the Airy phase. Figures 12.13 and 12.14 illustrate the wave field in a three-dimensional view as well as in a contour view, for Mach values of $M_\beta = 0.5$ for the subsonic range and $M_\beta = 2.0$ for the supersonic range. In the case of $M_\beta = 0.5$ the response resembles that of an impulse load at a fixed position. However, in the case of $M_\beta = 2.0$ the response changes drastically, to show a shock wave field making a Mach cone behind the moving load. The angle is determined by the expression $\theta = \sin^{-1}(c_R/c)$, in which c_R denotes the Rayleigh wave velocity (see Fig. 12.14). In order to see the effect of driving frequency f_0 in conjunction with motion at speed c, the associated response for $f_0 = 25$ Hz is depicted. In the subsonic situation, the wave field clearly indicates the Doppler phenomenon, in such a way that the wavelengths are shortened in the area in front

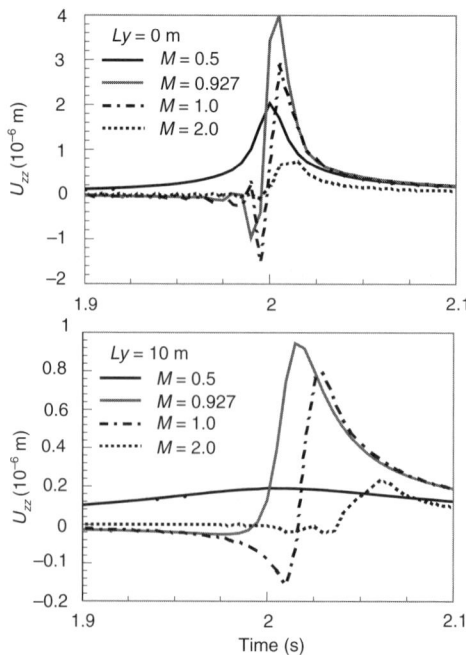

Fig. 12.15. Transient response of a track on a half-space to a unit moving load on the track, and the nearby ground response, $f_0 = 0$ Hz

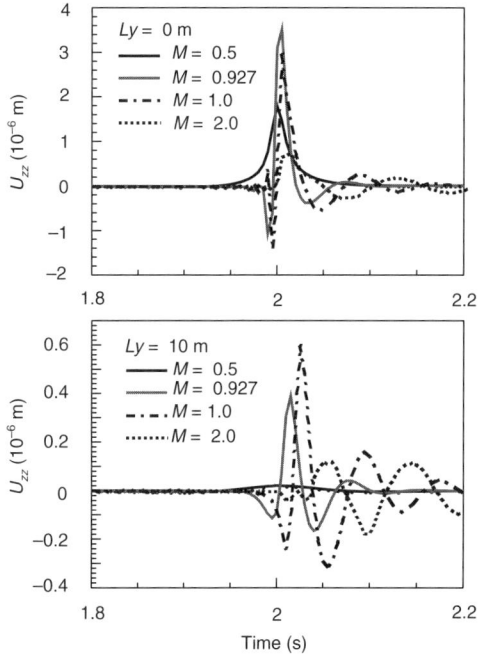

Fig. 12.16. Transient response of a track on a layer to a unit moving load on the track, and the nearby ground response, $f_0 = 0$ *Hz*

of the load, while they are lengthened in the area behind it. In the transonic and supersonic situations, the wave field exists only behind the load position. In these ranges, for $f_0 = 0$ Hz, a hyperbola-type wave field appears, while for $f_0 = 25$ Hz a parabola-type wave field appears. The wavelength in the former case may be interpreted with the help of Fig. 12.7(b), showing that the moving characteristic line has generated the first mode, which is enhanced by the Airy phase of the layer.

12.3.4. Response of track–ground system [12.23]

The track–ground model analysed here is described for the following structural dimensions: the bending rigidity EI is 1.3×10^5 kN m^2, the mass per unit length m is 7.5 t/m, the track width B is 6 m and the internal damping δ is 0.03 in the form of $E(1 + i\delta)$. The soil properties are given in Fig. 12.3. The ground is assumed to be either a half-space or a layer whose depth is fixed at $D = 20$ m.

Figure 12.15, for the half-space case, shows the vertical response of the track beam under the action of a constant force $P_0 = 1$ kN moving at a given speed. A drastic change of the response is noted with changes of the value of M_β. In case of $M_\beta = 0.5$, the track response appears like the static deformation, showing a symmetric nature before and after the arrival of the load at the location under consideration. However, in the case $M_\beta \geq 1.0$, a small response at high frequency appears just before the arrival of the load, a big impulsive response results at the time when the load passes by, and then a decaying response follows. Figure 12.16 corresponds to a layered ground. The track beam has a similar response before the arrival of the load, but afterwards a significant oscillation behaviour appears owing to wave trapping inside the layer. Figures 12.17 and 12.18 illustrate the results of a computation for a moving force with a driving frequency of $f_0 = 25$ Hz. In the response for $M_\beta < 1$ the Doppler effect can be clearly recognized before and after the load passes by, while for $M_\beta > 1$ the effect can be recognized only afterwards.

The investigation of the dynamic interaction between the track and the ground revealed that its effect on the response was negligibly small, so that a quasi-static force proportional to the static deformation may be used as an approximation.

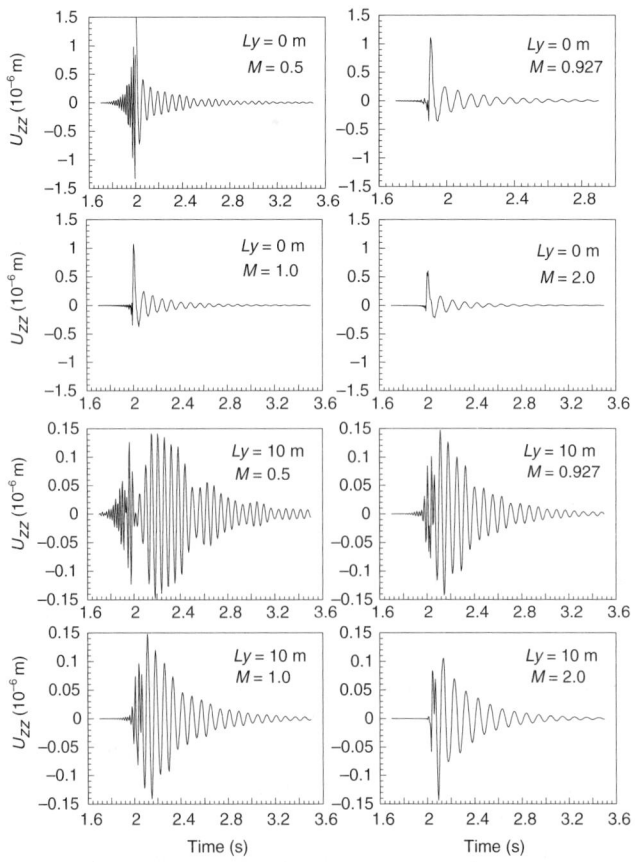

Fig. 12.17. Transient response of a track on a half-space to a unit moving load on the track, and the nearby ground response, $f_0 = 25$ Hz

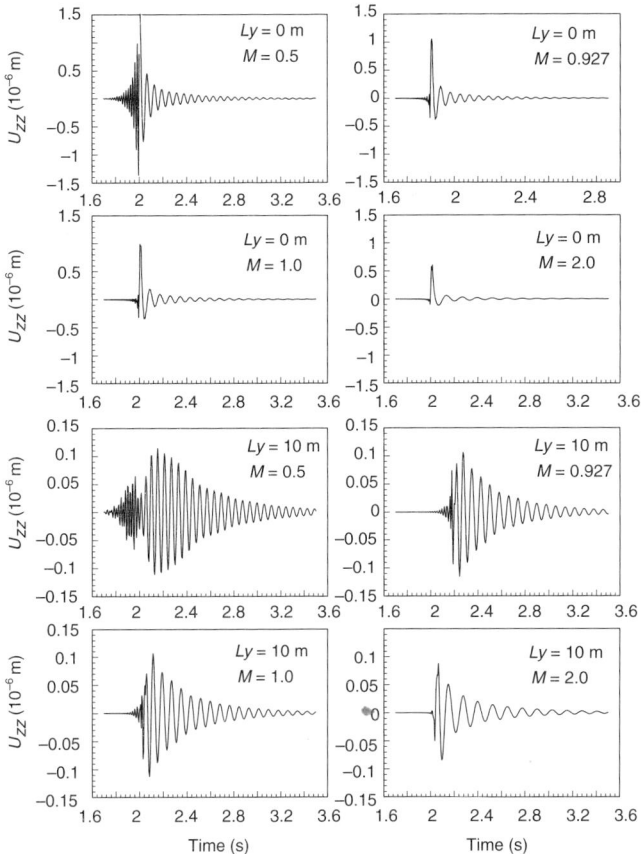

Fig. 12.18. Transient response of a track on a layer due to a unit moving load on the track, and the nearby ground response, $f_0 = 25$ Hz

12.4. Field measurements, theoretical prediction and mitigation

12.4.1. Measurement data [12.20, 12.21]

Since 1995, the Shinkansen trains have been classified into types 0, 100 and 300; their speed ranges are shown in Table 12.1. Field measurements of vibrations induced by these Shinkansen trains have been conducted. Here, the measured data are interpreted for a surface track on flat ground. Figure 12.19 shows a site with a layered soil where vibration measurements were conducted. The site condition is described in the figure, and the measurement locations are indicated by triangles. Figure 12.20 shows the recorded acceleration time histories, in which the top three records are from the down-bound train and the bottom two records are from the up-bound train. The time history on the ballast bed indicates very spiky vibrations with a pseudo-periodic nature. Actually, impact loads are imposed on the rails at the moment when wheels pass over them and are then transferred to the ballast track via the sleepers. The measurement records become spindle-shaped as the observation

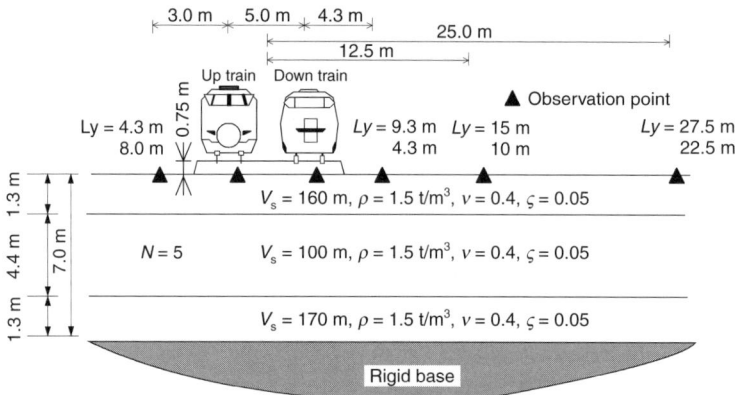

Fig. 12.19. *Conditions of flat track and ground*

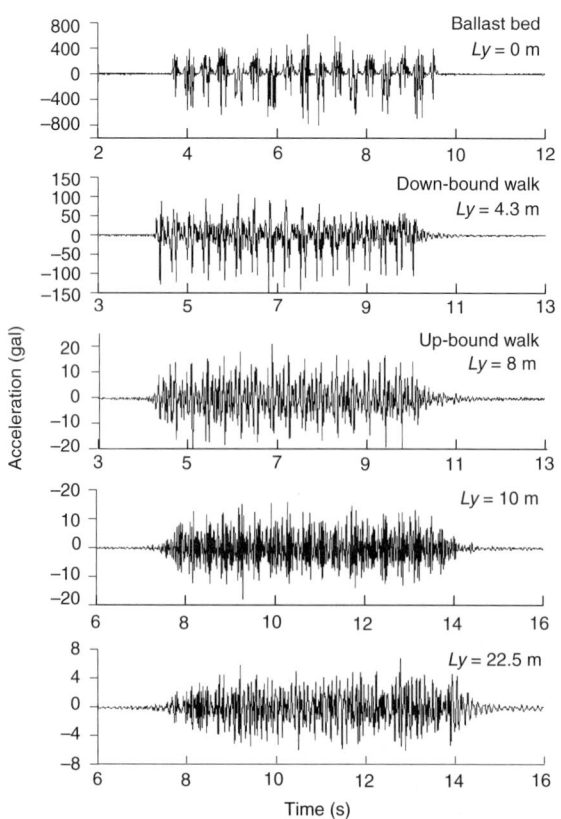

Fig. 12.20. *Measured acceleration time histories due to down-bound train (300), z component, c = 252 km/h*

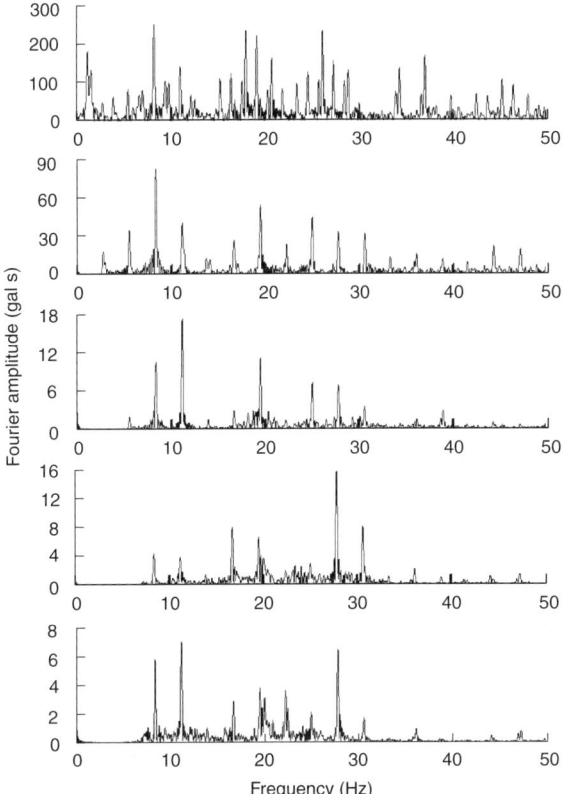

Fig. 12.21. Fourier amplitudes of measured accelerations

distance is increased in the transverse direction since the effect of all the moving axle loads arrive at the measurement location by wave propagation in the soil. For information in the frequency domain, the Fourier spectral density has been computed from the time histories and is depicted in Fig. 12.21. Note that at the ballast bed, peaks at very evenly spaced intervals appear over a wide range of frequencies. These peak frequencies almost coincide with those calculated from equation (12.31). It is interesting that the frequencies below several Hz and also higher frequencies disappear as the distance of the observation is increased. Therefore, the frequency spectrum contents are shifted in such a way that middle frequencies contents from 10 to 30 Hz are increased relative to other frequencies.

In order to investigate vibration transmission from the track outwards, the frequency transfer functions have been evaluated by dividing the Fourier transforms at distances of 10 m and 22.5 m by the Fourier transform at the track. The results are shown in Fig. 12.22. Here, a flat trend corresponds to the transmission of vibrations, while a sharp decline means no transmission. One can observe a clear cut-off frequency of vibration transmission/impediment at around 8 Hz. In the figure, regression lines are drawn; these were computed from the minimum mean square error with respect to the observed data, separately for the ranges below 8 Hz, between 8 and 20 Hz and above 20 Hz. Also depicted is an analytical simulation, the theoretical aspects of which are explained in the next section.

12.4.2. Wave propagation at the site [12.21]

The layered soils under the track shown in Fig. 12.19 have been analysed on the basis of the theory described in Section 12.2. In order to predict the dispersive wave field in these layered soils, the relationships between the wave velocity V and the frequency f, and between the frequency f and the wavenumber ξ were calculated and are depicted in Fig. 12.23. The group velocity gives information on the Airy phase at the local minimum values, which were detected at 8.0 and 19.5 Hz and are indicated in Fig. 12.23 by circles. According to the findings in Section 12.3, the lowest Airy phase frequency becomes the cut-off frequency, below which wave propagation is impeded in the layer.

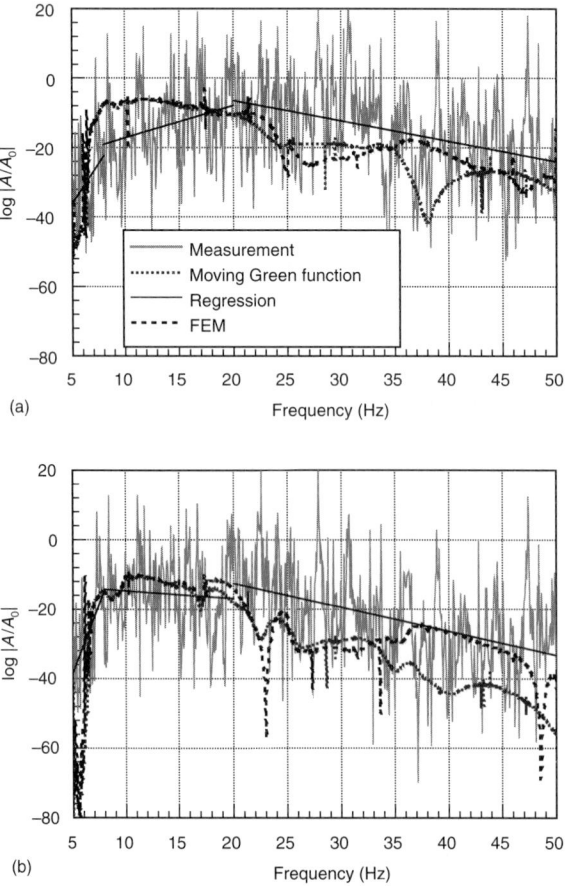

Fig. 12.22. Frequency transfer function: (a) 10 m and (b) 22.5 m from the loaded track

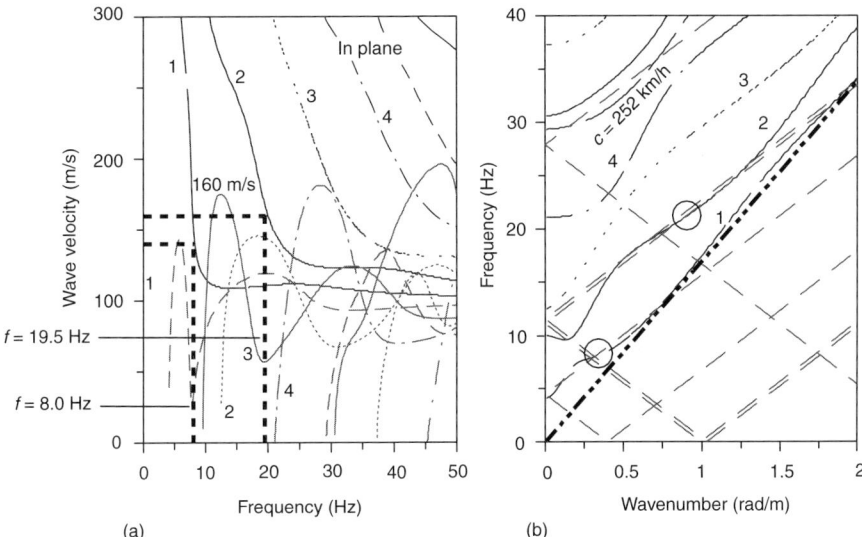

Fig. 12.23. Wave dispersion characteristics for a layered model: (a) wavenumber versus frequency; (b) frequency versus wavenumber. 1, first phase velocity; 2, second phase velocity; 3, third phase velocity; 4, fourth phase velocity; circles, Airy phase

The moving load, on the other hand, induces vibrations that depend on the velocity c and the driving frequency f_0. The characteristics of the moving source are described by $f = c\xi/2\pi \pm |f_0|$. When the characteristic lines pass through the Airy phase, then the associated modes can be induced as primary vibration modes. Referring to Fig. 12.23, at a train speed of 252 km/h the characteristic lines pass through the first and second Airy phases at frequencies of $f_0 = 4$ Hz and $f_0 = 12$ Hz. Then, there is a good possibility that the first mode will be induced in the former situation, and the second mode in the latter situation.

The theoretical interpretation will be given using the frequency transfer function to investigate the wave transmission and wave impediment. Adopting the models of equations (12.29) and (12.30) for the loading by the train and using the traction distribution of equations (12.33) and (12.34) with $q_c = 1.5$ m on the ground, we can simulate the ground motion induced by the motion of the train. Frequencies of generation were considered in the range below 50 Hz. These frequencies are obtained as $f_0 = 4.49, 8.33, 11.67, 14.58, 19.44$ and 29.17 Hz, for a sleeper spacing $d = 0.6$ m and a train speed of 252 km/h (70 m/s), for one axle load. Some additional frequencies, $f_0 = 26$ and 28 Hz, were also taken into account in view of the velocity characteristic curve of the load.

The load so defined was imposed on the surface of the ground and the ground vibrations were computed. The amplitude ratio of the frequency responses at distances of 10 m and 22.5 m with respect to that at the track was obtained and is drawn in Fig. 12.22 for comparison. This solution gives an excellent prediction of the trend, even though the source loading (a train) is replaced by a harmonic time function. The same trend was obtained with the measurement data, in which almost

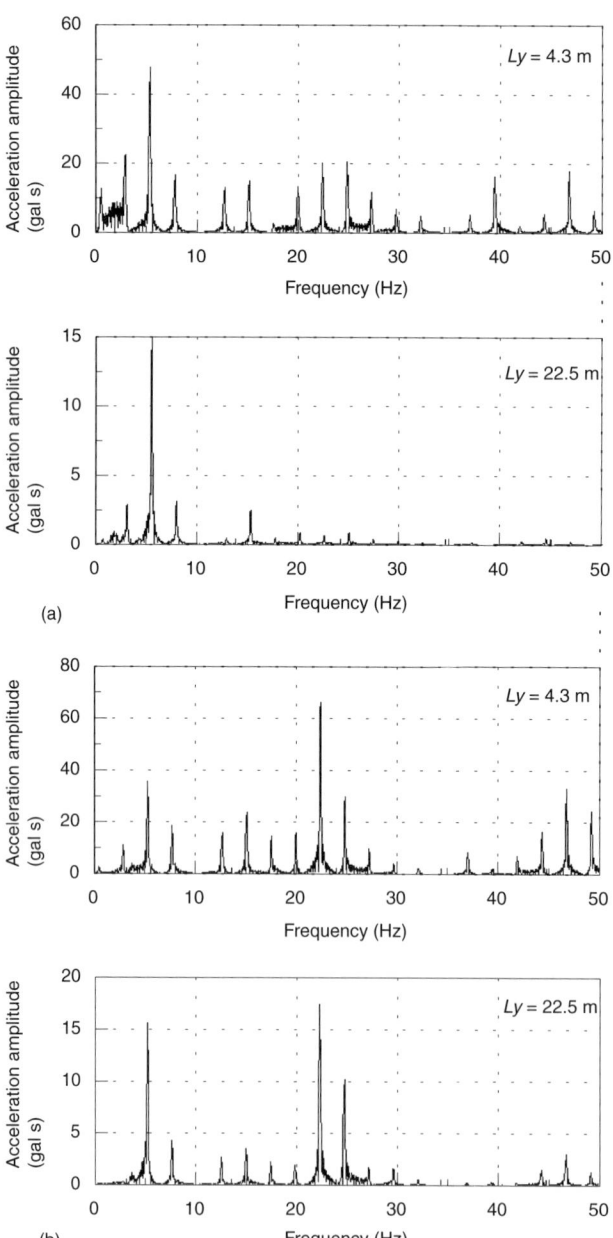

(a)

(b)

GROUND VIBRATIONS: PREDICTION AND MITIGATION

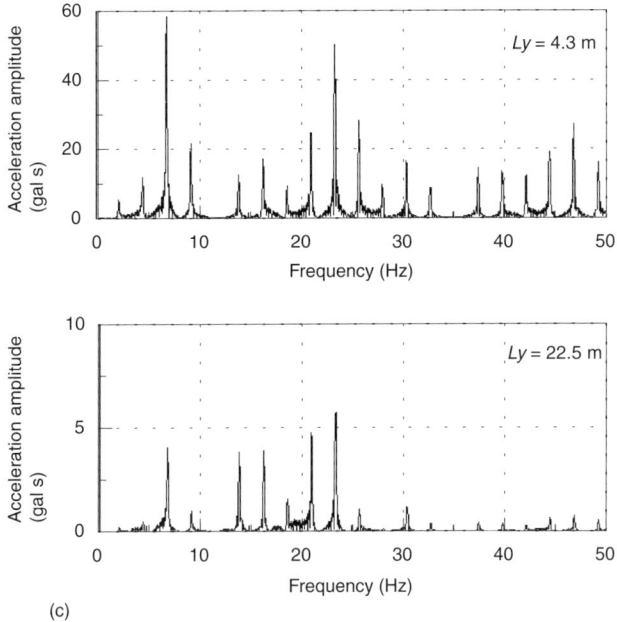

Fig. 12.24. Fourier amplitudes: (a) x *direction; (b)* y *direction; (c)* z *direction*

no vibration propagation exists below 8 Hz in the direction transverse to the track. Only higher frequencies above 8 Hz are transmitted, with a slight damping in the high-frequency range. This phenomenon is confirmed to have been caused by the lowest cut-off frequency of the layer, according to Fig. 12.23(a).

12.4.3. Prediction of ground motions [12.20]

The cause of frequency generation at the source is two-fold: one factor is the sleeper spacing and the other is the train geometry. The sleeper frequency and the train geometry frequency can be combined into one formula

$$f = \frac{L_d}{c} \qquad (12.86)$$

in which the L_d values are distances for the various source mechanisms, chosen to have a common multiple for the given train speed c. Therefore, the axle loading is prescribed by equation (12.36). The axle loads were referred to in Table 12.1. The contributing elements may be affected by the track condition, such as the unevenness, and by the track topography. The measurement data can be used to determine these coefficients $A_k(\omega_j)$, by solving the inverse problem. The acceleration time histories at the distances $L_y = 4.3$ m and 22.5 m were used in this study. The fitting of the simulation and measurements was conducted in the frequency domain in the sense of mean square error minimization:

$$\sum_{i}^{N_i} \left(O_k(y_i, \omega_l) - \sum_{\omega_j} A_k(\omega_j) C_k(y_i, \omega_j, \omega_l) \right)^2 \to \min \quad (12.87)$$

where $O_k(y_i, \omega_l)$ and $C_k(y_i, \omega_j, \omega_l)$ denote, respectively, the Fourier amplitudes from the measurement and from the computation for the referred N_i different frequencies ω_l at the distances y_i. Here, we imposed the condition at $\omega = 0$ of $A_x(0) = A_y(0) = 0$, $A_z(0) = 1$, under the normalized condition of the axle dead load.

The numerical computations have been carried out for the Shinkansen 0 type train travelling at a speed of 211 km/h. The frequencies have been calculated for the representative values of L_d listed in Table 12.2. The associated weighting coefficients $A_k(\omega_j)$ are also listed in Table 12.2. On the basis of these values $A_k(\omega_j)$, the ground response computation has been carried out for the track shown in Fig. 12.19, and the results are depicted in the form of the Fourier amplitudes in Fig. 12.24. These Fourier amplitudes compare well with the measure values in Fig. 12.21.

The maximum accelerations are plotted in Fig. 12.25 to demonstrate the attenuation with distance. These correspond to different types of Shinkansen, up-bound and down-bound, as characterized in Table 12.1. Vibration assessment is conventionally performed by use of the acceleration level, which is defined as

$$VAL = 20 \log_{10} \frac{A}{A_0} \quad (12.88)$$

with respect to a specific value of $A_0 = 10^{-5}$ m/s². Frequency modification is also included in equation (12.88) in order to account for the human perception of vibrations. The results are depicted in Fig. 12.25(a), labelled as 'modification' and 'before modification', the latter being for the original measurement data.

Table 12.2. Weighting coefficients A_k, flat train track (speed 211/245 km/h)

L_d (m)	Frequency (Hz), 211/245 km/h	Weight coefficients, A_k		
		X	Y	Z
∞	0.0/0.0	0.000	0.000	1.000
18.0	3.26/3.78	−0.580	0.003	0.341
8.4	6.98/8.10	0.086	0.196	0.008
3.6	16.28/18.91	0.015	−0.046	0.039
2.4	24.42/28.36	0.008	0.062	0.025
1.2	48.84/56.72	0.020	0.041	0.061

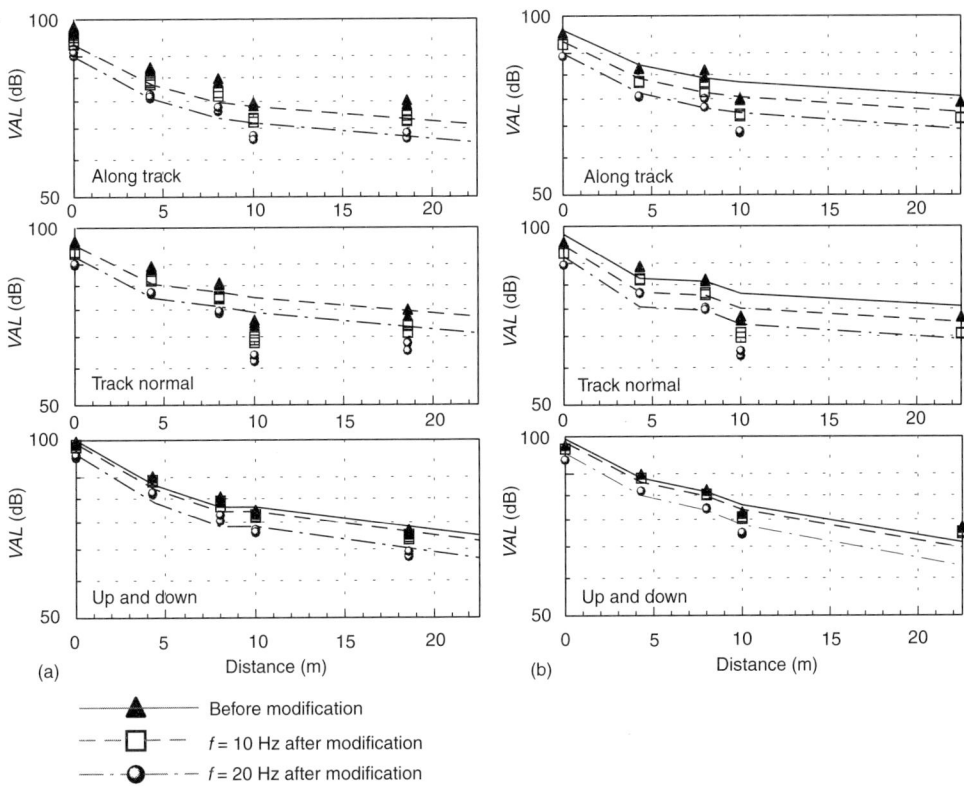

Fig. 12.25. Attenuation of maximum acceleration with distance, flat track: (a) Shinkansen type 0; (b) Shinkansen type 300

Once we obtain the weighting coefficients, we may use these values for assessing the vibrations for different types of Shinkansen trains on the same track. The maximum response thus obtained is shown in Fig. 12.25(b) for the type 300 Shinkansen running at a speed of 245 km/h.

12.4.4. Vibration mitigation measures – WIBs [12.21]

Vibration reduction measures should be undertaken when excessive vibrations come out of a source. Traditionally, trenches and walls have been employed for shielding against wave propagation. The present author has been engaged in the development of wave-impeding blocks or wave-impeding barriers (WIBs) for blocking waves propagating from a source. The wave-impeding block is based on the cut-off frequency of the layer, so that the soil layer depth is reduced by artificially adding a bottom layer with rigid, spread-out obstacles. The recent development is the idea of constructing X-configured wave-impeding barriers made of stiffened soil columns. These soil column systems work for constraining the soil motions due to the propagating waves. Design criteria on the depth of the crossings of X-columns and on the total depth must be fulfilled for the relevant modal wave. Therefore, the total depth is set equal to the wavelength and the crossing location is set to the centre of gravity of the mode shape. Figure 12.26 shows an FEM layout of the model used for the numerical analysis. The computational model has an additional extent of 45 m at both sides, in comparison with the meshes indicated in Fig. 12.26, in order to avoid wave reflection from the sides. Further, an internal damping ratio $y = 0.3$ is imposed in the farthest 10 m range. The finite-element solution method was used for a 2.5-dimensional analysis. In view of the soil layering and the change of stiffness with depth, a rigid base condition was assumed at the bottom of the top three layers shown in Fig. 12.19. Figure 12.27 indicates the effect of the X-WIB in comparison with the ground motion without any measures for vibration reduction. Figure 12.28 shows the ground surface motion around the first three cars due to a train of 16 cars, under the assumption of particular driving frequencies. For constant moving loads, only static deformation is observed under the axles. At the lower driving frequency the major wavelength is greater than the layer depth, so that wave propagation is

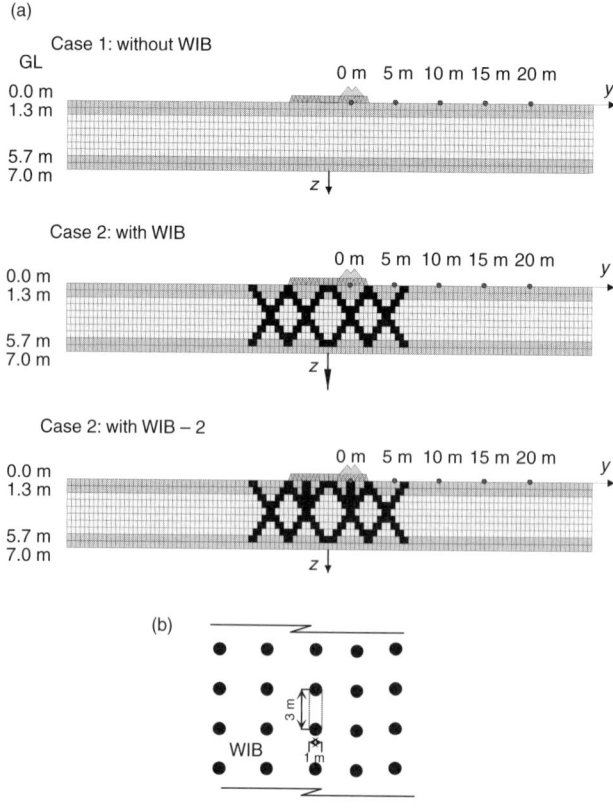

WIB, wave impeding barrier

Fig. 12.26. FEM model for analysis: (a) side view; (b) plan view

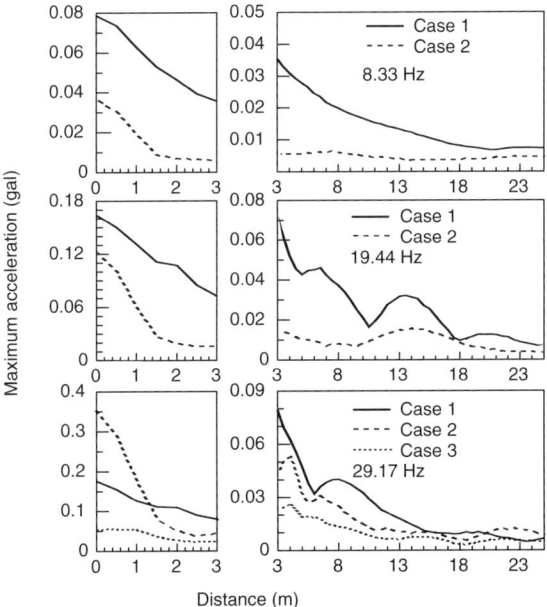

Fig. 12.27. Attenuation of the maximum response

observed only in front of the load, not behind it. At high driving frequencies wave interference occurs; Fig. 12.29 shows the maximum response as a function of depth. It is clearly noticeable that the installation of the X-WIB reduces the ground response significantly.

12.5. Conclusion

Prediction of vibrations alongside railway tracks is of importance from the point of view of track stability and of vibration assessment of the built environment. The method described in this chapter applies a theory of soil–structure interaction and wave propagation in soils.

Computer simulation has been developed on the basis of the track–ground model using pseudo-static loading. Features of the response can be predicted by assuming single or multiple axle loads, taking account of the train geometry and the carriage connection. Some uncertainty arises from the placement of the sleepers. Depending on the deviation of the sleepers in their interaction with the ballast, the sleeper frequency varies. This pseudo-periodicity is included in the source, but cannot be determined in a definite way. The measurement data may be of use for this purpose.

A magnetic-levitation transportation system is now under development for a supertrain in the near future. The vehicle–track interaction for such a system under high-speed running must be checked for safe operation. The environmental impact of this system should also be assessed before it is put into service.

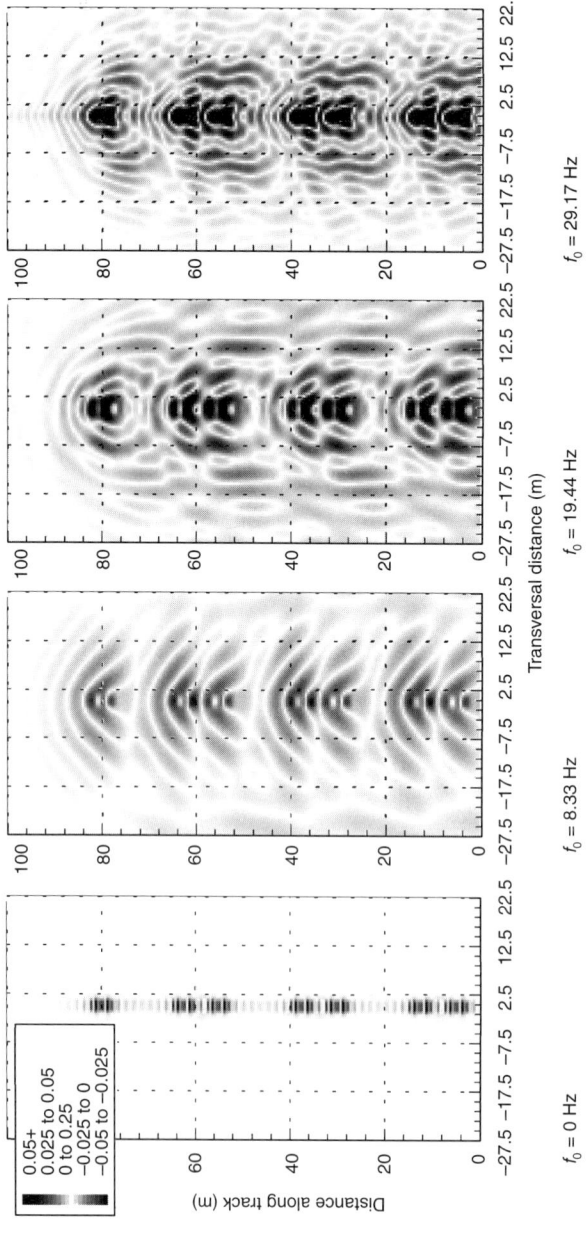

Fig. 12.28. Simulation of the maximum response due to a train travelling at a speed of 252 km/h

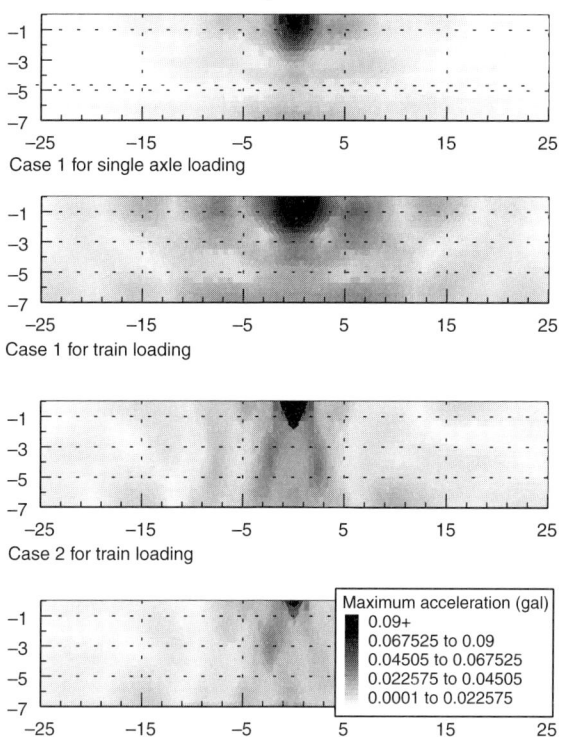

Fig. 12.29. Maximum response profile in a vertical section through the ground ($f_0 = 19.44$ Hz) (dimensions in m)

12.6. Appendix: layer stiffness matrix [12.19]

12.6.1. The layer stiffness matrix with respect to stresses acting on the z plane $\{\sigma_{12} \ \sigma_{22} \ \sigma_{32}\}$

The layer stiffness matrix for the SV and P wave fields is given below:

$$[K_{SV-P}^{AB}] = \begin{bmatrix} K_{11} & & & \text{sym} \\ K_{21} & K_{22} & & \\ K_{31} & K_{32} & K_{33} & \\ K_{41} & K_{42} & K_{43} & K_{44} \end{bmatrix} \qquad (12.89)$$

where

$$K_{11} = \frac{1}{D} \mu^c k_\alpha (\xi^2 - k_\beta^2)(k_\alpha k_\beta S^\alpha C^\gamma - \xi^2 S^\gamma C^\alpha)$$

$$K_{21} = \frac{1}{D} \mu^c \xi [k_\alpha k_\beta (3\xi^2 + k_\beta^2)(C^\alpha C^\gamma - 1) - (\xi^4 + \xi^2 k_\beta^2 + 2k_\alpha^2 k_\beta^2) S^\alpha S^\gamma]$$

$$K_{31} = \frac{1}{D} \mu^c k_\alpha (\xi^2 - k_\beta^2)(\xi^2 S^\gamma - k_\alpha k_\beta S^\alpha)$$

$$K_{41} = \frac{1}{D}\mu^c \xi k_\alpha k_\beta (\xi^2 - k_\beta^2)(C^\gamma - C^\alpha)$$

$$K_{22} = \frac{1}{D}\mu^c k_\beta (\xi^2 - k_\beta^2)(k_\alpha k_\beta S^\gamma C^\alpha - \xi^2 S^\alpha C^\gamma)$$

$$K_{32} = -K_{41}$$

$$K_{42} = \frac{1}{D}\mu^c k_\beta (\xi^2 - k_\beta^2)(\xi^2 S^\alpha - k_\alpha k_\beta S^\gamma)$$

$$K_{33} = K_{11}$$

$$K_{43} = -K_{21}$$

$$K_{44} = K_{22}$$

Here

$$D = k_\alpha k_\beta \left(-2\xi^2 + 2\xi^2 C^\alpha C^\gamma - \frac{k_\alpha^2 k_\beta^2 + \xi^4}{k_\alpha k_\beta} S^\alpha S^\gamma \right)$$

$$C^\alpha \equiv \cosh(k_\alpha h), \qquad C^\gamma \equiv \cosh(k_\gamma h)$$

$$S^\alpha \equiv \sinh(k_\alpha h), \qquad S^\gamma \equiv \sinh(k_\gamma h)$$

$$k_\alpha = \sqrt{\xi^2 - \frac{\rho \omega^2}{\lambda^c + 2\mu^c}}, \qquad k_\beta = \sqrt{\xi^2 - \frac{\rho \omega^2}{\mu^c}}$$

ρ is the mass density, ω is the frequency, h is the layer thickness, and λ^c and μ^c are the complex Lamé constants in the addressed layer.

In case of $\omega/\xi V_S^c \ll 1$, the following approximations can be used:

$$K_{11} = \frac{2}{D'}\mu^c \xi [\xi h(1-\varepsilon^2) - (1+\varepsilon^2) S^\xi C^\xi]$$

$$K_{21} = \frac{2}{D'}\mu^c \xi [\xi^2 h^2 (1-\varepsilon^2)^2 - \varepsilon^2 (1+\varepsilon^2)(S^\xi)^2]$$

$$K_{31} = \frac{2}{D'}\mu^c \xi [(1-\varepsilon^2) S^\xi - \xi h(1-\varepsilon^2) C^\xi]$$

$$K_{41} = -\frac{2}{D'}\mu^c \xi [\xi h(1-\varepsilon^2) S^\xi]$$

$$K_{22} = -\frac{2}{D'}\mu^c \xi [\xi h(1-\varepsilon^2) + (1-\varepsilon^2) S^\xi C^\xi]$$

$$K_{32} = -K_{41}$$

$$K_{42} = \frac{2}{D'}\mu^c \xi [(1+\varepsilon^2) S^\xi + \xi h(1-\varepsilon^2) C^\xi]$$

$$K_{33} = K_{11}$$

$$K_{43} = -K_{21}$$

$$K_{44} = K_{22}$$

Here

$$D' = \xi^2 h^2 (1-\varepsilon^2)^2 - (1+\varepsilon^2)^2 (S^\xi)^2$$

$$\varepsilon = \sqrt{\frac{\mu^c}{\lambda^c + \mu^c}}, \qquad C^\xi \equiv \cosh(\xi h), \qquad S^\xi \equiv \sinh(\xi h)$$

For the SH wave field,

$$[K_{SH}^{AB}] = \frac{k_\beta \mu^c}{\sinh(k_\beta h)} \begin{bmatrix} \cosh(k_\beta h) & -1 \\ -1 & \cosh(k_\beta h) \end{bmatrix} \tag{12.90}$$

12.6.2. The stiffness matrix for a half-space with respect to stresses acting on the z plane

The stiffness matrix for the SV and P wave fields is given below:

$$[K_{SV-P}^{C}] = \frac{k_\beta \mu^c}{\xi^2 - k_\alpha k_\beta} \begin{bmatrix} k_\alpha(\xi^2 - k_\beta^2) & \xi(\xi^2 + k_\beta^2 - 2k_\alpha k_\beta) \\ \xi(\xi^2 + k_\beta^2 - 2k_\alpha k_\beta) & k_\beta(\xi^2 - k_\beta^2) \end{bmatrix} \tag{12.91}$$

In the case of $\omega/\xi V_S^c \ll 1$, the following approximation can be used:

$$[K_{SV-P}^{C}] = \frac{2\xi \mu^c}{1+\varepsilon^2} \begin{bmatrix} 1 & \varepsilon^2 \\ \varepsilon^2 & 1 \end{bmatrix} \tag{12.92}$$

For the SH wave field,

$$[K_{SH}^{C}] = k_\beta \mu^c \tag{12.93}$$

12.7. References

12.1. EASON, G. The stresses produced in a semi-infinite solid by a moving surface force. *International Journal of Engineering Science*, 1965, **2**, 581–609.

12.2. GAKENHEIMER, D. C. and MIKLOWITZ, J. Transient excitation of an elastic half-space by a point load travelling on the surface. *Journal of Applied Mechanics, E*, 1969, **36**(3), 505–515.

12.3. HOOP, A. T. A modification of Cagniard's method for solving seismic pulse problems. *Applied Scientific Research B*, 1959, **8**, 349–356.

12.4. DIETERMAN, H. A. and METRIKINE, A. V. The equivalent stiffness of a half-space interacting with a beam. Critical velocities of a moving load along the beam. *European Journal of Mechanics A: Solids*, 1996, **15**(1), 67–90.

12.5. DIETERMAN, H. A. and METRIKINE, A. V. Steady-state displacements of a beam on an elastic half-space due to a uniformly moving constant load. *European Journal of Mechanics A: Solids*, 1997, **16**(2), 295–306.

12.6. JONES, C. J. C. and BLOCK, J. R. Prediction of ground vibration from freight trains. *Journal of Sound and Vibration*, 1996, **193**(1), 205–213.

12.7. KAYNIA, A. M., MADSHUS, C., HARVIK, L. and ZACKRISSON, P. Ground vibration from high speed railway lines. *Geotechnics*, 1998, **4**, 411–423.

12.8. KAUSEL, E. and ROESSET, J. M. Stiffness matrices for layered soils. *Bulletin of the Seismological Society of America*, 1981, **71**(6), 1743–1761.

12.9. YOSHIOKA, O. Some considerations on generating mechanism of vibration due to running trains. *Butsuri-Tanko [Geophysical Exploration]*, 1976, **29**(2), 23–33.

12.10. YOSHIOKA, O. A. Dynamic model on excitation and propagation of Shinkansen-induced ground vibrations. *Butsuri-Tansa*, 1995, **48**(5), 299–315.

12.11. KRYLOV, V. and FERGUSON, C. Calculation of low-frequency ground vibrations from railway trains. *Applied Acoustics*, 1994, **42**, 199–213.

12.12. Krylov, V. Generation of ground vibrations by superfast trains. *Applied Acoustics*, 1995, **44**, 149–164.

12.13. KRYLOV, V. Vibrational impact of high-speed trains. I. Effect of track dynamics. *Journal of the Acoustical Society of America*, 1996, **100**(5), 3121–3134.

12.14. KRYLOV, V. Effects of track properties on ground vibrations generated by high-speed trains. *Acustica–Acta Acustica*, 1998, **84**, 78–80.

12.15. LAMB, H. On the propagation of tremors over the surface of an elastic solid. *Philosophical Transactions of the Royal Society A*, 1904, **203**, 1–42.

12.16. TAKEMIYA, H. and GODA, K. 3-D transient response characteristics of soil stratum due to moving loads. *Proceedings of the Japan Society of Civil Engineers*, 1997, **563**(I-39), 137–148.

12.17. TAKEMIYA, H. and GODA, K. Wave propagation/impediment in a soil stratum over rigid base due to impulse/moving loads. *Proceedings of the Japan Society of Civil Engineers*, 1998, **605**(I-45), 161–169.

12.18. TAKEMIYA, H. and GODA, K. Prediction of ground vibration induced by high-speed train operation. *Proceedings of the 5th International Conference on Sound and Vibration*. Adelaide, 1997, vol. 5, pp. 2681–2688.

12.19. TAKEMIYA, H. and GODA, K. FEM–BEM analysis of vibration of an embankment track on layered soils for harmonic moving loads. *Proceedings of the Japan Society of Civil Engineers*, 1998, **605**(I-45), 199–152.

12.20. TAKEMIYA, H., GODA, K. and KOMORI, D. Computer simulation prediction of ground vibration induced by high-speed train running. *Proceedings of the Japan Society of Civil Engineers*, 1998, **619**(I-47), 193–201.

12.21. TAKEMIYA, H. and SHIOTSU, Y. Features of ground vibrations induced by high-speed train running and the counter vibration measure X-WIB. *Proceedings of the Japan Society of Civil Engineers*, submitted.

12.22. SHENG, X., JONES, C. J. C. and PETYT, M. Ground vibration generated by a load moving along a railway track. *Journal of Sound and Vibration*, 1999, **228**(1), 129–156.

12.23. TAKEMIYA, H. Simulation of high-speed train induced vibration and mitigation measure WIB. *Seminar on High Speed Lines on Soft Ground, Dynamic Soil–Track Interaction, Ground-borne Vibration*. Swedish National Rail Administration, Getheborg, 2000.

12.24. BOUCHON, M. and AKI, K. Discrete wave number representation of seismic source wave fields. *Bulletin of the Seismological Society of America*, 1977, **167**(2), 259–277.

12.25. GRUNDMANN, H., LIEB, M. and TROMMER, E. The response of a layered half-space to traffic loads moving along its surface. *Archive of Applied Mechanics*, 1999, **69**, 55–67.

12.26. LIEB, M. and SUDRET, B. A fast algorithm for soil dynamics calculations by wavelet decomposition. *Archive of Applied Mechanics*, 1998, **68**, 147–157.

12.27. TAKEMIYA, H. and JIANG, J. Q. Wave impeding effect by buried rigid block and response reduction of dynamically excited pile foundation. *Structural Engineering/Earthquake Engineering, Proceedings of the Japan Society of Civil Engineers*, 1994, **489**(I-27), 243–250.

12.28. TAKEMIYA, H. and FUJIWARA, A. Wave propagation/impediment in a stratum and wave impeding block (WIB), measured for SSI response reduction. *Soil Dynamics and Earthquake Engineering*, 1994, **13**, 49–61.

12.29. TAKEMIYA, H. and FUJIWARA, A. Installation of a wave impeding block (WIB) for dynamic response reduction of soil–structure system. *Proceedings of the Japan Society of Civil Engineers*, 1994, **489**(I-27), 243–250.

12.30. KENNY, J. T. Steady-state vibrations of beam on elastic foundation for moving load. *Journal of Applied Mechanics. American Society of Mechanical Engineering*, 1954, **21**, 359–364.

Part 5
Ground vibrations generated by underground trains

13. Prediction and measurements of ground vibrations generated from tunnels built in water-saturated soil

S. A. Kostarev,[*] S. A. Makhortykh[†] and S. A. Rybak[‡]
[*]Tunnelling Association, Moscow, Russia
[†]RAS Institute of Mathematical Problems in Biology, Pushchino, Russia
[‡]N. N. Andreyev Acoustics Institute, Moscow, Russia

13.1. Introduction

High-speed underground railways represent one of the major sources of vibrations and ground-borne noise in built-up areas. Prediction of vibrations from underground railways and the possible reduction of their excessively high levels involve solutions of the following problems:

- determination of sources of underground vibration and their physical characteristics
- development of reliable methods for calculation of transmission of vibrations to the ground surface
- evaluation of the influence of ground vibrations on nearby residential buildings
- development of practical measures for reduction of vibration levels.

In this chapter, a simplified model for calculating ground vibrations from underground trains is described. Using this model, the expected levels of generated ground vibrations are predicted and some measures are proposed for their control.

Section 13.2 describes the modelling of a tunnel as a thin cylindrical elastic shell buried in the ground, the latter being considered as liquid (the so-called 'liquid ground' approximation). It is shown that such an approximation works reasonably well for water-saturated soils characterized by a Poisson's ratio of 0.4–0.5. The influence of the first three oscillation modes of an elastic shell on the ground vibration field generated in the surrounding medium is analysed.

In Section 13.3 the transmission of generated ground vibrations to a free ground surface is considered, including a special case of waveguide propagation of generated vibrations. Section 13.4 deals with some vibration reduction measures, and Section 13.5 describes a simple method of field estimation of the elastic parameters and damping of layered ground.

13.2. Waves radiated by a cylindrical oscillating shell

The problem considered in this section concerns the properties of the first three oscillation modes of a tunnel wall (modelled as a cylindrical elastic shell) and their contribution to the total acoustic field generated in the surrounding ground by periodic forces applied to the shell's inner surface. The ground is considered as water-saturated and the so-called 'liquid ground' approximation is applied.

The character of the force modelling a train travelling underground can be of various kinds. In particular, it can be a time-dependent spatially distributed force applied to the interior surface of a shell. Depending on the character of the external force, various types of oscillation modes can be excited in the shell. In this context, it is essential to establish which modes are the most important and which can be neglected to reduce the complexity of the problem.

Let us consider a thin elastic cylindrical shell of thickness h and radius R placed in an unbounded ideal liquid (modelling the ground) and subjected to a dynamic pressure $p_0(x, \varphi, t)$ applied to its inner surface. The equations describing the oscillations of the shell and the equation for the acoustic pressure field p in the exterior medium (in cylindrical coordinates) [13.1–13.3] may be written as

$$\begin{aligned} L_{11}u + L_{12}v + L_{13}w &= S_1 \\ L_{21}u + L_{22}v + L_{23}w &= S_2 \\ L_{31}u + L_{32}v + L_{33}w &= S_3 + f \end{aligned} \tag{13.1}$$

and

$$\frac{\partial^2 p}{\partial t^2} - c^2 \Delta p = 0$$

$$\Delta = r^{-1}\frac{\partial}{\partial r}\left(r\frac{\partial}{\partial r}\right) + r^{-2}\frac{\partial^2}{\partial \varphi^2} + \frac{\partial^2}{\partial x^2} \tag{13.2}$$

Here u, v, w are equal to x, φ, r, the axial, azimuthal and radial components of the displacement vector of the shell surface, and the differential operators L_{ij} have the form

$$L_{11} = \frac{\partial^2}{\partial x^2} + \frac{1-\nu}{2R^2}\frac{\partial^2}{\partial \varphi^2} + c_1^{-2}\frac{\partial^2}{\partial t^2}$$

$$L_{12} = L_{21} = \frac{1+\nu}{2R}\frac{\partial^2}{\partial x\partial \varphi}$$

$$L_{22} = \frac{1-\nu}{2}\frac{\partial^2}{\partial x^2} + R^{-2}\frac{\partial^2}{\partial \varphi^2} + c_1^{-2}\frac{\partial^2}{\partial t^2} \tag{13.3}$$

$$L_{23} = L_{32} = R^{-2}\frac{\partial}{\partial \varphi}$$

$$L_{13} = L_{31} = \frac{\nu}{R}\frac{\partial}{\partial x}$$

$$L_{33} = a^2\left(R^2\nabla^2\nabla^2 + \frac{2}{R^2}\frac{\partial^2}{\partial\varphi^2} + R^{-2}\right) + R^{-2} + c_1\frac{\partial^2}{\partial t^2}$$

where

$$a^2 = \frac{h^2}{12R^2}$$

$$\nabla^2 = \frac{\partial^2}{\partial x^2} + R^{-2}\frac{\partial^2}{\partial\varphi^2}$$

The following notations are used here: $f = \gamma[p_0(x, \varphi, t) - p^+(x, \varphi, t)]$, where p_0 and p^+ are the pressure values on the internal and external shell surfaces, respectively; c_1 is the longitudinal elastic-wave velocity in the shell material; t is the time; $\gamma = (1 - \mu^2)/Eh$; and E and ν are the Young's modulus and Poisson's ratio. Also, the condition $h/R \ll 1$ is assumed. The right-hand side of equation (13.1) contains the terms S_j, which are expressed through the projections of the elastic response of the medium Σ_j, so that $S_j = \gamma\Sigma_j$. If the medium considered is liquid then all S_j are equal to zero. The expression for the shear response is then written as follows [13.3]:

$$\Sigma_1 = \mu\left(\frac{\partial w}{\partial x} + \frac{\partial u}{\partial r}\right)$$

$$\Sigma_2 = \mu\left(R^{-1}\frac{\partial w}{\partial\varphi} + \frac{\partial v}{\partial r} + \frac{v}{r}\right) \quad (13.4)$$

$$\Sigma_3 = (\lambda + 2\mu)\frac{\partial w}{\partial r} + \lambda\left(\frac{w}{r} + \frac{1}{R}\frac{\partial v}{\partial\varphi} + \frac{\partial u}{\partial x}\right)$$

Here λ and μ are the Lamé coefficients, and ρ is the mass density of the medium. The values of λ and μ are expressed through the known values of the Young's modulus and Poisson's ratio [13.3]. Equation (13.2) for the pressure field is valid in the case of zero shear elasticity.

The relationship between pressure and displacement in the exterior medium is given by the following expression:

$$\frac{\partial p}{\partial r} = -\rho_1\ddot{w} \quad (13.5)$$

Let the pressure applied to the interior surface, modelling a moving train, be expressed as a progressing wave:

$$p_0 = A\delta\left(\varphi + \frac{\pi}{2}\right)e^{i(k_x x - \omega t)} \quad (13.6)$$

Here $\delta(x)$ is the Dirac delta function, and ω and k_x are the angular frequency and the wavenumber in the axial direction, respectively. It is assumed in equation (13.6) that the area to which the dynamic pressure is applied is small compared with the diameter of the shell. This condition is normally satisfied for trains in underground tunnels. The periodicity with respect to the coordinate x is related to the periodic structure of the train. Let the function p_0 be represented as a Fourier series expansion [13.4]:

$$p_0 = \frac{A}{\pi}(\tfrac{1}{2} - \sin\varphi - \cos 2\varphi + \ldots)e^{i(k_x x - \omega t)} \tag{13.7}$$

In this case the solution of the system of equations (13.1) for the component u can be sought as

$$u = (u_1 - u_2 \sin\varphi - u_3 \cos 2\varphi + \ldots)e^{i(k_x x - \omega t)} \tag{13.8}$$

Similar expressions can be written also for the components v and w. It is not difficult to show that the pressure field in the surrounding medium, which satisfies the condition expressed by equation (13.5), is given by an expression describing waves radiating from the tunnel:

$$p = [p_1 H_0^{(1)}(kr) + p_2 H_1^{(1)}(kr)\sin\varphi + p_3 H_2^{(1)}(kr)\cos 2\varphi + \ldots]e^{i(k_x x - \omega t)} \tag{13.9}$$

where $k = \sqrt{(\omega^2/c^2 - k_x^2)}$, and the condition $\omega^2/c^2 - k_x^2 \geq 0$ holds. Here $H_i^{(1)}(x)$ is the Hankel function. If $\omega^2/c^2 - k_x^2 < 0$, then the pressure p is expressed through the modified Hankel functions $K_i(x)$ [13.5] and its magnitude decays exponentially with distance from the tunnel.

The expansion coefficients p_i can be expressed through the values w_i after the substitution of equation (13.9) into equation (13.5). The physical meaning of the expansion in equation (13.9) is clear: its terms correspond to the monopole, dipole, quadrupole and higher-order components of the field. Indeed, if one had substituted a multipole source ($\delta(z)$, $\delta'(z)$, $\delta''(z)$, …, where z is the vertical coordinate in a Cartesian basis) into the right-hand side of equation (13.1), then its solutions satisfying the radiation condition $p \to 0$ when $r \to \infty$ would look like the first, second and third terms of the expansion in equation (13.9). If g_i is the solution of equation (13.2) with $\delta^{(i)}(z)$ in the right-hand side, then the following relationship is valid:

$$\frac{\partial g_i}{\partial z} = a_i g_{i+1} f(\varphi)$$

where a_i is a constant and f is a cosine or sine function for even or odd i, respectively, $i = 1, 2, 3, \ldots$.

Thus, it is possible to write a linear set of equations for each component of the vector $\boldsymbol{u} = (u_i, v_i, w_i)$:

$$\|\mathbf{l}\|\boldsymbol{u} = \boldsymbol{b} \tag{13.10}$$

The matrix $\mathbf{l}$ is given by

$$l_{11} = -k_x^2 - \frac{1-\nu}{2R^2}n^2 - c_1^{-2}\omega^2$$

$$l_{12} = l_{21} = -\frac{1+\nu}{2R}nk_x$$

$$l_{22} = -\frac{1-\nu}{2}k_x^2 - R^{-2}n^2 - c_1^{-2}\omega^2 + \mu\gamma R^{-1}$$

$$l_{23} = iR^{-2}n - iR^{-1}n\mu\gamma$$

$$l_{32} = iR^{-2}n + \frac{2}{3R}i\mu\gamma n$$

$$l_{13} = ik_x\frac{\nu}{R} - i\mu\gamma k_x$$

$$l_{31} = ik_x\frac{\nu}{R} + \frac{2}{3}i\mu\gamma k_x$$

$$l_{33} = a^2\left(R(k_x^2 + R^{-2}n^2)^2 - n^2\frac{2}{R^2} + R^{-2}\right) + R^{-2} - c_1^{-2}\omega^2 + \gamma p_n^+$$

where $a^2 = h^2/(12R^2)$, and the right-hand side is given by $b = (s_1, s_2, s_3 + \gamma p_n)$. Here $p_n^+ w_n$ is the pressure on the external surface of the shell, and $p_0 = 1/2$, $p_1 = -1$, $p_2 = -1$, ... are the Fourier coefficients of the series in equation (13.7).

The system described by equation (13.10) has been investigated and checked numerically. Since the sound field in the surrounding medium is described by the expansion in equation (13.9), the first three terms of equation (13.9), corresponding to the zero, first and second modes of oscillation, give a close approximation to the total field at frequencies $f \leq f_c/5$, where $f_c = c_1/6\pi R$ is the radial shell frequency. Qualitative evaluations are valid for frequencies up to f_c. In practice, a cast-iron–concrete envelope of radius 2.75 m (the radius of the tunnels of the Moscow underground railway) would have a radial frequency of about 300 Hz. In this case, at frequencies below 60 Hz, there are three modes which give a good approximation to the total field, whereas at higher frequencies up to 300 Hz, only a qualitative description is possible.

Considering a realistic case of the excitation of the tunnel's inner surface allows for fewer terms in the expansion in equation (13.7), so that the three-mode approximation is quite sufficient. It is also known that the frequencies of vibration generated by underground trains are primarily confined to the range from 30 to 60 Hz, which supports the above-mentioned limitation to the first three modes of an elastic shell.

The behaviour of the ordinary and modified Hankel functions is well known [13.5]. The Hankel functions and the modified Hankel functions can be approximated at large distances as $[\sqrt{(2/\pi kr)}]e^{ikr}$ and $[\sqrt{(\pi/2kr)}]e^{-kr}$, respectively. The functions $\sin\varphi$ and $\cos 2\varphi$ have maximum values at $\varphi = \pi/2$; therefore it is enough to consider a point $r = r_1$ and $\varphi = \pi/2$, where r_1 is a fixed distance from the shell's axis.

The characteristic parameters of the inner layer of the circular cast-iron–concrete tunnel used for detailed numerical analysis were the following: $c_1 = 4000$,

$E = 3.3 \times 10^{10}$, $\mu = 0.25$ and $h = 0.2$. Let us consider the surrounding medium to be a sandy ground with $E_1 = 5.3 \times 10^9$ and $\mu_1 = 0.25$, where E_1 and μ_1 are the Young's modulus and Poisson's ratio, $c_1 = 600$ is the velocity of the longitudinal elastic waves and $\rho = 1800$ is the mass density. All these values are in SI units. Damping in the ground was neglected, and shear elasticity was taken into account only for determining the response on an envelope in equations (13.1).

Figure 13.1 shows the values $L_1 = 20 \lg |p_1/p_0|$ and $L_2 = 20 \lg |p_1/p_2|$ (in dB) as functions of frequency. Here p_0, p_1 and p_2 are the amplitudes of the zero-order, first- and second-order modes for $\varphi = \pi/2$, $k_x = 0.1$ and $r_1 = 2.75$ (on the external shell

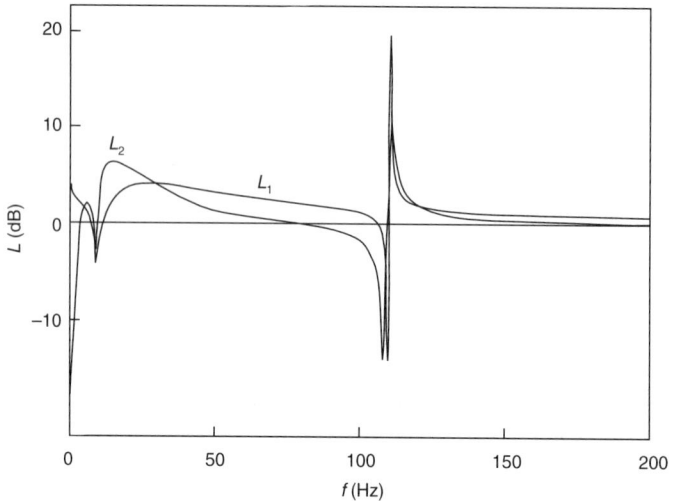

Fig. 13.1. Amplitudes L_1 and L_2 as functions of frequency: $\varphi = \pi/2$, $k_x = 0.1$, $r_1 = 2.75$

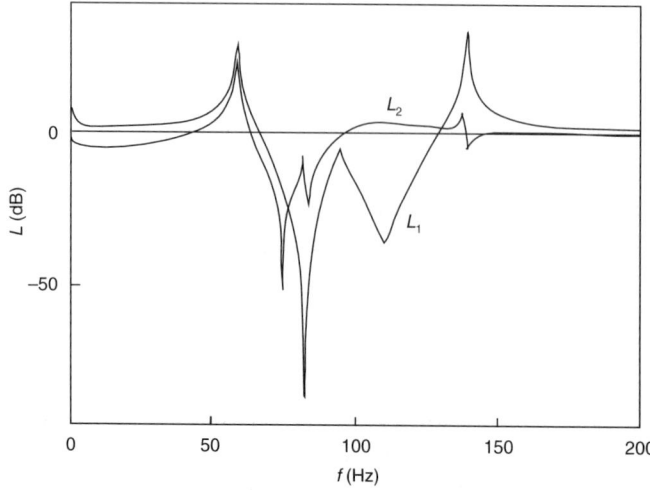

Fig. 13.2. Amplitudes L_1 and L_2 as functions of frequency: $\varphi = \pi/2$, $k_x = 1$, $r_1 = 2.75$

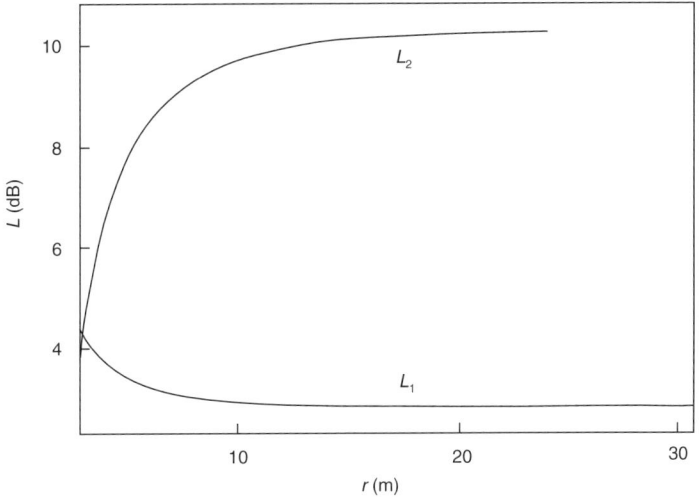

Fig. 13.3. Amplitudes L_1 *and* L_2 *as functions of the distance from the shell*

surface). Figure 13.2 shows the corresponding values for $\varphi = \pi/2$, $r_1 = 2.75$ and $k_x = 1$. Generally, the behaviour of the functions investigated does not change, and only for the lowest frequencies is some growth of the amplitude of the second mode observed. The computed values of L_1 and L_2 as functions of the wavenumber k_x show very complex behaviour owing to the appearance of numerous resonances. The values of L_1 and L_2 as functions of distance from the shell are shown in Fig. 13.3.

In the case of $r_1 = 2.75$ and $k_x = 0.1$ the condition $kr \gg 1$ is not satisfied, and it is necessary to consider the opposite asymptotic expressions for the cylindrical functions, namely

$$H_0^{(1)}(x) \to \frac{2}{\pi}\ln x$$
$$H_1^{(1)}(x) \to \frac{2}{\pi x} \quad (13.11)$$
$$H_2^{(1)}(x) \to \frac{4}{\pi x^2}$$

A common property of the frequency-dependent amplitudes L_1 and L_2 for the wavenumber $k_x = 0.1$, for different distances from the shell axis, is a maximum at a frequency around 115 Hz and a deep minimum at a frequency around 110 Hz. A dominance by mode 1 takes place in the range of frequencies from 20 to 80 Hz (this mode exceeds the levels of modes 0 and 2 by about 5 dB). The difference in the levels of modal components is related only to the excitation of the shell and does not depend on the distance. The dependence on the angular coordinate is defined by equations (13.7)–(13.9). To explain the behaviour of $L(k)$, it is necessary to consider the system described by equation (13.10). The shell has oscillatory eigenvalues which satisfy the equation

$$\det \|1\| = 0 \tag{13.12}$$

The controlling parameter in this case is k_x. It follows from equation (13.12) that each mode has only one resonance frequency ω^i_{res} (the solution of equation (13.12)) if the shear elasticity of the medium is neglected and $k_x \to 0$ (i is the mode index). Thus, we have $\omega^1_{res} \sim \omega^0_{res} \sim \omega^2_{res}$, which explains the emergence of maxima for modes 1, 2 and 3 at frequencies between 60 and 110 Hz. Shear elasticity leads to an increase of the eigenfrequencies for the first and second modes to 110 Hz. Equation (13.12) has three solutions for each mode for $k_x \sim 1$. This fact explains the complexity of the function $L(k)$ at frequencies from 60 to 150 Hz. The basic maxima of the first mode are at frequencies from 60 to 140 Hz, and the principal maxima of the zero and second modes lie at lower frequencies.

The level of the first acoustic mode is 2–7 dB higher than that of the other two modes at frequencies from 10 to 60 Hz, for $k_x \sim 0.1$. In some cases, the domination of the first mode at frequencies around 110 Hz increases this difference to 20 dB. The opposite case is that in which modes 0 and 2 dominate. This occurs at frequencies around 2 Hz and also from 100 to 110 Hz. The contribution of mode 2 at short distances from the shell is also increased. The amplitudes of modes 0 and 2 increase as $k_x \to 1$ (except at frequencies around 60 and 140 Hz, where mode 1 is the most important).

Often, the measured resonance frequencies differ from the predicted ones owing to damping and radiation of energy into the outer medium. This leads to the existence of an imaginary part in the solution of equation (13.12) for fixed k_x. The first mode dominates for $k_x < 0.2$ in the octave bands at 31.5 Hz and 63 Hz and also for the interval $0.5 < k_x < 1$. The functions $L_i(r)$ ($i = 1, 2$) are presented in Fig. 13.3 for a frequency of 31.5 Hz and $k_x = 0.1$ (r is the distance from the shell's axis). It is obvious that the relative contributions of the different modes do not depend on the distance r for larger values of r. The situation is different for smaller r, which can be explained by the different growth rates of Hankel functions of different orders for $r \to 0$ (see equation (13.11)). In this case, the contribution of mode 2 increases and that of mode 0 decreases for smaller distances.

Conclusions can now be drawn on the basis of the above. The main factors influencing the amplitudes of the vibrations generated in the surrounding medium are the amplitude of the force, its frequency f and its longitudinal wavenumber k_x, modelling the train speed $v = 2\pi f / k_x$. The first mode of vibration is dominant for values of $k_x \sim 0.1$ m^{-1} in the octave bands at 31.5 and 63 Hz. For $k_x > 0.2$ and $k_x > 0.5$ it is necessary to account for the zero- and second-order modes. The dependence of the relative contributions of the oscillation modes on the distance from the shell axis is important only when the inequality $kr = r\sqrt{(\omega^2/c^2 - k_x^2)} \ll 1$ holds. For a force with a longitudinal wavenumber exceeding 0.5 m^{-1}, the vibration field is expressed through the modified Hankel functions, which decay exponentially with distance.

It can be shown that the approximation of 'liquid ground' used in the above model works rather well for most water-saturated soils satisfying the condition $c_t/c_l \ll 1$, where c_t and c_l are the velocities of shear and longitudinal elastic waves.

To account for an arbitrary azimuthal force distribution $p_0(\varphi)$ one can use the Green's function approach. The Green's function can be obtained if the problem expressed by equation (13.6) is rewritten for the case of a point oscillating force $p_0 = \delta(\varphi - \psi)e^{i(kx - \omega t)}$. In this case the Green's function may be expressed in the form $G = p_1(\varphi, \psi)e^{i(kx - \omega t)}$. The solution of the wave equation (13.2) for an arbitrary force distribution $p_0(\varphi)$ can then be expressed as the integral

$$p = \exp(kx - \omega t) \int_0^{2\pi} p_1(\varphi, \psi) p_0(\psi) d\psi$$

Thus, the results obtained in this section can be easily extended to the case of an arbitrary force distribution over the perimeter of the shell's inner surface.

13.3. Transmission of vibrations to the ground surface

Ground vibrations generated from tunnels are transmitted to the surface through the surrounding medium, which can be described as a viscoelastic homogeneous or layered ground characterized by the complex elastic modulus $E_0(1 - i\eta)$ (we remind the reader that in the model considered here, shear elasticity is neglected). Here E_0 is real and $i\eta E_0$ is imaginary part of the modulus; these quantities describe the elastic and viscous properties, respectively, of the ground (typically, the loss factor $\eta < 0.3$). Note that in a model of fully elastic ground there are two components of the wave field: shear and longitudinal elastic waves. The velocities of these waves are denoted by c_t^s and c_l^s, respectively. The velocities of the longitudinal and shear waves are expressed through the Young's modulus E, Poisson's ratio ν and mass density ρ in the following manner:

$$c_l = \left(\frac{E(1-\nu)}{\rho(1+\nu)(1-2\nu)} \right)^{1/2}$$
$$c_t = \left(\frac{E}{2\rho(1+\nu)} \right)$$
(13.13)

If a plane longitudinal wave is incident onto a free boundary of a solid medium (where shear elasticity is now taken into account) the reflection coefficients into longitudinal and shear waves take the forms [13.6]

$$v_l = \frac{4\operatorname{ctg}^2 \theta \sqrt{n^2 + (n^2 - 1)\operatorname{ctg}^2 \theta} - [n^2 + (n^2 - 2)\operatorname{ctg}^2 \theta]^2}{4\operatorname{ctg}^2 \theta \sqrt{n^2 + (n^2 - 1)\operatorname{ctg}^2 \theta} + [n^2 + (n^2 - 1)\operatorname{ctg}^2 \theta]^2}$$
$$v_t = \frac{4\operatorname{ctg}^2 \theta [n^2 + (n^2 - 2)\operatorname{ctg}^2 \theta]^2}{4\operatorname{ctg}^2 \theta \sqrt{n^2 + (n^2 - 1)\operatorname{ctg}^2 \theta} + [n^2 + (n^2 - 1)\operatorname{ctg}^2 \theta]^2}$$
(13.14)

Here n is the ratio of the velocities c_l/c_t, which has a similar meaning to the refractive index. The angle θ is the grazing angle of the incident wave (angle between the direction of incidence and the layer). Note that the reflection coefficient into the longitudinal wave is close to -1, and the reflection coefficient into the shear wave becomes close to 0 as $\theta \to \pi/2$. If the ground has vertical stratification, the transmission matrix approach can be used to describe wave transmission and reflection (see, e.g., [13.7, 13.8]).

For shallow tunnels (depths less than 10 m), the tunnel wall oscillations can also generate Rayleigh waves on the ground surface (in the case of deep tunnels, the amplitudes of Rayleigh waves generated are negligibly small).

In some cases of vertical stratification of the ground, the condition for waveguide propagation of ground vibrations away from the surface may be realized, contrary to the typical situation of subsurface waveguide propagation caused by the soft top layer. For the analysis, we shall use the above-mentioned 'liquid ground' approximation applicable to water-saturated soils with $c_t \ll c_l$ and consider a typical situation in the Moscow area characterized by three geological layers: (1) an upper water-saturated soil layer (down to 2 m), (2) a soft sandy layer (2–8 m) and (3) a clay or limestone layer (>8 m).

An important property of the geological profile considered here is the relatively high sound speed in the upper layer (700–900 m/s). This fact is attributed to a large amount of melting of snow in some years and the resulting permanent presence of a water-saturated layer.

The scalar wave equation modelling the above situation should be solved in two steps: in the first step, the problem of free wave propagation should be considered to find the vertical and horizontal field distributions. In the second step, absolute values of the field, depending on the intensity of the source, should be calculated at various depths. Only the first step will be discussed below.

Let us seek the acoustic field of the longitudinal waves in the above-mentioned waveguide system in the following form:

$$p = A\varphi(z)\exp[i(\xi x - \omega t)] \qquad (13.15)$$

where A is the wave amplitude, $\xi = \xi_1 + i\xi_2$ is the complex wavenumber ($\xi_2 > 0$) and ω is the angular frequency. The velocity of longitudinal waves in the ground, $c(z)$ (for shortness we shall use the term 'sound velocity'), is considered to be a depth-

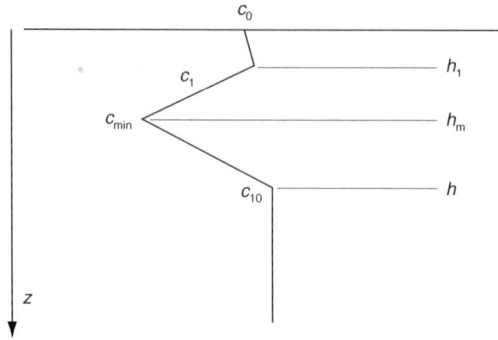

Fig. 13.4. Dependence of the longitudinal-wave velocity on depth z, responsible for waveguide propagation

dependent function. Then, the distribution function $\varphi(z)$ should satisfy the following equation [13.2]:

$$\frac{d^2\varphi}{dz^2} + \omega^2 \left(\frac{1}{c^2(z)} - \frac{1}{c_\phi^2} \right) \varphi = 0 \qquad (13.16)$$

where $c_\phi = \omega/\xi$ is a waveguide mode velocity, which has to be determined. In this equation the mass density variation with depth, $\rho(z)$, is neglected. The dependence of the sound velocity on the vertical coordinate z for typical ground conditions can be modelled as follows:

$$c(z) = \begin{cases} c_0 + z\dfrac{c_0 - c_1}{h_1}, & -h_1 < z < 0 \\[6pt] c_1 + \dfrac{h_1}{h_m - h_1}(c_{min} - c_1) + z\dfrac{c_{min} - c_1}{h_m - h_1}, & -h_m < z < -h_1 \\[6pt] c_{min} + \dfrac{h_m}{h - h_m}(c_{10} - c_{min}) + z\dfrac{c_{10} - c_{min}}{h - h_m}, & -h < z < -h_m \end{cases} \qquad (13.17)$$

It is assumed here that $h = 10$ m, $h_1 = 2$ m, $h_m = 4$ m, $c_0 = 700$ m/s, $c_1 = 900$ m/s, $c_{min} = 420$ m/s and $c_{10} = 910$ m/s (see Fig. 13.4).

Calculations for realistic cases must take into account also the effect of energy dissipation in the medium. Numerical solution of the problem expressed by equations (13.15)–(13.17) shows that the acoustic field can be captured by the slowest-velocity region in the soil. The dependence of the corresponding acoustic mode on the depth z is shown in Fig. 13.5 for octave bands with geometric-mean frequencies $f = 31.5$ and 125 Hz.

The first curve corresponds to the value dp/dz (this parameter is proportional to the z component of the vibration acceleration) and the second curve represents the elastic stress (acoustic pressure p). Calculations for the octave band with a central frequency of 16 Hz indicate that the waveguide effect does not occur in this case, and the acoustic field may be approximated using the model of a homogeneous medium.

The calculated results and experimental values for the vibration acceleration in the octave bands at 31.5 and 125 Hz are presented in Table 13.1. The measurements were carried out in city conditions. The two last lines of the table contain calculated values for a homogeneous medium and are given for comparison. The measured values of wave velocity and dissipation in the medium for this case and the corresponding values calculated using the above-mentioned waveguide model are quite close.

13.4. Two-level elastic system for vibration reduction

One effective measure for the reduction of vibration from underground trains is acoustic isolation of the railway tracks using elastic layers. As a result, a shift in the frequency of oscillation of the inner layer of the tunnel towards the lower part of the spectrum may occur. This leads to a reduction of the amplitudes of vibration in all

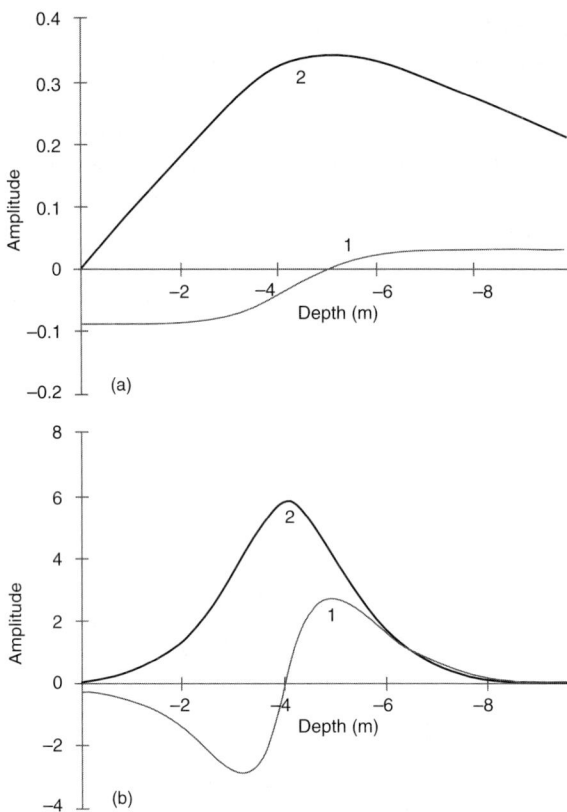

Fig. 13.5. Dependence of the vibration acceleration (curve 1) and pressure (curve 2) on depth (in relative units): (a) f = 31.5 Hz; (b) f = 125 Hz

Table 13.1. Calculations and measurements of the vibration acceleration under conditions of waveguide propagation

Frequency (Hz)	Level of x component of vibration acceleration (dB)		Level of z component of vibration acceleration (dB)	
	Waveguide case			
	Calculations	Measurements	Calculations	Measurements
31.5	22	23	24	23
125	24	26	30	28
	Homogeneous soil (calculation results)			
31.5	20		20	
125	22		23	

octave bands, except at the shifted resonance frequency. However, to obtain the required effect, careful consideration is necessary. Otherwise, a negative effect may be obtained.

In Fig. 13.6 some published results on the reduction of vibration are presented [13.9]. It can be seen that in the most important octave band, with a central frequency of 31.5 Hz, the measures previously considered can have a negative effect.

In this section, a two-unit system for vibration reduction is discussed [13.10]. A diagram of the system is shown in Fig. 13.7. Here m_1, m_2, m_3, m_4 and m_5 are the masses of the carriage, of the bogie, of the wheel pair and rail, of the ballast and of the concrete tunnel lining including the apparent additional mass of the soil, respectively; the two last parameters are given per unit length along the tunnel axis.

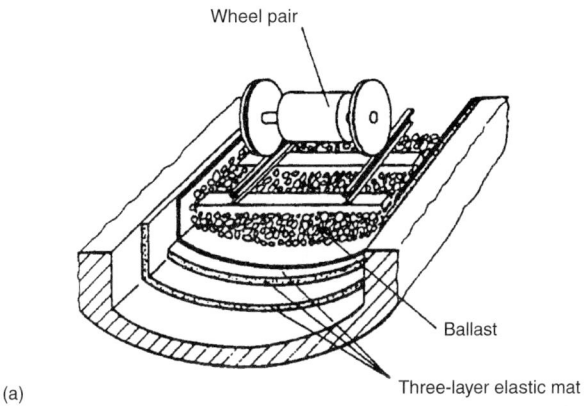

(a)

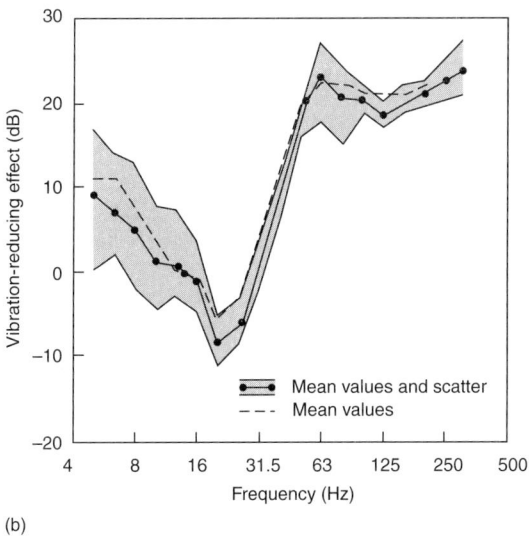

(b)

Fig. 13.6. A vibration-isolating system [13.10]: elastic mats under a ballast layer (a) and the frequency effect of their application (b)

The parameters k_1 and b_1, k_2 and b_2, k_3 and b_3, k_4 and b_4, and k_5 and b_5 are the elasticity and damping coefficients of the central suspension unit of the carriage, of the axle-box suspension unit, of the shock absorbers under the sleepers and of the ballast mat per unit length, respectively. The elastic parameters k_5 and b_5 were calculated using the known parameters of the soil and of the tunnel lining.

The added mass of soil and the elastic parameter of the surrounding medium per unit length are given by the following expressions [13.3]:

$$m_s = 4\pi \rho_s R^2$$
$$k^* = \frac{4\pi E(1-\nu)}{1+\nu} \tag{13.18}$$

Here ν and E are the Poisson's ratio and Young's modulus of the ground, ρ_s is the mass density of the soil and R is the tunnel radius.

In what follows, a one- and a two-level vibration reduction system (shock absorbers under the sleepers and elastic mats under the ballast layer) will be considered. The general equations describing oscillations in a system of this kind have the form

$$m_1 \ddot{x}_1 + b_a \dot{x}_1 + b_1(\dot{x}_1 - \dot{x}_2) + k_1(1 - i\eta_1)(x_1 - x_2) = F_1$$
$$m_2 \ddot{x}_2 + b_1(\dot{x}_2 - \dot{x}_1) + b_2(\dot{x}_2 - \dot{x}_3) + k_1(1 - i\eta_1)(x_2 - x_1) + k_2(1 - i\eta_2)(x_2 - x_3) = F_2$$
$$\vdots$$
$$m_j \ddot{x}_j + b_{j-1}(\dot{x}_j - \dot{x}_{j-1}) + b_j(\dot{x}_j - \dot{x}_{j+1}) + k_{j-1}(1 - i\eta_{j-1})(x_j - x_{j-1}) + k_j(1 - i\eta_j)(x_j - x_{j+1}) = F_j \tag{13.19}$$
$$\vdots$$
$$m_{n-1} \ddot{x}_{n-1} + b_{n-2}(\dot{x}_{n-1} - \dot{x}_{n-2}) + b_{n-1}(\dot{x}_{n-1} - \dot{x}_{n-2}) + k_{n-2}(1 - i\eta_{n-2})(x_{n-1} - x_{n-2}) + k_{n-1}(1 - i\eta_{n-1})(x_{n-1} - x_n) = F_{n-1}$$
$$m_n \ddot{x}_n + b_n(\dot{x}_n - \dot{x}_{n-1}) + k_{n-1}(1 - i\eta_{n-1})(x_n - x_{n-1}) + k_n(1 - i\eta_n)x_n = F_n$$

Here η_i is the imaginary part of the elastic modulus, F_i is the external force acting on the ith oscillator and b_a is the damping of the carriage.

The level of vibration reduction was calculated using the formula $\Delta L = 20 \lg(w_0/w)$, where w and w_0 are the vibration accelerations for the railway design with shock absorbers and for the basic construction (without shock absorbers), respectively.

The results of calculations of the reduction of vibration are shown in Fig. 13.8. Values of the parameters in the system of equations (13.19) were selected for realistic underground rolling stock and a realistic tunnel design. For example, m_1 was set equal to 40 and 55 t for an empty carriage and for a carriage with passengers, respectively, and the tunnel radius was 5.5 m. One can see that the system considered works quite well, with the exception of low frequencies.

GROUND VIBRATIONS GENERATED FROM TUNNELS

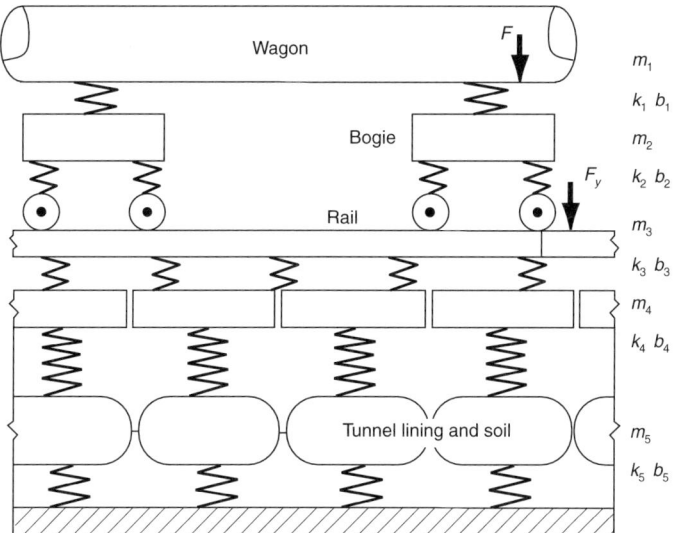

Fig. 13.7. Diagram of the two-level model system for reduction of vibration

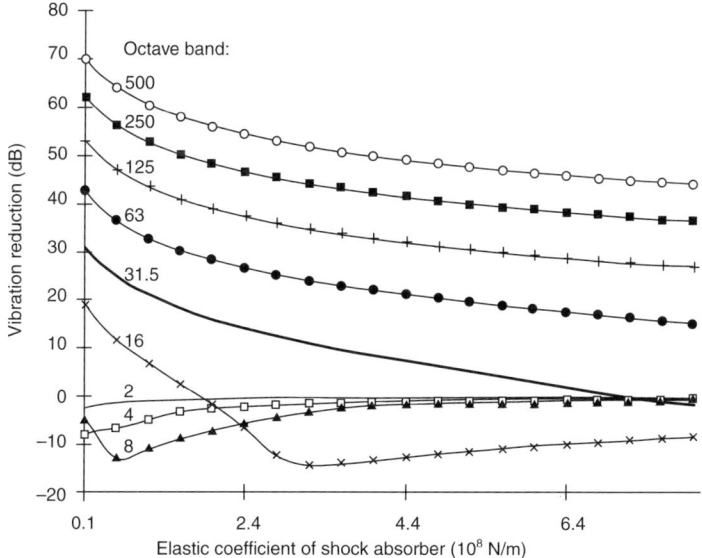

Fig. 13.8. Dependence of the vibration reduction on the rigidity of shock absorbers (per unit length) for the case of soft ground, with a velocity of longitudinal waves of 300 m/s

13.5. Method of estimation of the elastic parameters and damping of layered ground

The dynamic parameters of the ground required to predict the amplitudes of generated ground vibrations are usually determined by means of direct measurements or using available tabulated data (see, for example, Table 13.2). The

Table 13.2. *Typical wave velocities and values of damping parameter for some types of soil: the lower and upper limits of the velocity values correspond to soils with low and high mass densities, respectively*

Ground type	Density (kg/m³)	Longitudinal wave velocity (m/s)	Shear wave velocity (m/s)	Damping coefficient
Anthropogenic layer	1600	300	100	0.1
Coarse sand with water content $G < 0.8$	1700	500	150	0.1
Dense plastic loam	1700	600	250	0.15
Hard clay	1700	1500	350	0.15
Loess	1500	400	150	0.15–0.2
Peat	1000	200	80	0.2
Sandy silt	1500–1800	1100	300	0.2
Water-saturated soil with water content $G > 0.9$	2000	1750	250	0.1
Medium sandy soil	1400–1700	100–300	70–150	0.1–0.2
Gravelly sandy soil	1600–1900	200–500	100–250	0.1
Sandy loam	1600–2000	300–1200	120–600	0.1–0.15
Wet plastic clay	1700–2200	500–2800	130–200	0.2
Marl	1800–2600	1400–3500	800–2000	0.05–0.1
Friable sandstone	1800–2200	1500–2500	800–1700	0.1
Hard sandstone	2000–2600	2000–4300	1100–2500	0.05–0.1
Hard limestone	2000–3000	3000–6500	1500–3700	0.05
Argillaceous schist	2000–2800	2000–5000	1200–3000	0.05–0.1

latter frequently appear unacceptable because of a significant scatter of the parameter values for the chosen ground type. Therefore, it is recommended that one should use direct measurements whenever possible.

In what follows, the dynamic properties of the ground will be characterized by the velocities of longitudinal and shear elastic waves in the solid medium and by related parameters, such as the Young's modulus and Poisson's ratio. The dissipation properties will be characterized by the power damping factor η or by the amplitude factor β, related by $\beta = \eta/2$. The procedure described in this section enables the consideration of ground stratification with a limited number of layers ($1 \leq N_S \leq 3$). In this procedure it is assumed that the adjacent layers are characterized by essentially different physical properties. Note that a limit on the number of layers $N_S \leq 3$ has been imposed in view of practical requirements relating to the accuracy of the calculations. Numerous calculations and tests of the proposed method were carried out under real urban conditions in Moscow, St Petersburg and Ekaterinburg. It appears that a maximum number of layers of 3 is sufficient in most cases to reach the required accuracy of estimation (the maximum permitted error was 25%). Adaptation of the technique considered here for other types of geological structures may require some modification.

If the dynamic properties of two adjacent layers differ by a factor of less than 1.5 and the dissipative properties differ by a factor of less than 2, i.e.

$$0.67 < \frac{c^{i+1}}{c^i} < 1.5 \quad \text{and} \quad 0.5 < \frac{\beta_{i+1}}{\beta_i} < 2$$

the layers considered are merged into one with a combined thickness $h = h_i + h_{i+1}$, and the resulting average values of elastic-wave velocity and damping factor

$$c_{1,t} = \frac{c_{1,t}^i h_i + c_{1,t}^{i+1} h_{i+1}}{h_i + h_{i+1}}$$

$$\beta = \frac{\beta_1 h_i + \beta_2 h_{i+1}}{h_i + h_{i+1}}$$

(13.20)

are used. Here i and $i + 1$ are the numbers of the layers, and the indices 'l' and 't' correspond to longitudinal and shear waves. We shall consider only the upper part of the ground, above the depth

$$H = H_{\text{tun}} + 5 \text{ m} \qquad (13.21)$$

where H_{tun} is the distance from the ground surface to the lower boundary of the tunnel.

At the preliminary stage, it is desirable for reconstruction of the ground structure to be guided by a priori available geological information. An initial one-, two- or three-layer ground model can be formulated on the basis of such values (e.g. see Table 13.2) and the above rule for combining layers with close properties. In this case, the thickness of each layer is provided by available charts of geological layers, but the dynamic and damping parameters require further adjustment. If preliminary information about the geological structure of the ground is absent it is necessary to consider a full inverse problem with an unknown number of layers, an unknown thickness of each layer and unknown corresponding dynamic and damping properties. Because the mass density of the ground varies insignificantly (within the limits 1600–2000 kg/m³), it can be considered as constant, $\rho = 1800$ kg/m³. Poisson's ratio for typical water-saturated soils varies from 0.35 to 0.45. On this basis the value of Poisson's ratio adopted can be $\nu = 0.4$.

The solution to the inverse problem of evaluation of the soil properties can be obtained by direct simulation. The calculated levels of ground vibration are compared with the experimental data and their residuals are minimized. The calculation scheme has been briefly described in the previous sections and has been implemented in a computer code [13.11]. As the initial approximation, the lowest values of the velocity of longitudinal waves and of the damping factor for the ground considered are used (see Table 13.2). If the residual obtained is within the accuracy of the corresponding experimental data (typically within 25%), then the problem of estimation of the parameters is considered as solved. After estimation of the longitudinal-wave velocity, the velocity of shear waves is calculated using the expression

$$c_t = c_1 \left(\frac{1-2\nu}{2(1-\nu)} \right)^{1/2} \qquad (13.22)$$

for the value of Poisson's ratio ν adopted. If the required accuracy is not achieved in the first step, it is necessary to vary the initial values of the parameters. First, it is advisable to change the value of the damping factor. This value is modified at each subsequent iteration so that

$$\beta = \beta_{min} + i\Delta\beta, \qquad i = 1, 2, 3 \qquad (13.23)$$

where the increment $\Delta\beta$ is defined as follows:

$$\Delta\beta = \frac{\beta_{max} - \beta_{min}}{3} \qquad (13.24)$$

Here β_{max} and β_{min} are the maximum and minimum values of the damping factor for a given type of ground (see Table 13.2). At this iteration, the problem is considered as solved for the selected values of parameters if a satisfactory agreement with the experimental data is achieved. Otherwise, the velocity of longitudinal waves for each ground layer is varied as well. The value of the velocity at the ith iteration is set to

$$c_i = c_{min} d^i, \qquad i = 1, 2, 3, \ldots \qquad (13.25)$$

where factor D is selected on the basis of the value of the ratio c_{max}/c_{min} as presented in Table 13.3. Here c_{min} and c_{max} are the minimum and maximum velocities of longitudinal waves for the given type of ground.

If the required accuracy is achieved at the current step, the problem of the parameter characterization is considered as solved. The general flow chart for the procedure described above is presented in Fig. 13.9.

The approach described above is based on the use of experimental data for vibration levels at different distances from a standard source. To test this approach, a series of measurements of vibration levels transmitted to the ground surface from a standard impulsive source has been carried out. The results have shown that the approach provides satisfactory accuracy and can be applied to obtain a reliable evaluation of parameters of water-saturated soils with the aim of predicting ground vibrations from underground railways.

In the measurements, the vibrations on the ground surface were generated by calibrated explosions. The explosive power was 50 g of TNT equivalent (a preliminary assessment showed that under city conditions the power of the controlled burst must be within the interval 50–100). The source was placed in a borehole. The borehole was drilled in sandy soil and its upper part was strengthened with a concrete casing. The source depth was 10 m. The experiment is illustrated schematically in Fig. 13.10. The absolute values of the vertical and horizontal components of the vibration at the ground surface were measured at six points at distances of 0, 10, 20, 30 and 40 m from the borehole (points 0–5 in Fig. 13.10). Points 1 and 2 were located on the opposite sides of the borehole mouth. Measurements at point 1 were carried out for reference and were averaged later with vibration amplitudes obtained at point 2. The cross-section of the proposed tunnel is illustrated on the left-hand side of Fig. 13.10. The source was suspended on a rope at the depth of the proposed tunnel. The measurements were expected to provide vibration amplitudes in octave bands at 16, 31.5 and 63 Hz. In addition, a fast Fourier transform package was used for spectral analysis of the measured signal,

Table 13.3. Definition of factor D

c_{max}/c_{min}	D
≤ 3	$\sqrt{c_{max}/c_{min}}$
≤ 6	$\sqrt[3]{c_{max}/c_{min}}$
> 6	$\sqrt[4]{c_{max}/c_{min}}$

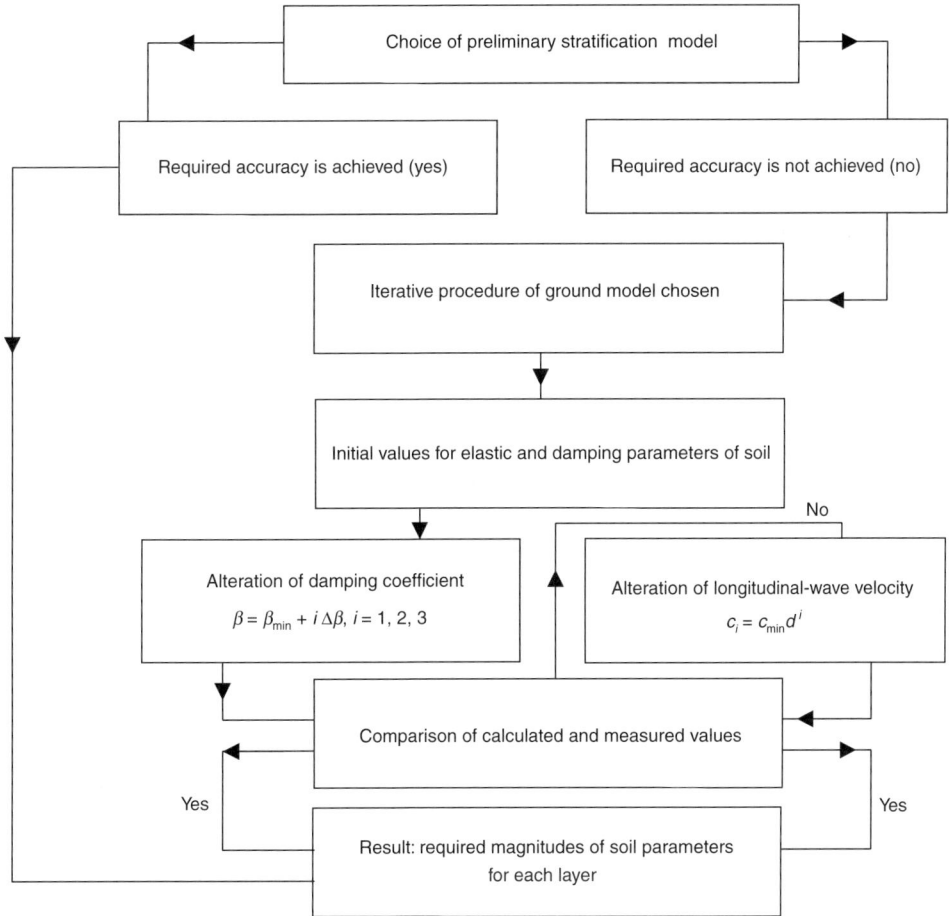

Fig. 13.9. Flow chart of the algorithm for estimation of elastic and damping parameters

with averaging over $\frac{1}{3}$ octave and octave bands. The length of the recorded file and the sampling frequency provided a resolution of 0.25 Hz. The averaged results of spectral analysis of the experimental data are shown in Tables 13.4 and 13.5.

A priori information on the ground geology was available at the time when the measurements were carried out. The preliminary data on the properties of soils are presented in Table 13.6. The initial minimal and maximal parameter values used in the calculations (columns 4–7) are presented in Table 13.2.

The criterion applied for the correct estimation of the ground parameters m_j was the requirement for a minimum of the following residual function:

$$F(m) = \sqrt{\sum_{i=0}^{N} W(x_i)[f(x_i, m_j) - \phi(x_i)]^2}$$

Here $\phi(x_i)$ are the experimental values at the points x_i, $f(x_i, m_j)$ are the results of direct calculations using the computer code [13.11], $W(x_i)$ is a weight function

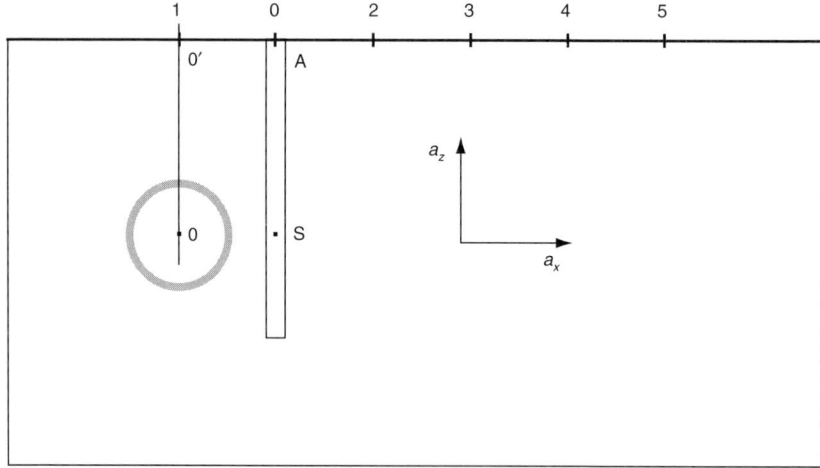

Fig. 13.10. Experimental set-up. Point O indicates the level of the tunnel axis; AS is a borehole in which an explosive source S is placed; $|AS| = |OO'|$, $|O'A| < 10$ m; 0, 1, ..., 5 are the measurement points; a_z and a_x are the measured amplitudes of the vertical and horizontal vibration components

Table 13.4. Measured values (mm/s²) of the vertical component of the vibration amplitude

Geometric frequency of octave band (Hz)	Distance from borehole (m)				
	0	10	20	30	40
16	15	15	15	8	5
31.5	30	28	15	8.2	4.5
63	56	51	27	14	7.5

GROUND VIBRATIONS GENERATED FROM TUNNELS

Table 13.5. Measured values of the horizontal component of the vibration amplitude (mm/s^2)

Mean geometric frequency of octave band (Hz)	Distance from the borehole (m)				
	0	10	20	30	40
16	17	17	17	15	3
31.5	34	34	30	13	5.3
63	67	53	59	26	8.4

Table 13.6. Preliminary data on properties of soils

Stratum number	Soil type	Stratum thickness (m)	c_{min} (m/s)	c_{max} (m/s)	β_{min}	β_{max}
1	Loam soil	6	500	2800	0.1	0.4
2	Sandy soil	20	500	500	0.05	0.2

Table 13.7. Results of the minimization procedure

Stratum number	Longitudinal-wave velocity, c (m/s)			Damping coefficient, β		
	16 Hz	31.5 Hz	63 Hz	16 Hz	31.5 Hz	63 Hz
1	800	900	1000	0.2	0.15	0.1
2	600	700	800	0.15	0.1	0.06

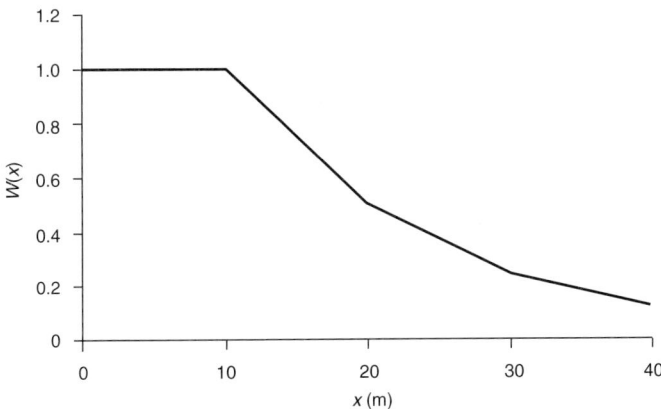

Fig. 13.11. Weight function $W(x)$ used to increase the accuracy of predictions in the vicinity of the source

defined to be positive, designed to improve the accuracy of the prediction of the vibration amplitudes at a given distance, and the subscript j defines the specific parameter evaluated: $m_1 = c_1, m_2 = \beta_1, m_3 = c_2, m_4 = \beta_2$.

The results of the minimization procedure are presented in Table 13.7. The weight function $W(x_i)$ used in the calculation is shown in Fig. 13.11. A comparison of the measured vibration levels and those calculated with the weight function $W(x_i)$ is presented in Fig. 13.12. This figure shows the dependence of the vertical acceleration component on the distance from the source in the octave bands at 16, 31.5 and 63 Hz. Vibration levels were calculated relative to a reference level of 10^{-6} m/s².

The analysis of the results presented in Fig. 13.12 shows that this procedure for estimation of parameters provides reliable predictions at relatively short distances. The accuracy of the prediction deteriorates at greater distances from the source. This phenomenon is clearly observed in the octave bands with centre frequencies of 16 and 31.5 Hz. To improve the accuracy in all frequency bands, a more complex model should be used to allow for more than a two-layer stratification. A three-layer model using data for the two upper strata presented in Table 13.6 was considered. The third layer was selected to fit the experimental data. The results of this fitting procedure are presented in Table 13.8 and Fig. 13.13. Comparison of Figs 13.12 and 13.13 suggests that the three-layer model is more suitable than the initial geological approximation.

Application of the simplified procedure described above allows one to estimate the ground parameters with a precision sufficient for vibration prediction in the vicinity of underground railways. A knowledge of a priori geological information is useful for reliable predictions but is not crucial. The results described here may be used for developing a more precise approach, which may be based on an adaptive orthogonal expansion of the recorded signals [13.12, 13.13].

13.6. Discussion

In this chapter, a simple approximation of elastic ground as a purely volumetric medium (the 'liquid ground' approximation) has been applied to solve the problem of generation and propagation of elastic waves from underground tunnels constructed in water-saturated soils. Calculations and measurements indicate that the approximation of 'liquid ground' does not result in significant errors in the predictions of generated ground vibration levels. The 'liquid ground' approximation has also been applied for the description of waveguide propagation of the generated ground vibrations.

The magnitudes of generated vibrations depend greatly on the tunnel design, which may be rather complicated. However, a simplified elastic-shell model can be used to include the effect of oscillations of the tunnel wall. A shell approximation allows one to investigate in detail the mode structure of the radiated wave field and its main components. As a result, the desired accuracy can be achieved with small computational effort.

GROUND VIBRATIONS GENERATED FROM TUNNELS

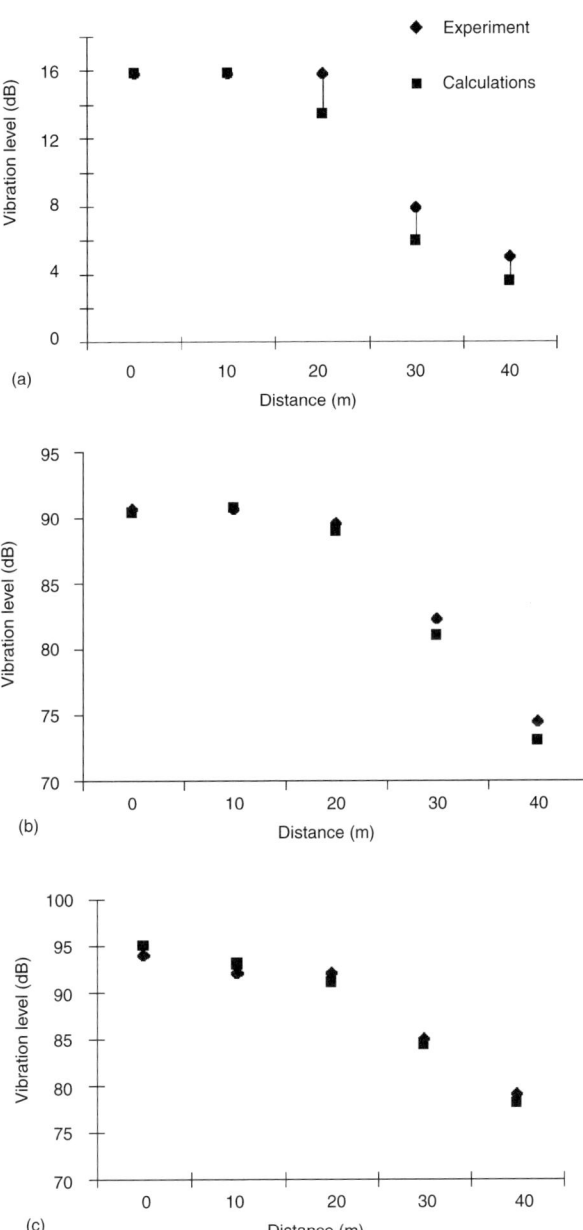

Fig. 13.12. Comparison of measured and predicted vibration levels for a two-layer model of the ground: (a) f = *16 Hz; (b)* f = *31.5 Hz; (c)* f = *63 Hz*

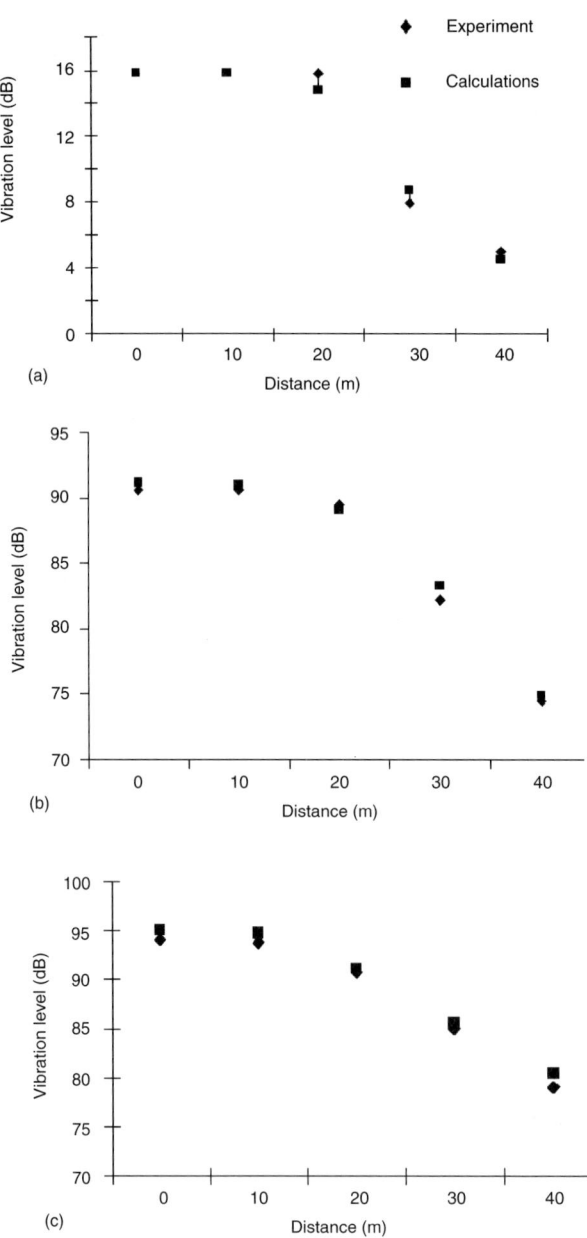

Fig. 13.13. Comparison of measured and predicted vibration levels for a three-layer model of the ground: (a) f = *16 Hz; (b)* f = *31.5 Hz; (c)* f = *63 Hz*

Table 13.8. *Results of the fitting procedure using the three-layer model*

Stratum number	Stratum thickness (m)	Longitudinal-wave velocity, c (m/s)			Damping coefficient, β		
		16 Hz	31.5 Hz	63 Hz	16 Hz	31.5 Hz	63 Hz
1	6	750	900	1000	0.2	0.15	0.1
2	6	600	700	800	0.15	0.1	0.06
3	14	1200	1250	1250	0.15	0.1	0.06

Application of the simplified procedure described above for estimating the elastic moduli and damping of layered ground allows one to obtain values of the parameters with satisfactory precision in most of the cases related to problems of predicting ground vibrations generated by underground railways. A knowledge of a priori geological information is useful for reliable predictions but is not crucial. The results described in this chapter have passed a number of practical checks on several sections of the Moscow underground.

13.7. Acknowledgements

We are grateful to Dr K. Horoshenkov for reading a draft of this paper and for useful comments. We also acknowledge financial support from the Russian Foundation for Basic Research (projects 98-02-16833 and 97-01-00526).

13.8. References

13.1. MUZYCHENKO, V. V. and RYBAK, S. A. Low-frequency scattering of a sound by bounded shells. Review. *Soviet Physics – Acoustics*, 1988, **34**(4), 325.
13.2. BREKHOVSKIKH, L. M. and GONCHAROV, V. S. *Mechanics of Continua and Wave Dynamics*. Springer, Berlin, 1985.
13.3. LANDAU, L. D. and LIFSHITZ, E. M. *Theory of Elasticity*. Pergamon, Oxford, 1959.
13.4. TOLSTOV, G. P. *Fourier Series*. Prentice Hall, Englewood Cliffs, 1962.
13.5. ABRAMOVITZ, M. and STEGUN, I. A. *Handbook of Mathematical Functions, with Formulas, Graphs, and Mathematical Tables*. Dover, New York, 1965.
13.6. ISAKOVICH, M. A. *General Acoustics*. Nauka, Moscow, 1973 [in Russian].
13.7. RYBAK, S. A. and TARTAKOVSKY, B. D. Some applications of the transition matrix to the theory of plane waves in the elastic layers system. *Soviet Physics – Acoustics*, 1962, **8**(1), 88–90.
13.8. TOMPSON, W. H. Transmission of elastic waves through a stratified solid material. *Journal of Applied Physics*, 1950, **21**(2), 89–93.
13.9. WETTSCHURECK, R. Ballast mats in tunnels – analytical model and measurement. *Proceedings of Internoise '85*. Munich, 1985.
13.10. KOSTAREV, S. A., MAKHORTYKH, S. A. and RYBAK, S. A. Two-unit elastic system for vibration reduction of underground railway. *Proceedings of Transport Noise '98*. Tallinn, 1998.
13.11. KOSTAREV, S. A., MAKHORTYKH, S. A. and RYBAK, S. A. Calculations of vibration, induced by underground sources in soil. Software package UNSONIC, version 2.1. *Proceedings of the 12th International FASE Symposium on Transport Noise and Vibration*. St Petersburg, 1996, pp. 79–82.

13.12. DEDUS, F. F., DEDUS, A. F., MAKHORTYH, S. A. and USTININ, M. N. Generalized spectral-analytical method: theoretical foundations. *Proceedings of SPIE*, 1995, **2363**, 109–112.

13.13. DEDUS, F. F., MAKHORTYH, S. A. and USTININ, M. N. Generalized spectral-analytical method: applications. *Proceedings of SPIE*, 1995, **2363**, 113–116.

14. Measures for reducing ground vibration generated by trains in tunnels

H. E. M. Hunt
Department of Engineering, University of Cambridge, UK

14.1. Introduction

All high-speed trains generate vibration in the ground, but trains in tunnels deserve special treatment for three main reasons. Firstly, tunnels are often found in densely populated areas and low vibration levels are desirable. Secondly, space and access restrictions limit the range of solutions that is available to the track designer. Thirdly, the flexural modes of the tunnel wall along with the waveguide characteristics of the tunnel and track itself lead to curious effects that call into question some of the traditional methods of vibration isolation. This chapter will address the effects of these tunnel modes and the interaction of the tunnel with a floating track slab.

Analytical modelling of a long, circular tunnel has a computational advantage. It is possible to consider the tunnel and track to be an infinitely long periodic structure and the train itself can be considered to be infinitely long. Periodically spaced axle loads may be treated in this way using random-process theory. The whole three-dimensional problem is thereby reduced to two dimensions. Furthermore, axial symmetry enables the cylindrical modes of tunnel vibration to be assembled as a Fourier series. This is possible provided that the free surface of the ground is not close to the tunnel and that the soil is homogeneous in the vicinity of the tunnel. These assumptions are good from the point of view of evaluating vibration countermeasures in the track and the tunnel, as only near-field effects are important. This approach to modelling complements alternatives using boundary element methods, but the latter cannot sensibly be used for design, owing to the lengthy computations involved.

The main conclusion reached in this chapter is that the effectiveness of vibration countermeasures in tunnels is strongly influenced by the dynamics of the tunnel itself. In some cases it may be better not to isolate the track, because this increases the level of transmitted vibration in some critical frequency range. The reasons are complex, involving analysis of the torsional modes of the track slab, the flexural modes of the tunnel, radiation efficiency and waveguide effects. These conclusions should be of interest to designers of new tunnels as well as to those responsible for maintaining and improving existing infrastructure.

14.2. Tunnels with floating slabs

Vibration generated by rail traffic is now a well-established field of research on account of the availability of computing power, by which it is possible to solve complicated three-dimensional wave propagation problems (e.g. see [14.1–14.11]). The first in-depth study of ground vibration from railways was published in the 1970s [14.2], and since then a wide range of methods have been established to control vibration propagation from the track. For tunnels, popular vibration countermeasures include the Cologne Egg, the Clouth system and the Eisenmann track, a form of floating-slab track [14.9]. At the same time, many different computational procedures have evolved to assess the performance of vibration countermeasures and also to make absolute predictions of vibration levels in structures adjacent to railways. Early analytical approaches for vibration from surface railways employ well-established results for an elastic half-space [14.10, 14.12] and there have been many attempts to fit empirical relationships to measured vibration levels in an attempt to circumvent the mathematical complexity of the problem [14.6]. However, the boundary element method (BEM) is now the most widely used method for calculations of vibration transmission because of its ability to deal with complex ground structures. Boundary element methods have been used to model such phenomena as soil–structure interaction [14.1], wave-impeding blocks [14.14], vibration from railways [14.13] and vibration in stratified soils, including stochastic effects of structures embedded in the ground [14.3].

A common method used to reduce vibration transmitted into nearby buildings is to 'float' the track slab. The track is fixed through rail pads to a massive concrete slab, which itself is isolated from the tunnel invert by means of rubber bearings or steel springs. Such an underground railway system is illustrated in Fig. 14.1. In conventional analyses of the performance of such systems it is usual to assume that

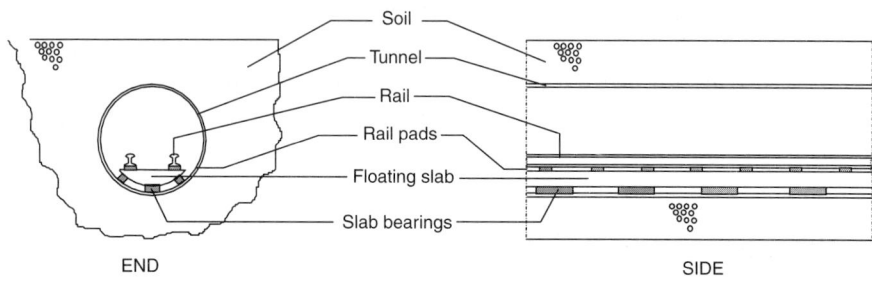

Fig. 14.1. Layout of an underground railway line and its various components

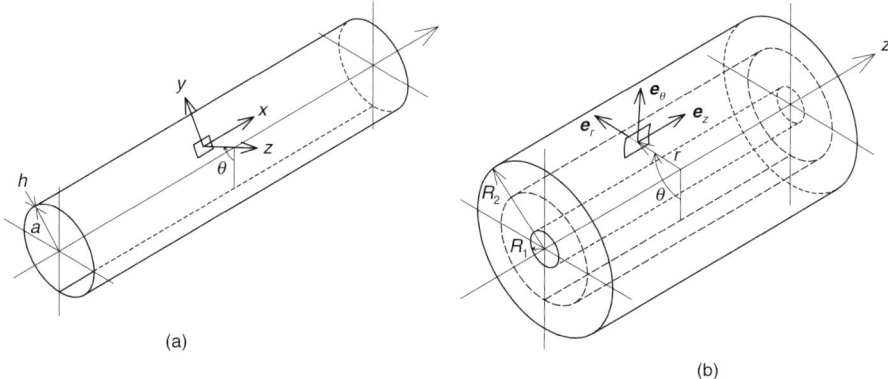

Fig. 14.2. (a) Cylindrical shell and (b) continuum used to model the tunnel and soil

the slab support is sufficiently soft and that it is decoupled from the tunnel invert, which can therefore be considered rigid. The analysis which usually then follows is based on Winkler beam theory. The force transmitted through the slab support springs is less than the axle force applied to the track by several tens of dB at frequencies well above the 'natural frequency' of the foundation. On the basis of these results, it is not unusual for a floating slab with a 30 Hz natural frequency to be used to attenuate underground-railway vibration, commonly centred around 80 Hz. It is, however, inappropriate to use the Winkler model for transmitted-force calculations, as can easily be demonstrated by calculating the total force transmitted to the ground by the beam [14.11]. It transpires that a simple single-degree-of-freedom mass-on-spring model gives exactly the same results as the more elaborate Winkler beam model. Another obvious limitation of the Winkler model is that the tunnel invert cannot be considered rigid. Interactions between tunnel vibration modes and track–slab modes are very significant and the computation of the overall attenuation of vibration becomes a complicated process.

Although researchers have in the past recognized the importance of interaction, the computational tools suitable for this purpose (usually the FEM and BEM) are expensive and not very flexible for investigative purposes. Two-dimensional methods are incomplete, as they cannot account simultaneously for both circumferential and longitudinal modes of the tunnel. For these reasons, an analytical track–tunnel–soil model has been developed. The model assumes that the tunnel and track are infinitely long and that there is no layering and no free surface to the soil. These are not serious limitations for an investigation of the effectiveness of floating track, because accurate modelling is required only for those elements nearest to the floating slab.

14.3. Vibration from railway tunnels

The essential element of the model used here is a pair of concentric thick-walled cylinders, as depicted in Fig. 14.2. For details of the model, the reader is referred to the work of Forrest [14.4]. In outline, the methodology involves three steps. The first is to establish a model for the coupling between an infinitely long circular tunnel and an infinite solid in which it is embedded, as depicted in Fig. 14.2. The solution for the motion of this cylinder follows the method employed by Gazis [14.7] to investigate the modes of thick-walled cylindrical shells and developed further by Köpke [14.8] to model the dynamics of buried undersea pipelines. The problem is formulated analytically in the frequency–wavenumber domain using a discrete Fourier decomposition of the cylindrical harmonics. The solution is obtained for particular boundary conditions to include the details of the applied load, the compatibility conditions at the tunnel–soil interface and a radiation condition for an infinite soil. In general, the loading applied to the tunnel will not be cylindrically harmonic, so a discrete Fourier decomposition of the loading is required.

The second stage is to produce a model for the track and slab as infinite beams that may be coupled to the base of the tunnel, as shown in Fig. 14.3. This model may include bending and torsional effects, and the coupling with the tunnel invert need not be along a single line at the base of the tunnel. In this way, the effect of torsional coupling between the slab and the tunnel may be investigated with a view to establishing new ways to localize vibrational energy within the tunnel structure itself. It has never been easy to conceptualize the three-dimensional interactions between the slab and the tunnel.

The final stage is to include a model for an infinitely long train and to incorporate random rail roughness as the mechanism for vibration generation, taking into account the various propagation paths (Fig. 14.4). The approach used is based on random process theory as applied to loads on the surface of a semi-infinite solid [14.10]. Correlation between adjacent loads is therefore taken into account. The input to the model is the random track roughness. The roughness of each rail is likely to be uncorrelated at high frequencies, reflecting local surface irregularities of the individual rails and correlated at low frequencies on account of irregularities of the slab.

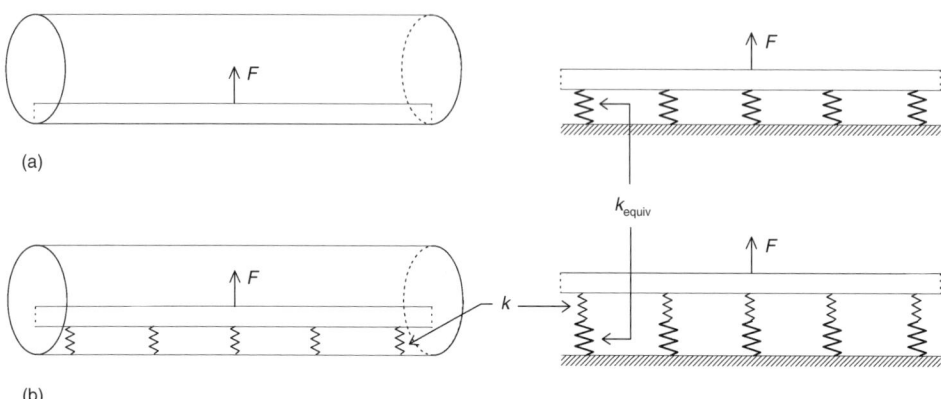

Fig. 14.3. A slab beam fixed (a) directly to the tunnel invert and (b) with a resilient layer between, to illustrate the concept of 'equivalent stiffness' k_{equiv}

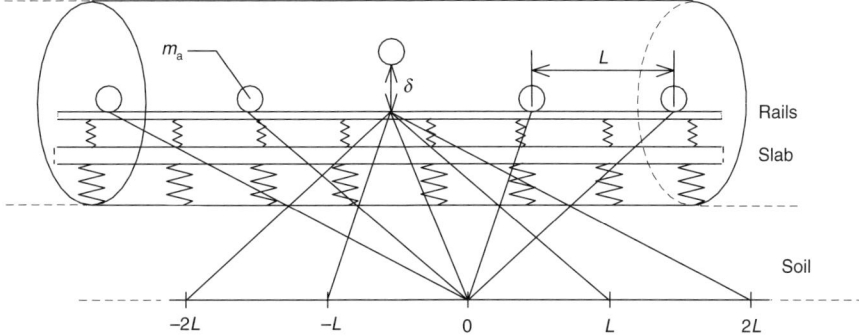

Fig. 14.4. Full track model with masses added to represent axle–wheel assemblies of a train excited by rail roughness δ. *For a model of infinite length, the responses at a single point in the soil to each axle input (paths shown) are equivalent to a line of separate responses to a single input at the middle axle (paths shown)*

A case study is presented below, for which the tunnel consists of concrete, with a diameter of 6 m and a thickness of 250 mm. The surrounding soil has compressive- and shear-wave speeds of 944 m/s and 309 m/s, respectively. The floating concrete slab has a flexural rigidity $EI = 1430$ MN m^2 and a mass per unit length of 3500 kg/m. The two rails have $EI = 10$ MN m^2 and a mass per unit length of 100 kg/m, and the rail pads have a distributed stiffness of 400 MN/m^2. Damping is incorporated by means of complex moduli, and the loss factors for concrete, steel, soil and rubber are taken as 0, 0, 0.06 and 0.3, respectively. The train has an axle mass of 500 kg, and the distance between axles is 20 m.

Central to the presentation of the results below is the concept of 'equivalent Winkler stiffness' k_{equiv}, as illustrated in Fig. 14.3(a). The parameter k_{equiv} is the Winkler subgrade stiffness required to give static deflections equal to the deflection calculated for a slab rigidly connected to the tunnel invert. The value of subgrade stiffness k required to give 'natural frequencies' of the slab of 60 Hz, 45 Hz and 30 Hz (as conventionally defined with reference to the Winkler model) can be applied in series with the equivalent stiffness, as shown in Fig. 14.3(b). In this study the Winkler equivalent stiffness $k_{equiv} = 821$ MN/m and the spring stiffness k in Fig. 14.3 is 1262 MN/m^2, 424 MN/m^2 and 147 MN/m^2 for 60 Hz, 45 Hz and 30 Hz slabs, respectively.

Two examples of ground vibration responses with a simple slab beam in the tunnel are shown in Fig. 14.5. Figure 14.5(a) shows the response 20 m away from the tunnel, and a classic isolation performance is observed. But at 40 m along the track from an applied load, Fig. 14.5(b), shows *higher* responses for the floating track. This can be explained by considering the fact that vibrational energy can be propagated along the slab as flexural waves, and slab motion is then transmitted into the soil. The vibration levels at these two points 40 m apart are comparable (both around −120 dB) and it is clearly important to incorporate the effect of multiple inputs from consecutive axles, and these will be seen below to result in large vibration levels in almost all areas around the tunnel.

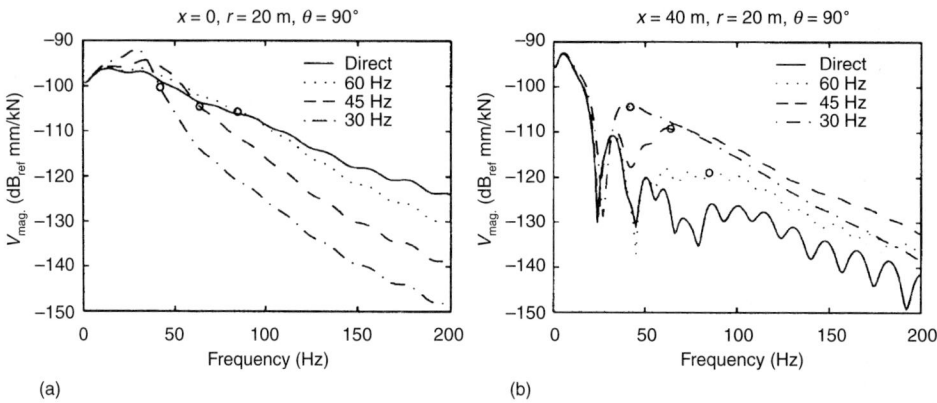

Fig. 14.5. Vertical soil displacement responses to a point load acting at x = 0 on a simple slab track: (a) 20 m horizontally out into the soil and (b) 20 m horizontally out and 40 m longitudinally parallel to the tunnel from the centre of the tunnel. The circles indicate the '√2 points' above which vibration isolation becomes effective using simple SDOF models

Taking a summation of inputs according to Fig. 14.4 and using random-process theory, the RMS vibration levels at different radii for different slab support 'frequencies' are given in Fig. 14.6. The RMS level was obtained by using standard rail roughness spectra [14.5] and by integrating the response over frequencies up to 200 Hz. It can be seen that up to 5 dB reduction can be obtained close to the tunnel, but that further away from the tunnel there is a marginal – and sometimes adverse – effect. The effect of doubling the speed of the train is to increase all vibration levels by 5 dB, and halving the train speed reduces all responses by 5 dB. There is no evidence of any consistent vibration reduction, and certainly nothing like the tens-of-dB reduction that many practitioners like to claim. It is only for floating-track systems with very low frequencies (say 15 Hz or lower) that some noticeable isolation performance may be obtained, and such systems are expensive. In other words, cheap floating-track systems are probably not worth having.

There are two further observations that may be made from Fig. 14.6. Firstly, the small beneficial effect of floating-slab systems is only noticeable directly below the tunnel ($\theta = 0$); to the sides and above the tunnel there is no observable effect. Secondly, the benefit of a floating slab is only felt very close to the track. These are important observations because buildings and their foundations generally pass through these unaffected regions, further calling into question the efficacy of floating-slab systems for controlling the transmission of vibration from railways into buildings. These observations are summarized graphically in Fig. 14.7, in which the regions where a floating slab leads to vibration reduction are shown in grey. A line representing a supposed ground level (the model itself has no free surface) indicates that there would be no benefit at ground level. Most of the benefit is close to the tunnel and at some distance directly beneath the tunnel.

The computational model described here suggests that predictions of insertion loss performance of above 5 dB are in many cases exaggerated and that the

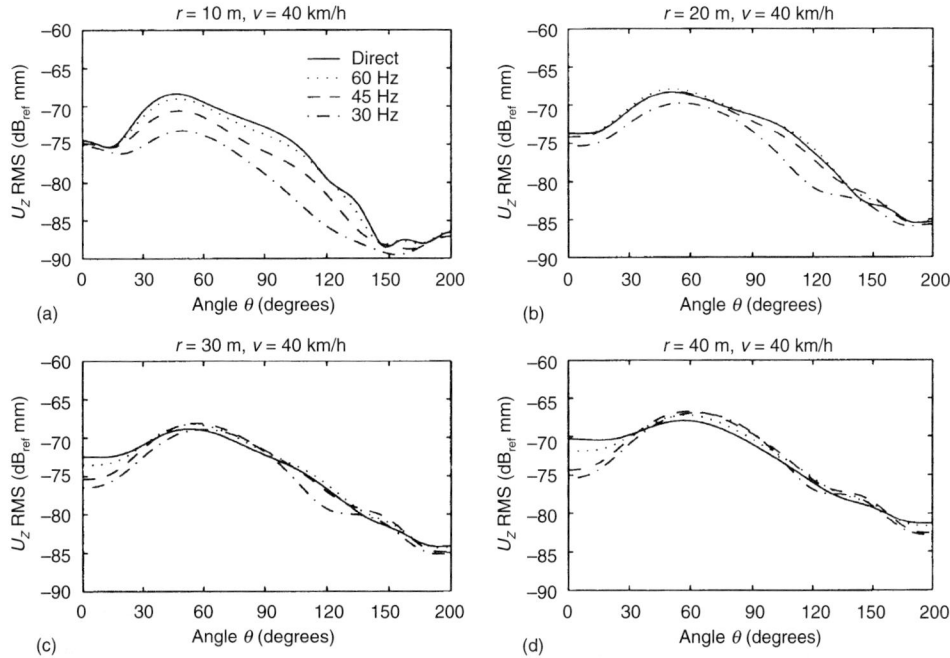

Fig. 14.6. RMS values of vertical soil displacement for a train speed of 40 km/h showing the effect of different stiffnesses of slab bearings for radii of (a) 10 m, (b) 20 m, (c) 30 m and (d) 40 m. Input spectrum for realistic rail irregularities

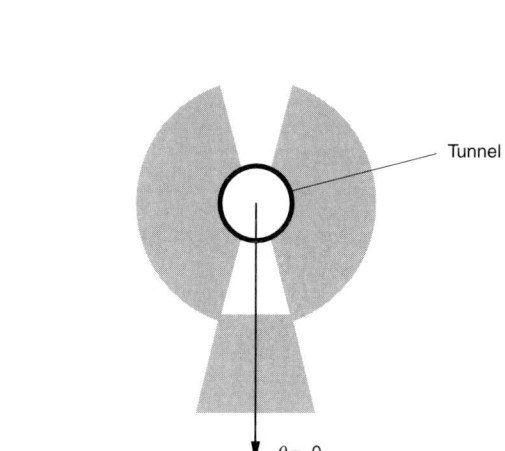

Fig. 14.7. Schematic illustration showing distribution of isolation performance around the tunnel. Regions showing vibration reduction are shown in grey. In the unshaded regions the floating slab has no effect

technique of floating the track slab may in fact cause increased transmission of vibration to surrounding buildings. These are predictions that cannot be deduced with the simple mass–spring models commonly used in the design of railway tracks for isolation of vibration.

14.4. Conclusions

The interactions between a floating-slab track, a tunnel and the soil are complex processes. Taking into account the effects of simultaneous excitation of vibration by all the axles of a long train makes it clear that there is little advantage to be obtained by using medium-frequency floating-slab track construction in railway tunnels. The beneficial effect of floating slab construction is only to be found very close to the tunnel or in the region directly beneath the tunnel.

14.5. References

14.1. AUBRY, D. and CLOUTEAU, D. A regularized boundary element method for stratified media. *Proceedings of the 1st International Conference on Mathematical and Numerical Aspects of Wave Propagation Phenomena, SIAM, Strasbourg.* 1991, pp. 660–668.

14.2. DAWN, T. M. and STANWORTH, C. G. Ground vibration from passing trains. *Journal of Sound and Vibration*, 1979, **66**(3), 355–362.

14.3. CLOUTEAU, D., SAVIN, E., ELHABRE, M. L. and AUBRY, D. Stochastic BEM for large structures embedded in an elastic halfspace. In: *Mathematical Aspects of Boundary Element Methods* (eds M. Bonnet, A.-M. Sändig and W. L. Werland). Chapman and Hall, London, 1999, pp. 91–103.

14.4. FORREST, J. A. Modelling of ground vibration from underground railways. PhD thesis, University of Cambridge, 1999.

14.5. FREDERICH, F. Die Gleislage – aus fahrzeugtechnischer Sicht. [Effect of track geometry on vehicle performance.] *Zeitschrift für Eisenbahnwesen und Vekehrstechnik – Glasers Annalen*, 1984, **108**(12), 355–362.

14.6. FUJIKAKE, T. A prediction method for the propagation of ground vibration from railway trains. *Journal of Sound and Vibration*, 1986, **111**, 357–360.

14.7. GAZIS, D. C. Three-dimensional investigation of the propagation of waves in hollow circular cylinders. I. Analytical foundation. *Journal of the Acoustical Society of America*, 1959, **31**(5), 568–573.

14.8. KÖPKE, U. G. Transverse vibration of buried pipelines due to internal excitation at a point. *Proceedings of the Institution of Mechanical Engineers, Part E: Journal of Process Mechanical Engineering*, 1993, **207**(E1), 41–58.

14.9. HEMSWORTH, B. Reducing groundborne vibrations: state-of-the-art study. *Journal of Sound and Vibration*, 2000, **231**(3), 703–709.

14.10. HUNT, H. E. M. Stochastic modelling of traffic-induced ground vibration. *Journal of Sound and Vibration*, 1991, **144**(1), 53–70.

14.11. HUNT, H. E. M. Floating slab track for vibration reduction: why simple models don't work. *Proceedings of the 7th International Congress on Sound and Vibration.* Garmisch-Partenchirchen, 2000.

14.12. JONES, D. V. and PETYT, M. Ground vibration in the vicinity of a strip load: a two-dimensional half-space model. *Journal of Sound and Vibration*, 1991, **147**(1), 155–166.

14.13 TAKEMIYA, H. and GODA, K. Prediction of ground vibration induced by high-speed train operation. *Proceedings of the 6th International Congress on Sound and Vibration.* Adelaide, 1997, pp. 2681–2688.

14.14. SHENG, X., JONES, C. J. C. and PETYT, M. Ground vibration generated by a load moving along a railway track. *Journal of Sound and Vibration*, 1999, **228**(1), 129–156.

Index

Page numbers in *italics* refer to figures

accelerometers in roughness measurement 45–6, *45*
acoustic shock waves 213–45
   dispersion 215
   generation 213, *213*
   pressure disturbances in tunnels 216–18
aerodynamic noise 120, 154
Airy phase 365, 368
antenna measurements 122–5, *124*
auxiliary equipment, noise from 120
average-roughness model 31
A-weighted average sound level 66, 86, 87, 146, 159

ballastless track ('slab track') forms 177
barrier tests 79–81
Bloch wave function 232
Bloch wavenumber 232
blocked force 30
bogie shields 180, *181*
boundary element method (BEM) 65, 94, 424
Bragg reflection 233–4
Burgers equation 227

Cagniard–de Hoop technique 348
circular frequency 9
computational-fluid-dynamics (CFD) models 22
cone penetration testing (CPT) 329
constrained layer damping 165–6, *165*
contact area increase 56–61, *57*
contact filter 17
contact patch 17
contact stiffness reduction 56–61, *57*

conventional diffraction theory 65
cross-hole method of determining shear wave velocity vs depth 328
curve and braking squeal 120
cut-off wavelength 43
cylindrical oscillating shell, waves radiated by 398–405

decibel level 86
directivity index 98
discrete wavenumber technique 348–9
dispersion of acoustic waves 215
displacement-based devices in roughness measurement 45, 46–8
distributed point-reacting spring (DPRS) model 17–18, 33–9, 60
   of roughness excitation 36, 37
down-hole method of determining shear wave velocity vs depth 328–9

Eisenmann track 424
equivalent radii 35
equivalent roughness *60*, 61
equivalent sound level 106–7
equivalent Winkler stiffness 427
Eurosabot 164
exterior noise measurement 120–58
   antenna systems 122–5, *124*
   diagnostics 122–5
   monitoring 144–9
      permanent emission monitoring 145–6, 146
      permanent perception monitoring 146
      special noise perception measurement techniques 146–9

non-acoustic factors influencing exterior
rail noise 149–58
  braking 158
  speed 153–8
  wheel roughness and rail roughness
    149–52
  simplified approach 125
  type testing 126–44
    average level vs maximum level 138–41
    distance between track and microphone
      142–3
    frequency weighting 142
    general measurement conditions 143
    rail noise directivity 132–8
    regulations and noise limits 143–4, 144
    sound pressure level vs sound power
      level 129–32
    time weighting 138
extra excess attenuation 79

far (pressure) field 218
  analysis 223–9
    evolution of pressure wave into a shock
      228–9
    formulation 223–6
    non-linear wave equation 226–8
FEM–BEM solution method 348
finite-element equations 94
force–deflection relationship
  predicted by DPRS model cf Hertz's
    theory 36, *36*
Fourier transform technique 348
free-field vibrations of Thalys high-speed
  train *see* Thalys high-speed train
full elastic-interaction model 40–5

geographical information system (GIS) 92
Green's function 219, 223, 256–61, 271, 277,
  336, 405
  effect of layered ground structure 258–61
  homogeneous elastic half-space 257–8
ground vibration boom 251–82
  calculation 262–3
    vibrations from complete train 262–3
    vibrations from single axle load 262
  quasi-static pressure mechanisms of
    generating ground vibrations
    252–6
    dynamic properties of the track 253–5
    forces applied from sleepers to the
      ground 255–6
  trans-Rayleigh trains 263–81

high-speed trains travelling
  underground 270–7
TGV and Eurostar trains 265–70
waveguide effects of embankments
  277–81
ground vibration reduction in tunnels
  423–30
  tunnels with floating slabs 424–5
  vibration from railway tunnels 425–30
ground vibrations alongside tracks 347–91

Hankel function 400, 401
Helmholtz resonators 215, *215*, 229
  effects on shock-free propagation 234–6
  experimental verification 241–4
horizontal directivity index 98, *98*
hush rail 174

interior noise measurement 158–60
  diagnostics 158–9
  type testing 160
ISVR 860F wheel 168

Korteweg–de Vries equation 238
Krylov's analytical prediction model 298–302

layer stiffness matrix 389–91
  for half-space with respect to stresses
    acting on the z plane 391
linear variable displacement transducers
  (LVDTs) 61
  in roughness measurement 45, 46–8
liquid ground approximation 397, 398, 404,
  406

Mach number 190
meta-models 94
MetaRail project 119, 150
method of dimensions 65
micropressure waves 187–210, 218
  generation 188, *188*, 189–91
  measures to decrease 203–10
    applied to Shinkansen trains 206–10
    applied to Shinkansen tunnels 204–5
    effect of train nose shape 206–10, *207–9*
    use of tunnel head 204–5, *205–6*, 214
  peak value 188
  pressure rise at wavefont 189, *190*
  propagation through a tunnel 192–8
  radiation out of tunnel portal 198–203
mode shape 6, 10
modified Hankel function 400, 401

natural frequency 9
Navier–Stokes equation 224
near (pressure) field 218
    analysis 219–23
        evaluation of pressure field 220–3
        linear acoustic theory 219–20
noise indicators 85–8
    annoyance 85–6
    basic indicators 87–8
    composite indicator 88
    long-time average sound level and equivalent sound level 87
    maximum sound level 87
    noise level and A-frequency-weighted noise level 86
    root mean square average 86–7
    statistical indicators 87
noise level, calculation of 106–9
    with monopole or dipole noise sources 107–9
noise prediction
    calculation 89–91
    input data 92–3
    location 88–9
    reason for 88
    sequence 93–4
    timing 91–2
noise prediction model
    accuracy 112–13
    definition 94
    general set-up 95, 95
    methodology 95–6
noise recertification test 128
noise source control 29
    rolling noise 163–82
    roughness and 53–61

OFWHAT experiment 164, 165, 168, 170, 173
overhanging barriers 73

pantograph noise 21
parabolic equations 94
perceived equivalent loudness 148, 148
permanent emission monitoring 145–6, 146
permanent perception monitoring 146
pinned–pinned mode 11
preload, effects of 13–14
propagation models 98–106
    absorption by the ground 101–2
    atmospheric absorption 101
    attenuation due to barrier/obstacle 102–4

geometrical spreading 100–1
    meterological correction 105–6
    reflections 105
    types of attenuation 104
propulsion noise 120
psychoacoustics 146, 159

radiation efficiency 19
radiation of sound 18–23
    aerodynamic sources 21–2
    contribution of various sources 22–3
    from rail 20
    from sleepers 21
    from wheel 18–20
radiation ratio 19
rail-grinding block 56
Rayleigh wave reflection 277–81
ray-tracing techniques 65
receptance 9, 15–16
reflections, calculation of 110, *111*
resiliently treaded wheel 57–9, *58*
Reynolds number 190
'roaring rail' 179
rolling damping 164
rolling noise 4
    flow chart for TWINS calculation 5–6, *6*
    generation 5, *5*
roughness 14–18, 177–80
    changes to contact zone 180
    characteristics of wheel and rail 48–53
    contact receptances 15–16
    definition 27
    effective damping of rolling wheel 18
    effectiveness in exciting noise 28
    effects of braking system 177–8
    equivalent *60*, 61
    measurement 45–8
    modelling 29–45
        average roughness model 31–3
        distributed point-reacting spring model 33–9
        full elastic-interaction model 40–5
    rail corrugation 179–80
    roughness modification at contact zone 17–18
    wheel and rail roughness 16–17
    wheel/rail interaction, equations of 14–15
roughness amplitude reduction 54–6
roughness velocity 30

scale modelling 65–73
    atmospheric absorption 70–1

barriers 73
ground plane 71–3
 ballast bed 71–2
 between barrier and track 73
 characteristics 71
 grassland 72
measurable quantities 66
railway noise 73–6
 acoustical 1:32 scale model of high-speed train 73–5
 acoustical 1:32 scale model of railway track 75–6
receiver 70
similarity 65–6
sound sources 66–70
 distribution of moving point sources 70
 pneumatic sound source 68–70
 railway track in light of I_{Aeq} 68
 train characteristics and source type 66–7
Schalloptimiertesrad 166
shielding 180
shock-free propagation 229–41
 effects of array of Helmholtz resonators 234–6
 linear dispersion characteristics 229–34
 suppression of shock formation 236–41
shunting noise 120
Silent Freight project 164, 168
Silent Track project 164, 173, 176
sine wheel tests 25
sleep disturbance 86
sound-absorbing barriers at scale of 1:32 76–9
 absorption extracted from excess attenuation 79
 reference absorption curve 76–9
sound-absorbing box 145, *147*
sound directivity 132–8, *133–7*
sound exposure level 106–7
sound pressure levels at 25 m from track 4, *4*
sound propagation paths, determination 109–12
sound, radiation of 18–23
sound ratio characteristics 98
source description model 96–8
specific model 94
spectral analysis of surface waves (SASW) test 288–90
stationary solution 254
structural radiation damping 164
sub-Rayleigh trains 263–4

subseismic moving-load problems 333
superseismic moving-load problems 333

TGV wheel, modes of vibration 6–8, *7*
Thalys high-speed train, free-field vibrations 285–313
 experimental results 291–3
 influence of train speed 293
 passage at speed of $V = 314$ km/h 291–3
 free-field response 291–3
 track response 291
 experimental set-up 290
 free-field response 306–12
 Green's functions 305–6
 in situ measurements 287–90
 soil 288–90
 track 288
 train configuration *286*, 287, 288
 Krylov's analytical prediction model 298–302
 force transmitted by sleeper due to single axle load 300–1
 forces transmitted by all sleepers due to train passage 301–2
 response of soil 302
 track response 303–5
track critical velocity 251, 254
track dynamics 11–14
 effects of preload 13–14
 frequency response functions 11–12
 models for 11
 propagation along the track 12–13
 sleeper response 13
track–embankment–soil response and vibration generation (Ledsgård) 315–43
 case study 315–23
 observations 317–23
 test site and test programme 317
 countermeasures 337–9
 dispersion characteristics of layers 363–4
 dynamic properties of soil and embankment materials 326–33
 elastodynamic analysis 357–63
 finite element-boundary element method 362–3
 multi-layered system 358–60
 three-dimensional wave motions 357–8
 environmental vibration 342–3
 field measurements, theoretical prediction and mitigation 377–87
 measurement data 377–80

INDEX

prediction of ground motions 383–5
wave propagation at the site 380–3
ground surface motions 374–5
ground vibration due to quasi-static moving load 355–6
measurements 323–6
modelling of a loading by train 354–5
numerical simulation 333–6
physical model 339–42
response of track–ground system 375–6
simulations and comparisons 336
solution method for a moving load 349–51
track–ground dynamic interaction 351–4
transient responses 364–73
wave-impeding barriers (WIBs) 385–7
track noise control 170–7
  ballastless track ('slab track') forms 177
  damping treatments 173
  rail pad stiffness 170–2
  rail shape optimization 174–6, *175*
  track mobility 176–7
track-on-ballast resonance frequency 254
transeismic moving-load problems 333
trans-Raleigh trains 263–81
T-shaped barriers 73
tuned absorbers 165, *165*
tuned resonance devices 164–5, *165*
TWINS (Track–Wheel Interaction Noise Software) model 22–3, 61, 164
  model validation 23–5

underground tunnels 397–421

estimation of elastic parameters and damping of layered ground 411–18
ground vibration transmission to surface 405–7
two-level elastic system for vibration reduction 407–11

validation 23–5
  experimental set-up 23
vertical directivity index 98, *99*
VibTrain 334, 337, 339

wave-impeding block/barrier (WIB) 349, 385–7
wheel belt grinder 56
wheel dynamics, railway 6–10
  effects of rotation 10
  frequency response functions 9–10
  modes of vibration 6–9, *8*
wheel noise control 164–70
  damping treatments 164–6, *165*
  reduced wheel radiation 169–70
  resilient wheels 168–9
  wheel shape optimization 166–8
wheel–rail interaction 27, *28*
wheel–rail rolling noise 120
  origin 149–50, *150*
  speed and 153–4
wheel, railway, structure 7, *7*
WiBaO2 (UIC standard freight wheel) 167, *167*
Winkler beam theory 425